Flat Panel Display Manufacturing

Wiley - SID Series in Display Technology

Series Editor:
Ian Sage, Abelian Services, Malvern, UK

The Society for Information Display (SID) is an international society which has the aim of encouraging the development of all aspects of the field of information display. Complementary to the aims of the society, the Wiley-SID series is intended to explain the latest developments in information display technology at a professional level. The broad scope of the series addresses all facets of information displays from technical aspects through systems and prototypes to standards and ergonomics.

Flat Panel Display Manufacturing
Jun Souk, Shinji Morozumi, Fang-Chen Luo, Ion Bita (Eds.)

Display Systems: Design and Applications
Lindsay W. MacDonald and Anthony C. Lowe (Eds.)

Electronic Display Measurement: Concepts, Techniques, and Instrumentation
Peter A. Keller

Reflective Liquid Crystal Displays
Shin-Tson Wu and Deng-Ke Yang

Display Interfaces: Fundamentals and Standards
Robert L. Myers

Colour Engineering: Achieving Device Independent Colour
Phil Green and Lindsay MacDonald (Eds.)

Digital Image Display: Algorithms and Implementation
Gheorghe Berbecel

Flexible Flat Panel Displays
Gregory Crawford (Ed.)

Polarization Engineering for LCD Projection
Michael G. Robinson, Jianmin Chen and Gary D. Sharp

Introduction to Microdisplays
David Armitage, Ian Underwood and Shin-Tson Wu

Mobile Displays: Technology and Applications
Achintya K. Bhowmik, Zili Li and Philip Bos (Eds.)

Photoalignment of Liquid Crystalline Materials: Physics and Applications
Vladimir G. Chigrinov, Vladimir M. Kozenkov and Hoi-Sing Kwok

Projection Displays, Second Edition
Matthew S. Brennesholtz and Edward H. Stupp

Introduction to Flat Panel Displays
Jiun-Haw Lee, David N. Liu and Shin-Tson Wu

LCD Backlights
Shunsuke Kobayashi, Shigeo Mikoshiba and Sungkyoo Lim (Eds.)

Liquid Crystal Displays: Addressing Schemes and Electro-Optical Effects, Second Edition
Ernst Lueder

Transflective Liquid Crystal Displays
Zhibing Ge and Shin-Tson Wu

Liquid Crystal Displays: Fundamental Physics and Technology
Robert H. Chen

3D Displays
Ernst Lueder

OLED Display Fundamentals and Applications
Takatoshi Tsujimura

Illumination, Colour and Imaging: Evaluation and Optimization of Visual Displays
Peter Bodrogi and Tran Quoc Khanh

Interactive Displays: Natural Human-Interface Technologies
Achintya K. Bhowmik (Ed.)

Addressing Techniques of Liquid Crystal Displays
Temkar N. Ruckmongathan

Fundamentals of Liquid Crystal Devices, Second Edition
Deng-Ke Yang and Shin-Tson Wu

Modeling and Optimization of LCD Optical Performance
Dmitry A. Yakovlev, Vladimir G. Chigrinov and Hoi-Sing Kwok

Flat Panel Display Manufacturing

Edited by

Jun Souk
Hanyang University, Korea

Shinji Morozumi
Crystage Inc, Japan

Fang-Chen Luo
AU Optronics, Taiwan

Ion Bita
Apple Inc, USA

This edition first published 2018
© 2018 John Wiley and Sons Ltd

All rights reserved. No part of this publication may be reproduced, stored in a retrieval system, or transmitted, in any form or by any means, electronic, mechanical, photocopying, recording or otherwise, except as permitted by law. Advice on how to obtain permission to reuse material from this title is available at http://www.wiley.com/go/permissions.

The rights of Jun Souk, Shinji Morozumi, Fan-Chen Luo, and Ion Bita to be identified as the the editorial material in this work has been asserted in accordance with law.

Registered Offices
John Wiley & Sons, Inc., 111 River Street, Hoboken, NJ 07030, USA
John Wiley & Sons Ltd, The Atrium, Southern Gate, Chichester, West Sussex, PO19 8SQ, UK

Editorial Office
The Atrium, Southern Gate, Chichester, West Sussex, PO19 8SQ, UK

For details of our global editorial offices, customer services, and more information about Wiley products visit us at www.wiley.com.

Wiley also publishes its books in a variety of electronic formats and by print-on-demand. Some content that appears in standard print versions of this book may not be available in other formats.

Limit of Liability/Disclaimer of Warranty
While the publisher and authors have used their best efforts in preparing this work, they make no representations or warranties with respect to the accuracy or completeness of the contents of this work and specifically disclaim all warranties, including without limitation any implied warranties of merchantability or fitness for a particular purpose. No warranty may be created or extended by sales representatives, written sales materials or promotional statements for this work. The fact that an organization, website, or product is referred to in this work as a citation and/or potential source of further information does not mean that the publisher and authors endorse the information or services the organization, website, or product may provide or recommendations it may make. This work is sold with the understanding that the publisher is not engaged in rendering professional services. The advice and strategies contained herein may not be suitable for your situation. You should consult with a specialist where appropriate. Further, readers should be aware that websites listed in this work may have changed or disappeared between when this work was written and when it is read. Neither the publisher nor authors shall be liable for any loss of profit or any other commercial damages, including but not limited to special, incidental, consequential, or other damages.

Library of Congress Cataloging-in-Publication Data applied for
HB ISBN: 9781119161349

Cover design by Wiley
Cover image: © Shannon Fagan/gettyimages

Set in 10/12pt WarnockPro by SPi Global, Chennai, India
Printed and bound in Singapore by Markono Print Media Pte Ltd

10 9 8 7 6 5 4 3 2 1

Contents

List of Contributors *xxi*
Series Editor's Foreword *xxv*
Preface *xxvii*

1 Introduction *1*
Fang-Chen Luo, Jun Souk, Shinji Morozumi, and Ion Bita
1.1 Introduction *1*
1.2 Historic Review of TFT-LCD Manufacturing Technology Progress *1*
1.2.1 Early Stage TFT and TFT-Based Displays *2*
1.2.2 The 1990s: Initiation of TFT-LCD Manufacturing and Incubation of TFT-LCD Products *2*
1.2.3 Late 1990s: Booming of LCD Desktop Monitor and Wide Viewing Angle Technologies *4*
1.2.4 The 2000s: A Golden Time for LCD-TV Manufacturing Technology Advances *4*
1.3 Analyzing the Success Factors in LCD Manufacturing *5*
1.3.1 Scaling the LCD Substrate Size *7*
1.3.2 Major Milestones in TFT-LCD Manufacturing Technology *9*
1.3.2.1 First Revolution: AKT Cluster PECVD Tool in 1993 *9*
1.3.2.2 Second Revolution: Wide Viewing Angle Technology in 1997 *9*
1.3.2.3 Third Revolution: LC Drop Filling Technology in 2003 *10*
1.3.3 Major Stepping Stones Leading to the Success of Active Matrix Displays *10*
References *11*

2 TFT Array Process Architecture and Manufacturing Process Flow *13*
Chiwoo Kim
2.1 Introduction *13*
2.2 Material Properties and TFT Characteristics of a-Si, LTPS, and Metal Oxide TFTs *15*
2.2.1 a-Si TFT *15*
2.2.2 LTPS TFT *16*
2.2.2.1 Excimer Laser Annealing (ELA) *17*
2.2.3 Amorphous Oxide Semiconductor TFTs *22*
2.3 a-Si TFT Array Process Architecture and Process Flow *22*
2.3.1 Four-Mask Count Process Architecture for TFT-LCDs *24*
2.4 Poly-Si TFT Architecture and Fabrication *27*
2.5 Oxide Semiconductor TFT Architecture and Fabrication *30*
2.6 TFT LCD Applications *32*
2.7 Development of SLS-Based System on Glass Display [1, 11, 14, 15] *33*
References *35*

3		Color Filter Architecture, Materials, and Process Flow	39
		Young Seok Choi, Musun Kwak, and Youn Sung Na	
3.1		Introduction	39
3.2		Structure and Role of the Color Filter	39
3.2.1		Red, Green, and Blue (RGB) Layer	40
3.2.1.1		Color Coordinate and Color Gamut	41
3.2.2		Black Matrix	44
3.2.3		Overcoat and Transparent Electrode	45
3.2.4		Column Spacer	46
3.3		Color Filter Manufacturing Process Flow	46
3.3.1		Unit Process	46
3.3.1.1		Formation of Black Matrix	46
3.3.1.2		Formation of RGB Layer	48
3.3.1.3		Overcoat (OC)	51
3.3.1.4		Formation of ITO Electrodes	53
3.3.1.5		Column Spacer (Pattern Spacer)	53
3.3.2		Process Flow for Different LC Mode	54
3.3.2.1		Color Filter for the TN Mode	54
3.3.2.2		Color Filter for the IPS Mode	54
3.3.2.3		Color Filter for the VA Mode	55
3.4		New Color Filter Design	55
3.4.1		White Color (Four Primary Colors) Technology	55
3.4.2		Color Filter on TFT	56
		References	57
4		Liquid Crystal Cell Process	59
		Heung-Shik Park and Ki-Chul Shin	
4.1		Introduction	59
4.2		Liquid Crystal Cell Process	59
4.2.1		Alignment Layer Treatment	61
4.2.2		Process of Applying PI Layers	62
4.2.3		Rubbing Process	63
4.2.4		Photo-Alignment Process	64
4.2.5		LC Filling Process	65
4.2.5.1		Vacuum Filling Method	66
4.2.5.2		End Seal Process	66
4.2.5.3		One Drop Filling (ODF) Method	67
4.2.6		Vacuum Assembly Process	68
4.2.7		Polarizer Attachment Process	69
4.3		Conclusions	70
		Acknowledgments	70
		References	70
5		TFT-LCD Module and Package Process	73
		Chun Chang Hung	
5.1		Introduction	73
5.2		Driver IC Bonding: TAB and COG	73
5.3		Introduction to Large-Panel JI Process	74
5.3.1		COF Bonding	75

5.3.1.1	Edge Clean	75
5.3.1.2	ACF Attachment	76
5.3.1.3	COF Pre-Bonding	77
5.3.1.4	COF Main Bonding	78
5.3.1.5	Lead Check	78
5.3.1.6	Silicone Dispensing	78
5.3.2	PCB Bonding	79
5.3.3	PCB Test	79
5.3.4	Press Heads: Long Bar or Short Bar	79
5.4	Introduction to Small-Panel JI Process	79
5.4.1	Beveling	80
5.4.2	Panel Cleaning	80
5.4.3	Polarizer Attachment	80
5.4.4	Chip on Glass (COG) Bonding	81
5.4.5	FPC on Glass (FOG) Bonding	81
5.4.6	Optical Microscope (OM) Inspection	81
5.4.7	UV Glue Dispense	82
5.4.8	Post Bonding Inspection (PBI)	82
5.4.9	Protection Glue Dispensing	82
5.5	LCD Module Assembly	83
5.6	Aging	84
5.7	Module in Backlight or Backlight in Module	85
	References	86
6	**LCD Backlights**	*87*
	Insun Hwang and Jae-Hyeon Ko	
6.1	Introduction	87
6.2	LED Sources	90
6.2.1	GaN Epi-Wafer on Sapphire	92
6.2.2	LED Chip	93
6.2.3	Light Extraction	94
6.2.4	LED Package	96
6.2.5	SMT on FPCB	97
6.3	Light Guide Plate	98
6.3.1	Optical Principles of LGP	98
6.3.2	Optical Pattern Design	99
6.3.3	Manufacturing of LGP	101
6.3.3.1	Injection Molding	101
6.3.3.2	Screen Printing	102
6.3.3.3	Other Methods	103
6.4	Optical Films	104
6.4.1	Diffuser	106
6.4.2	Prism Film	107
6.4.3	Reflector	108
6.4.4	Other Films	108
6.5	Direct-Type BLU	111
6.6	Summary	111
	References	112

7	**TFT Backplane and Issues for OLED** *115*	
	Chiwoo Kim	
7.1	Introduction *115*	
7.2	LTPS TFT Backplane for OLED Films *116*	
7.2.1	Advanced Excimer Laser Annealing (AELA) for Large-Sized AMOLED Displays *117*	
7.2.2	Line-Scan Sequential Lateral Solidification Process for AMOLED Application *120*	
7.3	Oxide Semiconductor TFT for OLED *122*	
7.3.1	Oxide TFT–Based OLED for Large-Sized TVs *123*	
7.4	Best Backplane Solution for AMOLED *125*	
	References *127*	
8A	**OLED Manufacturing Process for Mobile Application** *129*	
	Jang Hyuk Kwon and Raju Lampande	
8A.1	Introduction *129*	
8A.2	Current Status of AMOLED for Mobile Display *130*	
8A.2.1	Top Emission Technology *130*	
8A.3	Fine Metal Mask Technology (Shadow Mask Technology) *133*	
8A.4	Encapsulation Techniques for OLEDs *135*	
8A.4.1	Frit Sealing *135*	
8A.4.2	Thin-Film Encapsulation *136*	
8A.5	Flexible OLED technology *137*	
8A.6	AMOLED Manufacturing Process *137*	
8A.7	Summary *140*	
	References *140*	
8B	**OLED Manufacturing Process for TV Application** *143*	
	Chang Wook Han and Yoon Heung Tak	
8B.1	Introduction *143*	
8B.2	Fine Metal Mask (FMM) *144*	
8B.3	Manufacturing Process for White OLED and Color Filter Methods *147*	
8B.3.1	One-Stacked White OLED Device *149*	
8B.3.2	Two-Stacked White OLED Device *152*	
8B.3.3	Three-Stacked White-OLED Device *155*	
	References *157*	
9	**OLED Encapsulation Technology** *159*	
	Young-Hoon Shin	
9.1	Introduction *159*	
9.2	Principles of OLED Encapsulation *159*	
9.2.1	Effect of H_2O *160*	
9.3	Classification of Encapsulation Technologies *162*	
9.3.1	Edge Seal *163*	
9.3.2	Frit Seal *164*	
9.3.3	Dam and Fill *166*	
9.3.4	Face Seal *167*	
9.3.5	Thin-Film Encapsulation (TFE) *168*	
9.4	Summary *170*	
	References *170*	

10	**Flexible OLED Manufacturing** *173*	

Woojae Lee and Jun Souk

10.1	Introduction *173*	
10.2	Critical Technologies in Flexible OLED Display *174*	
10.2.1	High-Temperature PI Film *175*	
10.2.2	Encapsulation Layer *176*	
10.2.2.1	Thin-Film Encapsulation (TFE) Method *176*	
10.2.2.2	Hyrid Encapsulation Method *177*	
10.2.2.3	Other Encapsulation Methods *178*	
10.2.2.4	Measurement of Barrier Performance *179*	
10.2.3	Laser Lift-Off *180*	
10.2.4	Touch Sensor on F-OLED *181*	
10.3	Process Flow of F-OLED *181*	
10.3.1	PI Film Coating and Curing *181*	
10.3.2	LTPS TFT Backplane Process *183*	
10.3.3	OLED Deposition Process *183*	
10.3.4	Thin-Film Encapsulation *185*	
10.3.5	Laser Lift-Off *185*	
10.3.6	Lamination of Backing Plastic Film and Cut to Cell Size *185*	
10.3.7	Touch Sensor Attach *186*	
10.3.8	Circular Polarizer Attach *186*	
10.3.9	Module Assembly (Bonding Drive IC) *186*	
10.4	Foldable OLED *186*	
10.5	Summary *188*	
	References *189*	
11A	**Metal Lines and ITO PVD** *193*	

Hyun Eok Shin, Chang Oh Jeong, and Junho Song

11A.1	Introduction *193*	
11A.1.1	Basic Requirements of Metallization for Display *193*	
11A.1.2	Thin-Film Deposition by Sputtering *195*	
11A.2	Metal Line Evolution in Past Years of TFT-LCD *198*	
11A.2.1	Gate Line Metals *199*	
11A.2.1.1	Al and Al Alloy Electrode *199*	
11A.2.1.2	Cu Electrode *201*	
11A.2.2	Data line (Source/Drain) Metals *202*	
11A.2.2.1	Data Al Metal *202*	
11A.2.2.2	Data Cu Metal *203*	
11A.2.2.3	Data Chromium (Cr) Metal *203*	
11A.2.2.4	Molybdenum (Mo) Metal *203*	
11A.2.2.5	Titanium (Ti) Metal *204*	
11A.3	Metallization for OLED Display *205*	
11A.3.1	Gate Line Metals *205*	
11A.3.2	Source/Drain Metals *205*	
11A.3.3	Pixel Anode *206*	
11A.4	Transparent Electrode *207*	
	References *208*	

11B	Thin-Film PVD: Materials, Processes, and Equipment *209*
	Tetsuhiro Ohno
11B.1	Introduction *209*
11B.2	Sputtering Method *210*
11B.3	Evolution of Sputtering Equipment for FPD Devices *212*
11B.3.1	Cluster Tool for Gen 2 Size *212*
11B.3.2	Cluster Tool for Gen 4.5 to Gen 7 Size *213*
11B.3.3	Vertical Cluster Tool for Gen 8 Size *213*
11B.4	Evolution of Sputtering Cathode *215*
11B.4.1	Cathode Structure Evolution *215*
11B.4.2	Dynamic Multi Cathode for LTPS *217*
11B.4.3	Cathode Selection Strategy *217*
11B.5	Transparent Oxide Semiconductor (TOS) Thin-Film Deposition Technology *218*
11B.5.1	Deposition Equipment for TOS-TFT *218*
11B.5.2	New Cathode Structure for TOS-TFT *219*
11B.6	Metallization Materials and Deposition Technology *221*
	References *223*
11C	Thin-Film PVD (Rotary Target) *225*
	Marcus Bender
11C.1	Introduction *225*
11C.2	Source Technology *227*
11C.2.1	Planar Cathodes *227*
11C.2.2	Rotary Cathodes *229*
11C.2.3	Rotary Cathode Array *230*
11C.3	Materials, Processes, and Characterization *232*
11C.3.1	Introduction *232*
11C.3.2	Backplane Metallization *232*
11C.3.3	Layers for Metal-Oxide TFTs *234*
11C.3.4	Transparent Electrodes *236*
11C.3.5	Adding Touch Functionality and Improving End-User Experience *238*
	References *239*
12A	Thin-Film PECVD (AKT) *241*
	Tae Kyung Won, Soo Young Choi, and John M. White
12A.1	Introduction *241*
12A.2	Process Chamber Technology *243*
12A.2.1	Electrode Design *243*
12A.2.1.1	Hollow Cathode Effect and Hollow Cathode Gradient *243*
12A.2.1.2	Gas Flow Control *245*
12A.2.1.3	Susceptor *245*
12A.2.2	Chamber Cleaning *246*
12A.3	Thin-Film Material, Process, and Characterization *248*
12A.3.1	Amorphous Si (a-Si) TFT *248*
12A.3.1.1	Silicon Nitride (SiN) *248*
12A.3.1.2	Amorphous Silicon (a-Si) *253*
12A.3.1.3	Phosphorus-Doped Amorphous Silicon (n^+ a-Si) *257*
12A.3.2	Low-Temperature Poly Silicon (LTPS) TFT *258*
12A.3.2.1	Silicon Oxide (SiO) *259*
12A.3.2.2	a-Si Precursor Film (Dehydrogenation) *260*

12A.3.3	Metal-Oxide (MO) TFT	*263*
12A.3.3.1	Silicon Oxide (SiO)	*265*
12A.3.4	Thin-Film Encapsulation (TFE)	*269*
12A.3.4.1	Barrier Layer (Silicon Nitride)	*269*
12A.3.4.2	Buffer Layer	*271*
	References	*271*
12B	*Thin-Film PECVD (Ulvac)*	*273*
	Masashi Kikuchi	
12B.1	Introduction	*273*
12B.2	Plasma of PECVD	*273*
12B.3	Plasma Modes and Reactor Configuration	*273*
12B.3.1	CCP-Type Reactor	*274*
12B.3.2	Microwave-Type Reactor	*274*
12B.3.3	ICP-Type Reactor	*275*
12B.4	PECVD Process for Display	*276*
12B.4.1	a-Si Film for a-Si TFT	*276*
12B.4.2	a-Si Film for LTPS	*277*
12B.4.3	SiN_x Film	*278*
12B.4.4	TEOS SiO_2 Film	*279*
12B.5	PECVD System Overview	*279*
12B.6	Remote Plasma Cleaning	*279*
12B.6.1	Gas Flow Style of Remote Plasma Cleaning	*281*
12B.6.2	Cleaning and Corrosion	*281*
12B.7	Passivation Layer for OLED	*282*
12B.7.1	Passivation by Single/Double/Multi-Layer	*282*
12B.8	PECVD Deposition for IGZO TFT	*283*
12B.8.1	Gate Insulator for IGZO TFT	*283*
12B.8.2	Passivation Film for IGZO TFT	*284*
12B.9	Particle Generation	*284*
	References	*286*
13	*Photolithography*	*287*
	Yasunori Nishimura, Kozo Yano, Masataka Itoh, and Masahiro Ito	
13.1	Introduction	*287*
13.2	Photolithography Process Overview	*288*
13.2.1	Cleaning	*289*
13.2.2	Preparation	*289*
13.2.3	Photoresist Coating	*289*
13.2.4	Exposure	*289*
13.2.5	Development	*289*
13.2.6	Etching	*289*
13.2.7	Resist Removal	*289*
13.3	Photoresist Coating	*290*
13.3.1	Evolution of Photoresist Coating	*290*
13.3.2	Slit Coating	*290*
13.3.2.1	Principles of Slit Coating	*290*
13.3.2.2	Slit-Coating System	*291*
13.4	Exposure	*292*
13.4.1	Photoresist and Exposure	*292*

13.4.1.1 Photoresist 292
13.4.1.2 Color Resist 292
13.4.1.3 UV Light Source for Exposure 292
13.4.2 General Aspects of Exposure Systems 292
13.4.3 Stepper 293
13.4.4 Projection Scanning Exposure System 294
13.4.5 Mirror Projection Scan System (Canon) 296
13.4.6 Multi-Lens Projection System (Nikon) 296
13.4.6.1 Multi-Lens Optics 296
13.4.6.2 Multi-Lens Projection System 296
13.4.7 Proximity Exposure 297
13.5 Photoresist Development 300
13.6 Inline Photolithography Processing Equipment 301
13.7 Photoresist Stripping 302
13.8 Photolithography for Color Filters 303
13.8.1 Color Filter Structures 303
13.8.1.1 TN 304
13.8.1.2 VA 304
13.8.1.3 IPS 304
13.8.2 Materials for Color Filters 305
13.8.2.1 Black Matrix Materials 305
13.8.2.2 RGB Color Materials 305
13.8.2.3 PS (Photo Spacer) Materials 306
13.8.3 Photolithography Process for Color Filters 307
13.8.3.1 Color Resist Coating 307
13.8.3.2 Exposure 307
13.8.3.3 Development 308
13.8.4 Higher-Performance Color Filters 309
13.8.4.1 Mobile Applications 309
13.8.4.2 TV Applications 309
References 310

14A Wet Etching Processes and Equipment 311
Kazuo Jodai
14A.1 Introduction 311
14A.2 Overview of TFT Process 312
14A.3 Applications and Equipment of Wet Etching 313
14A.3.1 Applications 313
14A.3.2 Equipment (Outline) 313
14A.3.3 Substrate Transferring System 315
14A.3.4 Dip Etching System 316
14A.3.5 Cascade Rinse System 316
14A.4 Problems Due to Increased Mother Glass Size and Solutions 317
14A.4.1 Etchant Concentration Management 317
14A.4.2 Quick Rinse 317
14A.4.3 Other Issues 318
14A.5 Conclusion 318
References 318

14B	**Dry Etching Processes and Equipment** *319*	
	Ippei Horikoshi	
14B.1	Introduction *319*	
14B.2	Principle of Dry Etching *319*	
14B.2.1	Plasma *320*	
14B.2.2	Ions *321*	
14B.2.3	Radicals *321*	
14B.3	Architecture for Dry Etching Equipment *322*	
14B.4	Dry Etching Modes *323*	
14B.4.1	Conventional Etching Mode and Each Characteristic *324*	
14B.4.2	Current Etching Mode and Each Characteristic *325*	
14B.5	TFT Process *325*	
14B.5.1	a-Si Process *325*	
14B.5.2	LTPS Process *326*	
14B.5.3	Oxide Process *327*	
	References *328*	
15	**TFT Array: Inspection, Testing, and Repair** *329*	
	Shulik Leshem, Noam Cohen, Savier Pham, Mike Lim, and Amir Peled	
15.1	Defect Theory *329*	
15.1.1	Typical Production Defects *329*	
15.1.1.1	Pattern Defects *329*	
15.1.1.2	Foreign Particles *331*	
15.1.2	Understanding the Nature of Defects *332*	
15.1.2.1	Critical and Non-Critical Defects *332*	
15.1.2.2	Electrical and Non-Electrical Defects *333*	
15.1.3	Effect of Defects on Final FPD Devices and Yields *333*	
15.2	AOI (Automated Optical Inspection) *334*	
15.2.1	The Need *334*	
15.2.2	AOI Tasks, Functions, and Sequences *335*	
15.2.2.1	Image Acquisition *335*	
15.2.2.2	Defect Detection *336*	
15.2.2.3	Defect Classification *336*	
15.2.2.4	Review Image Grabbing *337*	
15.2.2.5	Defect Reporting and Judgment *337*	
15.2.3	AOI Optical Concept *337*	
15.2.3.1	Image Quality Criteria *338*	
15.2.3.2	Scan Cameras *339*	
15.2.3.2.1	Camera Type *339*	
15.2.3.2.2	Resolution Changer *339*	
15.2.3.2.3	Backside Inspection *339*	
15.2.3.3	Scan Illumination *339*	
15.2.3.3.1	Types of Illumination *339*	
15.2.3.4	Video Grabbing for Defect Review and Metrology *340*	
15.2.3.4.1	Review/Metrology Cameras *340*	
15.2.3.4.2	On-the-Fly Video Grabbing *340*	
15.2.3.4.3	Alternative to Video Images *340*	
15.2.4	AOI Defect Detection Principles *341*	

15.2.4.1 Gray Level Concept *342*
15.2.4.2 Comparison of Gray Level Values Between Neighboring Cells *342*
15.2.4.3 Detection Sensitivity *342*
15.2.4.4 Detection Selectivity *344*
15.2.5 AOI Special Features *344*
15.2.5.1 Detection of Special Defect Types *344*
15.2.5.2 Inspection of In-Cell Touch Panels *345*
15.2.5.3 Peripheral Area Inspection *346*
15.2.5.4 Mura Defects *346*
15.2.5.5 Cell Process Inspection *347*
15.2.5.6 Defect Classification *347*
15.2.5.7 Metrology: CD/O Measurement *349*
15.2.5.8 Automatic Judgment *350*
15.2.6 Offline Versus Inline AOI *350*
15.2.7 AOI Usage, Application and Trends *351*
15.3 Electrical Testing *352*
15.3.1 The Need *352*
15.3.2 Array Tester Tasks, Functions, and Sequences *353*
15.3.2.1 Panel Signal Driving *353*
15.3.2.1.1 Shorting Bar Probing Method *354*
15.3.2.1.2 Full Contact Probing Method *354*
15.3.2.2 Contact or Non-Contact Sensing *354*
15.3.2.2.1 Contact Sensing *355*
15.3.2.2.2 Non-Contact Sensing Methods *355*
15.3.2.3 Panel Image Processing and Defect Detection *355*
15.3.2.4 Post-Defect Detection Processes *355*
15.3.3 Array Tester System Design Concept *356*
15.3.3.1 Signal Driving Probing *357*
15.3.3.2 Ultra-High-Resolution Testing *357*
15.3.3.3 System TACT *358*
15.3.3.4 "High-Channel" Testing *358*
15.3.3.5 Advanced Process Technology Testing (AMOLED, FLEX OLED) *358*
15.3.4 Array Tester Special Features *359*
15.3.4.1 GOA, ASG, and IGD Testing *359*
15.3.4.2 Electro Mura Monitoring *359*
15.3.4.3 Free-Form Panel Testing *361*
15.3.5 Array Tester Usage, Application, and Trends *361*
15.3.5.1 Source Drain Layer Testing for LTPS LCD/OLED *362*
15.3.5.2 New Probing Concept *363*
15.3.5.3 In-Cell Touch Panel Testing *363*
15.4 Defect Repair *363*
15.4.1 The Need *363*
15.4.2 Repair System in the Production Process *364*
15.4.2.1 In-Process Repair *364*
15.4.2.2 Final Repair *364*
15.4.3 Repair Sequence *364*
15.4.4 Short-Circuit Repair Method *365*
15.4.4.1 Laser Ablation Concept *365*
15.4.4.1.1 Thermal Ablation *366*

15.4.4.1.2 Cold Ablation   *366*
15.4.4.1.3 Photochemical Ablation   *366*
15.4.4.2   Laser Light Wavelengths and their Typical Applications   *366*
15.4.4.2.1 Laser Matter Interaction   *366*
15.4.4.2.2 Using DUV Laser Light (266 nm) for Short-Circuit Defect Repair   *367*
15.4.4.2.3 Using Infrared Laser Light (1,064 nm) for Short-Circuit Defect Repair   *367*
15.4.4.3.4 Using Green Laser Light (532 nm) for Short-Circuit Defect Repair   *367*
15.4.4.3   Typical Applications of the Short-Circuit Repair Method   *367*
15.4.4.3.1 Cutting   *367*
15.4.4.3.2 Welding   *368*
15.4.5   Open-Circuit Repair Method   *369*
15.4.5.1   LCVD (Laser Chemical Vapor Deposition)   *369*
15.4.5.2   Metal Ink Deposition Repair   *370*
15.4.5.2.1 Dispensing   *370*
15.4.5.2.2 Metal Inkjet Deposition   *370*
15.4.5.2.3 LIFT (Laser-Induced Forward Transfer) Deposition   *371*
15.4.5.3   Main Applications of the Deposition Repair (Open-Circuit Repair)   *372*
15.4.6   Photoresist (PR) Repair   *372*
15.4.6.1   Main Applications of the Photoresist Repair   *373*
15.4.6.2   Photoresist Repair Technology   *373*
15.4.6.2.1 Using DMD for Patterning   *373*
15.4.6.2.2 Using FSM for Patterning   *373*
15.4.7   Special Features of the Repair System   *375*
15.4.7.1   Line Defect Locator (LDL)   *375*
15.4.7.2   Parallel Repair Mode for Maximum System Throughput   *375*
15.4.8   Repair Technology Trends   *376*
15.4.8.1   Cold Ablation   *376*
15.4.8.2   Full Automatic Repair Solution   *377*
15.4.9   Summary   *377*

16   LCM Inspection and Repair   *379*
         Chun Chang Hung   *379*
16.1   Introduction   *379*
16.2   Functional Defects Inspection   *379*
16.3   Cosmetic Defects Inspection   *381*
16.4   Key Factors for Proper Inspection   *383*
16.4.1   Variation Between Inspectors   *383*
16.4.2   Testing Environments   *385*
16.4.3   Inspection Distance, Viewing Angle, and Sequence of Test Patterns   *385*
16.4.4   Characteristics of Product and Components   *387*
16.5   Automatic Optical Inspection (AOI)   *388*
16.6   LCM Defect Repair   *388*
         References   *391*

17   Productivity and Quality Control Overview   *393*
         Kozo Yano, Yasunori Nishimura, and Masataka Itoh
17.1   Introduction   *393*
17.2   Productivity Improvement   *394*
17.2.1   Challenges for Productivity Improvement   *394*
17.2.2   Enlargement of Glass Substrate   *395*

17.2.2.1 Productivity Improvement and Cost Reduction by Glass Size Enlargement *397*
17.3 Yield Management *399*
17.3.1 Yield Analysis *399*
17.3.1.1 Inspection and Yield *399*
17.3.1.2 Failure Mode Analysis *401*
17.3.2 Yield Improvement Activity *404*
17.3.2.1 Process Yield Improvement *404*
17.3.2.2 Systematic Failure Minimization *404*
17.3.2.3 Random Failure Minimization by Clean Process *404*
17.3.2.4 Yield Improvement by Repairing *406*
17.4 Quality Control System *406*
17.4.1 Materials (IQC) *407*
17.4.2 Facility Control *408*
17.4.3 Process Quality Control *408*
17.4.3.1 TFT Array Process *409*
17.4.3.2 Color Filter Process *410*
17.4.3.3 LCD Cell Process *412*
17.4.3.4 Modulization Process *412*
17.4.4 Organization and Key Issues for Quality Control *413*
References *417*

18 Plant Architectures and Supporting Systems *419*
Kozo Yano and Michihiro Yamakawa
18.1 Introduction *419*
18.2 General Issues in Plant Architecture *420*
18.2.1 Plant Overview *420*
18.2.2 Plant Design Procedure and Baseline *422*
18.3 Clean Room Design *423*
18.3.1 Clean Room Evolution *423*
18.3.2 Floor Structure for Clean Room *424*
18.3.3 Clean Room Ceiling Height *424*
18.3.4 Air Flow and Circulation Design *427*
18.3.5 Cleanliness Control *428*
18.3.6 Air Flow Control Against Particle *428*
18.3.7 Chemical Contamination Countermeasures *431*
18.3.8 Energy Saving in FFU *433*
18.4 Supporting Systems with Environmental Consideration *433*
18.4.1 Incidental Facilities *433*
18.4.2 Water and Its Recycle *434*
18.4.3 Chemicals *436*
18.4.4 Gases *436*
18.4.5 Electricity *437*
18.5 Production Control System *437*
References *440*

19 Green Manufacturing *441*
YiLin Wei, Mona Yang, and Matt Chien
19.1 Introduction *441*
19.2 Fabrication Plant (Fab) Design *441*
19.2.1 Fab Features *441*

19.2.2	Green Building Design *442*	
19.3	Product Material Uses *443*	
19.3.1	Material Types and Uses *443*	
19.3.2	Hazardous Substance Management *444*	
19.3.3	Material Hazard and Green Trend *446*	
19.3.4	Conflict Minerals Control *446*	
19.4	Manufacturing Features and Green Management *447*	
19.4.1	The Manufacturing Processes *447*	
19.4.2	Greenhouse Gas Inventory *448*	
19.4.3	Energy Saving in Manufacturing *449*	
19.4.4	Reduction of Greenhouse Gas from Manufacturing *449*	
19.4.5	Air Pollution and Control *451*	
19.4.6	Water Management and Emissions Control *452*	
19.4.7	Waste Recycling and Reuse *453*	
19.5	Future Challenges *453*	
	References *454*	

Index *457*

List of Contributors

Amir Peled
Orbotech, Ltd.
Yavne, Israel

Chang Oh Jeong
Samsung Display Co., Ltd.
Youngin City, Gyeonggi-Do, Korea

Chang Wook Han
LG Display
Gangseo-gu, Seoul, Korea

Chiwoo Kim
Seoul National University
Daehak-dong, Gwanak-gu, Seoul, Korea

Chun Chang Hung
AU Optronics, Inc.
Taoyuan City, Taiwan

Fang-Chen Luo
AU Optronics, Inc.
Hsinchu, Taiwan

Heung-Shik Park
Samsung Display Co., Ltd.
Youngin City, Gyeonggi-Do, Korea

Hyun Eok Shin
Samsung Display Co., Ltd.
Youngin City, Gyeonggi-Do, Korea

Insun Hwang
Samsung Display Co., Ltd.
Youngin City, Gyeonggi-Do, Korea

Ion Bita
Apple, Inc.
Cupertino, USA

Ippei Horikoshi
Tokyo Electron Limited
Akasaka Minato-ku, Tokyo, Japan

Jae-Hyeon Ko
Hallym University
Chuncheon, Gangwondo, Korea

Jang Hyuk Kwon
Kyung Hee University
Hoegi-dong, Dongdaemun-gu, Seoul, Korea

John M. White
Applied Materials, Inc.
Santa Clara, California, USA

Jun Ho Song
Samsung Display Co., Ltd.
Youngin City, Gyeonggi-Do, Korea

Jun Souk
Hanyang University
Korea

Kazuo Jodai
SCREEN Finetech Solutions Co., Ltd.
Horikawa-dori, Kamigyo-ku, Kyoto, Japan

Ki-Chul Shin
Samsung Display Co., Ltd.
Youngin City, Gyeonggi-Do, Korea

Kozo Yano
Foxconn Japan RD Co., Ltd.
Yodogawa-Ku, Osaka, Japan

Marcus Bender
Applied Materials GmbH & Co.
Alzenau, Germany

Masahiro Ito
Toppan Printing Co., Ltd.
Taito-Ku, Tokyo, Japan

Masahi Kikuchi
ULVAC, Inc.
Chigasaki, Kanagawa, Japan

Masataka Itoh
Crystage, Inc.
Yodogawa-Ku, Osaka, Japan

Matt Chien
AU Optronics, Inc.
Hsinchhu, Taiwan

Michihiro Yamakawa
Fuji Electric Co., Ltd.
Osaki, Shinagawa, Tokyo, Japan

Mike Lim
Orbotech Pacific, Ltd.
Bundang-Gu, Sungnam City, Kyoungki Do, Korea

Mona Yang
AU Optronics, Inc.
Hsinchu, Taiwan

Musum Kwak
LG Display
Wollong-myeon, Paju-si, Gyeonggi-do, Korea

Noam Cohen
Orbotech, Ltd.
Yavne, Israel

Raju Lampande
Kyung Hee University
Hoegi-dong, Dongdaemun-gu, Seoul, Korea

Savier Pham
Photon Dynamics, Inc.
San Jose, California, USA

Shinji Morozumi
Crystage, Inc.
Yodogawa-Ku, Osaka, Japan

Shulik Leshem
Orbotech, Ltd.
Yavne, Israel

Soo Young Choi
Applied Materials, Inc.
Santa Clara, California, USA

Tae Kyung Won
Applied Materials, Inc.
Santa Clara, California, USA

Tetshuhiro Ohno
ULVAC, Inc.
Chigasaki, Kanagawa, Japan

Woojae Lee
E&F Technology
Yongin-si, Gyeonggi-do, Korea

Yasunori Nishimura
FPD Consultant
Nara, Japan

YiLin Wei
AU Optronics, Inc.
Hsinchu, Taiwan

Yoon Heung Tak
LG Display
Gangseo-gu, Seoul, Korea

Youn Sung Na
LG Display
Wollong-myeon, Paju-si, Gyeonggi-do, Korea

Young Seok Choi
LG Display
Wollong-myeon, Paju-si, Gyeonggi-do, Korea

Young-Hoon Shin
LG Display
Wollong-myeon, Paju-si, Gyeonggi-do, Korea

Series Editor's Foreword

Flat panel displays (FPDs) have transformed consumer electronics and are driving profound changes in the way people around the globe interact with one another and engage with markets and information. Although many technical disciplines have contributed to the impact of FPDs, the most direct source of these changes has arguably been in the manufacturing field where cooperation across a growing and globally based industry has allowed an unprecedented scale of technology introduction and provided the high quality, large numbers and reasonable unit costs needed for mass adoption of these display devices. An understanding of the many processes which underpin FPD manufacturing is essential, not only to those directly engaged in production but also to scientists and engineers developing new display technologies, to materials and component suppliers and to those seeking to devise new processing routes, equipment and fabrication concepts. In all these cases, compatibility with established manufacturing capabilities will be essential for the rapid adoption of innovations in displays aimed at consumer electronic products with global reach. The central role of manufacturing processes extends much further; the availability of devices in different sizes and form factors, and the economic and environmental impacts of the displays industry have their roots in the established, optimised norms of FPD fabrication.

In this latest volume of the Wiley-SID book series, the editors and authors have compiled a broad and detailed account of the device structures and corresponding processes and operations involved in the manufacture of the active matrix OLED and LCD devices which dominate the modern display market. The different chapters cover the full scope of processing from preparation of display substrates to the final assembly and testing of completed modules, with specialist sections covering diverse topics such as automated inspection and defect repair, colour filter fabrication, yield analysis, semiconductor deposition, lithography, etching and thin film coating techniques. Active matrix technologies based on amorphous and polycrystalline silicon are fully covered, and topics which have become important more recently including oxide semiconductor backplanes, drop-filling of LCDs, flexible OLED display fabrication and encapsulation and many more are also included. Alongside these accounts of device processing, the reader will find an overview of critical topics which support the manufacturing process, such as plant design and layout, and environmental considerations in display production. Although this book concentrates on the leading LCD and OLED display modes, much of the content is also relevant to alternatives such as electrophoretic or electrowetting displays which share the need for an active matrix backplane to realise a high information content, fast switching product.

The authors and editors of this book bring to its contents an outstanding track record and knowledge of display technologies. Their experience spans senior roles in display manufacturing companies, in equipment suppliers and in both industrial and academic research and development. I believe that under the guidance of the editors, the authors have assembled material which will provide an essential reference and learning resource for display engineers, scientists and practitioners. For all interested, it also offers a fascinating insight into this unique field of large-scale, high-technology, multifunctional manufacture.

Ian Sage
Malvern, 2018

Preface

Many display-related books can be found today. While there are a number of excellent books on display technologies that cover the fundamentals and applications in many different areas, currently there is no book dedicated to modern flat panel display panel manufacturing.

The Wiley-SID book series organizers and our editorial team believe that there is a strong need in the display field for a comprehensive book describing the manufacturing of the display panels used in today's display products. The objective of this book, entitled "Flat Panel Display Manufacturing," is to give a broad overview for the key manufacturing topics, serving as a reference text. The book will cover all aspects of the manufacturing processes of TFT-LCD and AMOLED, which includes the fabrication processes of the TFT backplane, cell process, module packaging, and test processes. Additionally, the book introduces important topics in manufacturing science and engineering related to quality control, factory and supporting systems architectures, and green manufacturing. We believe this text will benefit the display engineers in the field by providing detailed manufacturing information for each step, as well as an overall understanding of manufacturing technology. The book can serve as a reference book not only for display engineers in the field, but also for students in display fields.

One might think that flat panel display manufacturing is a mature subject, but the state of the art manufacturing technologies enabling today's high end TFT-LCDs and OLED displays are still evolving for the next-generation displays. Considering the rapid progress and evolution of display technology today, such as flexible OLED displays and VR/AR wearable displays, we expect the new display manufacturing technologies will continue to evolve so rapidly to make this book just the beginning in a series on modern display manufacturing.

The editorial team, a group of veteran display engineers including Jun Souk, ex- Executive VP, CTO of Samsung Display, Fan Luo, ex-CTO of AUO, Shinji Morozumi, ex-President, PVI, Executive VP at Hosiden, Seiko Epson, and Ion Bita, Display Manager of Apple, decided to work together and prepare this comprehensive display book. This team invited authors from major display manufacturers and experts in each manufacturing topic (equipment, processing, etc.). We are grateful to all the authors who collaborated with us and shared their in-depth knowledge and experience to produce high-quality chapters. This book could not be completed without their diligence and patience. Last but not least, we are also grateful to the Wiley-SID book series team, and especially to Dr. Ian Sage for help planning, reviewing the early drafts, and bringing this book to light.

(2018)

The editorial team
Jun Souk
Shinji Morozumi
Fang-Chen Luo
Ion Bita

1

Introduction

Fang-Chen Luo[1], Jun Souk[2], Shinji Morozumi[3], and Ion Bita[4]

[1] AU Optronics, Taiwan
[2] Department of Electronic Engineering, Hanyang University, South Korea
[3] Crystage Inc., Japan
[4] Apple Inc., USA

1.1 INTRODUCTION

Flat panel displays (FPDs) have greatly changed our daily life and the way we work. Among several types of FPDs, the thin-film transistor (TFT) liquid crystal displays (TFT-LCDs) are presently the leading technology, with 30 years of manufacturing history. Recently, TFT-LCDs reached over 95% market share across TVs, computer monitors, tablet PCs, and smartphones, and are still expanding into other application areas.

The tremendous progress in TFT-LCD technology has brought us to the point where display performance, screen size, and cost far exceeded most industry leaders' expectation projected at the time of initial TFT-LCD production, which started in the late 1980s. Owing to a sustained, enormous effort of display engineers around the world for the past three decades, the performance of TFT-LCD has not only surpassed the original leading cathode ray tube (CRT) in most areas, but also the cost barrier, initially considered to be prohibitive for mass adoption, has been reduced significantly by rapid advances in display manufacturing technology.

In this chapter, we briefly review the history and evolution of display technologies, focusing mainly on the manufacturing technology associated with TFT-LCD. With organic light emitting diode (OLED) displays in the form of mobile and TV becoming flexible, and growing and drawing a great deal of attention, the current status of active matrix driven OLED (AMOLED) display manufacturing technology is explored in sections 8A, 8B and 10.

1.2 HISTORIC REVIEW OF TFT-LCD MANUFACTURING TECHNOLOGY PROGRESS

Counted from the early stage production of TFT-LCD notebook panels that happened in the late 1980s, display manufacturing technologies have evolved enormously over the past three decades in order to meet the market demand for applications and different pixel technologies for notebook, desktop PC, LCD-TV, and mobile devices. The progress did not come easily, but it was the result of continued innovations and efforts of many engineers across the industry that enabled adoption of new technologies, process simplifications, and increased automation in order to achieve such economics of scale. In this section, we review the historic evolution in TFT-LCD manufacturing technology from notebooks to LCD-TV application.

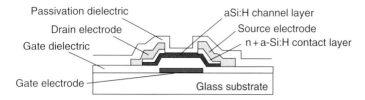

Figure 1.1 Schematic cross-section of an inverted-staggered a-Si:H TFT.

1.2.1 Early Stage TFT and TFT-Based Displays

TFTs, more precisely, insulated-gate TFTs, have been critical enablers for the development of flat panel displays. Figure 1.1 shows a schematic cross-sectional view of the most common type of TFT device, a so-called inverted staggered TFT. In this device, there is a gate electrode on the bottom, which is covered with an insulator, followed by the active semiconductor material and a top passivation insulator. The passivation insulator is etched back to allow fabrication of source and drain contacts to the semiconductor.

P. K. Weimer at RCA reported first working TFT devices by using CdS as the semiconductor material in 1962 [1]. Various active materials have been developed in addition to CdS: CdSe, polysilicon, amorphous silicon (a-Si), and so on. Among these, a-Si remains presently the most widely used due to its practical advantages over other materials.

The first a-Si TFT was reported by LeComber et al. [2] in 1979, and considered a major milestone in TFT history from a practical standpoint. The characteristics of a-Si are well matched with the requirements of liquid crystal driving and provide uniform, reproducible film quality over large glass areas using plasma enhanced chemical deposition (PECVD).

The active matrix circuit incorporating a field-effect-transistor and a capacitor in every pixel element for LC display addressing, still widely used today, was first proposed by Lechner in 1971 [3]. Fisher et al. [4] reported on the design of an LC color TV panel in 1972. The first attempt for a TFT-LC panel was reported in 1973 by the Westinghouse group led by Brody [5], which demonstrated the switching of one row of pixels in a 6×6-inch 20 line-per-inch panel. In 1973 and 1974, the group reported on an operational TFT-EL (electroluminescence) [6] and a TFT-LC panel [7] respectively, all using CdSe as the semiconductor. In the early 1980s, there were active research activities working on a-Si and high temperature polycrystalline silicon TFTs. Work on amorphous and poly silicon was in its infancy in the late 1970s and nobody had succeeded in building a commercial active matrix display using these materials. In 1983, Suzuki et al. [8] reported a small-size LCD TV driven by a-Si TFT and Morozumi et al. [9] reported a pocket-size LCD TV driven by high-temperature poly Si. The 2.1inch 240×240 pixel LCD-TV introduced by Seiko Epson (Epson ET-10) is regarded as the first commercial active matrix LCD product. It prompted the Japanese companies to intensify their efforts to build large-screen color TFT-LCDs. In 1989, Sera et al. reported on the low-temperature poly Si (LTPS) process [10] by recrystallization of a-Si film using pulsed excimer laser.

1.2.2 The 1990s: Initiation of TFT-LCD Manufacturing and Incubation of TFT-LCD Products

In 1988, Sharp produced first 10.4-inch a-Si based TFT-LCD panels for notebook PC application, which launched the TFT-LCD manufacturing. In 1992, DTI (a company jointly owned by IBM Japan and Toshiba) introduced a 12.1-inch SVGA panel that was used for the first color laptop computer introduced by IBM (Figure 1.2).

Until that time, the product yield of 10.4-inch or 12.1-inch LCD panels stayed in a very low level, leading to very expensive panel prices. The industry was not yet fully convinced that the large-screen sized TFT-LCD panels could enter a mass production scale with a proper production yield and meet the cost criteria. Nonetheless, the demand of the full-color TFT-LCD portable laptop computer was very high despite its high price tag (the initial price of IBM CL-57SX was almost $10,000) and thus was able to accommodate the

Figure 1.2 12.1-Inch TFT-LCD introduced in 1992 and the first color laptop computer by IBM.

unusual high price of notebook display panels (LCD panel cost was nearly 70% of that of laptop computer), opening the door for the expansion of LCD mass production.

Triggered by the demonstration of high image quality large-screen 12.1-inch LCD panels, competition on larger LCD panels for laptop computers continued from 1994 to 1998. At the same period, determining an optimum mother glass size that can produce "future standard size" laptop screens was a critical issue in the LCD industry. Starting with 1995, three Korean big companies, Samsung, LG, and Hyundai, also entered the TFT-LCD business.

Figure 1.3 shows the landscape of LCD companies adopting different mother glass sizes to manufacture notebook display panels. The mainstream LCD panel size increased from 9.4 inches to 14.1 inches from 1993

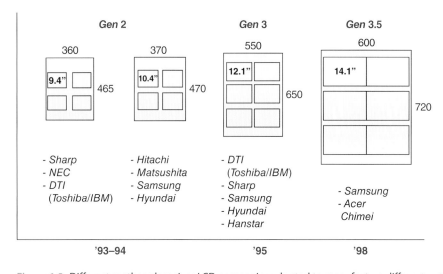

Figure 1.3 Different mother glass sizes LCD companies adopted to manufacture different notebook panel sizes.

to 1998, therefore, mother glass sizes increased accordingly from Gen 2 sizes (360×475 or 370×479 mm) to Gen 3.5 size (600×720 mm).

The panel size competition settled when NEC introduced 14.1-inch XGA panels in 1998. The expansion of TFT-LCD so rapidly progressed that it finally overturned the competitive super twisted nematic (STN) passive matrix LCD market in 1996, which has been used for a long time as the main display panel for laptop computers. This was possible due to the cost down effort and the superior resolution and color performance of TFT-LCD. Since then, between 1990 and 2010, the mother glass sizes of TFT-LCD plants have continuously increased to higher generation every two or three years.

Increasing the mother glass size is the most efficient way to reduce panel cost and meet the trend of increasing panel size. Furthermore, the LCD manufacturing technology progressed with numerous technology innovations such as reduced process steps and enhanced productivity with the introduction of new concepts for process equipment.

In the early 1990s, TFT-LCD panels were primarily used for notebook PC applications. In 1997, 15-inch diagonal panels were produced for initial LCD desktop monitor applications.

1.2.3 Late 1990s: Booming of LCD Desktop Monitor and Wide Viewing Angle Technologies

After TFT-LCD penetrated laptop computer screen by major portion reaching near 25 million units in 2000, the TFT-LCD industry had plans to enter the desktop monitor market, which at that time was 100% CRT. However, TFT-LCD was far behind in optical performance especially for viewing angle characteristics. In order to compete with CRT, which has a perfect viewing angle by its emissive display nature, viewing angle improvement became the most urgent requirement for the TFT-LCD industry before launching into the desktop monitor market. In that regard, wide viewing angle was no longer considered as a premium technology at that time, but was regarded as a standard feature.

In 1993, Hitachi announced the development of IPS (in plane switching) technology. In 1995, M. Ohta and a group at Hitachi built the first 13.3-inch color IPS panel [11]. IPS technology was commercially initiated by Hitachi and the technology demonstrated itself excellent viewing angle capabilities due to the nature of horizontal (in plane) movement of liquid crystal molecules with respect to the substrate plane. A few years later in 1998, Fujitsu announced the development of MVA (multi-domain vertical alignment) technology based on VA technology [12], in which building protrusion shapes on each pixel electrodes generates a fringe field for LC molecules and widens the viewing angle. However, the two technologies showed a large difference in device structure, process, and display panel characteristics. In order to establish a dominant market share over CRT monitors, a number of cost reduction features were rapidly pursued for incorporation into the production flow. In this regard, it was considered very important to ensure practical wide viewing technology that provides both a wide viewing performance as well as a high productivity. The alignment layer rubbing-less feature of VA resulted in key advantages for easiness in processing over larger mother glass, reduced a process step, and offered a wide process margin for screen uniformity. Throughout the LCD industry's growth, IPS and VA technology groups historically competed with each other. This healthy competition promoted the progress of each technology. Owing to the rapid advance of wide viewing angle technologies, CRT monitor replacement has grown steadily since 1998. With the appearance of 17-inch and 18.1-inch LCD monitors equipped with PVA (patterned VA) and IPS technology, respectively, desktop monitor sales have doubled every year between 1998 and 2001 (Figure 1.4).

1.2.4 The 2000s: A Golden Time for LCD-TV Manufacturing Technology Advances

Followed by the rapid LCD desktop monitor market expansion, the LCD industry was knocking on the door of the TV market. However, in the late 1990s, the biggest mother glass size was Gen 4 sizes (730×920 mm) and the manufacturing technology for larger panel size was not ready at that time. In addition, color performance, contrast ratio, and motion picture quality of TFT-LCD were not sufficient for the LCD-TV application. Since then, intensive efforts were assembled in order to overcome these handicaps, with especially motion picture

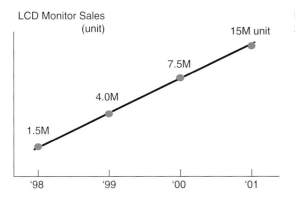

Figure 1.4 TFT-LCD sales doubled every year during 1998 and 2001 due to the rapid expansion of LCD desktop monitors.

quality being the most urgent item to be improved for target TV applications. From late 1990s to early 2000s, numerous new technologies for LCD TV were developed in the area of high transmittance, high contrast ratio, high color gamut panel fabrication, wide viewing angle technology, and motion picture enhancement technology. During this period, TFT-LCD technology as well as manufacturing technology advanced rapidly. This period was regarded as a golden time for LCD manufacturing that led the LCD technology to today's mature technology.

In 2001, a prototype of a 40-inch HD grade (1280×768) TFT-LCD TV panel was introduced by Souk et al. [13] The panel demonstrated not only a size breakthrough at the time, but also it demonstrated a 76% NTSC color gamut, 12 msec response time, screen brightness 500nits, and PVA wide viewing angle technology that triggered the large-size TV market.

In 2005, the first Gen 7 size factory (1870×2250 mm) was built by Samsung Electronics and began to produce TV panels. One mother glass could produce 12 panels of 32 inches, 8 panels of 40 inches, or 6 panels of 46 inches as shown in Figure 1.5.

The flexibility to produce multiple display panel sizes from the available mother glass substrates initiated the full-scale production of LCD-TV followed by a rapid panel size increase from 40-inch to 42-inch, 46-inch, and 52-inch TV applications as shown in Figure 1.6.

The successful launching of Gen 7 size lines for large-size TV panels triggered the opening of the LCD-TV market. From the period of 2005 to 2010, the average TV panel size increased 2.5 inches every year. In October 2006, Sharp Corp. started the first Gen 8 size factory (2160×2460 mm), capable of producing 8 units of 52-inch LCD TV panels per glass. Through the rapid increases in volume production of LCD-TV panels, LCD-TV surpassed CRT TV volume in 2008 (Figure 1.7).

1.3 ANALYZING THE SUCCESS FACTORS IN LCD MANUFACTURING

The rapid growth of TFT-LCD market backed by the rapid reduction of panel cost was possible due to the fast LCD manufacturing technology progress. The advance of LCD manufacturing technology led to productivity enhancements and panel cost-down, that in turn contributed to the fast expansion of LCD market. The progress of LCD manufacturing technology was attributed to optimizing each individual process in the TFT-LCD process flow, as well as the advancement of process equipment and fab layout. The key factors that contributed to the rapid cost down were productivity enhancements by:

(1) moving to larger mother glass size.
(2) process simplification effort, such as four mask step TFT process and LC drop filling process.
(3) efficient fab layout in conjunction with equipment technology advance that enabled higher throughput and reduced standby time.

Figure 1.5 First Gen 7 size mother glass factory, allowing fabrication of 8 units of 40-inch TV panels per substrate (2005).

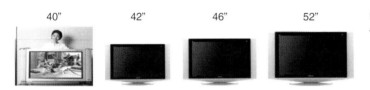

Figure 1.6 LCD-TV panel sizes increase from 40 inches to 46 inches fabricated in Gen 7 lines.

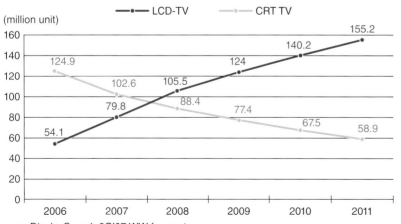

Figure 1.7 The volume of LCD-TV exceeded CRT TV in 2008.

Source: DisplaySearch 3Q'07 WW forecast

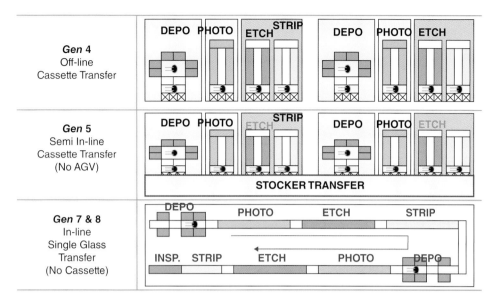

Figure 1.8 Fab layout evolved to enhance glass handling efficiency, from transferring individual cassettes to the single glass substrate transfer system.

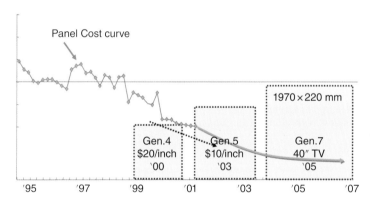

Figure 1.9 Average panel cost of $20 per diagonal inch dropped to $10 per diagonal inch when the industry moved from Gen 4 size to Gen 5 size.

Fab layouts evolved from configurations based on process equipment using individual cassette transfer systems, to use of inline single glass substrate transfer systems to enhance glass transfer efficiency as shown in Figure 1.8.

The transition to larger mother glass sizes demonstrated to be the most effective way to reduce the panel cost. Figure 1.9 showed the average panel cost $20 per diagonal inch, which dropped to $10 per diagonal inch when the industry moved from Gen 4 size to Gen 5 size substrates.

1.3.1 Scaling the LCD Substrate Size

The initial mother glass size started with Gen 1 size (300×400 mm) in the mid-1980s, and was used as TFT-LCD pilot lines. Since then, the glass size evolved to Gen 2 size, Gen 3 size, Gen 4 size, and so on, as shown in Figure 1.10, with the largest to date reaching Gen 10 size (2880×3130 mm) at the factory built by Sharp Corp. in October 2009. The competition on the larger mother glass size doesn't seem to end here. BOE

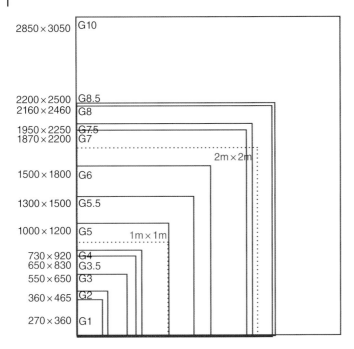

Figure 1.10 TFT-LCD substrate sizes for each generation that increased from Gen 1 size to Gen 10 size (Source: AU Optronics Corp. 2012).

Group is building its Gen 10.5 size LCD panel fabrication plant in Hefei, China. With the construction of the Hefei fab underway, which is scheduled for mass production in 2018, BOE will be capable of processing glass substrates that reach 3370 × 2940 mm. In addition to the construction of Gen 10.5 size fab, China Star Optoelectronics Technology (CSOT), a subsidiary of TCL Group, will kick off its construction project of the world's largest Gen 11 size LCD panel fabrication plant in Shenzhen, China.

The productivity of LCD panels, measured by the number of display panels produced per mother glass, naturally increases as the mother glass size increases from Gen 3.5 size to Gen 8.5 size, as shown in Figure 1.11.

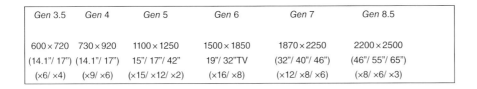

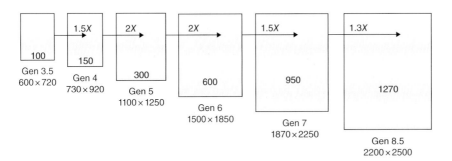

Figure 1.11 Panel productivity and size increase as mother glass size evolves from Gen 3.5 size to Gen 7 size.

1.3.2 Major Milestones in TFT-LCD Manufacturing Technology

There have been numerous technology evolutions during the 30 years of TFT-LCD manufacturing history, from the late 1980s to the present, which can be found in every process step, materials, and equipment technology advances; these combined efforts contributed to make TFT-LCD the dominant flat panel display nowadays. Nonetheless, we select three most significant technology revolutions, based on the consideration that without these revolutionary technologies, TFT-LCD would not become a commodity product and the industry would not able to grow as much. These three selected technologies are:

(1) AKT cluster PECVD tool
(2) Wide viewing angle technology
(3) Liquid crystal (LC) drop filling technology

1.3.2.1 First Revolution: AKT Cluster PECVD Tool in 1993

Scientists and engineers in CVD technology believed that a high vacuum process should be used for a-Si TFT fabrication in order to prevent ppm level contamination of oxygen and carbon in the films. The initial PECVD tools were the inline high vacuum system. The overall process time for depositing a-Si TFT took a long time and the PECVD process itself was regarded as a particle generation process. The cleaning of each chamber wall took nearly half a day. In early stage of TFT-LCD production, PECVD process was the most time consuming and troublesome process. In 1993, AKT introduced a new concept PECVD tool, equipped with mechanical pumps only without high vacuum pumps, fast glass handling cluster type chamber arrangement, and a convenient in situ chamber cleaning method. In situ chamber cleaning with NF_3 gas greatly reduced CVD film particle problem and maintenance time (Figure 1.12).

The appearance of this innovative concept tool significantly reduced the burden of a-Si TFT process.

1.3.2.2 Second Revolution: Wide Viewing Angle Technology in 1997

Throughout the LCD industry's growth, wide viewing angle technologies, IPS and VA (Figure 1.13), contributed the most in replacing CRT monitors in desktop monitors and TVs. Without the wide viewing angle technology advance, the widespread adoption of LCDs would not happen.

Early stage Anelva/ Shimazu in-line high vacuum PECVD tool

AKT Cluster PECVD tool

Figure 1.12 Cluster type PECVD tool from AKT.

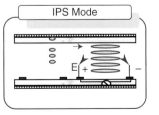
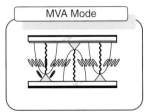
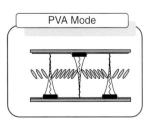

Figure 1.13 The structure of three wide viewing angle modes: IPS, MVA, and PVA.

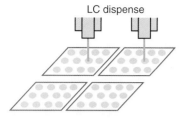

Figure 1.14 Schematic picture of LC drop filling process. The accurately measured amount of LC is dispensed directly onto each TFT panel.

1.3.2.3 Third Revolution: LC Drop Filling Technology in 2003

Before the invention of the liquid crystal drop fill method, the traditional liquid crystal filling process was vacuum filling process that has been used for a long time in the industry. This process was time consuming, by filling the liquid crystal into the cell gap of the assembled TFT and CF glass by capillary action. For example, the LC fill process itself took a few days for a 40-inch LCD panel in early stage of LCD-TV prototype in the early 2000s.

The appearance of LC drop filling technology, as well as the accurately measured amount of LC is dispensed onto a TFT panel directly (Figure 1.14), greatly reduced the bottleneck process and contributed significantly to the growth of LCD-TV market that started in 2003.

1.3.3 Major Stepping Stones Leading to the Success of Active Matrix Displays

Besides the abovementioned three technology revolutions, there were numerous major technology contributions that led to the success of TFT-LCD and AMOLED. We list other major stepping stone technologies that led to the success of active matrix displays [14].

- 1962 First working TFT using CdS, RCA
- 1971 Active matrix circuit, RCA
- 1974 First working TFT-LCD panel, Westinghouse
- 1976 First TFT-EL/TFT-LCD video panel, Westinghouse
- 1979 First working a-Si TFT, U of Dundee
- 1983 First working a-Si TFT-LCD, Toshiba
- 1983 First working poly-Si TFT-LCD, Seiko Epson
- 1984 9.5-inch 640 × 400 CdSe TFT-LCD, Panelvision
- 1988 First 9-inch a-Si color TFT-LCD for avionics applications, GE
- 1989 First 14.3-inch a-Si TFT-LCD for PC, IBM, Toshiba
- 1990 First large-scale Gen 1 TFT-LCD manufacturing, Sharp
- 1995 IPS, Hitachi
- 1997 14.1-inch notebook panel production, Samsung
- 1997 14-inch and 15-inch monitor TFT-LCD panel production
- 1998 MVA, Fujitsu
- 1999 First commercial AMOLED by Pioneer
- 2000 LED backlight, IBM Research
- 2001 40-inch TFT-LCD TV, Samsung
- 2003 First 52-inch TFT-LCD TV
- 2003 First color AMOLED product in camera, Kodak/Sanyo
- 2005 First 82-inch TFT-LCD TV
- 2006 First 102-inch TFT-LCD TV
- 2012 Large-area AMOLED TV panels, Samsung and LGD

References

1. P. K. Weimer, Proc. IRE, 50, p. 1462 (1962).
2. P. G. LeComber, et al., Electron Lett., 15, p. 179 (1979).
3. B. J. Lechner, Proc. IEEE, 59, p. 1566 (1971).
4. A. G. Fisher, et al., Conf. Rec. IEEE Conf. On Display Devices, p. 64 (1972).
5. T. P. Brody et al., IEEE Trans. Electron Devices, ED-20, p. 995 (1973).
6. T. P. Brody, et al., IEDM, Washington DC (1973).
7. F. C. Luo, et al., WESCON (1974).
8. K. Suzuki, et al., SID '83 Digest, p. 146 (1983).
9. S. Morozumi, et al., SID '83 Digest, p. 156 (1983).
10. K. Sera et al., IEEE Trans. Electron Devices, ED-36, p. 2868 (1989).
11. M. Ohta and K. Kondo, Asia Display 95, p. 707 (1995).
12. A. Takeda, et al., SID '98 Digest, p. 1077 (1998).
13. J. Souk, et al., SID'02 Digest, p.1277 (2002).
14. Fan Luo, Keynote Talk, "Major Stepping Stones Leading to the Success of Active Matrix Displays" IDMC, Taiwan (2013).

2

TFT Array Process Architecture and Manufacturing Process Flow
Chiwoo Kim

Seoul National University, Daehak-dong Gwanak-gu, Seoul, Korea

2.1 INTRODUCTION

Liquid crystal display (LCD) technology has evolved over the years and presently holds a dominant position in the flat panel display market. Amorphous silicon (a-Si)–based thin-film transistor LCDs (TFT-LCDs) were developed to adopt large glass substrates. a-Si TFT-LCD grew rapidly in the 1990s through the adoption in laptop and notebook PC displays. The size of the glass substrates has been continuously enlarged to improve productivity. Since the a-Si TFT-LCDs have cost competitiveness as well as scalability, they are currently being used from small-size mobile devices to large size TV screens.

Recently, as the display resolution in both mobile devices as well as in large-size TVs has been continuously increasing, it is necessary to develop next-generation TFTs with higher electron mobility, electrically stable, and uniform performance. Smartphones, which are the fastest-growing products among mobile devices, require resolutions up to WQHD (2560×1440) or even higher. In addition to resolutions, the mobile devices require reduced power consumption by increasing the transmittance of each pixel on the TFT-LCD backplanes and low-voltage driving. The maximum aperture ratio of the pixel can be obtained by using fine design rule, minimizing the storage capacitor and TFT sizes. To achieve that, high-resolution photo-lithography, high-mobility TFTs and self-aligned TFT structure are required.

The conventional a-Si TFT structure has large parasitic capacitance and low electron mobility. With a-Si TFT backplane, it is impossible to handle the high-resolution mobile LCD displays. Even worse, the bias stability of the a-Si TFT is not good enough for use in active matrix organic light emitting diode (AMOLED) display. Alternatively, poly-silicon (poly-Si) TFT with thin gate insulator and self-aligned structure is the ideal backplane solution for producing high-resolution and low-power consumption TFT-LCDs as well as AMOLED displays. Poly-Si has an electron mobility about 200 times higher than a-Si, due to its crystalline atomic structure (Figure 2.1) [3]. For the mass production of poly-silicon TFT-LCDs, it is necessary to develop low-temperature manufacturing techniques for compatibility with alumina borosilicate large glass substrates, rather than costly and small-sized quartz wafers. The development of excimer laser annealing (ELA) for the crystallization of the a-Si precursor was a breakthrough for the use of glass substrates, enabling the mass production of low-temperature poly-silicon (LTPS) LCDs. The high mobility of LTPS brings many benefits to TFT-LCDs. The dimension of LTPS TFT is smaller than the a-Si TFT, which directly leads to higher aperture ratio pixels and narrower panel border width. The high aperture ratio reduces backlight

Flat Panel Display Manufacturing, First Edition. Edited by Jun Souk, Shinji Morozumi, Fang-Chen Luo, and Ion Bita.
© 2018 John Wiley & Sons Ltd. Published 2018 by John Wiley & Sons Ltd.

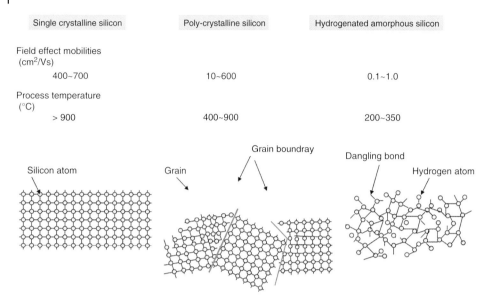

Figure 2.1 Comparison of single crystalline, poly crystalline, and hydrogenated amorphous silicon.

power consumption. The small TFT size further reduces the driving power consumption because the electrode overlapped parasitic capacitance becomes smaller. In addition, LTPS enables much higher pixel density without increasing the bezel form factor compared to a-Si because LTPS allows integrating the gate drivers and the data multiplexing circuits on the glass substrate [3]. This circuit integration capability enables the interconnections within small area.

Presently, the LTPS demand is significantly increased due to the requirement of high-resolution displays for smartphones. In addition, the AMOLED display is considered the next-generation display technology after TFT-LCD because it has higher optical performance, lower power consumption, and thinner form factor than LCD. The OLED pixel is operated by electric current, whereas liquid crystals in LCDs are driven by electric voltage. For high-resolution mobile AMOLED displays, LTPS TFTs are thus the primary choice since high electron mobility and high device stability are required. It is expected that, eventually, when the OLED material cost is reduced through mass production and the production yield is matured, the production cost of mobile OLED displays can be lower than LTPS based TFT-LCDs and the OLED displays dominate the mobile display market.

In 2004, the amorphous oxide semiconductor (OS) indium-gallium-zinc-oxide (IGZO) was introduced as a new TFT channel material [2]. Since then, many groups started research and development of OS materials for display applications. IGZO has relatively high carrier mobility and low leakage current while using a-Si like manufacturing processes. OS TFT's potential for high pixel density, low power consumption, and high-performance display-integrated touch screen in display applications were reported [2, 3]. Mass production of IGZO technology has already started for high-resolution TFT-LCD applications, although the volume is not yet comparable with conventional silicon-based backplane technologies. Furthermore, there are possibilities to use other OS materials for display TFT applications besides IGZO. For example, the potential use of Hf–In–Zn–O, which has higher mobility than IGZO, was reported by another group [4]. Since the manufacturing process of OS TFTs is compatible with a-Si infrastructure, the production cost of OS TFT is lower than the LTPS while offering reasonable electron mobilities. Thus, OS TFT is regarded as a good candidate for AMOLED TV backplanes as well [5].

2.2 MATERIAL PROPERTIES AND TFT CHARACTERISTICS OF a-Si, LTPS, AND METAL OXIDE TFTS

2.2.1 a-Si TFT

The semiconductor and dielectric layers in the TFT device structure are normally deposited by plasma enhanced chemical vapor deposition (PECVD). For a-Si TFT, the PECVD process is used for the silicon nitride (SiN_x) gate dielectric layer, a-Si intrinsic semiconductor layer, the phosphorus doped n+ a-Si layer for source and drain ohmic contact regions, and for the passivation SiN_x layer. For the back channel etch (BCE) TFT structure, three layers composed of SiN gate dielectric, a-Si intrinsic, and n+ a-Si are deposited in sequence without breaking vacuum in PECVD cluster tools. This is the key process step for BCE a-Si TFT. Using this tri-layer sequential deposition within the same tool, the process flow can be significantly simplified and the impurities between three layers can be minimized for achieving stable TFT performance. To reduce cross-contamination between layer materials, these deposition processes are normally carried out in separate chambers of the cluster tool [6]. For the etch stopper (ES) type TFT, SiN_x, a-Si, and ES-SiN_x layers are deposited sequentially. After patterning ES and a-Si layers, n+ a-Si layer is deposited. The pre-deposition cleaning process of n+ a-Si deposition is critical for ohmic contact.

Key requirements for PECVD systems include good uniformity over large areas and a relatively low deposition temperature (<350°C) for compatibility with large glass substrates. A representative PECVD deposition chamber structure is illustrated in Figure 2.2 [6]. It is a parallel plate reactor housed in a vacuum chamber, and consists of a grounded heated substrate holder (susceptor) and a gas shower head that distributes the flow of reactant gases for uniform material growth. The shower head is powered by a 13.56 or 27.1 MHz RF generator. The basic reactant gas for a-Si is silane (SiH4), with phosphine (PH_3) added for depositing n+ a-Si layers, or with ammonia (NH_3) added for depositing SiN_x. In addition, carrier gases such as hydrogen or nitrogen can also be added. The key parameters of the deposition process are the substrate temperature, the RF power, the parallel plate spacing, the gas partial pressure and composition, and the pumping speed.

The atomic structure of PECVD Si films only includes short range order from the fourfold Si coordination. This short range order leads to corresponding weak bonds within the material. Hydrogen plays a key role of defect passivation within the material during the a-Si film deposition [7, 8]. The hydrogen atoms passivate the Si dangling bonds (DBs) forming stable Si–H bonds. The DB density of hydrogenated amorphous silicon is typically reduced from $10^{20}/cm^3$ to $10^{16}/cm^3$. The lack of a long-range atomic potential results in both localized as

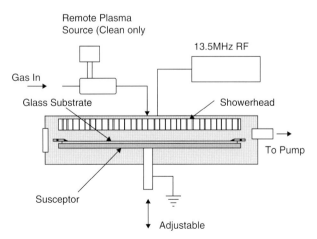

Figure 2.2 A plasma CVD system used for depositing a-Si, n+ a-Si, and SiN_x films. Silane (SiH4) gas is dissociated in an RF chamber to deposit a-Si, with ammonia and nitrogen gases added to deposit the SiN_x layer. Phosphine (PH_3) can be added to deposit heavily doped p+ a-Si [6].

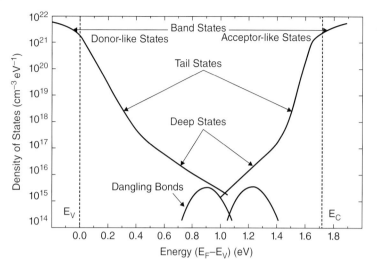

Figure 2.3 Density of states distribution of a-Si [7].

well as extended, electron states. The localized states are characterized by a distribution of band tail states extending from both the conduction and valence bands into the electronic band gap, resulting in a "mobility gap" of 1.85eV for a-Si [8]. The density of states (DOS) distribution in a-Si:H is shown in Figure 2.3 [7] illustrating both the band tail states and deep states, the density of which is given by the summation of the DB densities.

Silicon nitride films are used for both gate dielectric as well as passivation layer of a-Si TFTs. The preference for SiN_x in a-Si:H TFTs is due to the reduced DOS resulting from the positive charge in the nitride [7].

2.2.2 LTPS TFT

The current smartphone displays are yet behind in respect to the resolution of human eye and the a-Si TFT is way behind in performance to meet the current display requirements [1]. To enable high-resolution TFT-LCD or AMOLED, high mobility and electrically stable TFT backplane technologies are vital. Since the pixel density is in a trade-off relationship with power consumption, low power consumption technologies like high aperture pixel design and low power driving are critical. For slim border design, the driving circuits like shift register and data multiplexer need to be integrated on the display panel itself. These circuits have traditionally been located external to the panel, in silicon chips, requiring bulky fan-out pads on the panel edge and costly added module assembly processes.

Polycrystalline silicon has electron mobility 200 times higher than that of a-Si, typically ~100cm^2/Vs versus ~0.5cm^2/Vs. This allows poly-Si to be usable for peripheral driving circuitry, which requires higher speeds and current carrying capabilities than possible with a-Si. Poly-Si even provides CMOS capability that is missing for a-Si due to its poor p-type device performance. Poly-Si CMOS allows the fabrication of low power-consumption circuits. As a result, poly-Si can provide both the functionality and the performance necessary for low-power driver circuits on large area glass substrates. Highly integrated poly-Si TFT-LCDs are already in mass production.

In the early stages, poly-Si films were directly deposited by low pressure chemical vapor deposition (LPCVD) process. LPCVD with deposition temperatures of ~620°C allowed direct growth of poly-Si with grains of about 100nm. The electron mobility was below 10cm^2/Vs [9]. The LPCVD poly-Si process was replaced by the low-temperature a-Si precursor of PECVD at 350°C combined with subsequent crystallization processing, since the LPCVD deposition temperature was too high and the resulting TFT performance was not good enough.

Crystallization technologies allowing conversion of a-Si precursor films to poly-Si film can be classified into two groups related to furnace annealing and laser annealing, as shown in Figure 2.4. Solid-phase crystallization

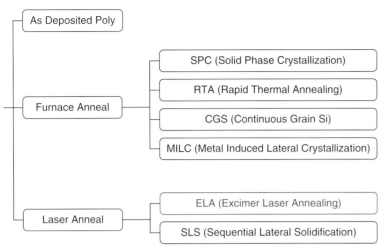

Figure 2.4 Crystallization technologies to a-Si precursor film to poly-Si film.

(SPC) has been developed for cost-competitiveness and scalability to large area displays [10]. Nonetheless, SPC requires long annealing times to crystallize a-Si film at high temperature of 600°C, and thus researchers have utilized metal impurities to promote crystallization kinetics. To reduce the annealing temperature, metal impurities are added before annealing. Metal-induced crystallization (MIC), metal-induced lateral crystallization (MILC), and super grain silicon (SGS) technologies have been studied in depth as strong candidates for SPC methods. Metal impurities can effectively reduce the process time and annealing temperature significantly. The minimization of metal impurities in the crystallized Si film is the key process and SGS has the advantage because the SGS process includes a capping layer on top of the a-Si precursor. During the annealing process, only a small amount of Ni atoms diffuses through the capping layer to form Ni-silicide nuclei that allow converting a-Si to poly-Si. Still, it was extremely difficult to produce high-quality displays because of the leakage current associated with the remaining Ni impurities and the non-uniformity associated with the low crystallinity.

For the melt-mediated crystallization, many researchers and equipment companies have attempted to develop various types of laser systems. The Excimer laser annealing (ELA) process [1, 12, 13], in which a-Si film is irradiated several times to obtain uniform poly-Si film, is widely used in flat panel display manufacturing. Another laser based technology is the sequential lateral solidification (SLS) process [1, 11, 14, 15], which includes (1) complete melting of pre-determined area of a-Si film, resulting in controlled super-lateral growth (C-SLG), (2) substrate movement with respect to the previous laser beam and re-irradiation of the panel, and (3) iteration of the above steps. These melt-mediated crystallization processes require relatively expensive laser system and operation cost.

In Figure 2.5, three types of poly crystal films are compared. As-deposited poly shows small grain structure and the electron mobility is the lowest. MIC shows relatively big grain structure and high electron mobility, but bad uniformity. ELA shows the big grain structure and also the electron mobility is the highest among them. ELA processed poly-Si shows best TFT performance and uniformity, but the process cost is highest.

2.2.2.1 Excimer Laser Annealing (ELA)

LTPS processes normally use laser-induced melt-mediated crystallization of thin silicon films. Polycrystalline films are grown on glass or even on plastic substrates. The excimer laser is widely used as the laser source because it can offer desirable characteristics of high pulse energy, strong absorption, and a low degree of coherence. ELA has found the preferred method of processing LTPS films due to its balance of functionality, yield, cost, and scalability to large substrates. Excimer lasers are gas lasers operating in the ultraviolet wavelength range, from 193 to 351nm depending upon the gas mixture. For crystallization of a-Si, the preferred

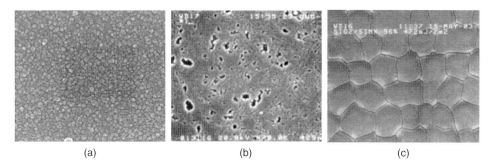

(a) (b) (c)

Figure 2.5 Images of film morphology of (a) as-deposited, (b) MIC, and (c) ELA poly-Si films.

gas mixture is XeCl giving a wavelength of 308nm, which being a longer wavelength is less damaging to the optical components in the beam path. Pulsed lasers with a typical pulse duration of 28ns and a maximum frequency of 600Hz can deliver from 350mJ/cm^2 to 600mJ/cm^2 in order to optimize the microstructure of the poly-Si films [12, 13]. Few years ago, the largest glass size for LTPS mass production was Gen 4 size (730mm × 920mm). Today, Gen 6 size (1500mm × 1850mm) LTPS process is under mass production. Even Gen 8 size (2200mm × 2500mm) process is available [12, 13]. To support the larger glass, ELA equipment has made great progress in throughput, yield, and cost efficiency. It has been confirmed that ELA line beam scaling is not a fundamental issue, and does not limit the application of LTPS to small glass sizes. The development of a very high power excimer laser in combination with a unique line beam optics system now plays a pivotal role in LTPS production.

A schematic illustration of an ELA crystallization system is shown in Figure 2.6 [13], where the key components are an optical unit for controlling beam intensity, the homogenizer and beam shaper to produce the line-beam, and a condensing lens to focus the beam on the underlying plate. The plate is mechanically swept through the short axis of the beam at a rate of typically 10 to 30 shots per point for better uniformity of the poly-Si. So the plate translation distance between shots is typically in the range 12 to 50μm for short axis width of 350 to 500μm.

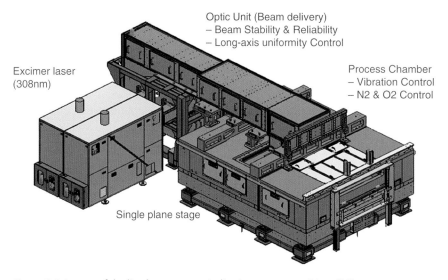

Figure 2.6 Layout of the line beam system indicating sensors positions [13].

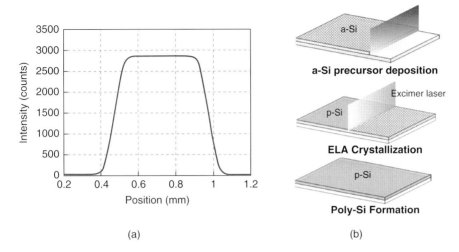

Figure 2.7 (a) Short-axis cross section of laser beam line, (b) ELA process steps.

The raw pulse shape is semi-Gaussian (Figure 2.7(a)), and beam shaping optics are used to produce a highly elongated line-beam, whose dimensions can be up to 1500mm in the long axis and 400μm for the short axis. The steep edges in the short axis profile have led to the beam shape being referred to as a flat-top beam. The up-scaling of laser power and line beam length to enable high throughput on large glass substrates is achieved. The laser can overlap the pulses from the two oscillators to provide a high energy of 2 Joules per pulse or it can trigger the oscillators in an alternating scheme leading to 1 Joule per pulse at 1200Hz [3]. Using multiple oscillators has also led to a breakthrough in the lasers stability, which is a challenging task at high power level when using conventional technology. The concept of using multiple oscillators provides further scalability of the power. Four or six oscillator systems that deliver a total of 4 or 6 Joules are available [13].

ELA process is shown in Figure 2.7(b). First, bare glass is covered by blocking SiO_2 layer to protect the Si from impurities from the glass. Then, the a-Si precursor film is deposited on top of the blocking layer. a-Si precursor of 400Å to 500Å thick film is deposited by PECVD. Film thickness control and the thickness uniformity are important because the poly-Si grain size is directly related to the film thickness and the laser energy [16, 17]. ELA step is next. Laser beam irradiated several times to obtain uniform poly-Si film. Glass is moving continuously with respect to the previous laser beam with the speed of the corresponding multi-shot distance.

The single laser beam crystalized a-Si precursor film is shown in Figure 2.8. The melt depth of the film is the key parameter in determining the outcome of the process [16, 17]. At 308nm wavelength, the optical absorption depth in a-Si is 7.6nm. The incident energy is strongly absorbed in the a-Si film, resulting in intense heating. If the incident energy density is high enough, this will heat the film to its melting point of 1420°C. When the top-flat shape 28ns laser beam in Figure 2.8(a) is irradiated on the a-Si precursor film, a-Si is instantly melt and crystalized. The grain shapes on the line beam edge region is shown in Figure 2.8(c). There are three melting regions, that is, partial melting region A, near complete melting region B, and complete melting region C. In the partial melting region A, crystallized film has a vertically grown appearance, with mid-sized grains (100nm). When the film is fully melted in region C, small solid islands remain at the back of the film. Between C and A region, where laser energy density is just matched to melt the a-Si precursor film precisely, big and uniform grains are obtained.

It is apparent that the optimum energy range is B region. The energy density above and below this region result in smaller grain poly-Si as shown in Figure 2.9. This process was identified by Im et al., [16, 17] and given the B region's name super-lateral growth (SLG). This SLG region is the crystallization

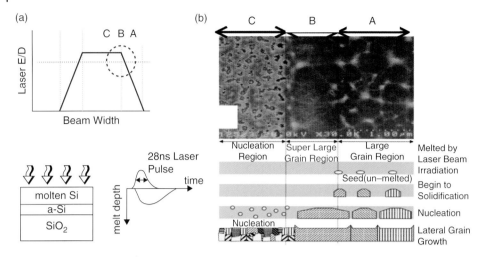

Figure 2.8 Single laser beam crystalized a-Si precursor film. (a) 28ns flat-top laser beam's short axis profile and the vertical stack up of the irritated a-Si precursor film on the blocking layer (b) Crystallized film structures of the region A, B, and C.

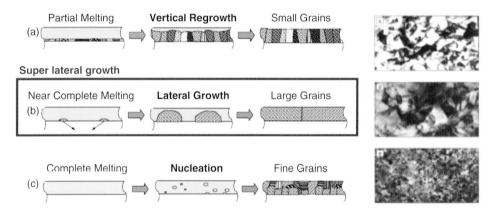

Figure 2.9 Phase transformation of a-Si for various laser energy densities.

regime, which yields high-quality TFTs. For the fully melted film in Figure 2.9(c), the seeding centers responsible for the SLG growth are lost. And due to the random nucleation, only fine grain film is grown in this complete melting region. Unfortunately, as shown in Figure 2.9(b), the SLG regime occurred over a very limited energy density range of about 10mJ/cm^2. These special results were observed in single shot irradiations (Figure 2.8), and qualitatively identical results were obtained from multi-shot irradiations, apart from a growth in SLG grain size with increasing shot number. The narrow energy density window of the SLG process is the major hurdle for the implementation of the crystallization process to mass produce high performance TFTs [16, 17].

To obtain high carrier mobility devices, the film needs to be crystallized within the SLG regime. But, there is limited accuracy in the precise setting of the laser energy density and the pulse-to-pulse fluctuations mean that the panels will be exposed to higher intensity irradiations, and the TFT characteristics such as V_{th} or mobility becomes non-uniform [13, 18]. In addition, a-Si precursor film's thickness uniformity is not perfect, either. Usually, the uniformity of the CVD a-Si precursor film is higher than 5% and this film thickness distribution affects the long-range non-uniform crystal growth especially at the substrate edge region. When an anomalously high-intensity pulse has fully melted the region, the next pulse will overlap most of this poorly crystallized region, and the fine grain material will be re-melted, and converted back to the large grain SLG material.

Nevertheless, there remains a thin stripe region (equal to the plate translation distance, 10 to 30μm), which will not be fully overlapped by the top-hat region of the next pulse. To get better TFT uniformity, slower plate translation is preferable, but it reduces the process throughput. Plate translation speed (number of shot overlap) is the most important parameter to control the shot induced mura. This shot mura is also related to the laser energy density, and the pixel design parameters such as pixel pitch, transistor's size, and so on. Both the optimum laser energy density and the tolerance of the ELA process window can be obtained by irradiating the sample with various laser energy densities, and examining the crystal size and the uniformity. This sampling evaluation can be done by highly experienced optical and SEM analysis. Figure 2.10 shows the grain size variation and the mobility change for various shot overlaps. Typical plate processing uses a 20 shot (for 400μm beam width, 20μm/shot translation) process, yielding an electron mobility of 100cm^2/Vs. The peak-to-peak pulse energy stability should be reduced to below 1 % for AMOLED application, which is more sensitive to laser energy fluctuation than AMLCDs [12, 13].

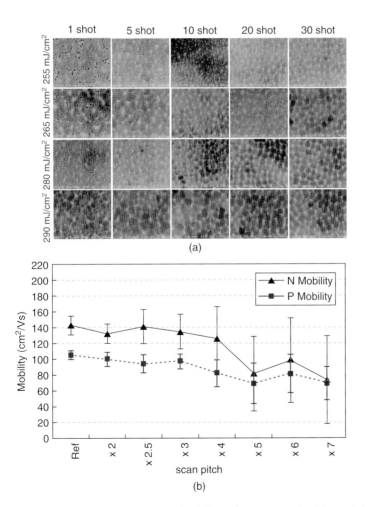

Figure 2.10 (a) Grain size variations for different laser energy densities and shot overlaps. (b) Scan pitch versus mobility [1].

2.2.3 Amorphous Oxide Semiconductor TFTs

There have been many studies in amorphous oxide semiconductor (OS) TFT for applying to the next generation AMOLEDs and AMLCDs [20–22]. Kimizuka et al. synthesized $InGaZnO_4$ (IGZO) for the first time in 1985. In 1987, they proposed its use as a semiconductor element [19]. Transparent amorphous oxide semiconductors (TAOSs) were proposed by Kamiya and Hosono et al., who claimed that the amorphous structure is ideal [20]. This interest comes from the particular properties of OS TFT, which make them well suited to the FPD application. The electron mobility of OS TFT is about 20 times higher than that of a-Si TFT. The high electron mobility improves the charging performance, and the low leakage current improves the charge holding ratio of the TFT-LCD. These advantages enable the high-resolution TFT-LCD display with low-power consumption.

The properties of oxide semiconductor are highly dependent on the oxygen content, since oxygen vacancies provide the free carriers. The content of oxygen controls its film properties in applications from conductor to insulator. In the ionic bonding configuration of OS materials, due to charge exchange between the metal cation and the oxide anion, the outer s-states of the metal ion are empty, and the outer p-states of the oxygen ion are filled [23]. Its amorphous nature suggests that OS has better uniformity than poly-crystalline materials. OS TFT has the advantages for large size AMOLED TVs, whereas poly-Si is presently used in small size AMOLED displays.

The OS material can be doped via oxygen vacancy. The vacancy concentration is determined by the oxygen partial pressure during the deposition. A low level of free carrier density is required in TFT material to ensure a low off-current, and Ga plays an important role in this respect, because its strong Ga-O bond reduces the O-vacancy density. For this reason, instead of IZO, IGZO is the preferred composition for TFTs, even though it is not the highest mobility material [23].

OS TFT technology is still under development, and there is no well-established fabrication process yet [24–26]. The OS material can be deposited by magnetron sputtering at temperatures down to room temperature, and they can be fabricated with a simple contact metallization scheme. However, the mobility distribution of the sputtered oxide films is large over the deposited substrate [26]. The reason for this uniformity problem is related to the spatial distributions of the number and energies of the Ar ions, negative oxygen ions, and positive oxygen ions reaching the substrate. Even worse, the chemical composition of sputtered a-IGZO films is different from that of the target [27], which potentially causes difficulty in obtaining uniform atomic composition ratio of a-IGZO films for large substrates. Uniform sputtering of the OS film is the key for mass production.

The dielectric films have another issue. When PECVD SiNx was used in the a-IGZO TFTs, the gate modulation of the channel current was lost [28]. To avoid these effects, SiOx is the preferred etch stopper and passivation layer. There is ongoing research to identify the optimum gate dielectric for OS TFT fabrication on low-temperature, flexible substrates. To improve the device performance and uniformity of the a-IGZO TFT, annealing is normally used [29]. One to two hours at 250°C to 350°C post-deposition annealing is applied.

C-axis-aligned crystalline IGZO (CAAC-IGZO) and nano-crystalline IGZO (nc-IGZO) were also reported, whose crystal morphologies differ from those of single-crystal and amorphous IGZO [30]. The CAAC- and nc-IGZOs have stacked nano-crystals (pellets) structure that are 1–3nm wide and 0.7–0.8nm thick. Cross-sectional transmission electron microscopy (TEM) of CAAC-IGZO films shows that the structure is an atomic arrangement parallel to the substrate surface. CAAC-IGZO TFTs have an extremely small off-state current because of their low carrier density [31].

2.3 a-Si TFT ARRAY PROCESS ARCHITECTURE AND PROCESS FLOW

Inverted staggered structure is the simplest and the most common structure of a-Si TFT. The inverted means the gate is beneath the channel, and the staggered means the source and drain contacts are not co-planar to the channel but located on the upper side of the channel. There are two different inverted staggered TFT architectures shown in Figure 2.11. The most widely used inverted staggered structure is the back channel

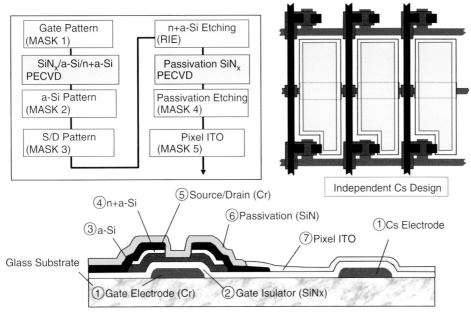

(a) The etch-back type TFT-Array structure and its process flow-chart

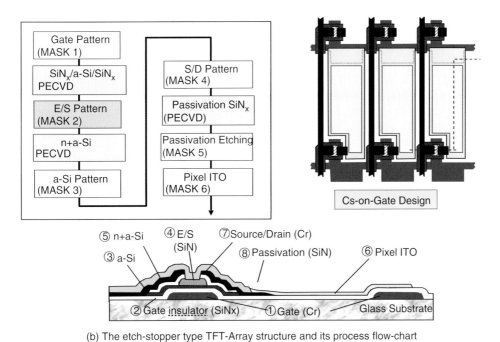

(b) The etch-stopper type TFT-Array structure and its process flow-chart

Figure 2.11 The TFT structures of (a) BCE and (b) ES types.

etch (BCE) type TFT, because BCE TFT has the simplest fabrication process for mass production. Second one is the etch-stopper (ES) type TFT, which has SiNx etch stopping layer on top of the TFT channel. The ES type TFT requires additional photolithography and CVD steps for ES layer. The pixel ITO layer in TFT-LCD can be located after a-Si is patterned or after the passivation is patterned.

Table 2.1 Typical TFT process steps and metals [32–35].

Step	5M	6M	7M
1	G	G	PXL	G
2	A	E/S	G	A
3	S/D	S/D	E/S	C
4	PA	PA	Via	S/D
5	PXL	PXL	S/D	PA1
6			PA	PA2
7				PXL
Gate metal	AlNd/Cr	AlNd	AlY, AlCu	TiN/Al/Ti
S/D metal	Cr	Mo/Al/Mo	Mo/Al/Mo	Al/Ti

G: gate line; A: a-Si TFT channel; SD: data line; PA: passivation open; PXL: pixel electrode; ES: etch stopper; C: contact hole open.

To fabricate the BCE type TFT panel, gate bus-lines, a-Si islands, data bus-lines, passivation and contact hole open, and pixel electrodes should be defined by using different photo masks, that is, four different patterns with different materials with at least one via hole are required. The minimum number of mask counts for the TFT-array fabrication is five. In Figure 2.11, (a) BCE and (b) ES type TFT fabrication processes are shown. The film thicknesses for commercial TFTs are typically 300nm SiN$_x$, 150nm a-Si, and 5nm n+ a-Si [38, 39]. The contacts to the TFT are n+ doped a-Si source and drain regions, which are contacted by the low resistant source-drain metal lines.

In Table 2.1, the existing five- to seven-mask processes used for notebook PC and monitor applications are summarized [32–35]. Different companies use their own unique choices of materials and processes depending on their application and display quality. For larger displays like monitors and TVs, aluminum or copper alloys are used to produce low-resistance bus lines. Aluminum or copper alloys of source-drain electrodes require under- and capping layers for ohmic contacts and low resistive contacts with pixel electrodes. Therefore, Mo/Al/Mo, Ti/Al/Ti or Cu alloy structures are widely used. But these sandwich structures make the TFT process even more complicated. It is difficult to design a proper metal structure without sacrificing the simple TFT structures. Nevertheless, because of the yield and the productivity, it is even more critical to use simple TFT structures. For TFT performance, normally, the leakage current (I_{off}) is in the sub-pico amp range, and the charging current (I_{on}) is in the few μA range, respectively. The pixel design is directly related to the aperture ratio of the panel, that is, the panel transmittance. By adopting the low dielectric constant interlayer insulator between the data line and pixel electrode, the coupling between the two electrodes can be effectively reduced [37]. A key aspect of the a-Si:H TFT process is its high manufacturing throughput. This is due, in part, to the low mask count, of four or less photolithography stages, needed to fabricate it, and with an extra mask needed for an AMLCD. Compared to BCE, ES structure has a SiN$_x$ etch stopping layer on top of the TFT channel that removes the need for critical time control for etching n+ aSi.

2.3.1 Four-Mask Count Process Architecture for TFT-LCDs

Major targets for the mass production of the TFT-LCDs are the low production cost and the high yield. They are closely related to the complexity of the fabrication process. By reducing the number of photolithography steps, the production capacity as well as the production yield is increased, and the process time is reduced. The minimum number of mask counts for the conventional TFT-array fabrication is five. It is even more difficult to reduce the mask count from five to four. To achieve four-mask count process any two layers must be defined by using the same photo mask or one material should be used for two layers, as in the case of the pixel electrode material, normally ITO, being used as the data bus-line at the same time [33–37]. However there is a limit of either using ITO as data bus-line or requiring a multi-layer simultaneous etching process. These approaches are not adequate for large size or high-resolution TFT-LCDs.

The main approach for four-mask structure is to pattern multiple thin-film layers simultaneously by using a single photomask. With the slit photolithography technique, a-Si channel layers and S/D layers are patterned simultaneously. The slit (or gray-tone) photolithography technology is introduced to create the partially exposed photoresist (PR) pattern. The parameters of the double slit used in this simulation are shown in Figure 2.12(a). Geometrical slit parameters are simulated and optimized to control the exposure energy at the slit region where the exposure energy is smaller than that of the normally exposed regions. Figure 2.12(b) shows the light distribution curve through the slit. By using the slit simulation mentioned above, a three-dimensional spatial-light intensity profile is obtained as shown in Figure 2.12(c). When the TFT channel area is exposed through the slit, the photoresist at the TFT channel will partially remain after development. The PR profile at the TFT channel region is directly related to the slit design parameters and the exposure energy. By using this photolithographic technology, the TFT channel can be defined without using an extra photo-lithographic step. Figure 2.12(d) shows the final PR pattern at the slit exposed region. Normal pattern and slit exposed pattern have significantly different PR thickness profile. Instead of using additional photolithography process, a simple photoresist etch-back process can define the TFT channel efficiently.

The slit photolithography process is shown in Figure 2.13. After photoresist coating, the data bus line and the TFT channel region are exposed by using the same mask. But the TFT channel region of the mask is slit patterned. The data bus line is normally exposed, whereas the TFT channel region is partially exposed. Figure 2.12(d) shows the focused-ion-beam image of the photoresist at the TFT channel region shown in Figure 2.14(a).

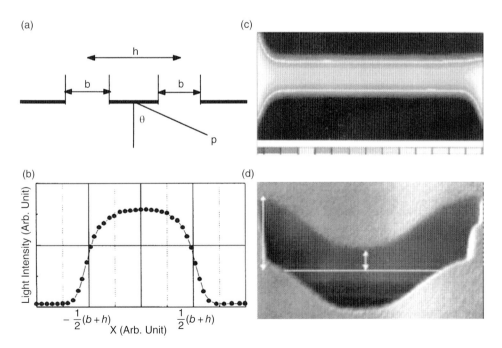

Figure 2.12 (a) Double slit used in the calculation, (b) Light intensity distribution curve through the slit (c) Simulation of the spatial light intensity profile for the optimized double-slit design (d) Focused-ion-beam image of the photoresist pattern at the slit exposed region. High and short arrows show the photoresist thickness of the unexposed and slit-exposed region, respectively.

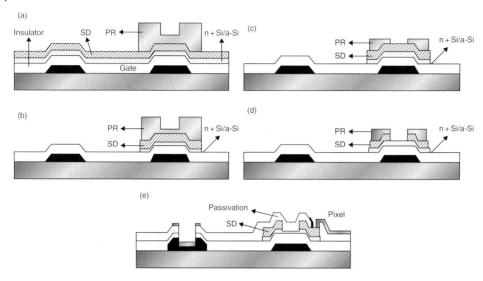

Figure 2.13 Slit photography process steps.

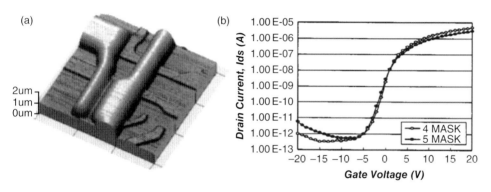

Figure 2.14 (a) AFM image of the TFT channel and data bus line photoresist pattern. (b) Comparison of the TFT transfer characteristics fabricated by four versus five-mask processes.

Figure 2.14(a) shows the photoresist profile (AFM image) of the data bus line and TFT channel region. In the TFT channel region, the partially exposed thin PR layer can be seen. With this PR pattern and using the metal-etching process, the data bus metal line is patterned. The partially exposed PR pattern at the channel region protects the channel during the data line etching process. Figure 2.13(b) shows the data bus-line formation after the data metal etching process. Then, the n+ a-Si/a-Si layers are patterned by a dry-etching process with the same PR pattern. The partially exposed photoresist at the TFT channel region is removed by using the etch-back process that is shown in Figure 2.13(c). After the TFT channel is defined by the etch-back process, the source and drain electrodes of the TFT are finally patterned by removing the metal layer at the channel region, as shown in Figure 2.13(d). The TFT characteristic is finally obtained by removing the n+ a-Si layer from the TFT channel region using the etch-back process. The passivation layer and pixel electrode are patterned successively to complete the TFT formation. Figure 2.13(e) shows the final TFT structure.

In Figure 2.14(b), the transfer characteristic of the TFT fabricated by the slit photolithography process is compared to that of the conventionally processed five-mask TFT where the TFT channel is defined by using the source-drain photolithographic process (see Figure 2.11(a)). The field-effect mobility ranges from 0.6 to

1.0cm^2/V-sec. The I_{on}/I_{off} ratio, threshold voltage, and sub-threshold characteristics are almost the same as that of conventionally processed TFTs. Self-alignment of the source-drain and a-Si layer at the TFT channel region as well as good ohmic contact between n+ a-Si and data metal guarantee a low leakage current.

2.4 POLY-Si TFT ARCHITECTURE AND FABRICATION

The fabrication cost of the LTPS LCD is higher than that of the a-Si LCD due to the increased number of process steps and masks. Normally, seven-mask CMOS process is used for LTPS LCD mass production. By using CMOS architecture, the power consumption is reduced and the circuit integration on the panel is available. But, in order to achieve the cost competitiveness, the CMOS fabrication process needs to be simplified. The five-mask PMOS and six-mask CMOS process architecture adopting the pixel-data electrode co-planar structure is introduced. In Table 2.2, the normal seven-mask CMOS process is compared to PMOS five-mask and CMOS six-mask process [40–47].

The cross-sectional view of TFT fabricated using the seven-mask CMOS process architecture is shown in Figure 2.15(a). The I-V curve of the seven-mask CMOS and six-mask PMOS TFT are shown in Figure 2.15(b). Detailed process steps are shown in Figure 2.16.

As shown in Figure 2.16, the first step is the glass pre-compaction. (a) LTPS process temperature is higher than a-Si or color filter process and the annealed glass at 650°C is used for total pitch matching in TFT-CF assembling process. Next, blocking SiO_2 layer is deposited on the annealed glass to protect the impurities from the glass. (b) The a-Si precursor is deposited. (c) The a-Si pre-cursor film normally contains a few percent of hydrogen atoms. If this high hydrogen content film is exposed to a laser beam, there will be an ablation of the a-Si film due to the hydrogen. To reduce the hydrogen contents to below 1%, thermal anneal at 400–450°C is required after the PECVD a-Si precursor film is deposited [88]. Low hydrogen content a-Si precursor is laser crystallized. (d) As mentioned in "ELA process control issue," laser energy setting is critical for obtaining uniform SLG poly-Si film. SLG poly-Si film is patterned with active mask so that the TFT channel and contact area is defined. (e) The low-dose boron ion doping is sometimes used to compensate for intrinsic electron richness in the crystallized poly-Si film, which is due to the positive charges in the oxide films. The threshold voltage is normally negatively shifted. Boron doping is used to compensate this shift.

The gate dielectric layer is deposited to cover the active pattern, as shown in Figure 2.16(f). For a-Si TFT, SiN_x is used for gate dielectric layer however for poly-Si TFT the positive charges in the nitride material can cause large negative shifts in threshold voltage. The nitride is also susceptible to gate bias induced trapping instabilities. Thus, for the LTPS process SiO_2 is used instead for the gate dielectric layer. The quality of the gate oxide is critical. The oxide film needs to have low leakage current, low densities of

Table 2.2 Comparison of various LTPS TFT-LCD fabrication process flows [44].

Process	CMOS (7 mask)	CMOS (6 mask)	PMOS (5 mask)
Active	0	0	0
Gate P	0 (P+)	0 (P+)	0 (P+ /P−)
Gate N	0 (N+ /N−)	0 (N+ /N−)	
ILD	0	0	0
S/D electrode	0	0	0
Passivation	0	0	0
Pixel electrode	0		

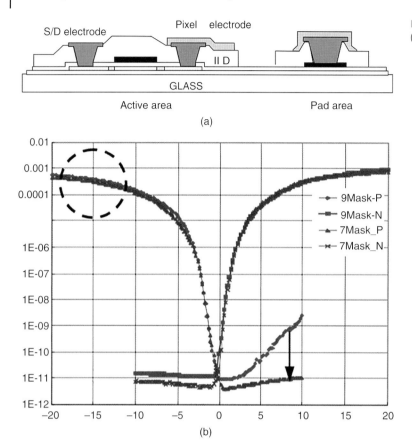

Figure 2.15 (a) CMOS 7 TFT structures (b) I-V curves of CMOS and PMOS TFTs.

fixed charges and interface states, high breakdown field, low pinhole density, and good bias-stress stability. In addition, the oxide deposition temperature should be below 450°C considering the large areas CVD system with good uniformity. PECVD from TEOS (tetra-ethyl-ortho-silicate) or SiH_4 gas are widely used for obtaining good quality SiO_2 film [51, 52]. Good quality oxide film can be obtained at low deposition rates through the dilution with oxygen or helium carrier gas [49, 50]. In some cases, SiO_2/SiN_x double layer gate dielectrics are used instead of the SiO_2 single layer. To prevent the electro-static breakdown during fabrication process as well as in operation, 800 Å to 1200 Å thick films are used for high-performance poly-Si TFTs.

Ion doping process is used to adjust the threshold voltage (channel doping: p−) as well as to form the ohmic contact (p+, n+) and the LDD (n−) at the source/drain region. The ion doping system delivers a rectangular (or ribbon-shaped) beam of ions. The beam is ion mass separated to remove hydrogen and any other unwanted ions from the beam. To achieve CMOS TFT, both PMOS and NMOS TFTs are required. PMOS TFT can be obtained by patterning the gate P metal and then doping with p+ ion (Boron) with gate P metal pattern as doping mask. (g) NMOS TFT with lightly doped drain (LDD) region is the next step. (h) By using gate N photoresist (PR) pattern gate N metal is patterned. This time gate N pattern is over etched about 1 um as shown in Figure 2.16(i). After the over etch, n+ doping (phosphorus ion) is processed with the N PR. Over etched region is not n+ doped because of the PR. Then PR is stripped and the n− doping is applied. (i) This process is called PR LDD. By using PR LDD process, additional LDD photolithography step is simplified. This is the CMOS LDD doping process. The self-alignment between the edges of the gate and the edges of the source and drain regions is achieved by using the gate electrode pattern as the ion doping mask and the dopants are implanted in the poly-Si film by passing through the 1000 Å SiO_2 gate insulator film.

2.4 Poly-Si TFT Architecture and Fabrication

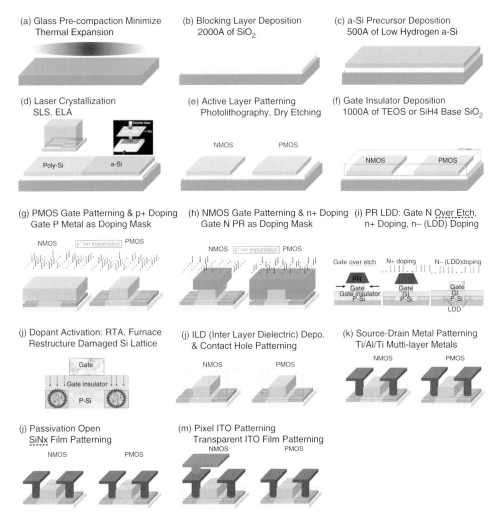

Figure 2.16 Seven-mask CMOS LTPS fabrication process.

When poly-Si TFTs are exposed to a drain-bias stress, I_{off} is increased and I_{on} is reduced [48–52]. The field at the drain/channel junction needs to be reduced, and it can be accomplished by the use of lightly doped drain (LDD) regions (i). It is known that hot carrier stress damages N-TFT more than P-TFT. LDD process is required only for N-TFT. High-dose ion doping, particularly the heavier phosphorus n+ ion doping, damages the poly-Si crystal structure. To activate the implanted dopant at the source/drain region, high-temperature activation process is necessary and RTA (rapid thermal anneal) or furnace anneal process is used. (j) The RTA process is carried out around 600°C to 650°C in short time to avoid the warpage of the large size substrate. Furnace activation is also used for dopant activation. Several hours of anneal at 400°C to 450°C is used to restructure the ion-damaged lattice by the undamaged poly-Si at the bottom of the film [53]. Next is the inter layer dielectric (ILD) deposition and contact open steps (j). PECVD SiO_2/SiN_x double layer is used for ILD. Thick ILD layer is preferred to reduce the gate-data coupling. Data metal is deposited and patterned (k). Low resistivity metal is used to minimize the coupling. Ti/Al/Ti triple metal is normally used. Ti protect Al from hillock formation as well as ITO contact. After the data patterning, passivation SiN_x is patterned (l). Finally, pixel ITO is patterned (m).

Since the doping process of CMOS TFT is rather complicated for OLED applications a simplified PMOS process is normally used. In this case only channel doping and p+ doping processes are used. In the case of five-mask PMOS device, comparing with the normal seven-mask CMOS architecture, the gate-N process and via (or passivation) process are skipped. PMOS structure does not need LDD process because PMOS is not affected by hot carrier stress. Instead, initial reverse bias is necessary to suppress the high leakage current.

2.5 OXIDE SEMICONDUCTOR TFT ARCHITECTURE AND FABRICATION

The advantage of the metal oxide semiconductor (OS) is the process similarity between a-Si TFT and OS TFT device fabrication. For inverted staggered type TFT, by replacing the a-Si active layer to OS, OS TFT can directly be achieved as shown in Figure 2.17(a). Even the fabrication equipment are almost the same. Just DC sputtering system is needed for OS deposition. The existing SiN_x/a-Si/n+ a-Si three layer PECVD system can be used for dielectric layer depositions. These similarities of processes and equipment are the major driving force of the OS TFT. Even the existing a-Si factory can be converted to OS factory or even both TFTs can be manufactured at the same time.

At present, the seven-mask ES a-IGZO pixel structure is commonly used for the mass production of AH-IPS LCDs. Although the inverted staggered type is suitable for mass production as shown in Figure 2.17(a), the preference is lower than the ES architecture because of the exposure of the back channel during fab processes [55–57]. The damage to the unprotected back surface of the a-IGZO film in the inverted staggered TFT resulted in almost order of magnitude degradation in device performance. ES TFTs in Figure 2.17(b) and (d) have stable back-channel structure by etch-stopper layer and stable light shield either by dual gate metal (2.17(b)) or light shield layer (2.17(d)). Compared to the a-Si BCE TFT, OS inverted staggered ES TFT with light shield have complicated fabrication process (seven-mask) and large parasitic capacitance (Cgs) due to the overlap area between gate and ES layer, which causes to increase ΔV_p. ΔV_p is a major cause of the image quality degradation (e.g. flicker, vertical crosstalk) in LCD. In order to minimize ΔV_p, it is important

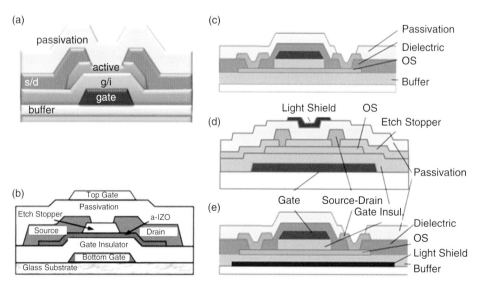

Figure 2.17 Various types of oxide TFT structures (a) Inverted staggered type (b) Inverted staggered etch stopper (ES) dual gate (c) Self-aligned coplanar (d) Inverted staggered ES with light shield (e) Self-aligned coplanar with light shield.

to maintain a large storage capacitance, which inevitably reduces the aperture area due to increasing storage area in pixel area. And it is also difficult to achieve the shorter channel length under 10μm in conventional inverted staggered ES TFT structure due to photo and etch process margin. This also reduces aperture ratio of the panel.

A new five-mask ES a-IGZO pixel structure is introduced by Yang [54], where the conventional inverted staggered ES OS TFT is seven-mask process. Unlike the conventional ES structure, a short channel length (under 5μm) TFT was achieved using a damage preventing layer (DPL) structure as shown in Figure 2.18. DPL pattern is self-aligned and can be realized by 3-D patterning technology such as contact hole filling process (CHF) [5, 6]. In addition, the overlap between gate and ES layer is eliminated by self-aligned DPL pattering process. Cgs could be reduced more than 30%. Figure 2.20 shows the new five-mask ES TFT structure. In new five-mask ES TFT structure, using the self-aligned S/D contact metals, TFT channel length can become shorter and Cgs can be reduced by eliminating overlap area between ES and S/D. Using the proposed DPL ES structure, two mask steps can be eliminated. The gate and active mask are combined and also the data line and common electrode mask are merged. DPL structure is very important to achieve a 4 to 5μm channel length TFT and reduce Cgs.

A typical process flow for the five-mask ES TFT with DPL consists of the following steps: first, a 300nm-thick Cu/Mo gate was deposited on a glass substrate. Then, 350nm of SiO_2 (gate insulator) and 50nm of a-IGZO (active) were directly deposited onto the substrate by PECVD and DC magnetron sputtering, respectively. The gate and active PR patterns are defined using a H/T mask (Mask #1). 200nm of SiO_2 (etch-stopper layer) is deposited and dry etched by using Mask #2. The CHF process is applied to form self-aligned DPL. 50nm of a-ITO is directly deposited on the substrate without removing the first PR. The second PR is then added, followed by the PR ashing process. The exposed a-ITO was removed by wet etch using 3% oxalic acid (100 sec). The substrate is then annealed at 230 degrees for 30 min in a convection oven to facilitate the poly-ITO formation. The S/D (200nm, Cu/Mo) and pixel (50nm, a-ITO) layers were fabricated simultaneously using a H/T mask. The fabrication of the passivation (200nm, SiN_x) and Vcom (50nm, a-ITO) layers followed.

Figure 2.19 shows the transfer curves of the proposed five-mask ES a-IGZO TFT with 4 and 5 um channel lengths, respectively. The linear field effect mobilities are 10.4 and 12.0 cm^2/Vs, respectively. The on and off currents were obtained at Vg = 15V, and –5V when V_{ds}=10V was applied. For 4 um short-channel length TFT, drain-induced-barrier-lowering (DIBL) effect is observed due to the channel contamination through the process. Also, on current is relatively low because of the triple contact layers' high resistance.

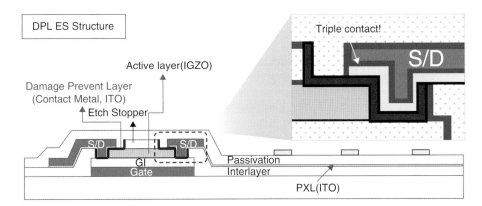

Figure 2.18 Five-mask ES TFT with a damage prevention layer (DPL).

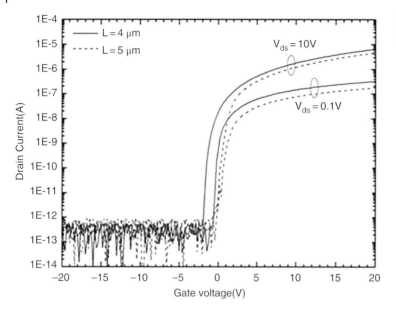

Figure 2.19 Transfer characteristics and Ig-Vg characteristic for five-mask ES a-IGZO TFT.

2.6 TFT LCD APPLICATIONS

Mobile displays require high resolution, narrow bezels, and a large aperture ratio pixel design. The high electron mobility of the LTPS TFT enables the slim bezel and the higher aperture ratio because the transistors can be smaller than the existing a-Si or OS TFTs. In addition, low parasitic capacitance and low driving voltage are obtained. This means LTPS consumes less backlighting power. LTPS is the backplane solution for high-resolution mobile displays. Other example of LTPS's advantage is the touch sensor integration into display (in cell touch sensor). In smart devices, a capacitive touch sensor is the key input device. Conventionally, a discrete touch sensor device is attached to the display panel. For better display quality and thin form factor, integration of the discrete capacitive touch sensor into the display panel has been developed [58]. A touch-integrated display is possible by both LTPS and OS, though the border size depends on the technology (see Figure 2.20).

To realize ultra-low power color and movie-capable displays, memory-in-pixel (MIP) technology was developed, which embeds static random access memory (SRAM) circuitry into each sub-pixel with the use of

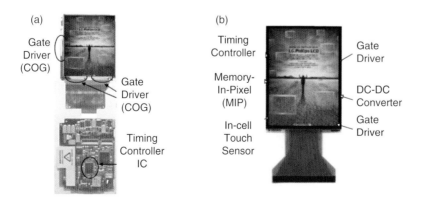

Figure 2.20 Comparison of the (a) conventional panel and the (b) system on panel (SOP).

LTPS. Since image data is stored in the SRAM in each sub-pixel, an MIP LCD can operate at ultra-low power for still images [59]. IGZO, on the other hand, utilizes its low off-current leakage to reduce display power consumption by reducing the refresh rate of the LCD [60].

2.7 DEVELOPMENT OF SLS-BASED SYSTEM ON GLASS DISPLAY [1, 11, 14, 15]

The so-called system on glass (SOG) TFT-LCD with LTPS (low-temperature polycrystalline silicon) technology enables to spare the silicon driver ICs and to enhance the productivity by reducing the module process steps. Several attempts to integrate the driver circuits on the glass substrate have been reported [61–65]. Comparing with the conventional a-Si TFT-LCD using external silicon driver ICs, the SOG-LCD requires high-performance TFTs, with high field effect mobility (μ_{FE}), low threshold voltage (V_{th}), and small sub-threshold slope to allow high-speed driving circuits that enable good display quality and small form factors. The SLS (sequential lateral solidification) technique provides sufficient TFT characteristics for the SOG-LCD, because the silicon grains can be grown like a single crystal in one direction, as shown in Figure 2.21 [66]. To meet the requirements of high throughput for mass production, a new laser crystallization technique entitled "TS-SLS" (two-shot sequential lateral solidification) technology is introduced.

The process scheme of the TS-SLS technique is as follows. Laser pulses are irradiated sequentially on an a-Si film while the substrate is loaded on a continuously moving stage. Meanwhile, a mask with patterned windows screens the laser pulse so that only the a-Si in selected areas is completely melted and crystallized. Right after the laser irradiation, lateral growth of silicon grains starts from the edge of the molten region to the center of it [66]. The length of the silicon grains that is typically 1.5μm, can be extended up to 4μm when the pulse duration of the laser is increased. When the second laser pulse is irradiated through the same mask with the stage translation as far as the width of the patterned window, the residual a-Si regions between the poly-Si regions are also exposed by the laser pulse as shown in the Figure 2.21. As a result of the second laser irradiation, the residual a-Si region is completely melted and crystallized continuously from the silicon grains that have been previously crystallized by the first laser shot. The whole area of the a-Si substrate can

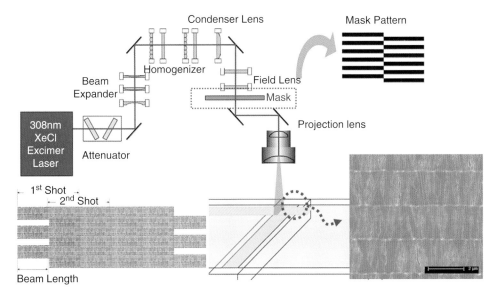

Figure 2.21 Artificially controlled super lateral growth method (sequential lateral solidification, SLS).

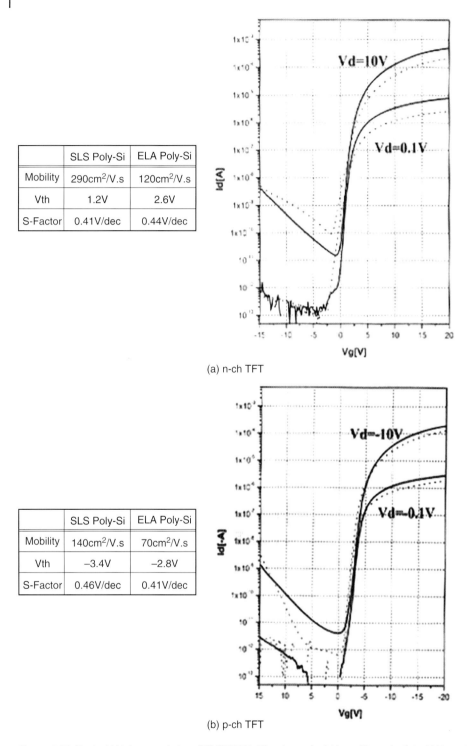

Figure 2.22 Typical I-V characteristics of TS-SLS TFTs. The channel width and length of the TFT are 20μm and 6μm, respectively.

be crystallized by further repetition of the laser irradiation with moving substrate. Since the SLS process uses the complete melting of a-Si film, it has wider process window than the conventional ELA (excimer laser annealing).

TS-SLS requires only two laser shots per region for the completion of a-Si re-crystallization. But the grain size of the TS-SLS poly-Si film is limited to 4μm, while the lateral grain growth with the conventional SLS process has no limit if the throughput is not an issue. Although the grain size is small, the grain size of the TS-SLS poly-Si film is at least six times larger than that of the conventional ELA. In addition, throughput of the SLS is two times higher than that of ELA where the laser beam is irradiated typically with 95% overlap (20 shots per region) due to the intrinsic narrow process margin with respect to the laser energy fluctuation. SEM images in Figure 2.21 show the microstructures of the Secco-etched TS-SLS poly-Si films after the first and the second laser irradiation. By the first laser irradiation, the grains are nucleated at the interface of the molten silicon and the solid a-Si. The nucleated grains grow toward the center of the molten silicon region. After the second laser irradiation, the lateral grain growth is initiated from the silicon grains previously crystallized by the first laser shot and continuously solidifies.

Figure 2.22 shows the typical transfer characteristics of TS-SLS TFTs. N-channel TFT has 1μm long LDD (lightly doped drain) to reduce the off current and to enhance reliability, while the p-channel TFT has no LDD. μ_{FE} and V_{th} of n-channel TFT are 181cm^2/Vs and 1.6V, respectively. μ_{FE} and V_{th} of p-channel TFT are 107cm^2/Vs and −2.4V, respectively. Still the performance of TS-SLS TFT is superior to that of the conventional ELA TFT, and it can be further improved by process optimization. Thanks to the superb characteristics of the TS-SLS TFTs, the de-multiplexing driving scheme using transmission gates (TGs) enables to integrate the source driver in the high-resolution LCD panel. Although the operation time of each amplifier is significantly decreased due to the de-multiplexing driving scheme, the TS-SLS SOG-LCD is successfully turned on. Push-pull amplifier with differential input stage guarantees the sufficient current driving capability under high-frequency operation. In addition, the layout techniques as well as the additional offset-canceling circuit effectively compensate the variations of the TFT characteristics.

References

1 C.-W. Kim, J.-G. Jung, J.-B. Choi, D.-H. Kim, C. Yi, H.-D. Kim, Y.-H. Choi, J. Im, Proc. SID Invited Paper, 59, p. 1, LTPS backplane technologies for AMLCDs and AMOLEDs (2011).
2 K. Nomura, H. Ohta, A. Takagi, T. Kamiya, M. Hirano, H. Hosono, Nature, 432, pp. 488–492, Room-temperature fabrication of transparent flexible thin-film transistors using amorphous oxide semiconductors (2004).
3 P. Semenza, Information Display, 27(9), pp. 30–32, Large TFT-LCD panels shift into high resolution (2011).
4 K. Ghaffarzadeh, A. Nathan, J. Robertson, S. Kim, S. Jeon, C. Kim, U.-I. Chung, J.-H. Lee, Appl Phys Lett, 97, p. 143510, Persistent photoconductivity in Hf–In–Zn–O thin film transistors (2010).
5 T. Arai, N. Morosawa, K. Tokunaga, Y. Terai, E. Fukumoto, T. Fujimori, T. Sasaoka, J SID, 19(2), pp. 205–211, Highly reliable oxide-semiconductor TFT for AMOLED displays (2011).
6 Y.-T. Yang, T. K. Won, S. Y. Choi, T. Takehara, Y. Nishimura, J. M. White, IEEE J Disp Tech 3(4), pp. 386–391, The latest plasma-enhanced chemical-vapor deposition technology for large-size processing (2007).
7 M. Hack, M.S. Shur, J. G. Shaw, Trans IEEE ED, 36(12), pp. 2764–2769, Physical models for amorphous-silicon thin-film transistors and their implementation in a circuit simulation program (1989).
8 R. A. Street. IEEE Trans ED, 36(12), pp. 2770–2774, Thermal equilibrium electronic properties of a-Si:H (1989).

9. S. D. Brotherton, J. R. Ayres, N. D. Young, Solid-State Electron, 34(7), pp. 671–679, Characterisation of low temperature poly-Si thin film transistors (1991).
10. S. D. Brotherton, N. D. Young, M. J. Edwards, A. Gill, M. J. Trainor, J. R. Ayres, I. R. Clarence, R. M. Bunn, J. P. Gowers, Proc SID IDRC pp. 130–133, Low temperature furnace processed poly-Si AMLCDs (1994).
11. C. W. Kim, K. C. Moon, H. J. Kim, K. C. Park, C. H. Kim, I. G. Kim, C. M. Kim, S. Y. Joo, J. K. Kang, U.J. Chung, Proc. SID, 35, p. 868, 21.4: Development of SLS-based system on glass display (2004).
12. M. Choi, S. Kim, J.-m. Huh, C. Kim, H. Nam, Proc. SID, 2014, 3.4: Advanced ELA for large-sized AMOLED displays (2014).
13. M. Sobey, K. Schmidt, B. Turk, R. Paetzel, Proc. SID, 2014, 8.2: Status and future promise of excimer laser annealing for LTPS on large glass substrates (2014).
14. J. S. Im, M. A. Crowder, R. S. Sposili, J. P. Leonard, H. J. Kim, J. H. Yoon, V. V. Gupta, H. Jin Song, H. S. Cho. Phys Stat Sol (a), 166(2) pp. 603–617, Controlled super-lateral growth of Si films for microstructural manipulation and optimisation (1998).
15. H. J. Kim, J. S. Im, Mat Res Soc Symp Proc, 321pp. 665–670, Multiple pulse irradiations in excimer laser-induced crystallisation of amorphous Si films (1994).
16. J. S. Im, H. J. Kim, M. O. Thompson, Appl Phys Lett, 63(14), pp. 1969–1971, Phase transformation mechanisms in excimer laser crystallization of amorphous silicon films (1993).
17. J. S. Im, H. J. Kim, Appl Phys Lett, 64(17), pp. 2303–2305, On the super lateral growth phenomenon observed in excimer laser- induced crystallization of thin Si films (1994).
18. P. C. van der Wilt, Proc. 13.1 SID, 2014, Excimer-laser annealing: microstructure evolution and a novel characterization technique (2014).
19. N. Kimizuka, T. Mohri, Journal of Solid State Chem. 60, pp. 382–384 (1985).
20. K. Nomura K, Ohta H, Takagi A, Kamiya T, Hirano M, Hosono H, Nature, 432, pp. 488–492, Room-temperature fabrication of transparent flexible thin-film transistors using amorphous oxide semiconductors (2015).
21. Y. G., Mo, M. Kim, C. K. Kang, J. H. Jeong, Y. S. Park, C. G. Choi, H. D. Kim, S. S. Kim, J SID 19(1), pp. 16–20, Amorphous-oxide TFT backplane for large-sized AMOLED TVs (2011).
22. H.-H. Lu, H.-C. Ting, T.-H. Shih, C.-Y. Chen, C.-S. Chuang, Y. Lin, SID'10 Technical Digest, pp. 1136–1138, 32-inch LCD panel using amorphous indium-gallium-zinc-oxide TFTs (2010).
23. T. Kamiya, K. Nomura, H. Hosono, J Display Technology, 5(7), pp. 273–288, Origins of high mobility and low operation voltage of amorphous oxide TFTs: Electronic structure, electron transport, defects and doping (2009).
24. T. Matsuo, S. Mori, A. Ban, A. Imaya, Proc. SID 2014, 8.3: Advantages of IGZO oxide semiconductor (2014).
25. T. Goto, S. Sugawa, T. Ohmi, Proc. SID, 2014, 3.2: Application of Rotation Magnet Sputtering Technology to a-IGZO Film Depositions (2014).
26. K. Ono, N. Ohshima, K. Goto, H. Yamamoto, T. Morita, K. Kinoshita et al., Jpn. J. Appl. Phys. 50, p. 023001 (2011).
27. K. Ide, K. Nomura, H. Hiramatsu, T. Kamiya, H. Hosono, J. Appl. Phys. 111, 073513 (2012).
28. M. Kim, J.-H. Jeong, H. J. Lee, T. K. Ahn, H. S. Shin, J.-S. Park, J. K. Jeong, Y.-G. Mo, H. D. Kim, Appl Phys Lett, 90, p. 212114, High mobility bottom gate InGaZnO thin film transistors with SiO_x etch stopper (2007).
29. K. Nomura, T. Kamiya, H. Ohta, M. Hirano, H. Hosono, Appl Phys Lett 93, p. 192107, Defect passivation and homogenization of amorphous oxide thin-film transistor by wet O_2 annealing (2008).
30. S. Yamazaki et al., SID Symposium Digest 43, 183 (2012).
31. S. Yamazaki, T. Matsuo, Proc. SID, 2015, 45.1: Future possibilities of crystalline oxide semiconductor, especially c-axis-aligned crystalline IGZO (2015).
32. C. W. Kim, C. O. Jeong, et al., Pure Al and Al-alloy gate-line processes in TFT-LCDs, SID Intl Symp Digest Tech Papers, pp. 337–340 (1996).
33. H. Kinoshita et al., High-resolution AMLCD made with a-Si:H TFTs and a five-mask Al-gate process, SID Intl Symp Digest Tech Papers, pp. 736–739 (1999).

34. K. Schleupen et al., High-information-content Color 16.3-in. desk-top-AMLCD with 15.7 million a-Si:H TFTs, Asia Display '98, pp. 187–90 (1998).
35. S. Nakabu et al., The development of super-high aperture ratio with low electrically resistive material for high-resolution TFT-LCDs, SID Intl Symp Digest Tech Papers, pp. 732–735 (1999).
36. C.-W. Kim, C.-O. Jeong, J.-H. Song, H.-G. Kim, Journal of the SID 9/3, Current manufacturing technologies for TFT-LCDs (2001).
37. J. H. Kim et al., High-aperture-ratio TFT-LCD using a low dielectric material, AMLCD, 97, pp. 5–8 (1997).
38. K. Sujuki, High aperture TFT array structures, SID Intl Symp Digest Tech Papers, p. 167 (1994).
39. C. W. Kim, Y. B. Park, D. G. Kim, et al., A novel four-mask-count process architecture for TFT-LCDs, SID Intl Symp Digest Tech Papers, p. 1016 (2000).
40. Pi-Fu Chen et al., Four photolithography process amorphous-silicon thin-film transistor array, SID Intl Symp Digest Tech Papers, pp. 1011–1013 (2000).
41. C. W. Han et al., A TFT manufactured by four-mask process with new photolithography, Asia Display, 98, pp. 1109–1112 (1998).
42. K. Ono et al., A simplified four-photomask process for 24-cm-diagonal TFT-LCDs, Asia Display, 95, pp. 693–696 (1995).
43. A. Ban et al., A simplified process for SVGA TFT-LCDs with single-layered ITO source bus-lines, SID Intl Symp Digest Tech Papers, pp. 93–96 (1996).
44. D. H. Kim, P. M. Choi, K. W. Chung, T. Uemoto, C. W. Kim, Proc. SID 03 Digest, P-39: development of 5" Poly-Si transflective panel by 7-mask process, operated with P-MOS circuit (2003).
45. Y. M. Ha. SID Digest, p. 1116 (2000).
46. Y. Aoki, T. Lizuka, S. Sagi, M. Karube, T. Tsunashima, S. Ishizawa, K. Ando, H. Skurai, T. Ejiri, T. Nakzono, M. Kobayashi, H. Sato, N. Ibaraki, M. Sasaki, N. Harada, SID 99 digest, pp. 176–179 (2015).
47. M. Itoh, Y. Yamamoto, T. Moita, H. Yoneda, Y. Yamane, SS. Tsuchimoto, F. Funada, K. Awane, SID 96 digest, pp 17–20 (2015).
48. N. Hirashita, S. Tokitoh, H. Uchida, Jpn J Appl Phys, 32(4), pp. 1787–1793, Thermal desorption and infrared studies of plasma-enhanced chemical vapor deposited SiO films with tetraethylorthosilicate (1993).
49. S. Asari, T. Kurata, T. Kikuchi, M. Hashimoto, K. Saito, Proc IDMC'02, Evaluation of gate insulation film for large-scale substrate in low temperature poly-Si TFTs. (2002).
50. J. Batey, E. Tierney, J Appl Phys, 60(9), pp. 3136–3145, Low-temperature deposition of high-quality silicon dioxide by plasma-enhanced chemical vapor deposition (1986).
51. J. R. Ayres, S. D. Brotherton, D. J. McCulloch, M. Trainor, Jpn J Appl Phys, 37(4a)D, pp. 801–1808, Analysis of drain field and hot carrier stability of poly-Si TFTs. (1998).
52. N. D. Young, IEEE Trans, ED-43(3), pp. 450–456, The formation and annealing of hot-carrier-induced degradation in poly- Si TFTs, MOSFETs, and SOI devices, and similarities to state-creation in a-Si:H (1996).
53. K. Yoneda, Proc SID-IDRC'97, pp. M40–M47, State-of-the-art low temperature processed poly-Si TFT technology (1997).
54. J.-Y. Yang, S.-H. Jung, C.-S. Woo, J.-Y. Lee, M. Jun, I.-B. Kang. Proc. SID 2015, 21.3: A novel 5-mask etch-stopper pixel structure with a short channel oxide semiconductor TFT (2015).
55. M. Kim, J. H. Jeong, H. J. Lee, T. K. Ahn, H. S. Shin, J.-S. Park, J. K. Jeong, Y.-G. Mo, H. D. Kim, Appl Phys Lett, 90, p. 212114, High mobility bottom gate InGaZnO thin film transistors with SiO_x etch stopper (2007).
56. T. Arai, Y. Shiraishi, Manufacturing issues for oxide TFT technologies for large-sized AMOLED displays, SID 2012 Digest (2012).
57. S. Ishikawa, T. Miyasako, H. Katsui, K. Tanaka, K. Hamada, C. Kulchaisit, M. Fujii, Y. Ishikawa, Y. Uraoka, IDW 2014 Digest, Reliability of amorphous InGaZnO thin-film transistors with low water-absorption passivation layer (2014).
58. K. Noguchi, Y. Kida, K. Ishizaki, T. Takeuchi, Euro Display, 2011, Novel in-cell capacitive touch panel technology (2011).

59. Y. Fukunaga et al., SID Digest, 44, pp. 701–704, Low power, high image quality color reflective LCDs realized by memory-in-pixel technology and optical optimization using newly developed scattering layer (2013).
60. Y. Kataoka et al., IDW, 2013, pp. 12–15, IGZO technology for the innovative LCD (2013).
61. C.-W. Kim, K.-C. Moon, H.-J. Kim, K.-C. Park, C.-H. Kim, I.-G. Kim, C.-M. Kim, S.-Y. Joo, J.-K. Kang, U.-J. Chung, SID, 04 Digest, 21.4: Development of SLS-based system on glass display (2004).
62. T. Nakamura, M. Karube, H. Hayashi, K. Nakamura, N. Tada, H. Fujiwara, J. Tsutsumi, T. Motai, Journal of the SID, 10/3 (2000).
63. 63.Y. Kida, Y. Nakajima, M. Takatoku, M. Minegishi, S. Nakamura, Y. Maki and T. Maekawa, Proceedings of Eurodisplay, p. 831 (2002).
64. S.-S. Han, K.-M. Lim, J.-S. Yoo, Y.-S. Jeong, K.-E. Lee, J.-K. Park, D.-H. Nam, S.-W. Lee, J.-M. Yoo, Y.-H. Jung, H.-S. Seo, C.-D. Kim, Proceedings of SID, p. 208 (2003).
65. H. Haga, H. Tsuchi, K. Abe, N. Ikeda, H. Asada, H. Hayama, K. Shiota, N. Takada, Proceedings of SID, p. 690 (2002).
66. R. S. Sposili, J. S. Im, Appl. Phys. Lett. 69, p. 2864 (1996).

3

Color Filter Architecture, Materials, and Process Flow

Young Seok Choi[1], Musun Kwak[2], and Youn Sung Na[1]

[1] *Process Development Division, LG Display, 245, LG-ro, Wollong, Paju, Gyeonggi, Republic of Korea*
[2] *Panel Performance Division, LG Display, 245, LG-ro, Wollong, Paju, Gyeonggi, Republic of Korea*

3.1 INTRODUCTION

The LCD (liquid crystal display) was initially used as a display device for digital watches or calculators in its early days. Back then, it was able to display information only in black and white (or at the on/off status). It was only after the introduction of the color filter on LCD and the development of the TFT (thin-film transistor)-LCD that LCD finally became capable of reproducing colors as they are found in the real world. The color filter is the key component in the LCD that is required to reproduce colors. For the LCD to be able to function as a display, by adjusting the amount of light passing through each pixel, it requires a TFT array, a liquid crystal (LC) layer, and a polarizing plate. These are not sufficient, though, to reproduce natural colors: for this, LCD has to have an additional layer in the form of a color filter (CF). The color filter enables the reproduction of millions of colors by mixing three primary colors: red, green, and blue.

Figure 3.1 shows the locations of CF layer, black matrix (BM), overcoat (OC) layer, LC layer, and TFT array layer inside LCD. The two glass panels face each other, with the liquid crystal material filling the gap between them, thereby completing a TFT-LCD structure capable of reproducing full colors. The CF layer is located above the TFT panel, facing toward the viewer, and it can also block the reflection from the metal layers in TFT structure, therefore, further improving the display image quality. The color affects the overall elegance of the display surface as well as its color reproduction performance. The characteristics of the visual presentation that are viewable to the users vary depending on the liquid crystal, which controls the passage of the backlight according to the control signals fed to the TFT. This is how the users can see the lights filtered through the color filter.

3.2 STRUCTURE AND ROLE OF THE COLOR FILTER

The primary role of the color filter is to reproduce colors, but it also works as the top panel of the LCD (i.e., the TFT is also called "bottom panel"). For this reason, numerous other layers are added to the CF panel. First, the black-matrix (BM) layer is inserted to prevent the RGB colors from mixing with one another, and light leakage from the backlight as shown in Figure 3.1. The black matrix is also the device that realizes the pitch-black quality of the display when seen from the front side with its power turned off. Figure 3.2 is a microscopic

Flat Panel Display Manufacturing, First Edition. Edited by Jun Souk, Shinji Morozumi, Fang-Chen Luo, and Ion Bita.
© 2018 John Wiley & Sons Ltd. Published 2018 by John Wiley & Sons Ltd.

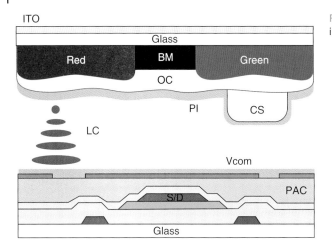

Figure 3.1 Cross-sectional structure of the TFT-LCD including color filter layer.

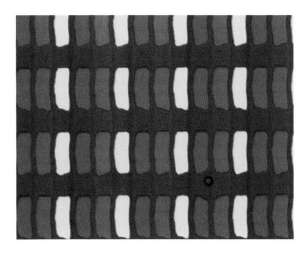

Figure 3.2 Microscopic image of the color filter applied to commercial products.

image of the color filter that was applied to commercial products. The black grid patterns among the red, green, and blue color cells represent the black matrix. The grid pattern is why it is called "matrix."

The color filter array is coated with an overcoat layer to protect liquid crystal, preventing organic foreign matters generated during the photochemical reaction of the RGB layer from infiltrating the liquid crystal layer. "OC" in Figure 3.1 stands for "overcoat layer."

The column spacer layer included in the CF panel keeps the cell gap between the TFT and the color filter layer at a constant level around 4 μm. In earlier days of LCD manufacturing, microscopic plastic balls with identical diameters were sprayed onto the panel to form the cell gap, but the practice phased out as it causes irregular light diffraction, which degrades the contrast ratio of the screen. In current LCD manufacturing, photolithographic patterning is used to form the column spacer on the CF glass.

Presently, the key components of the CF panel are the black matrix, three primary color layers, transparent electrodes (mainly ITO), and the column spacers. Figure 3.3 shows the cross-section of color filter structure.

3.2.1 Red, Green, and Blue (RGB) Layer

The different color patterns formed on the CF glass transmit selectively the intended light spectrum while absorbing the undesirable wavelengths. For example, the red material absorbs strongly in the short and

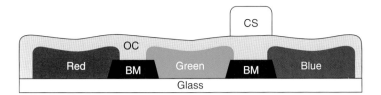

Figure 3.3 Cross-section of the color filter array.

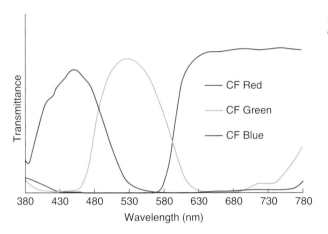

Figure 3.4 Transmission spectrum of red, green, and blue color filter materials.

medium wavelengths and transmits only the long wavelengths of the visual spectrum. Similarly, the blue material transmits the short wavelengths and absorbs all others. The various proportions of the three primary colors, R, G, B can generate the natural colors produced by today's LCDs. Figure 3.4 shows the transmission spectrum of red, green and blue color filter materials. Narrower spectrum band-width indicates higher purity of each color.

3.2.1.1 Color Coordinate and Color Gamut

The color coordinate is used to represent quantitatively the viewer perception of the color purity produced by the color filter. Globally accepted color standards became necessary to accelerate the industrial utilization of colors. The standards most commonly adopted by the display industry are the CIExy and CIEu'v' coordinates, which were established by the Commission Internationale de L'eclairage (CIE) in 1931 and 1976, respectively. Figure 3.5 shows example diagrams for these "CIE1931" and "CIE1976" color spaces.

Of the two aforementioned color coordinates, CIE1931 is the more commonly used, mainly for habitual reasons. The standard, however, is classified as a non-uniform color coordinate due to its low agreement with the actual colors perceived by human eyes in real life. The ellipses drawn on each coordinate system in Figure 3.5 represent the areas perceived by human eyes as actual colors. Green has a bigger space share in the CIE1931 coordinate system, whereas blue has a relatively smaller share [1]. The distances shown in the coordinate system suggest that the differences of colors as recognizable by human eyes are not uniform. Therefore, a more rational color coordinate system, which is required for the discussion of colors, became necessary, thereby resulting in the creation of CIE1976. In CIE1976, the distances of the colors in the ellipse are relatively similar in length [2], suggesting that the distances in the color coordinate system represent the color differences relatively better.

CIE1976 is a color coordinate system that was conceived to granulize the color differences within the color scheme by mathematically converting the theoretical primaries XYZ color system, because the distances in the CIE1931 coordinate system show some substantial deviations from the actual colors perceived by human eyes. CIE1976 is classified as a uniform coordinate system because it allows the calculation of the color differences by using the differences in the color coordinate system as proxies.

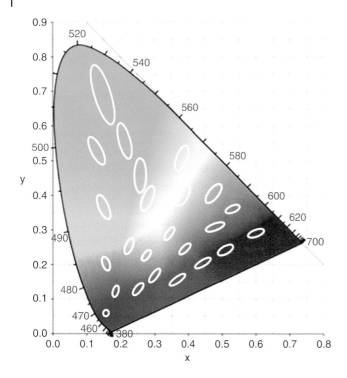

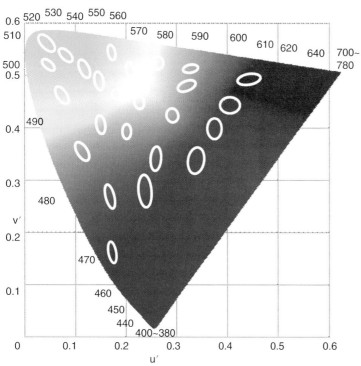

Figure 3.5 The CIE1931 (top) and CIE1976 (bottom) color coordinate systems. MacAdam ellipses proportional to the hue and saturation discrimination in the CIE 1931 chromaticity diagram, the same ellipses in CIE 1976.

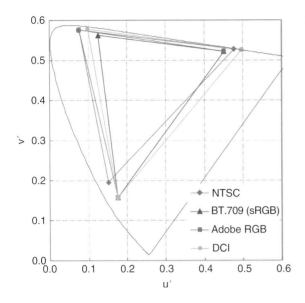

Figure 3.6 Leading types of color gamut for display devices.

Regardless of the improvement in color expression in CIE1976 as explained above, however, the industry is still using CIE 1931.

The color gamut area signifies the fraction of visible colors that can be reproduced. For a particular display design, the gamut area is defined by connecting the coordinates corresponding to the R, G, and B primary colors produced by emission through the CF array. The respective display is capable of producing any of the colors contained in the triangular area by suitable primary color mixtures. The larger the area of the triangle is, the better it is for the device to reproduce a pure color impression, which implies that it is better to move the R, G, and B dots to the far corners of the color coordinate system to realize this goal.

The types of color gamut commonly adopted by display devices are NTSC, BT709, sRGB, Adobe RGB, and DCI (Figure 3.6 and Table 3.1), with each system corresponding to a particular target of color performance and image content (see table below).

NTSC (National Television System Committee) is a system for analogue color television image reproduction established by the National Television System Committee in the United States, which was specified for CRT displays. Although this system is no longer in use in the signal processing domain, as it is not appropriate for digital TVs, it is still often used as a general reference concept to describe the characteristics in the color gamut. For instance, LCDs for notebook computers demonstrate a color gamut that is about 65% of the NTSC standard, which implies that the display is capable of reproducing colors up to 65% of the total area covered by NTSC.

BT.709 is a broadcasting specification for digital HDTV. Its color scheme is established in accordance with the international agreement on the standards on primary colors.

sRGB was proposed by HP and Microsoft in 1996 as a signal standard for desktop monitor displays. A display specification corresponding to BT.709, it is used mainly to minimize the errors in the color reproduction across different displays.

Adobe RGB was proposed in 1998 by Adobe System as a signal specification for display monitors. It is used mainly to simulate printed colors on a computer monitor screen.

DCI (Digital Cinema Initiative) is a specification for digital cinema that covers various specifications associated with the entire cinema cycle, from digital filming to digital projection.

3 Color Filter Architecture, Materials, and Process Flow

Table 3.1 Leading types of color gamut for display devices.

Name of CG	Size of CG	Main Characteristic
NTSC[1] (1953)	0.158	1. Based on the coordinates of the ideal phosphor 2. Abolished in the 1980s (ideal phosphor undeveloped)
BT.709 (1990)	0.112	1. ITU-R[2] Recommendation BT.709 2. Standard of HD TV
sRGB (1996)	0.112	1. HP and Microsoft to establish cooperation standards 2. Used at the monitor, printer, and Internet 3. Using the same BT.709 color gamut
Adobe RGB (1998)	0.151	1. Developed for the printer's color gamut (Adobe) 2. Used in electronic publishing 3. Green areas extended (compared to the BT.709 color gamut)
DCI[3] (2002)	0.152	1. Established by Digital Cinema Initiative 2. Digital cinema technical regulations and quality assurance standards 3. Green areas extended (compared to the BT.709 color gamut)

1) National Television System Committee
2) International Telecommunication Union-Radio Communication Sector
3) Joint venture established by Disney, Fox, Paramount, Sony Pictures, Universal, and Warner Bros. Studio

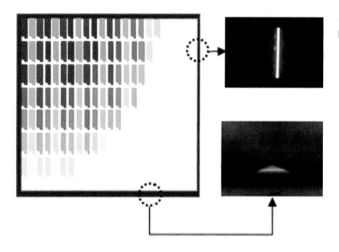

Figure 3.7 An example of leaked lights found in the panel edge.

3.2.2 Black Matrix

In general, the light emitted from the backlight unit could leak through the non-functional areas between pixels, that is, locations that display no image or where the liquid crystal is not modulated. Therefore, it is necessary to block such light leakage by placing an absorber layer in the leakage path to prevent degradation of the display image. As the lights also leak through the bezel area surrounding the panel pixel area, the bezel area should also be blocked appropriately. Figure 3.7 shows some real production examples of the lights spilling all over the bezel, often observed when the black matrix either disappears or becomes thinner. The direction of the liquid crystal cannot be controlled in those parts as there is no pattern or electric field that would otherwise block the light. The light spills over on the aforementioned parts because there is no black matrix in the first place, or it becomes thinner. As such, the black matrix prevents such unnecessary lights from being emitted through the panel edge.

The materials used for the black matrix should possess some degree of desirable properties: sufficient light blocking performance, a thickness less than $2\mu m$ and low reflectivity. A thicker material will block the lights

Table 3.2 Materials for the black matrix.

Material	Thickness (μm) for OD=3.5	Reflection
Chrome metal (Cr)	~0.2	~50%
Chrome oxide (CrOx)	~0.2	~4%
Carbon resin	~1.0	~2%
Graphite	~0.4	~7%

effectively but will have a detrimental impact on the subsequent manufacturing process and the overall display performance. Therefore, here, the key is narrowing down the choices until one finds some right material that will be thin enough but will still demonstrate excellent light blocking performance. The index that is often used to measure the light blocking performance is OD (optical density). This index is obtained by calculating the ratio of the transmitted light (I_{out}) out of the total incoming light (I_{in}) incident on the black matrix, as shown in equation 3.1 below.

$$OD(\text{Optical Density}) = -\log I_{out} / I_{in} \tag{3.1}$$

For instance, OD value 3 signifies that only 1/1000 (0.1%) of the total incoming light penetrated the black matrix, with the remainder having been absorbed. If a display panel with a contrast ratio higher than 1,000 is to be designed, the amount of transmitted light should be suppressed to below 0.1% of the "white" level across the whole display. In general, materials with a high optical density are needed for display panels, but the resulting increase in the thickness of the black matrix is not appropriate. Good black matrix candidate materials should have an optical density higher than 3.5 in <2 μm thick films, and also <5% reflectivity. Table 3.2 lists the candidate materials that could be considered for use as the black matrix in LCD panels.

Metallic thin films have excellent optical density and can be produced with conventional patterning methods, but their drawback is their high reflectivity. Ceramics or organic materials demonstrate lower reflectivity than metallic materials, but they are known to have difficulties in patterning. In the early days of LCD, when notebook computer panels were the primary product, chrome/chrome oxide multilayers were widely used for the black matrix due to its desirable properties, including its excellent optical density, reflectivity, and thinness. The chrome composite materials, however, are disappearing fast from the industry with the increasing demand for the green materials. In this chapter, we will focus on the process employing carbon-based synthetic resin.

3.2.3 Overcoat and Transparent Electrode

The thickness of the individual RGB layer often varies as the main focus is precision-controlling the color coordinate rather than controlling the thickness of the RGB layer. Such a variation in the thickness will alter the cell gap on each color cell and will even affect the rubbing effect in case of a serious gap difference. A transparent material is thus coated to eliminate these topographical irregularities, called overcoat layer. The layer also serves as a barrier for the organic impurities that were not eliminated completely during the RGB layer process. Cross-linked acrylic resins are commonly used for the transparent overcoat material.

The LCD panel (TN mode) can display images only when the liquid crystal sandwiched between the two transparent electrodes facing each can be modulated by electrical fields. One pixel electrode is connected to the drain terminal of the transistors on the TFT panel while the counter electrode is formed on the color filter side. Thus, a transparent conductive layer such as an ITO (indium thin oxide) or IZO (indium zinc oxide) layer needs to be included in the CF panel.

3.2.4 Column Spacer

LCD requires a precise management of the light pathway because the display relies on optical anisotropy, which implies that the cell gap should be managed precisely. In the early days of the LCD industry, plastic balls with diameters measuring up to the rough distance of the cell gap (4–5 μm) were sprayed onto the CF panel to maintain the cell gap uniformly over the glass. When the liquid crystal was vacuum filled into the inner space of the assembled glass panel with the ball spacers tucked between the two panes, no significant movement of the ball spacer was observed. The process was later changed to the "one-drop filling" method, where liquid crystal is dispensed in the vacuum chamber followed by the pressing two glass panes together, thereby causing the ball spacer to move and congregate together. As a result, some light leaks were observed in those areas adjacent to the spacers that lead to the deterioration of contrast ratio. To solve this problem, the column spacer patterning method was adopted, which allowed spacer fabrication with precise control of their locations, replacing the ball spray method. With the majority of the current LCD displays designs require a contrast ratio over 1,000, the spacer is now located on the black matrix so that it would not be visible to the users even when light leaks through it.

3.3 COLOR FILTER MANUFACTURING PROCESS FLOW

The color filter has various structures depending on the product type. Depending on their respective liquid crystal driving mode, LCDs are classified as TN, which is popular among notebook computers; IPS, which is mainly used for computer monitors or tablet computers; or VA for the television screens. The corresponding CF should have a structure optimized for the target LCD type. In this section, the unit process will be explained first, followed by a discussion of the process sequence for the selected structure.

3.3.1 Unit Process

3.3.1.1 Formation of Black Matrix

A negative-type photoresist containing carbon resin is used to produce the black matrix pattern. The manufacturing process of the black matrix is shown in Figure 3.8. First, the surface of the glass pane was cleaned before it was coated with a resin black-matrix solution. Spin coating has been widely used when the glass pane is less than 1 m, but slit coating method is used for glass larger than Gen 5 glass size. With the use of slit coating, it became feasible to apply a uniform organic coating on the larger than 2 m glass panes. Most of solvent

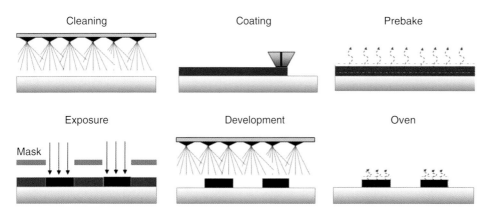

Figure 3.8 Process flow for black-matrix patterning.

Table 3.3 Key ingredients and their roles in the black matrix resin material.

	Components	Characteristic	Main function
Resin	Additives	surfactant, adhesive dispensing *etc.*	Coating, pattern adhesion, storage stability
Resin	Photo Initiator	Oxime ester Triazine based organic acids	Radical initiated polymerization
Resin	Multifunction Monomer	Photo-crosslinking acrylic monomer	Pattern formation by photo-crosslinking reaction
Resin	Polymer	Cardo based	Layer, Viscosity form, Alkali develop-ability, Mechanical properties of a given strength and elasticity.
Mill-base	Dispersant	Polyurethane, Polyester *etc.*	Dispersion stability
Mill-base	Carbon Black	Carbon / Organic Black (BM PR occupies the entire 60%)	Electrical properties _ High and low resistance, Control of optical density

left on the coated glass pane evaporates during the following prebake process, and the patterns are formed by the UV exposure through the photo mask followed by developing process. The type of exposure device to be used is determined by the resolution of CF pixel. In the case of television panels, the proximity type exposure device, which exposes light by proximity contact to the CF panel, will be sufficient to form the pattern. The projection type exposure device often adopted by the TFT manufacturing process is frequently used for smartphone and tablet PC panels because of their high resolution. During the photo process, the coated area exposed to UV light becomes hardened by the photochemical reaction with the exposed pattern remaining on the glass panel after the chemical develop process. The residual solvent will completely evaporate in the succeeding steps, while the hardening process is performed to produce a chemically stable layer. The hardening process is typically a thermal treatment performed for about 20 minutes at 230°C.

The black matrix material is typically a carbon-based synthetic resin and also a photoresist material, containing mainly the mill base and resin matrix. They have key ingredients, characteristics, and roles, as shown in Table 3.3. Resins such as additives, photo initiators, multifunction monomers, and polymers enable the patterning process and preserve the pattern after the completion of the exposure process. The carbon and dispersant in the domain of the mill base are mainly to increase the optical density and electrical resistance, which are the key functions of the black matrix. The microscopic structure of the carbon particles is shown in Figure 3.9 [3, 4]. The key required properties of the carbon particles are the particle diameter, cohesion structure, surface characteristic, and extent of dispersion. As it will be difficult to coat the glass panes if the dispersion performance drops, the dispersion performance is closely related with the overall performance of the process. To have a closer look, the black-matrix photoresist is made up in liquid form, in which the distributed carbon black is combined with resin and with solvent. As the carbon mill base content accounts for over 60% of the total amount, the material properties of the carbon black is considered critical. The key to the

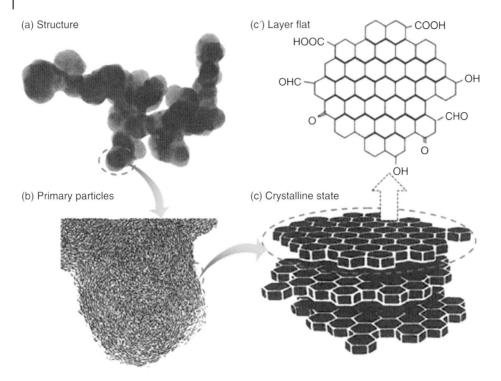

Figure 3.9 Overview of the carbon black structure.

satisfactory black-matrix photoresist is the selection of carbon black and how effectively the carbon black particles distributed and kept that way. As the RGB layer is similar in the dispersant and the resin, it will be explained in detail in the next section (3.3.1.2). Since carbon black is a conductive material, it cannot be used for those cases where the dielectric properties is required.

In general, some dielectric filler, such as black carbon or metal oxide powder, are used to impart the dielectric property to the polymer, and the black carbon demonstrates an excellent dielectric property. As the range of the required dielectric property may vary, however, depending on the applied product, it should be set to suit individual use. Recently, a high-resistance black matrix with a high dielectric property is required for the narrow-bezel and in-cell touch structure adopted by the latest mobile devices. To increase the optical density, which is the key property of the black matrix, it will be advantageous to raise the carbon content, but it will increase the conductivity, therefore, it is imperative to come up with a material design that suits each different product.

3.3.1.2 Formation of RGB Layer

The method of forming the RGB array layer on a glass pane is identical to the one applied to the black matrix: cleaning, spraying of color pigments, exposure, development, and reinforcement with oven heating. This method is generally called "pigment-dispersed method" [5], which is the method that is most commonly adopted by LCD manufacturers. The pigment is manufactured in the form of a photoresist. The photoresist is negative type, which preserves those areas exposed to the UV light by the exposure device. It is essential to disperse pigments in such forms that can be applicable to the photoresists to produce materials that can be hardened with exposure to UV light. As the amount of diffracted light drops with the decreasing size of the pigment, it possesses an excellent material properties for LCDs.

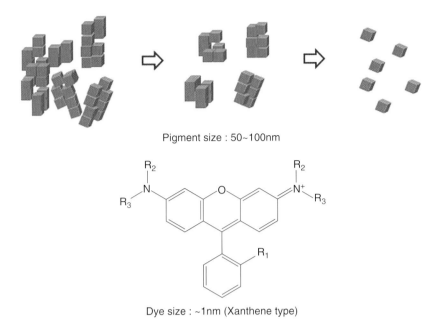

Figure 3.10 Structures of the pigment and the dye.

The base material for the RGB layer is the mill base, which is composed of key materials required to realize the optimal colors. The mill base contains colorants such as dyes, which are easily soluble in solvents, and pigments, which are not easily soluble (Figure 3.10). As dyes primarily consist of single molecules characterized by a delicate molecular structure, they are usually more vulnerable in terms of reliability and light stability compared to pigments. However, the industry is continuously investigating the use of advanced dyes with a sufficient heat resistance and light stability.

The pigments are formed into finer particles from larger agglomerates during the dispersion process (Figure 3.11). Because they are not easily soluble in solvents, it is necessary to grind large pigments and then wetting them by solvent before dispersing the wet particles with the diffuser. Once ground down to nm size, the pigments tend to coagulate easily due to the surface energy. Therefore, dispersant is mixed during the dispersion process to prevent re-coagulation. The dispersant is the key ingredient that maintains the repulsive force of the finely ground pigments, and a small dose of synergist, which is the surface treatment reagent for the pigments, is added into the process to accelerate the performance of the dispersants. Moreover, a binder polymer is added to the process to control the dispersion stability. The end product is called "mill base (MBS)," which is the same material as the resin black matrix explained earlier.

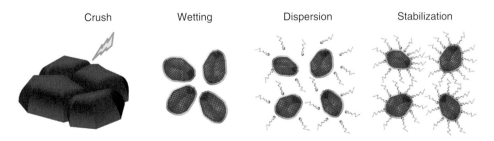

Figure 3.11 Pigment dispersion process flow.

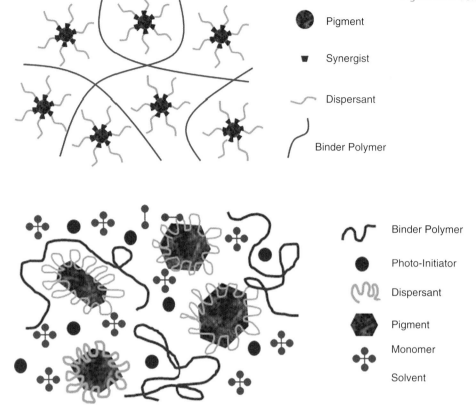

Figure 3.12 Formation of the mill base.

Figure 3.13 Composition of the color photoresist formulation.

The mill base in the RGB layer is composed of pigments, a synergist, a dispersant, and a binder polymer (Figure 3.12). The materials of the mill base cannot yet be considered photo-resistant but the pre-stage before the resist. Materials necessary for realizing the photo-resistant property should be added to make appropriate pattern.

The color photoresist is made by adding materials to the existing mill base, with the list of added materials shown in Figure 3.13. The photo-initiator, a material that generates radicals upon UV light exposure, is the key component required to create the pattern selectively during photolithography. The key required properties of the photo-initiator are high sensitivity, solubility in solvent, and stability. The monomer is the reactive molecule capable of reacting with the photo-generated radicals and subsequent polymerization. The binder polymer is the ingredient that helps coating uniform thin films on the substrate, and also enhances the stability of the pigment dispersion. Solvents are used to control the adhesion and enables coating by reducing the viscosity of the mill base solution. Some additives are also included in the mix either to help flatten the surface or to improve the adhesive power.

The red, green, and blue-colored photoresists produced in this manner are processed in the order desired, for example, R, G, and then B (Figure 3.14). The unit process for R, G, and B is identical to the process employed to produce the black matrix as described earlier (Figure 3.8). Basically, the process can be performed in almost identical equipment because it takes the form of a negative photoresist and is similar to the black-matrix process in terms of the adhesion. Proximity-type exposure tools are used typically, because the R, G, and B pattern is simpler than the black-matrix pattern.

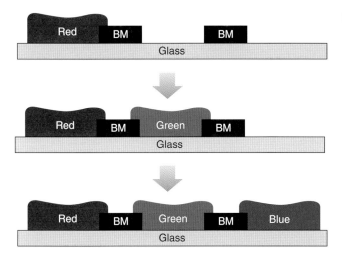

Figure 3.14 R,G,B color photoresist processing.

3.3.1.3 Overcoat (OC)

The overcoat layer does not require any patterning process. Depending on the product, sometimes the OC is removed around the panel edges to improve the adhesion with the seal. If patterning is not necessary, a heat-hardened overcoat will be applied. Additionally, some materials have been developed that do not require a separate exposure process, which can be hardened by evaporating solvents in the oven after the coating process. They are used in many products because they do not need a photoreaction or require a complex process. Meanwhile, the photo-hardening overcoat has the same basic processing order as those used for the black matrix or RGB layer. In case over coating is performed on the front side while performing photo hardening at the same time, the photoreaction is initiated by UV light exposure to the front side without using a mask.

The key roles of the overcoat layer are protection and planarization. The overcoat layer passivates the CF array and protects the pixels from the wet etchant used in the subsequent ITO etching process. It also plays the role of planarizing the RGB array to prevent variations of the LC cell gap and affecting pixel operation.

The ingredients and functions of the thermal-hardening and photo-hardening materials used as overcoat are listed in Table 3.4. The most significant among them is the difference in the composition of the photo initiator, which stimulates photoreaction. The overcoat turns to an acrylic family transparent membrane once it is completed, but its component elements will vary due to the differences in the process. It should be noted that the photo hardening type would not necessarily negate the need of thermal hardening. The photo energy will play the key role in the hardening process, but the thermal treatment process is still required to remove the residual solvents and to increase the density of the layer. Materials are designed to go through the same thermal treatment process as the one applied to the black-matrix or RGB process.

The simplicity of the associated equipment and process cannot be ignored in the manufacturing process. Due to such benefits, most LCD manufacturers adopted the thermal hardening type. Therefore, this chapter will be dedicated to explaining the thermal hardening type.

The thermal hardening method mainly involves the epoxy resin family, which uses acrylic polymer, and the polyimide family. In general, the epoxy resin family is being applied as the main material because polyimide requires additional care and material management, such as the installation of dedicated anti-corrosive stainless steel pipes because of the use of NMP (n-methyl pyrrolidone) solvent. On the contrary, epoxy resin has excellent adhesive performance and durability as well as high-quality insulation and mechanical property. For this reason, epoxy does not generate volatile substances during the hardening process. Moreover, it possesses high adhesive power on the surface. Epoxy has been used as an adhesive sealant to assemble the CF and TFT panels together, as well as for the CF OC material.

Table 3.4 Key elements and functions of two typical overcoat materials.

Components	Overcoat (OC) type		Function
	Thermoset OC	Photo-curing OC	
Binder	Epoxy resin or Polyimide	Alkali-soluble resin	Coating formation
Multifunction Monomer	Apply or Unapplied	Apply	Layer formation by crosslinking reaction
Photo Initiator	Unapplied	Apply	Radical initiated polymerization
Epoxy compound (Low molecular)	Apply or Unapplied	Unapplied	Adhesion, Planarization
Adhesive preparations	Apply or Unapplied	Apply	Adhesion
Surfactants	Modified silicone series or Fluorine based polymer		Coating properties
Solvent	Thinner type (PGMEA etc.)		Coating, Wetting

Table 3.5 Hardening mechanism by OC type.

Type	Thermoset OC		Photo-curing OC
	Epoxy resin	Polyimide precursor	Photosensitive acrylic
Curing mechanism	Epoxy (ring → OH)	Poly (amic acid) → Polyimide (−H$_2$O)	Acryl + Photoinitiator
Curing component	Epoxy, Hardener	amic acid	Acrylic, Photoinitiator

Details of the OC hardening mechanism are compared in Table 3.5. In an epoxy type, the hardening reaction is initiated with the cleavage reaction of the epoxy ring, which was caused by the potential acid (-OH) in the polymer, before it eventually forms a membrane-type (network structure) hardened film as a result of the crosslinking reaction between the multifunctional oligomers (M/M: multifunctional monomer) present in the photoresist.

The following properties of OC materials should be considered; easy to coat, mura-free film formation, re-workability, and fume generation. In particular, the easy-to-coat property is most critical in the role of a planarization as mentioned earlier. To improve the OC coating property, it is important to select the right surfactant to control the surface tension of the OC solution. The types of surfactant include the hydrocarbon family, the fluorine family, and the silicone family surfactant.

3.3.1.4 Formation of ITO Electrodes

ITO is commonly used as a transparent electrode because of its transparency and low electrical resistivity. The ITO film is commonly formed on OC/CF layer by sputtering thin-film deposition. In PVA LCD mode, ITO patterning is required to form a fringe field for the liquid crystals. As deposited ITO films are typically amorphous, but since it is an oxide compound it is difficult to etch it without damaging the underlying organic materials due to the ITO etchant strength. For this reason, IZO (indium zinc oxide), which is relatively easy to etch, is sometimes used instead of ITO. As IZO maintains an amorphous phase even after the metalizing process, it is relatively easy to etch. To form a pattern on ITO or IZO film, the photolithography process described in the TFT manufacturing section is used.

3.3.1.5 Column Spacer (Pattern Spacer)

The method of forming the column spacer (CS) is basically the same as that with the black matrix, which is formed by the exposure device. The photoresist that is used to fabricate column spacer is also a negative-tone type, and both proximity- or projection-type exposure tools can be used depending on the required resolution and alignment accuracy. The required material is acrylic, similar to the overcoat, but it is much more challenging to combine the two as the former is required to meet the stringent specifications for the latter, including the precision size, height, and taper angle. No additional description of the CS process will be included here because it is identical to other processes.

Recently, an exposure technology that allows forming a column spacer with varying heights within a single exposure step was developed and used commercially (Figure 3.15). The taller spacers (CS1) play the role of maintaining the cell gap, whereas those with a low height (CS2) are used to support the mechanical strength. Such diverse spacers with varying heights are formed with a single exposure by employing the method called "half-tone exposure" in the LCD industry. Unlike the mask generally used by the industry, the aforementioned exposure technology has a certain pattern that allows only partial light penetration. The mask with a half-penetrating layer is called "half-tone mask." The column spacer will have a relative hardening condition in those areas that are less exposed to light because it uses the negative-type photoresist. Once the material is finished with the development process in such condition, the thickness of the resulting material is changed. Moreover, a column spacer with varying heights will be formed after an additional hardening step in the oven.

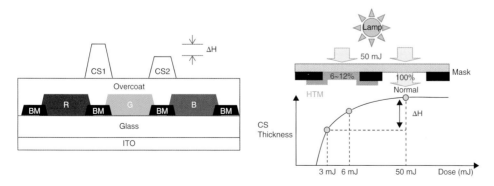

Figure 3.15 Half-tone spacer and exposure process.

Figure 3.15 shows the varying reactions of the resist for the column spacer, depending on the photo energy. The height of the column spacer can be controlled to a certain extent by adjusting the light penetration rate of the half tone in the mask. The difference in height between the two types of spacers formed as such is ΔH, and it has significance in the design of the spacer.

3.3.2 Process Flow for Different LC Mode

Multiple types of color filters are used in commercial products, with a process flow matching each of the applications. The diverse color filters operating in the TN, IPS, and VA modes, respectively, will be explained next.

3.3.2.1 Color Filter for the TN Mode

TN mode is the most commonly used mode for notebook computer displays. In the TN mode, a common electrode facing the pixel electrodes formed on the TFT panel is required. Toward this end, transparent electrodes (usually ITO) were used to cover the panel through the sputtering technique after the patterning of the black matrix and RGB layer (Figure 3.16). Overcoat was used in the TN mode, but it is actually no longer found in the notebook computers available in the market. The process is completed by forming a column spacer to create a cell gap.

3.3.2.2 Color Filter for the IPS Mode

IPS mode is frequently adopted by tablet computer displays. As it is easy to realize a high resolution when using such mode, the mode is often used in mobile phones with an LCD screen. No common electrode is necessary in the IPS mode. Instead, a transparent ITO layer is required on the back side of the CF glass to prevent electrostatic charges infiltrating from outside. The ITO layer is formed via sputtering followed by the formation of a black matrix layer and an RGB layer. The process is completed with the formation of the overcoat and column spacer on the panel. Figure 3.17 shows the process flow for manufacturing color filters in the IPS mode.

Since recently, the glass thinning technology has been widely used to design lightweight products. The technology makes the cell thinner by dissolving both sides of the glass pane by over half with hydrofluoric acid.

For that case where the glass pane should be etched as such, the ITO process is performed after thinning. In other words, CF process is performed without ITO sputtering.

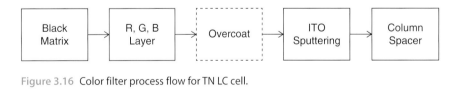

Figure 3.16 Color filter process flow for TN LC cell.

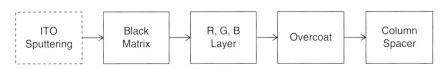

Figure 3.17 Color filter process flow for IPS LC cell.

Figure 3.18 Color filter process flow for VA LC cell.

3.3.2.3 Color Filter for the VA Mode

VA mode is usually found in television displays. A common electrode facing the pixel electrodes on the TFT glass is also required in the VA mode, but the common electrodes need be patterned to control the falling directions of the liquid crystal in the VA mode. Moreover, the overcoat is also essential for achieving precise control of the LC orientation. Figure 3.18 shows the process flow of the VA process.

3.4 NEW COLOR FILTER DESIGN

The color filter has evolved in terms of both its structure and function, along with the changing trends of the LCD technology. Among the changes, the RGBW color and the color filter on the TFT technology, two leading commercial technologies in the industry, will be introduced in this section.

3.4.1 White Color (Four Primary Colors) Technology

In principle, the optical transmission of the color filter cannot exceed 33% per pixel. To overcome such limitation, a new structure where pixels have a higher transmission was developed and adopted for commercial production. As shown in Figure 3.19, white color sub-pixels were added to the color filter forming a RGBW array. In theory, the pixels allow light transmission across the entire visible spectrum. For such reasons, the pixel has an advantage in expressing bright images. In the case when images with high chromaticity should be expressed, the pure colors are presented by turning off the white pixels. As this technology uses more lights emitted from the backlight, it consumes less energy and is used primarily in the condition where high brightness is required.

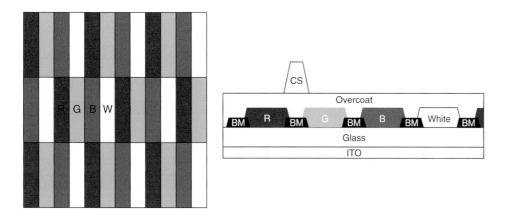

Figure 3.19 RGBW color pixel layout and cross-sectional view.

The manufacturing process usually involves the formation of an RGB layer followed by the formation of a white (transparent) layer. The materials used for this purpose are usually photosensitive transparent materials. The image on the right in Figure 3.19 is the cross-sectional and enlarged view of a CF with white pixels. The process is identical to the one used for RGB color filters, as explained earlier, which is then followed by the white-pixel process after the formation of a blue layer. Although it is a transparent resist, the applicable process is identical to the one applied to the RGB layer. Overcoat is applied after RGBW patterning, and the same process is followed as introduced earlier.

3.4.2 Color Filter on TFT

This technology is used to form a color filter on the TFT structure. For this technology to work, each color pixel in the RGB layer should touch each other with minimum overlap and should be accurately aligned with the corresponding TFT pixels. The technology enables the TFT and CF to accurately align each other on the same glass and reduces the light transmittance loss that occurs due to the CF glass/TFT glass misalignment. The alignment error occurred during the assembly of the CF and TFT glass panels is usually much higher than the misalignment of the CF and TFT mask layers when processed on the same substrate. As a fundamental solution to solve the large mis-alignment issue in the glass to glass alignment, a technology for forming color filters directly on the TFT structure was developed for commercial application. The color filter, however, should be formed on the panel by avoiding high temperatures, because throughout the TFT process the temperature is maintained at over 300°C. The color filter may be formed after the TFT is completed, but such alternative will raise the driving voltage of the liquid crystal. A feasible alternative will be to form transparent electrodes after the color filter formation and overcoat process. As such, the light transmittance ratio will go up while minimizing color mixes due to the alignment error, thereby resulting in bright and high color purity screen.

Figure 3.20 shows a cross-sectional view of the representative color filter on the TFT structure. The color filter and PAC (photo-acryl) layers are inserted between the data line and the pixel electrodes (labelled "PXL" in the drawing) on the TFT surface.

The role played by the overcoat in the conventional color filters is performed here by the photo-acryl. As it is difficult to realize intricate via holes with the conventional overcoat substance, an alternative material that enables the formation of via holes as well as planarization was developed.

This section has so far explained the structure of the color filter, the associated process, and the materials. Using this color filter technology, display users came to see the almost real colors through the LCD. The industry, however, will have to work on advancing the material properties and associated processes to be able to form a color filter on various display structures.

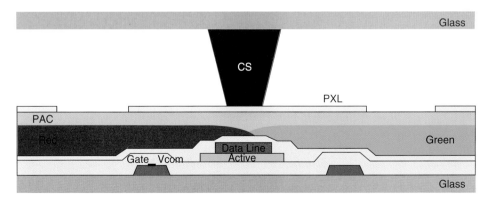

Figure 3.20 Color filter on the TFT (COT) structure.

References

1 CIE chromaticity diagram showing discrimination ellipses derived from Stiles' line element, Wyszecki, Stiles, p. 521 (1967).
2 R. W. G. Hunt, The Reproduction of Colour (Fountain Press, Surrey, UK), Chapter 7 (1995).
3 S. H. Maron, E. G. Bobalek, S. Fok, J. Colloid Sci., 11, p. 21 (1956).
4 A. Kitaha, Bull. Chem. Soc. Japan, 30, p. 586, Application Examples of Carbon Black, Mitsubishi Chemical, http://www.carbonblack.jp/en/index.html (1957).
5 R.W. Sabnis, Display, 20, p. 119 (1999).

4

Liquid Crystal Cell Process

Heung-Shik Park and Ki-Chul Shin

Next Generation R&D Team, Samsung Display Co., Youngin City, Gyeonggi-Do, 446-711, Korea

4.1 INTRODUCTION

Thin-film transistor (TFT) liquid crystal displays (LCDs) have many advantages related to size, resolution, lightness, thinness, and power consumption, which can meet the requirement of various applications (television, desktop computer monitor, laptop computer, and mobile phone). Various LCD operation modes, such as twisted nematic (TN) [1], in-plane switching (IPS) [2], fringe-field switching (FFS) [3], and vertical alignment (VA) [4, 5], have been investigated to meet the diverse requirements for different display applications.

In particular, IPS and VA modes have expanded the range of LCD applications to TVs because of their enhanced viewing angle properties. The LCD TV market has accelerated the need for large-sized LCD panels, with the average size of LCD TV becoming recently larger than 40 inches. Therefore, LCD panel makers have focused on producing large-sized panels with low cost, which drove an increase of the production speed and the use larger mother glass substrates. The size of the mother glass has been increased in order to improve efficiency and productivity of LCD panels as shown in Figure 4.1. For example, four 40-inch panels can be produced from the mother glass of Gen 6 size line (~1,500 × 1,850 mm). However, eight panels are obtained from Gen 8 size (~2,200 × 2,500 mm) and 18 panels from Gen 10 size (~2,880 × 3,130 mm) mother glass. Most global LCD panel makers have been operating production lines with Gen 8 mother glass size.

In this chapter, we explain the manufacturing process of the liquid crystal cell fabrication using large mother glasses, including the latest fabrication techniques of large size VA or IPS LCD.

4.2 LIQUID CRYSTAL CELL PROCESS

The cell fabrication process is the sum of the following steps:

(1) Application of LC alignment layers, such as thin polyimide (PI), on both TFT array and CF glass substrate
(2) Application of sealant on the edge of glass substrates
(3) Filling liquid crystal material inside the seal line
(4) Assembling TFT array and CF array glass substrates
(5) Cutting assembled substrates into LCD cells.

Flat Panel Display Manufacturing, First Edition. Edited by Jun Souk, Shinji Morozumi, Fang-Chen Luo, and Ion Bita.
© 2018 John Wiley & Sons Ltd. Published 2018 by John Wiley & Sons Ltd.

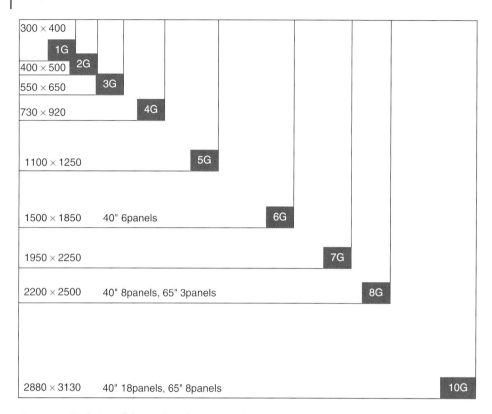

Figure 4.1 Evolution of the mother glass size. Each new generation allowed fabrication of larger displays with high production efficiency enabled by fitting more panels per mother glass.

TFT array patterns are generated on the glass substrate by photolithography processes, where micropatterns designed in a photo mask are transferred into a target material film through the steps of film deposition, photoresist coating, exposure and developing, film etching, and photoresist stripping. Color filter array glass substrates are also fabricated by photolithography on what will become the top glass substrate, while the TFT array process is performed on the bottom glass substrate of the LC cell.

The color filter process usually also includes forming column spacers that define a uniform cell gap thickness between the top and bottom glass substrates, as well as the black matrix pattern of black lines separating the RGB color sub-pixel patterns. The black matrix is fabricated first, helps shield the metal electrode lines and TFTs, and also prevents color mixing and light leakage at the edges of the sub-pixels. Each CF pixel is formed by sequentially patterning red, green, and blue color pigment diffused photo resists. A new color filter process, namely COA (color filter on array) with color filters patterned directly on the TFT substrates, has been recently developed and widely used. Even though COA process requires more patterning steps on the TFT array, it helps to increase the aperture ratio by reducing the overlay margin between the top glass substrate and TFT array substrate.

Figure 4.2 shows a typical fabrication process for LC cells. The first step deals with forming the LC alignment layers on the top substrate and on the TFT array substrate after electrodes fabrication. Metals or transparent conductive materials such as indium tin oxide (ITO) are used for pixel and common electrodes on a TFT array substrate, or on both TFT array substrate and top substrate glasses depending on the LCD switching mode. VA mode needs common electrodes on top substrate and pixel electrodes on the bottom TFT array substrate [4, 5], while for IPS and FFS modes the pixel and common electrodes are patterned only on the TFT substrate [2, 3].

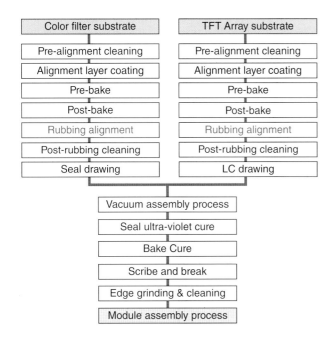

Figure 4.2 Typical process flow for LC cell fabrication.

4.2.1 Alignment Layer Treatment

Liquid crystal molecules can be uniformly oriented on an alignment layer, such as PI (polyimide). Figure 4.3 describes the typical orientations of liquid crystals: homogeneous, homeotropic, tilt, and hybrid orientations. The LC director, that is, the average direction of the liquid crystal molecules, aligns parallel to the substrate plane in homogeneous orientation, while the director aligns perpendicular to the substrates in homeotropic orientation. The tilt orientation is in the intermediate state between homogeneous and homeotropic orientation. The director in the hybrid orientation has homogeneous orientation on one substrate and homeotropic orientation on the other substrate, as shown in Figure 4.3(d).

PI materials have been typically used for the alignment layer not only because of their good aligning ability for LC molecules but also because of their optical and electrical properties, which are compatible to LCs. PIs with long side chains can align LC molecules perpendicular to substrates, therefore, they are practically used for VA LCD mode. In order to obtain the homogeneous alignment required for IPS or FFS modes, additional processes such as alignment layer rubbing or UV treatments with a particular direction are required for inducing LC alignment [2, 3]. The rubbing method and the UV treatment for photo-alignment technique are explained in detail in sections 4.2.3 and 4.2.4.

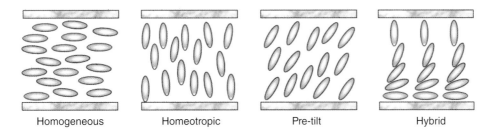

Figure 4.3 Four different types of LC orientations; (a) homogeneous, (b) homeotropic, (c) pre-tilt, and (d) hybrid.

4.2.2 Process of Applying PI Layers

In order to prepare the PI coating pre-mixture, soluble polyamic acid, dianhydride, and diamine are mixed in a polar aprotic solvent, such as N-methylpyrrolidone (NMP), with the addition of gamma-butyrolactone (γ-BL) to enhance wetting properties. The solid content of the PI pre-mixture is selected based on the method used to coat PI on the substrate glass.

There are two main methods for applying PI on the substrate glass. In the roll printing method, which has been used from the early stage of LCD industry, a PI pre-mixture with solid content of around 5% to 6% is used. The PI pre-mixture dropped on an Anilox roll is scraped onto a mesh-shaped polymer printing plate on a printing roll with a uniform thickness controlled by a doctor blade. Then, as shown in Figure 4.4, the PI pre-mixture is transferred on the substrate from the polymer printing plate. Organic solvents from the PI pre-mixture are evaporated during the prebake step at a temperature of about 65–75 °C for around 100 sec. The polyamic acid in the mixture is imidized to polyimde at high temperature around 180–220 °C.

Recently, ink jet printing methods have become widely used in the PI process, as shown in Figure 4.5. PI pre-mixture droplets are ejected from a reservoir through nozzles in inkjet print heads by acoustic pulses generated piezo-electrically. Compared with the conventional roll printing method, the inkjet printing method has many advantages: simpler process, easy modifications of the design and size of printing area, and a high coating accuracy. It allows obtaining uniform and precise alignment layers, controls the thickness of alignment layer easily, and reduces particle contamination because of its non-contact nature. Cost-wise, it can reduce alignment layer material consumption by ~50% and minimize the amount of cleaning solvent used for polymer printing plates compared to the roll printing method.

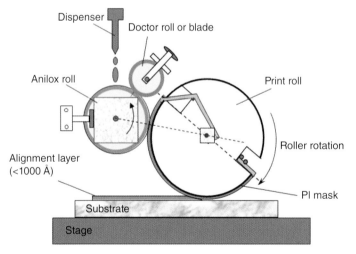

Figure 4.4 Schematic diagram of the roll printing method for PI layer coating.

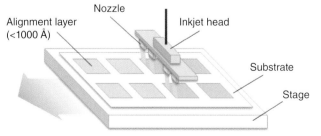

Figure 4.5 Schematic diagram of the PI inkjet printing method.

The key areas in PI coating/drying process are to achieve the uniformity of PI layer thickness and elimination of "mura" in the PI layer. The uniformity and thickness can be controlled by printing speed, the gap between doctor blade and Anilox roll, and the amount of PI solution dispensed. The typical thickness of PI layer ranges from 700 to 1000 Å. Pre-cleaning and post-cleaning processes used before and after PI layer coating are important since they are closely related to the particles and defects found after the PI coating/printing process. The cleaning processes typically used include ultrasonic, megasonic, and DI water.

4.2.3 Rubbing Process

The rubbing method has been the primary aligning process since Mauguin reported it in 1911 [6]. Nevertheless, the LC alignment mechanism has still been under debate since the finding of the alignment phenomena. According to the Berreman model, the elastic distortion energy is minimized when liquid crystal molecules are aligned parallel to the direction of grooves, which are formed on the PI surface during the rubbing process [7]. Geary et al. explained that the interaction between liquid crystal molecules and polymer chains elongated along the rubbing direction is the main reason for alignment [8].

The rubbing method is still widely used in LCD manufacturing because it is a simple and inexpensive process for mass production and it can easily control the rubbing properties. As shown in Figure 4.2, the process flow for the rubbing method alignment consists of PI material coating, pre-baking and post-baking steps, followed by the rubbing step, and a final cleaning step.

In general, the rubbing motion delivers energy to the PI layer by the friction between the rubbing cloth and PI layer, however, it is difficult to measure this energy. The empirical rubbing strength (S) defined by Uchida et al. [9] depends on the number of rubbing cycles (N), the radius of the rubbing roll, the rotation number of the roll (n), rubbing depth (l), and the shift speed of the roll (v), as shown in Equation 4.1 and Figure 4.6, where γ is a coefficient related to the rubbing pressure and the properties of the rubbing cloth.

$$S = \gamma N \cdot l \cdot (1 + 2\pi n / 60v) \tag{4.1}$$

Uchida et al. explained that the anchoring strength is linearly proportional to the rubbing strength in the weak rubbing regime, but that as the rubbing strength increases the anchoring strength shows a tendency to saturate [9].

The rubbing depth (l) is the most important parameter determining the alignment strength, however, a trade-off has been found between the alignment strength and defect level of the PI layer. The rotation number of the roll (n) is usually set between 500 and 1000 rpm for TN mode alignment. The number of rubbing cycles is normally chosen between 1 and 2.

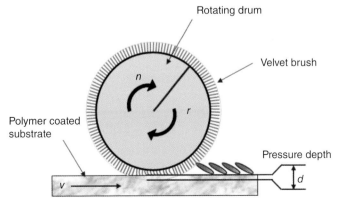

Figure 4.6 Schematic diagram of the PI layer rubbing process.

For IPS mode alignment applications, the rubbing directions for TFT and CF substrates are reversed from each other, with the rubbing direction around 10° with respect to the data line.

The VA mode usually does not require a rubbing process or any surface treatment after forming vertical alignment PI layer on the substrates [4, 5]. In VA mode, multi-domains are obtained by using pixel electrode patterns or pixel structures to control the azimuthal direction of LC molecules along the applied field. However, in-plane switching modes, such as IPS and FFS, and TN modes require the use of a PI rubbing process step to direct the LC molecules homogeneously on planar alignment PI layers [1–3].

4.2.4 Photo-Alignment Process

Another important method for inducing a homogeneous orientation of the LC molecules, parallel to the PI layer surface, is the photo-alignment process. Photo-alignment of LCs is a technique in which an alignment layer is irradiated with linear polarized light (usually ultraviolet (UV) or near-UV light depending on the alignment material) to introduce an anisotropy in the surface properties and induce the desired orientation of the LC molecules, as shown in Figure 4.7.

The photo-alignment techniques have been studied for a long time since Gibbons et al. presented their results in 1991 [10, 11], however, only very recently these techniques have been adopted for mass production by LCD panel makers in Korea and Japan. Nevertheless, the adoption of photo-alignment in production is still limited presently. For example, UVA^2 (ultraviolet-induced multi-domain vertical alignment) was used for LCD-TV by Sharp, who claimed that it was the world's first photo-alignment LCD-TV from a Gen 10 size factory and greatly improved the panel transmittance, contrast ratio and response time [12]. For IPS applications of photo-alignment, LG Display recently announced the use of the photo-alignment technique in the production of mobile display IPS panels.

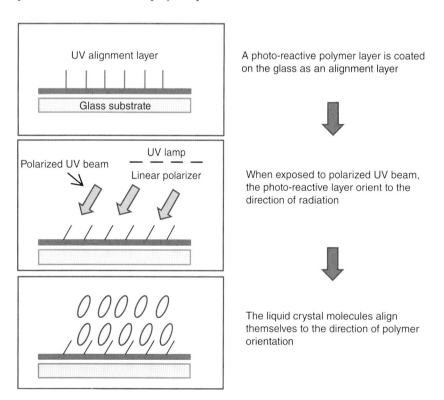

Figure 4.7 Overview of photo-alignment process.

Compared with the conventional rubbing method, the photo-alignment method has several advantages. Since photo-alignment is a non-contact process, it minimizes introduction of contamination particles and reduces the TFT electrostatic damage, which can easily occur during the rubbing process. Photo-alignment further induces a very low pre-tilt angle of < 0.1°, which enhances the uniformity of pre-tilt angle distribution and allows for maximizing the contrast ratio of the panel. In IPS mode, the photo-alignment technique also eliminates the variations of pre-tilt angles and contribute to a higher contrast ratio than that obtained with the rubbing method, where the pre-tilt angle varies about 1 to 2°. This technique is also very attractive for VA mode because it can make multi-domain pixel structures with different LC director orientations that increase the viewing angle. For example, some manufactures using VA technology for their large size panel production are in production or developing multi-domain VA structures by use of four different angle UV beams to induce four different pre-tilt angle domains.

The typical fabrication process for the photo-alignment method is similar to the conventional rubbing process except for the surface treatment step. After substrate cleaning, a photo-alignment material is coated on both color filter and TFT array plates followed by pre-baking near 100 °C for around 100 sec to evaporate solvent and then post-baked at a higher temperature near 200 °C. In order to be able to control the LC direction, the baked substrates are then irradiated with polarized UV light with a wavelength dependent on the photoreactive functional component of the alignment materials. The overall throughput of the photo-alignment process is comparable to the rubbing process because processing times for both methods are largely dependent on the alignment layer coating and baking steps, where similar processes are used in both cases.

Photo-alignment techniques can be generally categorized into three types based on the photo-chemical reaction and materials used: photoisomerization, photodimerization, and photo-decomposition.

Photoisomerization: this photo-alignment method controls the LC orientation due to an optically induced conformational change between cis- and trans-isomerized components of the alignment layer, such as azobenzene functional units, as shown in Figure 4.8(a) [13, 14]. The direction of LC alignment can be easily altered by exposure to light because the photoisomerization is reversible by changing the irradiation wavelength. Therefore, the resulting photo-alignment has poor long-term stability without further chemical processing, such as polymerization or cross-linking of the LC alignment material. An azobenzene-mixed polyimide material is typically used, with photoisomerization during exposure to 365 nm linearly polarized UV irradiation, as shown in Figure 4.8, which induces LC alignment perpendicular to the polarization direction.

Photodimerization: this photo-alignment method allows aligning LC molecules by a photo-induced dimerization chemical reaction using photo cross-linkable materials, such as polyvinyl cinnamate with linearly polarized UV light with 313nm wavelength. In this case, LC molecules align spontaneously perpendicular to the optical axis of polarized light, as shown in Figure 4.8(b) [14].

Photodecomposition: this method is attributed to the anisotropic photodecomposition of polyimide layer upon exposure to polarized UV light, where the LC alignment direction is also perpendicular to the axis of the polarized light as shown in Figure 4.8(c) [15, 16]. In this method, rinsing after photoreaction is the critical process because decomposed particles from the alignment layer can be generated. The second post bake step after photoreaction also enhances the LC alignment ability and increases the LC order parameter. One of the materials used with this method is a polyimide containing cyclobutadiene moieties, which decompose upon irradiation with 254nm UV light, recently adopted by several panel makers for IPS mode applications.

4.2.5 LC Filling Process

There are two main methods used for the LC filling process step. One is the vacuum filling method and the other one is so called one drop filling (ODF) method.

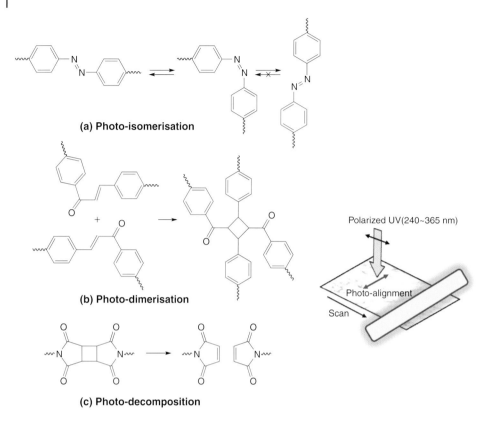

Figure 4.8 Three main types of photo-alignment materials: (a) photoisomerization, (b) photodimerization, and (c) photo-decomposition.

4.2.5.1 Vacuum Filling Method

The vacuum filling method had been used to fill LC material into small-sized cells formed between assembled CF and TFT substrates, using a small opening left at one side of the cell while the remaining peripheral area is sealed with sealant glue. The opening of the cell is contacted to the LC reservoir in a vacuum chamber and the cell fills with LCs by capillary force action when the pressure of chamber increases as shown in Figures 4.9 and 4.10. When the pressure of chamber is in balance with the outside, the cell is completely filled and then the opening of the cell is closed with end-sealant [17].

The process of vacuum filling method generally consists of four steps: printing the seal line with thermally cured sealant, cell cutting, LC filling, end sealing and UV curing as shown in Figure 4.10.

4.2.5.2 End Seal Process

The end seal process is important to control the final cell gap. During the end seal process, a uniform pressure is applied to the surface of LC filled panel in order to squeeze out excess amount of LC and also to fix the cell

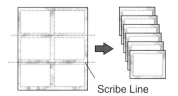

Figure 4.9 The assembled TFT and CF plates are scribed into small-sized cells used for vacuum filling process.

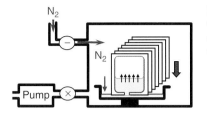

Figure 4.10 Vacuum filling process consists of (1) evacuate LCD panel and the vacuum chamber, (2) dip the open end of the panel to the LC reservoir, (3) vent the chamber with N_2, and (4) LC slowly fills in the LCD panel due to the pressure difference between inside LCD panel and the chamber.

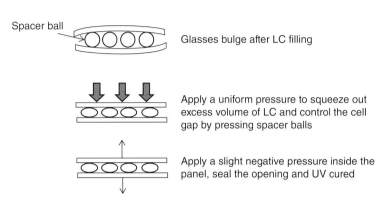

Figure 4.11 Schematic illustration of the end seal process.

gap. This process is illustrated in Figure 4.11. Then the opening that was used for LC filling is sealed and cured by UV.

The vacuum filling method had been used from the early stage of LCD industry. However, this method cannot be used for large sized LCD panels because the LC filling time is proportional to the cell size. For example, the vacuum filling method requires several days to completely fill up LCs into a 40 inch panel. Another issue of the vacuum filling method is that a large amount of LCs is wasted in the process.

4.2.5.3 One Drop Filling (ODF) Method

The ODF technique, which was developed in relatively recent years, can reduce the process time and save significant amounts of LC materials with a critical role for the mass production of large-sized LCD panels. Using this method, panel makers can produce large size panels for monitors and TVs with short lead time and lower costs.

Figure 4.12 shows the main steps of the ODF process. First, the exact amount of LCs is dropped or dispensed onto the active area of one side of substrate, usually the TFT array substrate in the ODF process. Then, seal lines are drawn with sealant (thermal and photo active polymer materials) on the other substrate, followed by bonding the two substrates together.

The total volume of LC materials needed to fill into a panel can be calculated by the following equation:

$$V_{total} = x \times y \times h \tag{4.2}$$

where $x \times y$ is the active area of the panel defined by the seal drawing area, and h is the height of the column spacer.

Therefore, the amount of one drop is V_{total}/N_{total}, where N_{total} is the number of the total drops from the LC dispenser. If an excessive amount of LC drops are dispensed relative to the exact amount needed for a panel, the edges of the panel will look brighter than the center of the panel. Alternatively, an insufficient amount of LCs can lead to air bubbles in the panels. The volume of LCs should be exactly checked by weighing tens of drops together with an electronic scale before ODF process because the amount of one drop is usually less than one milligram.

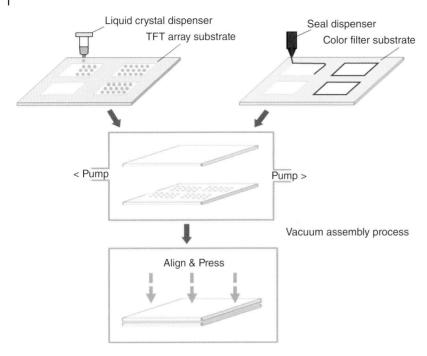

Figure 4.12 Schematic diagram of the ODF process.

The sealant, one of the most important materials for ODF process, is used not only for bonding the two substrates together but also for preventing moisture or impurities coming into the cells. The sealant for ODF can be dispensed and cured rapidly by UV light, followed by thermal curing [18]. For this reason, the sealant material typically consists of a mixture of epoxy and acryl resins. The acryl resins are polymerized under UV irradiation first, which prevents separation of the assembled substrates through fast curing. Then, the epoxy resins are cured at high temperature around 100 °C, which results in a strong and reliable panel assembly [19].

Recently, LCD panels with narrow bezels for TVs and PIDs (public information display) have become very popular. For these applications, the sealant must have a high adhesion strength to allow for narrow seal line width smaller than 1mm. The control of dispensing pressure and the position of dispenser are very important to obtain a uniform seal line width, especially at the panel corners. For reliability reasons, the sealant should also not be able to contaminate the LC material because it remains in direct contact with the LCs in the cell. The LC mixtures used for TFT-LCDs are purified and have a very high resistivity around 10^{13} to $10^{14}\,\Omega\text{cm}$. The contamination of LCs can lower the VHR (voltage holding ratio), one of the important parameters for driving TFT-LCD panels, which usually deteriorates image quality and reliability (such as image sticking problem).

4.2.6 Vacuum Assembly Process

After dispensing the arrays of LC droplets on the TFT substrate, the TFT and CF substrates are aligned in a vacuum chamber and bonded together as shown in Figure 4.13.

The TFT substrate with dispensed LC droplets on its surface is placed on the stage of the vacuum assembly system (VAS) chamber, and the upper substrate with printed sealant is handled without touching the lower substrate, then aligned using alignment marks on each corner with high accuracy. The upper substrate is then pressed down for bonding to the lower substrate.

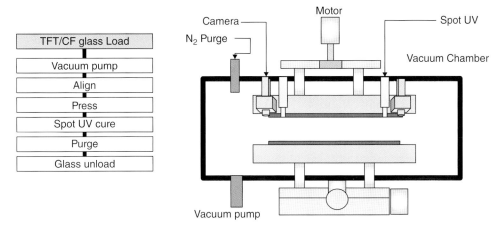

Figure 4.13 The procedure of vacuum assembly and schematic diagram of vacuum assembly system chamber.

VASs can hold substrates in the vacuum chamber using typically one of two different methods: electrostatic chuck (ESC) method and physical sticky chuck (PSC) method. Though ESC method can control easily the adhesion force, it often causes electrostatic discharge problems that can induce device damage. PSC method using polymer PSC sheets has been widely used because it has a stable adhesive force and can minimize the deformation after bonding the two substrates by maintaining uniform gap between substrates.

The assembled substrate stack is then cut into individual cell units, which are then rinsed under high-pressure pure water to remove cullet. Finally, two pieces of polarizer are attached on the individual cell and then electric components for driving panels and back lights are mounted and connected in the module assembly process.

4.2.7 Polarizer Attachment Process

As the final stage of TFT-LCD cell manufacturing process, two polarizer sheets are attached on each individual LCD cell during the module assembly process.

Linear polarizers are commonly used in LCDs, which transmit light with a particular polarization direction and absorb the other polarization components. They are generally made by polyvinyl alcohol (PVA) dyed with dichroic dye materials, such as iodine, which stretched to achieve uniaxial alignment of the light absorber materials. Therefore, the dichroic ratio defined based on the absorption coefficients for the two orthogonal polarizations becomes ~100, which provides excellent extinction ratio and polarization efficiency. After this, triacetyl cellulose (TAC) films are laminated onto both sides of the stretched PVA film in order to enhance the mechanical strength. The polarizing films can be further treated to get additional functions controlling light leakage and reflection, such as anti-glare (AG), low-reflection (LR), anti-reflection (AR), and hard coatings (HC). Figure 4.14(a) shows the structure of a conventional sheet polarizer with a protective film on the top surface, and a release film on the bottom surface.

Before attaching the sheet polarizers onto the LCD cell, the surface of the cell is cleaned to remove any glass cullets and particles, typically with high pressure pure water. The sheet polarizers are pressed down onto the LCD cell by using a roller after detaching the release film from the adhesive side to allow polarizer lamination as shown in Figure 4.14 (b).

Finally, the electric components necessary for driving display panels, such as display driver ICs, tape carrier package (TCP), printed circuit board (PCB), and backlights units are mounted and connected to the electrodes of LCD cells.

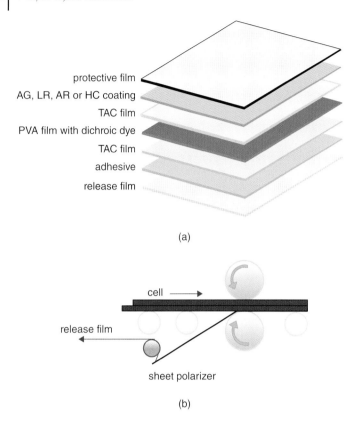

Figure 4.14 (a) Typical stack structure of the sheet polarizer and (b) schematic diagram of the polarizer attachment process.

4.3 CONCLUSIONS

We reviewed in this chapter the manufacturing process of liquid crystal cell fabrication. The cell fabrication process includes the steps for applying alignment layer material on a substrate, forming alignment layers for controlling the direction of liquid crystals molecules, applying sealant on the glass substrate and filling liquid crystal materials, assembling the TFT array and CF array glass substrates, and cutting the assembled substrates into LCD cells.

LCD panel makers continue to focus on how to produce large-sized panels from large mother glass in order to improve efficiency and to increase the productivity of LCD panels, lowering the manufacturing cost. We hope that this review provides not only a basic understanding of the LC cell fabrication process but also updated knowledge on the fabrication techniques for large-sized LCD panels.

Acknowledgments

Authors thank Mr. Jaesoo Jang and Mr. Kyeongjong Kim for their kind support.

References

1 M. Schadt, W. Helfrich, Voltage-dependent optical activity of a twisted nematic liquid crystal. Appl. Phys. Letters 18, p. 127 (1971).

2. M. Oh-e, K. Kondo, Electro-optical characteristics and switching behavior of the in-plane switching mode. Appl. Phys. Letters 67, p. 3895 (1995).
3. S. H. Lee, S. L. Lee, H. Y. Kim, Electro-optical characteristics and switching principle of a nematic liquid crystal cell controlled by fringe-field switching Appl. Phys. Letters 73, p. 2881 (1998).
4. K. H. Kim, K. H. Lee, S. B. Park, J. K. Song, S. N. Kim, J. H. Souk, Domain divided vertical alignment mode with optimized fringe field effect. Proceeding of Asia Display'98, pp. 383–386 (1998).
5. A. Takeda, S. Kataoka, T. Sasaki, H. Chida, H. Tsuda, K. Ohmuro, T. Sasabayashi, Y. Koike, K. Okamoto, A super-high-image-quality multi domain vertical alignment LCD by new rubbing-less technology. SID'98 Symposium Digest, pp. 1077–1080 (1998).
6. C. Mauguin, On the liquid crystal of Lehmann. Bull. Soc. Fr. Min. 34, p. 71 (1911).
7. D.W. Berreman, Solid surface shape and the alignment of an adjacent nematic liquid crystal. Phys. Rev. Lett., 28, p. 1683 (1972).
8. J. M. Geary, J. W. Goodby, A. R. Kmetz, J. S. Patel, The mechanism of polymer alignment of liquid crystals. J. Appl. Phys., 62, p. 4100 (1987).
9. Y. Sato, K. Sato, T. Uchida, Relationship between rubbing strength and surface anchoring of nematic liquid crystal. Jpn. J. Appl. Phys., 31, p. 579 (1992).
10. H. Zocher, Optical anisotropy of selective absorptive substances and mechanical generation of anisotropy. Naturwissenschaften, 13, p. 1015 (1925).
11. W. Gibbons, P. Shannon, S. Sun, B. Swetlin, Surface-mediated alignment of nematic liquid crystals with polarized laser light. Nature 351, 49 (1991).
12. K. Miyachi, et al., 2010 SID Sym. Digest Tech. Papers, 41, p. 5790582 (2010).
13. K. Ichimura, Photoalignment of liquid-crystal systems. Chem. Rev. 100, p. 1847 (2000).
14. Y. Kawanishi, T. Tamaki, M. Sakuragi, T. Seki, Y. Suzuki, K. Ichimura, Photoregulation of liquid crystal alignment by Langmuir-Blodgett Layers of azobenzene polymers. Langmuir 8, p. 2601 (1992).
15. M. Schadt, K. Schmitt, V. Hozinkov, V. Chigrinov, Surface-induced parallel alignment of liquid crystals by linearly polymerized photopolymers. Jpn. J. Appl. Phys., 31, p. 2601 (1992).
16. M. Nishikawa, J. L. West, Mechanism of generation of pretilt angles on polyimides with a single exposure to polarized ultraviolet light. Jpn. J. Appl. Phys. 38, p. 5183 (1999).
17. W. C. Yang, W. Y. Chien, M. S. Chen, W. M. Huang, Manufacturing an ultra slim LCD. SID Symposium Digest of Technical Papers, 42, pp. 649–651 (2011).
18. W. den Boer, Active matrix liquid crystal displays, Newnes, Oxford, p. 76 (2005).
19. M. Bajpai, V. Shukla, A. Jumar, Film performance and UV curing of epoxy acrylate resins. J. Prog. Org. Coat, 44, p. 271 (2002).

5

TFT-LCD Module and Package Process

Chun Chang Hung

AU Optronics, Taoyuan City, Taiwan, ROC, 32543

5.1 INTRODUCTION

In this chapter, the back-end manufacturing processes of TFT-LCD modules (LCMs) are described. As shown in Figure 5.1, the back-end process includes polarizer attachment, chip on film (COF) or chip on glass (COG) bonding, which are called collectively the JI process, module assembly, aging, and packing for shipment. The acronym "JI" comes from Japanese pronunciation and means "bonding." These processes are, in general, finished in a so-called module assembly (MA) factory. For some TFT-LCD module makers, the polarizer attachment may be done in a front-end cell fab and the MA factories do only the JI and remaining processes.

As a display device, a TFT-LCD panel doesn't emit light by itself. A bright and uniform illumination source is needed to produce an image, with the TFT-LCD panel acting as a spatial modulator. As shown in Figure 5.2, in addition to the light source, the TFT-LCD module includes several optical elements. In order to simplify the assembly of the TFT- LCD module, these optical components can be preassembled as a backlight module, which is subsequently used for final LCM assembly in another factory. Often, this makes the MA process as simple as just stacking together the display panel and the backlight module.

This chapter introduces in sequence the TFT-LCD back-end processes, starting from polarizer attachment to packing for shipment. For small and large panels the processes are similar but not identical. Process introductions are in separated sections. Furthermore, since the backlight module plays an important role in an LCM (TFT-LCD module), optical components and assembly process are described in a separate chapter. The status of art and trends of automation in the assembly process are included too. At the end, representative business models and material flow in LCM manufacturing are introduced.

5.2 DRIVER IC BONDING: TAB AND COG

TAB (tape automatic bonding) and COG (chip on glass) are two different processes for bonding driver ICs with the TFT-LCD cell. The process used depends on the configuration of the LCD panel. For TAB process, driver ICs are bonded onto flexible printed circuit boards (FPC), called COF (chip on film), and then COFs are bonded to the TFT-LCD cell glass. For COG process, driver ICs are bonded directly onto the TFT-LCD cell glass. In general, TAB process is better for production yield but is more expensive because of the FPC. Furthermore, without IC on glass, cell thickness can be thinner. On the other hand, the COG process will be more cost effective without using FPC. Since the ICs are bonded on glass, they are protected from deformation

Flat Panel Display Manufacturing, First Edition. Edited by Jun Souk, Shinji Morozumi, Fang-Chen Luo, and Ion Bita.
© 2018 John Wiley & Sons Ltd. Published 2018 by John Wiley & Sons Ltd.

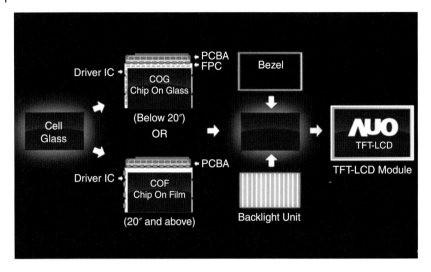

Figure 5.1 Typical LCM processes for a TFT-LCD panel maker [1].

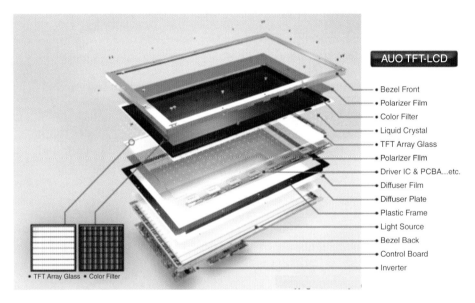

Figure 5.2 Components of a typical TFT-LCD module [1].

and damage. Furthermore, because of thinness in configuration, the TFT-LCD module can be designed to be smaller.

5.3 INTRODUCTION TO LARGE-PANEL JI PROCESS

Most of the large panels utilize COF for panel driving. Driver ICs are mounted on FPC instead of the cell glass. COFs are used as packed components. The JI process doesn't include IC bonding. As shown in Figure 5.3, JI

5.3 Introduction to Large-Panel JI Process

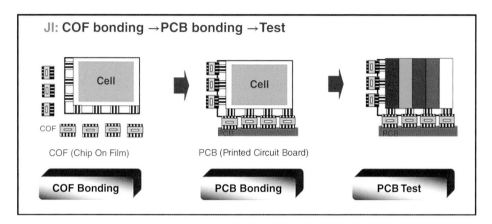

Figure 5.3 Illustration of large-panel JI process [1].

process includes three major sub-processes. They are COF bonding, PCB bonding, and PCB test. They are described as follows:

5.3.1 COF Bonding

This process bonds COF onto the cell glass, as shown in Figure 5.4. The adhesive used is ACF (anisotropic conductive film). Figure 5.5 shows details of the bonding process.

5.3.1.1 Edge Clean

To ensure good bonding quality, the area for cell-COF bonding has to be well cleaned to avoid the presence of any particles and residues on the bonding pad surfaces. Four processes are included:

- **Dry clean:** to remove large foreign particles from the cell surface.
- **Wet clean:** supersonic waves are to remove particles stuck on cell surface.
- **Wiper with solvent:** to wipe out particles on cell surface
- **Plasma:** to remove organic contamination from the cell surface.

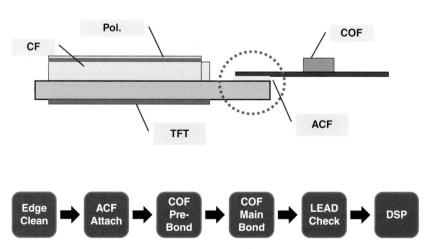

Figure 5.4 COF bonding structure.

Figure 5.5 COF bonding procedure.

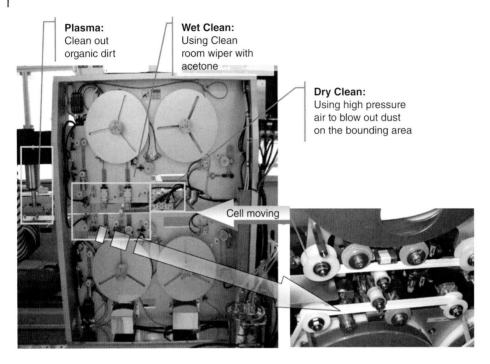

Figure 5.6 Edge clean [1].

Cleaning results of this process should be carefully monitored. Any particle left on the surface may induce panel fracture during the subsequent bonding process. Figure 5.6 shows an example of the edge cleaning equipment stages.

5.3.1.2 ACF Attachment

ACF is a double-sided adhesive tape coated with a mixture of adhesive binder and electrically conducting particles, as shown in Figure 5.7. The conducting particles in ACF are discrete and dispersed uniformly. These particles can form an electrical connection between two adhered metal parts pressed together. ACF is typically used in cases where high temperature soldering cannot be applied.

Figure 5.8 shows a basic diagram of ACF bond structure. When COF is attached to the glass and pressed together, the conducting particles between two electrodes are squeezed and they break through the insulating polymeric matrix to contact the electrode pads. As shown in the figure, the conduction is unidirectional in the vertical direction. Those nonsqueezed particles remain insulated and therefore no conduction occurs in the horizontal direction.

Prior to bonding, the ACF tape needs to first be laminated as shown in Figure 5.9. First, the carrier film is removed and then the ACF film is laminated onto the cleaned cell glass surface with controlled pressure,

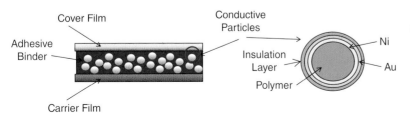

Figure 5.7 Anisotropic conductive film structure [1].

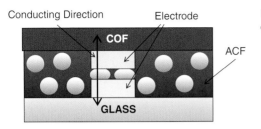

Figure 5.8 ACF bonding structure with electrical contact only in vertical direction [1].

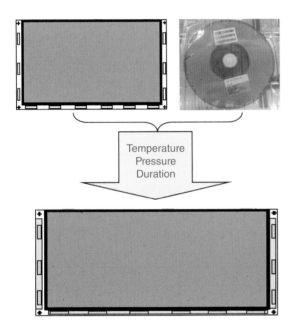

Figure 5.9 ACF attachment.

temperature, and duration. After the lamination is completed, the ACF protection is removed for the next process step.

5.3.1.3 COF Pre-Bonding

In this process, COFs are first taken out from COF tape by a punch die and then placed on the cell glass at the designated location by imaging positioning. After alignment, a small pressure is applied on the COF to fix in place. All the COFs are attached to the cell glass and pressed one by one (Figure 5.10).

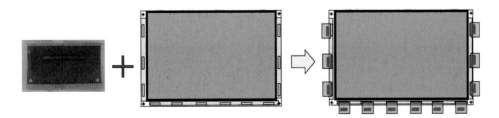

Figure 5.10 ACF pre-bonding.

5.3.1.4 COF Main Bonding

After pre-bonding, the cell is transferred to the next stage. At first, the position of the cell will be adjusted during an optical inspection process. Then, a heated bonding head will fall down and apply localized pressure on the COF at the target pressure, temperature, and duration. A buffer material between the bonding head and the COF is used to improve COF attachment. In general, the buffer film is made of teflon or silicone, as is shown in Figure 5.11.

5.3.1.5 Lead Check

This is an in-line optical testing process that follows right after the main bonding step, aimed at two key objectives. One is to verify the alignment between the COF and cell electrode leads, for proper connections. The other is to check if enough number of conducting particles have been pressed between each electrode for achieving a proper electrical resistance of the bond (Figure 5.12).

5.3.1.6 Silicone Dispensing

At this stage, silicone resin is dispensed on the cell lead bonding area to protect cell leads from corrosion and ambient moisture (Figure 5.13).

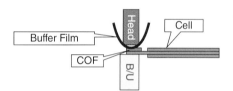

Figure 5.11 ACF main bonding step.

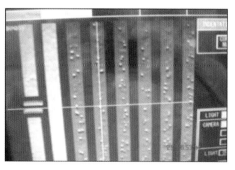

Figure 5.12 Lead check [2].

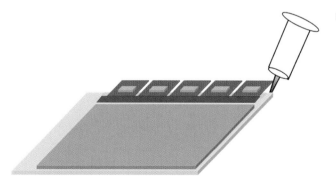

Figure 5.13 Silicone dispensing.

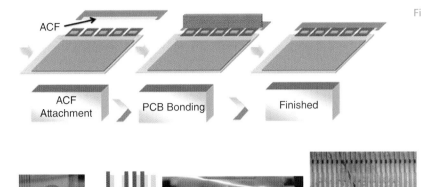

Figure 5.14 Procedures for PCB bonding.

Figure 5.15 Some typical defects found by PCB test: (a) poor alignment, (b) poor flatness, and (c) particle contamination.

5.3.2 PCB Bonding

During this process, the PCB is attached to the COF surfaces. As before, ACF is first laminated onto the PCB electrode pad array, and then the COFs are bonded on to the PCB by a pressing head, as shown in Figure 5.14.

5.3.3 PCB Test

This process checks both functional and visual quality of PCB bonding to prevent unqualified panels flowing down to the module assembly line. Mostly, it is tested by a sampling protocol. In recent years, one kind of business models is to sell PCB-bonded panel (open cell) to customers directly without mechanical parts assembly. For the open cell business, SKD (semi knock down, means sub-module) test will be done instead of PCB test. Instead of sampling, SKD test will do 100% checking with additional test items and images. Typical major defects found during this process are non-proper PCB bonding, particles, line defects, and so on. Figure 5.15 shows some typical defects found by PCB test.

5.3.4 Press Heads: Long Bar or Short Bar

Two kinds of press heads are used for ACF bonding: long bar and short bar. In the case of long bar, a single pressing head is used for the entire side of the panel, allowing simultaneous bonding of multiple COFs. It is convenient to use with the benefits of no limitation on the number of COFs and on the COF pitch. The main drawback, however, is that it is difficult to adjust the flatness of the bar for obtaining uniform pressing forces on each COF. Using a head with long bar also induces more wasting of ACF tape.

On the other hand, bonding heads with a short bar can more easily achieve uniform pressure for each COF, but have limitations on the pitch and the number of COFs. In general, the structure of the short bar machine is more complicated compared with the long bar machine, as shown in Figure 5.16.

5.4 INTRODUCTION TO SMALL-PANEL JI PROCESS

For small TFT-LCD panels, the chip-on-glass (COG) process is generally used. JI processes for small panels are similar to large panels, with a typical flow shown in Figure 5.17.

5 TFT-LCD Module and Package Process

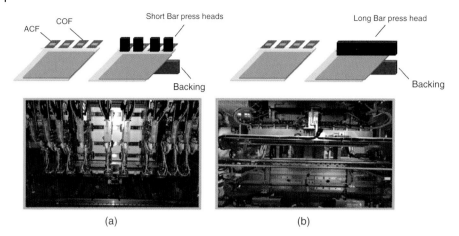

Figure 5.16 COF bonding: press head options: (a) array of short bars for each COF, and (b) long bar for pressing across entire panel edge [1].

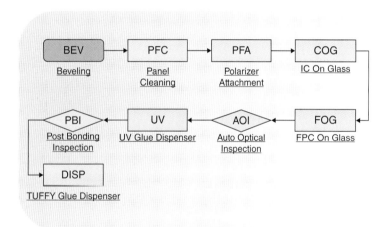

Figure 5.17 Typical process flow for small-panel JI process [1].

5.4.1 Beveling

To avoid glass chipping during handling, the JI process usually requires beveling all the glass corners at the beginning. In some occasions, beveling on edges is also required.

5.4.2 Panel Cleaning

After glass beveling, a cleaning step is required to prepare for subsequent polarizer attachment and COG bonding processes. DI water and clean room cloth are typically used. Sometimes solvent is used for more thorough cleaning.

5.4.3 Polarizer Attachment

The JI process for small panel LCM usually includes polarizer attachment, as shown in Figure 5.18. This process attaches polarizer sheets to the top (color filter) and bottom (TFT) surfaces of the TFT-LCD panel. Particle control is important for this process to ensure successful bonding. Particles may originate from

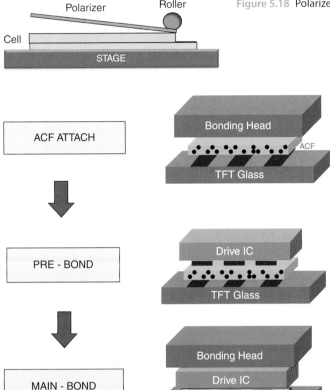

Figure 5.18 Polarizer attachments.

Figure 5.19 Chip on glass (COG) bonding.

facility environment, clean cloth, glass chipping, polarizer material itself, and inner parts of the assembly machine.

5.4.4 Chip on Glass (COG) Bonding

This process bonds driver ICs onto the cell glass using ACF as well. As shown in Figure 5.19, the procedures for ACF attachment are similar to large panels. At first, ACF film is laminated onto the connecting electrodes on the glass. Then ICs are loaded and pre-bonded for precise positioning and temporary bonding. Finally, the main bond process is performed at particular temperature, pressure, and time duration. During this process, the alignment and missing of driver IC are the main inspection goals.

5.4.5 FPC on Glass (FOG) Bonding

This process bonds FPC onto TFT glass. As shown in Figure 5.20, similar to large panels, three sub processes are included: ACF lamination, FPC pre-bond, and FPC main bonding.

5.4.6 Optical Microscope (OM) Inspection

This procedure checks the bonding status of IC and FPC. Detected bonding failures usually are misalignment, insufficient conducting balls between electrodes, improper press-crack of conducting particles, contaminant particles trapped in the bond, and so on (Figure 5.21).

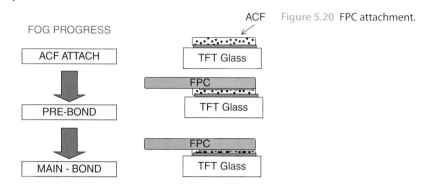

Figure 5.20 FPC attachment.

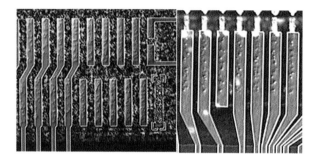

Figure 5.21 Example of OM inspection [3].

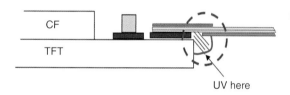

Figure 5.22 Example of glue dispensing.

5.4.7 UV Glue Dispense

As shown in Figure 5.22, UV or other kind of glue is dispensed between FPC and the edge of panel glass to increase mechanical robustness of the bond and prevent FPC from damage.

5.4.8 Post Bonding Inspection (PBI)

This process inspects functional defects and sorts out failed panels before assembling with the backlight unit. Defects found here usually are dot pixel defects, line defects, abnormal display, and so on (Figure 5.23). This test requires connecting the module to equipment with customized interface for providing power and testing signals. In addition, panels with cosmetic defects like cracks, scratches are also sorted out. For some applications, JI-finished panels will be shipped to customers directly without backlight. In that case, more detailed inspections have to be done before packing. The inspections are similar to those introduced in Chapter 16.

5.4.9 Protection Glue Dispensing

This process dispenses protection glue onto the top of exposed conductors of TFT and ICs. Fast curing silicone glue is usually used. Purpose for the glue dispensing is to prevent ICs and circuits from moisture corrosion, dust, and also for shielding leakage light (Figure 5.24).

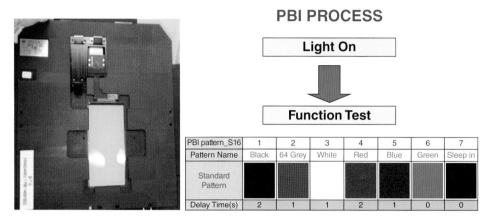

Figure 5.23 Example of PBI [3].

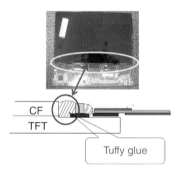

Figure 5.24 Protection glue dispensing.

5.5 LCD MODULE ASSEMBLY

The module assembly process in itself is not complicated. Figure 5.25 shows the assembly procedures for small panels (smart phone), while Figure 5.26 shows the same for large panels (TV).

Since backlight particles represent one of the major defects of LCD module, cleaning and double-checking the BLU quality at the beginning of the assembly process is sometimes necessary. Furthermore, the environment cleanness, especially at the stage of combing BLU with LCD cell, is one of the key factors for successful module assembly. When removing the protection film on both LCD panel and backlight unit, the speed to rip off the film is also critical—if too fast, it may induce high-static electric voltage and damage the panel circuit or electronic parts. The induced static electric voltage may also attract the particles in the air or on any objects nearby and cause particle defects.

In the early stage for some LCD makers, fully automatic module assembly lines had been used for years. In recent years, thin and narrow border frame design becomes a stylish trend to display modules. Some of the automatic processes, screw fastening for example, cannot meet the requirements for variation and flexibility in assembly by design. They are replaced by human hands and the assembly line becomes semi-automatic.

On the other hand, the small panel assembly lines at most module assembly makers used to be almost entirely based on manual labor in the past. However, mobile phone LCMs have evolved to narrow border or borderless designs as required by customer, and thus combining the LCD panel with BLU presently requires precise positioning and assembly. Assembly by human hand and eyes become critical and not achievable. Thus, such LCM assembly lines have become more and more automatic.

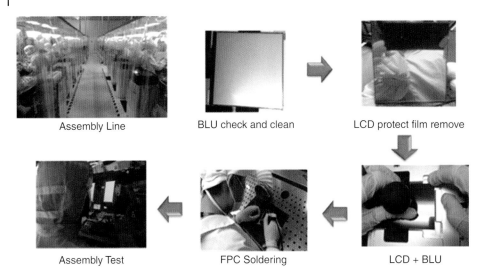

Figure 5.25 Small panel module assembly process [1].

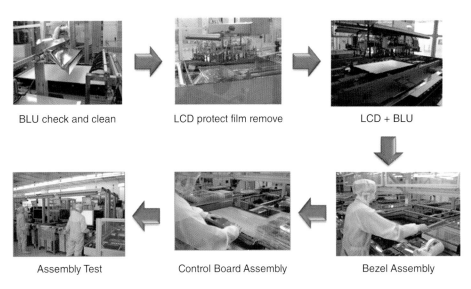

Figure 5.26 Large-panel module assembly process [1].

5.6 AGING

Aging is an optional process after LCD module assembly and before final test for shipping. The assembled LCD modules are placed in a high temperature environment for several hours (e.g., 60 °C for 5 hours).

The purpose for aging is to screen out potential defects of circuit parts, potential problems of mechanical and optical parts, and to stabilize the liquid crystal distribution. The process is costly (electricity usage) and time consuming. Since the design and manufacturing of LCD modules have been mature for recent years, this process has become optional for some customers. Figure 5.27 shows some types of aging lines. The design of the aging line depends on panel size and aging condition requirements.

Figure 5.27 Examples of aging lines [1].

5.7 MODULE IN BACKLIGHT OR BACKLIGHT IN MODULE

In recent years, due to the continuously dropping street price of display modules, the profit margin for LCD and component makers becomes critical. It causes all the makers reconsidering the possibility of redistribution of the value chain.

Figure 5.28 shows the LCD module production flow with different combinations. Originally, LCD panel makers produce LCM by themselves with backlight modules input from BLU makers. The BLU makers

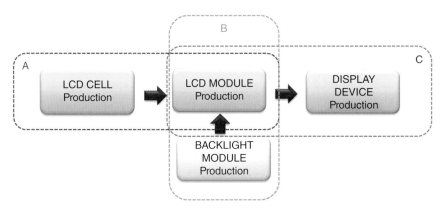

Figure 5.28 Different value chain combinations in LCM manufacturing.

procure all the components including optical parts and mechanical parts to assemble into backlight modules. This kind of business model induces additional cost that can be avoided. For example, repeated loading and unloading, repeated packing and unpacking, repeated inspection, cost of shipment, cost of over specification, and so on.

To eliminate the extra costs and enlarge profit margins, in recent years makers have tried to combine the process in either upstream or downstream. Some LCD makers procure all backlight parts and assemble LCMs combining with the JI process. Some backlight makers procure open cells from cell makers as one of the components and manufacture LCMs. Some TV makers own backlight and mechanical parts (plastic, metal) manufacturing and make TV sets.

References

1 Unpublished Internal training document of AU Optronics.
2 Unpublished TV product inspection SOP of AU Optronics (Xiamen).
3 Unpublished Mobile phone product inspection SOP of AU Optronics (Xiamen).

6

LCD Backlights

Insun Hwang[1] and Jae-Hyeon Ko[2]

[1] *Samsung Display Co. Ltd, Youngin City, Gyeonggi-Do, 446-711, Korea*
[2] *Department of Physics, Hallym University, Chuncheon, Gangwondo, 24252, Korea*

6.1 INTRODUCTION

Unlike other display devices, TFT-LCD is a non-luminescent device. Therefore, a flat light source module, which is called as backlight unit (BLU), is required and must be assembled under the LCD panel (Figure 6.1(a)). The BLU supplies the LCD panel with bright, two-dimensionally uniform white light, thanks to which necessary output beam profile and uniform light distribution are realized over the display area.

The BLU consists of several optical components, such as LED light source, light guide plate (LGP), and multiple optical films as shown in Figure 6.1(b) [1, 2]. Fluorescent lamps have been used as light sources in BLU during the past decades, but presently most LCD devices use LEDs. Depending on the location of LED, there are edge-lit type and direct-lit type BLU structures. Figure 6.1(b) shows the edge-lit type BLU, where several LEDs are located on the edge of LGP. In this chapter we will explain in detail the BLU design and its manufacturing process based on the edge-lit structure. In this section we describe the BLU architecture and main parameters important for optical performances of BLU conceptually when its components are assembled, and then in the following sections we will deal with important characteristics and manufacturing processes of LED, LGP, and optical films in sequence.

The first step in designing BLU is to determine the thicknesses of LED and LGP. The thickness of the LCD device products has become thinner and thinner, and the thin form factor is now one of the most important characteristics of LCD products. Accordingly, the thickness of each component must be thinner, too. With a given LCD panel thickness, the rest of the total thickness should be shared by the mold frame, LGP, and film stacks. In case of a thicker LGP, the thicker LED can be used, which may improve the efficiency of the LED. On the other hand, the thinner optical films may suffer from reliability failure such as film wrinkle due to high humidity and high temperature environment. Therefore, the thicknesses of individual components in BLU should be optimized.

The second step is to determine the number of LEDs. It is determined by the total luminous flux necessary for satisfying the luminance specification of the LCD device, where optical losses by other optical elements should be considered. The total flux also depends on the target color coordinates, because they change as the light passes through the LCD panel. In order to reduce material costs the number of LED should be minimized. In this case, the power consumption of individual LED should increase, and the distance from each LED to the active area, which is necessary for mixing light from individual LEDs, increases inevitably. Therefore suitable number of LEDs should be determined under these constraints.

Flat Panel Display Manufacturing, First Edition. Edited by Jun Souk, Shinji Morozumi, Fang-Chen Luo, and Ion Bita.
© 2018 John Wiley & Sons Ltd. Published 2018 by John Wiley & Sons Ltd.

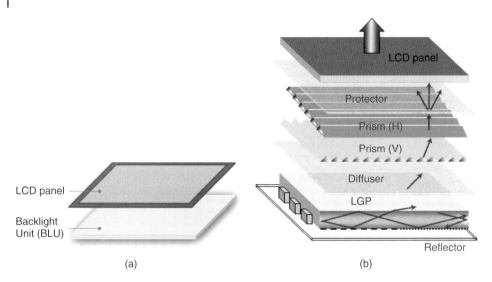

Figure 6.1 A schematic view of a backlight unit (a) and its components (b).

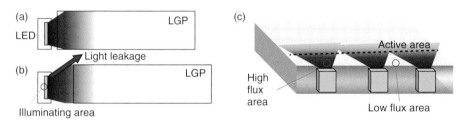

Figure 6.2 Some design parameters in LED BLU. (a) The case of adequate LED-LGP gap and (b) the case of wider LED-LGP gap under which some light from LED leaks and does not enter into LGP. (c) Light flux difference with respect to the positions of LEDs.

Once the number of LEDs is determined, the locations of LEDs can be fixed. We can then design the optical pattern of LGP based on the thickness of the LGP and the locations of LEDs in BLU. At this point it should be noticed that the gap between LED and LGP is an important factor for enhancing coupling efficiency of the light generated from LED into LGP. If the LGP-LED gap is wide (Figure 6.2(b)), some amount of light from LED cannot enter into LGP and can be a cause of light leakage failure. Besides, even though BLU is assembled with the zero-gap condition between LED and LGP initially, the gap may increase after thermal cycle test. It is due to the fact that metal chassis, mold frame, and LGP are made of different materials, and thus have different thermal expansion coefficients. These points should be taken into consideration when designing the product.

LEDs usually form bright regions, denoted as high-flux areas, near the incident surface of the LGP corresponding to their locations as shown in Figure 6.2(c). Dark regions (denoted as low-flux areas) are formed periodically between these high-flux areas. This kind of periodic modulation of light flux near the edge of the LGP can be a cause of non-uniformity failure during the test for liquid crystal modules (LCMs), which is called hot spots. Apparently, the hot spot looks similar to the light leakage phenomenon, which was mentioned earlier. However, the origins and the solutions of these two phenomena are quite different from each other. The non-uniformity caused by hot spot formation may be improved by adjusting the density of the optical patterns on/under the LGP. The optical patterns under the high flux areas should be smaller and/or lower density, while the optical patterns under the low-flux areas should be larger and/or higher density.

However, if there is no or much lower light flux in the dark regions, we cannot achieve satisfactory brightness uniformity only by optimizing the optical pattern on/under LGP. The positional and directional distributions of light inside LGP depend on the emitting profile of LEDs, the LED-LGP gap structure and the surface morphology of LGP for the light incident from LED. By optimizing these factors, we can widen the active area, reduce the dead space, and make a narrow-bezel product. However, the realization of enough luminous uniformity should not be sensitive to the manufacturing process of BLU, and thus the optimization of the optical structure of LGP should be carried out by considering all these viewpoints.

Figure 6.3 shows the cross-sectional view of the edge-lit BLU and the intensity distribution on each optical component. The light escaping out of LGP is usually inclined at large angles of 60 to 80° with respect to the normal direction of LGP as shown in Figure 6.3(b) bottom. As the light passes through the film stack, its direction is gradually turned along the normal direction. A diffuser, a vertical and a horizontal prism film, a protection film stacked on the LGP, and a reflection film under the LGP comprise a typical BLU configuration. There are various combinations of optical films according to the applications and required specifications. Diffuser scatters light and shades the image of the optical pattern on/under the LGP (Figure 6.3(c), bottom). It also changes the direction of the rays emitted from the upper surface of the LGP toward the normal direction via refraction. The prism film enhances the brightness of the LCD device along the normal direction by refraction at the prism lenses. The apex angle of the prism grooves can have various angles, but most of commercial prism films have an apex angle of 90° because it is easy to control the haze of the diffuser for optimizing the characteristics of films. Prism films of shaper apex angle can be damaged easily by handling while those with wider apex angle show lower optical gain. If the refractive index of the resin for fabricating prism grooves becomes higher, the optical gain can be increased. But UV-cured layer of higher refractive index becomes more brittle and tends to show yellowish color during long-term operation of BLUs. A protection film may be put on the upper prism film. The function of the protection film is to protect the prim grooves from any damage. If the sharp edge of the prism film is damaged, it can be perceived easily and become a failure. Since this protection film has no optical function, we do not include it in Figure 6.3(a). The change in the intensity distribution on each film can be seen from Figure 6.3(b) and (c). The final beam profile of BLU is shown from the top figures in Figure 6.3(b) and (c), which demonstrates a well-collimated intensity distribution.

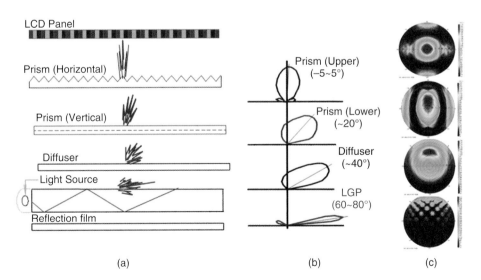

Figure 6.3 Beam profiles on LGP and optical films. (b) and (c) show the directional profile of light on LGP and sheets. The profiles in (b) are the vertical section diagrams of figures in (c).

There are other types of BLUs, which have optical patterns of LGP and optical film configurations different from those described above. However, the main subject of this book is related to the display manufacturing processes, and we think it is enough to discuss the manufacturing process of BLUs in relation to the most typical design. Therefore, we will not deal with other BLU designs in this chapter except for briefly mentioning the structure of direct-lit BLU at the end. So far, we have described the principle and key parameters in the design of BLU focusing on the light path in the edge-lit BLU system beginning from LED to the LCD panel. In the following sections, we explain how to make individual BLU components in more detail.

6.2 LED SOURCES

LED is a *p-n* junction light-emitting diode, which emits quasi-monochromatic light when a forward bias is applied. Under the forward bias voltage, electrons in the conduction band in the *n*-type region and holes in the valence band in the *p*-type region flow to each other and are recombined at the *p-n* junction. Their recombination energy is released as photon, the energy of which is equal to the energy gap between the conduction and the valence band. This light-generation mechanism is called electroluminescence. In order to use these LEDs as a light source of LCD BLU, LED packages should be manufactured according to carefully controlled processes: epitaxial growth on a substrate by chemical vapor deposition (CVD), making chips, packaging, surface mounting technology (SMT) on flexible printed circuit board (FPCB) processes, and so on. A schematic sequence of these processes is shown in Figure 6.4 [3].

LCD needs white light source with an appropriate color balance, which means individual primary colors of red, green, and blue colors should be well-balanced. There are several kinds of method to make white LEDs. The simplest one is to apply yellow phosphor on blue LED chips. In this case, phosphor particles are excited by blue photons from LED chips and emit yellow light characterized by a broad spectrum covering green and red. Figure 6.5(a) shows a schematic cross-section of this white LED and Figure 6.6(a) shows its spectrum with typical transmittance curves of color filters in LCD panel. This type of white LED exhibits the highest luminous efficiency compared to other types since yellow light contributes to luminance more than other colors. But the red component above 600 nm is insufficient and the green component has a broad width. Thus, there are limitations on supplying pure red and green colors on the LCD panel. These are main reasons for low-color gamut of LCDs combined with this type of white LED. In order to improve the color gamut, a mixture of red and green phosphors is used instead of the yellow phosphor material (dashed line in Figure 6.6(a)). However, there are different emitting characteristics between LED chips and phosphor materials. Therefore, white color tends to be deviated from its balanced one as temperature, driving current, or driving time changes. Alternative methods are to combine R, G, B primary lights emitted from phosphors coated over UV-LEDs (Figure 6.5(b)) or tri-color LED chips (Figure 6.5(c)). Unfortunately, the luminous efficacy of UV LED is quite low at the moment; this will be explained in the next sub-section. Figure 6.6(b) shows the light spectrum of the LED with R, G, B chips. It can give very wide color gamut. However, blue and green lights are generated from InGaN-based chips, while red light comes from AlGaAs- or AlInGaP-based chips. Because of the difference in the base materials, dependences of emitting properties of these three LED chips on temperature and current density are substantially different from one another.

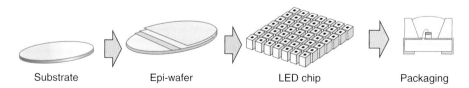

Figure 6.4 Process flow for fabricating LED packages.

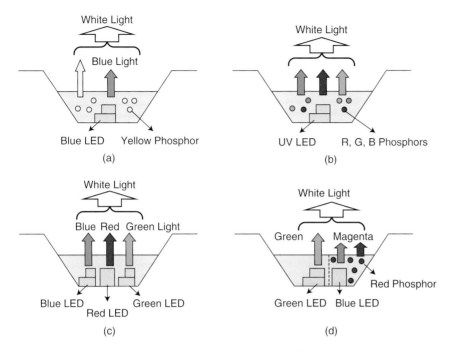

Figure 6.5 Several configurations of white LEDs consisting of LED chips and phosphor materials.

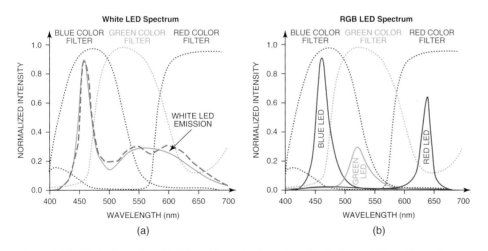

Figure 6.6 White LED spectra of (a) blue chip and yellow phosphor (solid line), blue chip and red-green phosphor (dashed line) system (Figure 6.6(a)) and (b) R, G, B LED system (Figure 6.6(c)) [4]. Typical transmission curves of LCD color filters are also plotted.

There are other combinations to make white LEDs. Figure 6.5(d) shows one example, where there is no phosphor material around the green LED chip while red phosphor is coated around the blue LED chip. This design may have some advantage because the luminous efficacy of the green chip is much lower than that of the blue chip, which is called green gap problem. It will be dealt with in the next sub-section. Besides, this LED uses only InGaN-based chips, so the change of emitting properties with respect to driving time and temperature is smaller than that of the LED consisting of R, G, B chips. The right side of the white LED shown in

Figure 6.5(d) emits magenta color, which is a mixture of blue and red colors and is the complementary color of the green light emitted from the left side.

In general, the white LED consisting of blue LED chips and phosphor particles (yellow or green/red mixture) shown in Figure 6.5(a) is mainly used in the market because of its high efficiency and low cost.

6.2.1 GaN Epi-Wafer on Sapphire

Metal-organic chemical vapor deposition (MOCVD) process is the most important process for manufacturing LED chips. As a substrate a sapphire (Al_2O_3) wafer is used in general. The sapphire substrate is very reliable from the viewpoints of thermal, chemical, and mechanical stability. It has corundum phase, similar to the wurtzite phase of nitrides. However, there is a large mismatch of lattice constants between the sapphire substrate and the nitride materials (see Table 6.1) [5]. So the hetero-epitaxial growth process of the GaN epi-layer results in a highly stressed layer where threading and edge dislocations at the density 10^8 to 10^9 cm^{-2} exist. Due to the large defect density the operating lifetime shortens and the internal quantum efficiency drops substantially at higher current density.

In order reduce dislocations in GaN epi-layer grown on a sapphire substrate several processes have been developed and applied. A technique, which is used commonly, is "epitaxial lateral overgrowth" method. Figure 6.7 shows a schematic cross-sectional view showing the sequence of this process. When the GaN epi-layer is grown on the substrate, it usually has lots of dislocations. Then some dielectric pattern using silicon nitride or silicon oxide with small apertures is formed on the first layer, and the GaN layer is regrown. If GaN can grow on small apertures between the pattern, there is no nucleation on the patterned layer and the epi-layer grows laterally over the pattern as shown in Figure 6.7 [6, 7]. This technique is relatively easy to apply compared to other techniques, but it still requires different types of growth equipment and photolithography process. An alternative method is the two-step deposition process involving growth of a low temperature (LT) GaN buffer layer. The nanoporous phase of the initial nucleation layer and faulted zone are formed generally

Table 6.1 Lattices constants and thermal expansion coefficients of GaN, AlN, InN, and Al_2O_3.

	a(Å)	c(Å)	$\alpha_a(10^{-6} K^{-1})$	$\alpha_c(10^{-6} K^{-1})$
GaN	3.189	5.185	5.59	3.17
AlN	3.111	4.98	5.3	4.2
InN	3.545	5.703	5.7	3.7
Al_2O_3	4.758	12.99	7.5	8.5

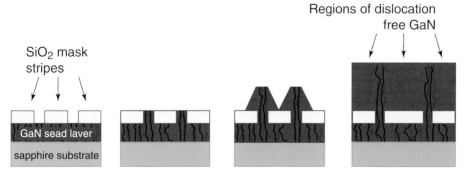

Figure 6.7 Lateral overgrowth of GaN epi-layer.

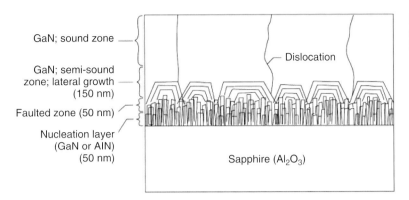

Figure 6.8 Dislocation structure of GaN epi-layer on a sapphire substrate. The initial layers are grown at about 500 °C [9].

Table 6.2 Various substrate materials for growing GaN LED.

Substrate	Cost	Size	Crystallinity	GaN epi-growth	Heat Conductivity
GaN	□	~4-inch	×	◎	2.1 Wcm^{-1}K^{-1}
SiC	×	~4-inch	△	○	4.9 Wcm^{-1}K^{-1}
Sapphire	△	~6-inch	○	○	0.4 Wcm^{-1}K^{-1}
Si	◎	~12-inch	◎	×	1.5 Wcm^{-1}K^{-1}

at a LT of 500 °C, and high dislocation density is self-annihilated through heat treatment as shown in Figure 6.8 [8, 9]. GaN epi-layer grown after self-annihilation has relatively low dislocation density. In this case, there are several advantages, for example, the number of process steps can be reduced. This is the reason why this technique has attracted great attentions [10, 11].

In order to reduce the dislocation density radically, a lattice matched substrate is necessary. GaN substrate is a perfect solution for the growth of the GaN epi-layer since there is no lattice mismatch. But it is much more expensive than the sapphire substrate, and wafers of large diameter are not yet available. GaN grown on another substrate SiC is also more expensive than multi-step processed GaN on sapphire. Silicone is the cheapest attractive substrate, but it is not easy to develop the epi-growth technique for buffer layer that offers a better match to the silicon lattice and then to gradually transpose the buffer layer into GaN. In addition, silicon is a good absorber of light, so the architecture of LED must be designed to minimize the absorption loss of the light emitted from quantum wells (QWs) in LED in the silicon substrate [12]. In Table 6.2 we compare the basic properties of theses substrate materials.

6.2.2 LED Chip

To adopt a complex growth system is essential to produce GaN-based LEDs consisting of multiple layers of "epitaxial" material, which must be deposited in sequence with accurate compositions and thicknesses. A simple LED structure is shown in Figure 6.9(a). The active region in the LED denotes ultra-thin QW layers consisting of the ternary InGaN alloy. Electrons and holes are confined in the QW structure, and thus their radiative recombination efficiency is increased.

Most GaN-based LEDs are grown on the *c*-plane of the wurtzite crystal. An important physical property of these nitride structures is the formation of spontaneous and piezoelectric polarization within the heterostructure caused by the crystal lattice mismatch between the layers with different compositions. The

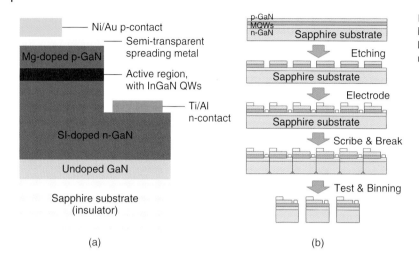

Figure 6.9 (a) A simple structure of GaN based LED, (b) Processes for making a LED chip after epi-layer growth (MQWs: multiple quantum wells).

polarization-related electric fields distort the energy bands, act upon the electron and hole distributions at the interfaces and reduce radiative recombination efficiency. To eliminate the effect of the fixed interfacial or surface polarization charges, alternative crystal planes other than the c-plane can be used to grow the GaN-based LEDs [13].

The wavelength of the light emitted from LED depends on the indium (In) composition in the InGaN layer and its thickness. At present, c-plane GaN crystals are widely used to fabricate violet and blue LEDs. However, achieving efficient operation of green devices at longer wavelengths (>500 nm) is challenging as the large polarization fields and defects such as In segregation in the active region reduce their radiative recombination rates as can be seen in Figure 6.10. The LED with UV wavelength also shows low efficacy. To shorten the emitted wavelength of the LED, the In composition must be decreased and, accordingly, the effect of composition variations is reduced. When the In composition is high, electrons and holes rapidly recombine (before getting captured by a dislocation); however, as the In composition decreases, the charges interact more actively with dislocations. In UV-LEDs emitting UV light of a wavelength of 365 nm, much portion of the UV light is absorbed in the GaN buffer layer grown on the sapphire substrate (the band gap energy of GaN is 3.4 eV), resulting in a rapid decrease in the light-extraction efficiency (LEE). This problem can be solved by using AlGaN as a buffer layer. However, this would increase the Al composition and in turn degrade the crystallinity [14]. After epitaxial growth, etching, electrode formation, scribing and break processes are followed to make a chip (Figure 6.9(b)).

6.2.3 Light Extraction

Similar to the light in LGP, which will be explained in Section 6.3.1, much portion of light is trapped inside the LED chips by the total internal reflection (TIR). The refractive index of GaN is about 2.5 and the corresponding critical angle θ_c is 25 degrees. The trapped light is easily absorbed by the substrate or internal defects, and is dissipated as heat. If the side cuts of the die are tilted, outcoupling efficiency can be improved. Light that internally reflects from the top face may reach the tilted facets at an angle smaller than the critical angle and then be out-coupled to air (e.g., see Figure 6.11(a)). The best tilt angle for efficient outcoupling is about 35 degree in AlGaInP LED material, and its LEE is about twice that of cuboid type chip (LEE ~52%) [16]. In the case of thin GaN external quantum efficiency can reach up to 75%.

Another method to improve LEE is surface texturing. There are two interfaces where a certain texture can be formed, that is, the interface between the sapphire surface and the p-GaN surface in addition to that between the top and the bottom surfaces of LED epi-layers, as shown in Figure 6.11(b) and (c). Figures 6.11(d)

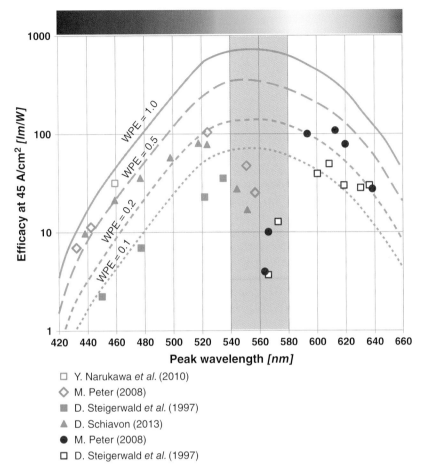

Figure 6.10 Efficacy of III-Nitride [e.g., InGaN] (green [blue-gray] data points) and III-phosphide [e.g., AlInGaP] (red data points) LEDs with different wavelengths. Marked in pink is the green-yellow range, which is not adequately covered by either the III-nitrides or the III-phosphides. This is the essence of the green gap problem [15].

and (e) show the SEM (scanning electron microscope) images of a wet-etched p-GaN layer surface and a plasma-etched sapphire surface, respectively. By using these textured structures LEE can be improved by 15 to 18% [17]. It is also possible to form textures on both surfaces, but the individual effects are not simply added. Besides, the number of process steps increases, and efficiency loss might occur due to damage in p-GaN layer. Therefore, the textured sapphire substrate prepared by plasma etching process, which is called the patterned sapphire substrate (PSS), is mainly used for LEE improvement recently.

If mirror layers are formed on both sides of active epi-layers and if its resonant wavelength is matched with that of the emitting light from the multiple-quantum well, the light output along the normal direction is increased by the Fabry-Perot resonance condition. As a result, wavelength dispersion is narrowed and LEE is also improved. It is called resonant cavity LED (RCLED). In addition to the modifications of LED structure and surface morphology, LEE can also be improved by the reduction of light absorption loss in LED chip and current spreading by electrode structure.

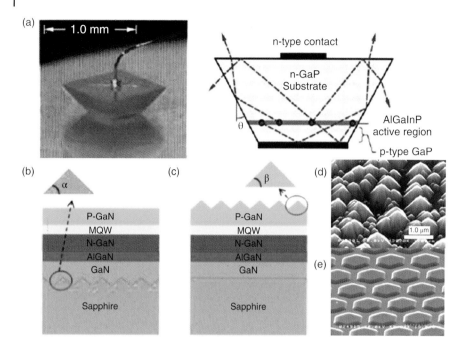

Figure 6.11 Examples of modified LEDs for the improvement of light extraction efficiency. (a) LED chip with inclined side-cut. Schematic diagrams of (b) a patterned sapphire substrate and (c) a roughened p-GaN surface. The SEM images of (d) a roughened surface by wet etching process and (e) a patterned sapphire substrate by plasma etching process.

6.2.4 LED Package

One of the most important factors that should be considered for the design of the LED package is related to the efficient heat dissipation by lead frame and heat-resistant molding resin. A lead frame is made by using an insert molding method where metal frame is inserted in a molding frame and molding resin is injected into the frame. An LED chip is attached on the metal frame, which is designed as large as possible for efficient heat dissipation. The metal frame also plays a role of a reflecting layer. After bonding with gold wire, transparent or phosphor-dispersed resin is dispensed on the chip. In the case of phosphor-dispersed resin inorganic beads are also included for efficient color mixing via scattering. Figure 6.12 shows the cross-sectional and top view of the LED package.

There are two methods for attaching an LED chip on the metal frame, that is, lateral bonding and flip chip bonding methods as shown in Figure 6.13. In the case of flip-chip bonding, there is no need for wire bonding and it is easy to dissipate heat generated from the emitting layer. In spite of these merits, it is costly to flip up a chip and bond on a patterned circuit. Recently a new LED architecture, where a phosphor pre-formed layer

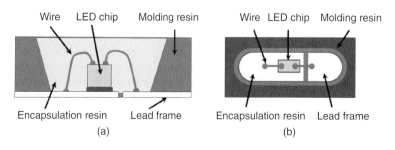

Figure 6.12 LED package design. (a) Cross section and (b) Top view.

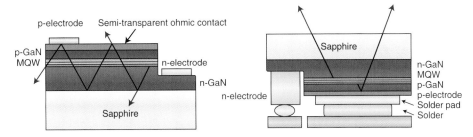

Figure 6.13 Lateral bonding LED (left) and flip chip bonding LED (right).

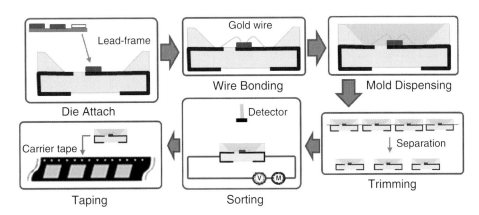

Figure 6.14 Process flow for the fabrication of LED package.

is applied directly on the flip-chip LED, is announced in the SID keynotes address [18]. In this case, white molding resin, gold wire bonding and phosphor resin dispensing process are not necessary, which makes the LED packaging process very simple. This type LED can be used in direct-lit BLU that we will explain in section 6.5.

When the resin is dispensed, it should not be more protrusive than the white molding resin. When you light up the LED the temperature of the dispensed resin becomes higher and expands more than the other part. If this encapsulation resin is in direct contact with LGP, it may cause the failure of light leakage. The manufacturing processes of the LED package are shown in Figure 6.14. The variation of LED characteristics, such as threshold voltage, brightness, and color coordinates, is large in general. Therefore, test and sorting processes are important for binning the LED packages. Standardization is also necessary for increasing the yield of LED packages.

Besides the dispensing method, there are compressive molding, transfer molding, and vacuum molding methods for lens molding. According to the lens shape and material an appropriate method should be chosen. In the case of high power LED polymer resin does not have sufficient heat resistivity. Ceramic and ceramic-polymer composite materials are used for LED package [19].

6.2.5 SMT on FPCB

Surface mounting technology (SMT) is a well-known process. Solder paste is printed on a flexible printed circuit board (FPCB), and LED package is attached on it. Finally, it is fixed by thermal reflow process. In this process the control of the tilting angle of LED package is important. The tilting angle of the LED package is affected by the outer lead shape, printing past pattern, and its thickness. After checking out the condition, the tilting angle of the LED package can be adjusted through rework process once it is failed.

When mounting LEDs on a FPCB, LEDs of neighboring ranks are placed alternately within allowed limitation, which may be determined by its effect on human eyes, in order to widen usable rank range. This is called a "rank mixing technique." The larger the neighboring LEDs are overlapped on the illumination area, the wider LED ranks we can combine for packaging.

Instead of FPCB, metal-core printed circuit board (MCPCB) and chip on board (COB) technologies can be used to dissipate heat from LED for high-brightness devices. An adequate design for heat dissipation is important for commercialization of high-power LED packages.

6.3 LIGHT GUIDE PLATE

The uniformity of LCD, a non-emitting device, depends on the pattern of the LGP in case of edge-lit backlights. LED is a point source, which means that additional optical components should be adopted to spread and homogenize the light from LEDs in order to achieve a two-dimensional uniform illumination condition. LGP plays a decisive role in this process. LGP is usually made from transparent polymeric material. If there is no optical pattern on/under the LGP, no light can be extracted through the top surface of the LGP as will be explained below.

Transparent polymeric materials such as PMMA (poly(methyl methacrylate)) or PC (polycarbonate) are used to fabricate LGP. There are two major processes for making LGP, screening printing on extruded plate and injection molding. These processes will be explained in detail in Section 6.3.4.

6.3.1 Optical Principles of LGP

If the LGP has mirror-like surfaces on both upper and lower sides, all the light entering the LGP through one or two of the four side surfaces is guided in the LGP to the opposite surfaces due to the total internal reflection (TIR). The critical angle θ_c of an PMMA LGP, the refractive index (n) of which is 1.49, is 42.2° according to Equation (6.1).

$$\theta_c = \sin^{-1}\left(\frac{1}{n}\right) \tag{6.1}$$

Let's consider one ray incident on the entrance surface of the LGP with an incident angle of θ_1 as shown in Figure 6.15(a). Its refraction angle θ_2 is given by the Snell's law. This ray travels in LGP and encounters, for example, the top surface of the LGP with an incident angle of θ_3, which is given as a complementary angle of θ_2 given by Equation (6.2).

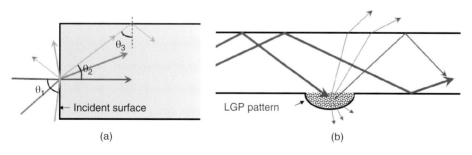

Figure 6.15 Light tracing in LGP.

$$\theta_3 = \frac{\pi}{2} - \theta_2 \tag{6.2}$$

The maximum angle for θ_1 is 90°, and the corresponding refraction angle θ_2 is 42.2°. In this case, θ_3 is calculated to be 47.8°, which is larger than the critical angle θ_c. All possible incident angles toward the upper or lower surface of the LGP are larger than 47.8°, which indicates that all optical rays are reflected back into the LGP via TIR at both upper and lower surfaces of the LGP and cannot escape from it. Hence, if there is no optical pattern on either the upper or the lower surface of the LGP and thus no light scattering function inside LGP, the light entering one side surface is guided along the LGP and escapes through the opposite surface. This is the reason why appropriate optical patterns should be formed on the bottom and, if necessary, the upper surface of the LGP for light extraction toward the LCD panel. On the way light travels in LGP, some rays encounter an optical pattern at the top/bottom surface, are scattered, and some of them may be directed toward the directions the angles of which are smaller than the critical angle. These rays can be extracted from the LGP toward the LCD panel (Figure 6.15(b)). Therefore, by controlling the size and/or density of optical patterns on the surface of LGP the amount of light extraction at each position can be adjusted. This makes it possible to adjust and optimize the overall luminance uniformity on the active area of BLU. The details of the design of optical patterns on LGP are described later.

6.3.2 Optical Pattern Design

Optical patterns on the bottom surface of LGP are usually formed in terms of diffusing elements as shown in Figure 6.16. This figure shows that the light hitting the optical pattern may undergo either diffuse reflection or diffuse transmission. The diffusing property of optical pattern varies depending on various process parameters. This is the reason why it is difficult to carry out exact modeling and optical simulation. In many cases, optical pattern is modified and optimized according to the difference between the target uniformity profile

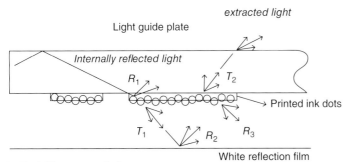

Figure 6.16 Light propagation in the LGP with printed ink dots on the bottom surface.

T_1, T_2 : Diffuse transmissions
R_1, R_2, R_3 : Diffuse reflections

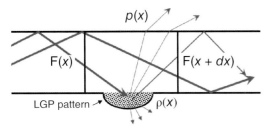

Figure 6.17 A schematic figure of a part of the LGP showing the light propagation and the resulting flux redistribution in the one-dimensional LGP model.

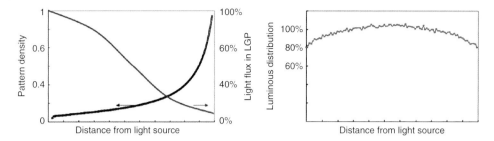

Figure 6.18 The positional dependence of the pattern density and the light flux in the LGP (left) and desirable luminous distribution on the LGP as a function of the distance from the light source (right).

and the tried uniformity result, which is fed back in the next optimization step. However, in order to reduce the number of trials, exact and reliable modeling is needed. For a simple explanation we suppose a one-dimensional LGP. At a point x, which denotes the distance from the light source, the extracted light flux $p(x)$ is proportional to the total flux in LGP and the areal pattern density (Equation 6.3 and Figure 6.17),

$$p(x) = \alpha \, F(x) \rho(x) \tag{6.3}$$
$$F(x + dx) = F(x) - p(x) \tag{6.4}$$

where ρ is the pattern density from 0 to 1, $F(x)$ the light flux at x, and α the coefficient of light extraction. α is related with diffusivity of the optical pattern. $F(L)$, where L is the length of LGP, should be minimized to reduce the light loss at the opposite side edge. Figure 6.18(a) shows the light flux in the LGP and pattern density with respect to the distance from the light source. The available light power in the LGP decreases as the distance from the edge source increases, because the guided light is extracted to the LCD panel and absorbed by the LGP material as it propagates along the LGP. Accordingly, the pattern density should increase with increasing x in order to compensate for the decreasing light power. Figure 6.18(b) shows the positional dependence of the extracted luminous flux on the LGP. This luminous distribution (which may also represent the on-axis luminance) on BLU is a typical one which looks uniform to our perception. If the brightness is constant regardless of position, we perceive the centre area to be dimmer than the surrounding area.

Propagation of light can be investigated by using optical simulation based on a ray-tracing technique. In general, the actual position and/or size of dot patterns can be determined via optimization process adopted in the simulation software. The initial distances between the patterns may be arranged in a linear or nonlinear manner. After one pattern configuration is obtained, the difference between trial result and target profile is evaluated, fed back to the optimization process to get improved pattern configuration. This feedback and optimization processes are iterated until satisfactory uniformity is obtained over the whole area of the LGP. The size of each dot can be adjusted and optimized while their locations are fixed and vice versa for this purpose [20]. For carrying out reliable optical simulation, exact scattering properties of the dot pattern, such as the bidirectional scattering distribution function, should be measured and incorporated in the simulation process [21]. The inclusion of diffusing element in the simulation usually makes the simulation process time-consuming. Fast calculation algorithms were proposed to speed up the optical simulation [22]. In some approaches, molecular dynamics method was used to generate a random dot pattern, which is more favorable to prevent the Moire formation compared to regular dot patterns [23]. If some optical structures other than printed dot patterns are adopted on bottom/top surfaces of the LGP such as prism patterns, the apex angle, the pitch, and the optical correlation between the two structures on the two surfaces are important factors that should be optimized to achieve uniform and collimated beam profile on the LGP.

Table 6.3 Representative polymeric materials for LGP.

	PMMA	PC (polycarbonate)	PS (polystyrene)	COP (cyclo-olefin polymer)
Refractive index	1.49	1.58	1.59	1.53
Density (g/cm^3)	1.18	1.20~1.22	0.96~1.04	1.02
Transmission (%)	93	87	89	91
Saturated water absorption (%)	0.3	0.2	0.05	0.01
Glass transition temperature (°C)	105	145	100	136-163

6.3.3 Manufacturing of LGP

Table 6.3 shows important physical properties of representative polymeric materials for LGP. The most common material for LGP is PMMA due to its high transmission, good durability against ultraviolet light and chemicals, satisfactory mechanical strength, and low optical loss. However, PC is mainly used for LGP of mobile phones and tablets where high thermo-stability is required. As we can see from Table 6.3, the glass transition temperature of PC is much higher than that of PMMA. There are several fabrication methods for LGP, but we will limit our description to the most conventional methods, that is, injection molding method and extrusion molding method combined with screen printing for pattern formation. Other techniques such as roll stamping and other non-printing methods will be dealt with briefly.

6.3.3.1 Injection Molding

One of the representative methods for fabricating LGP is the injection molding method. Figure 6.19 shows a schematic drawing showing the important part of the injection molding machine. In case of PMMA LGP, the MMA (methyl methacrylate) monomers or PMMA particles are used as raw starting materials. The MMA monomers should be mixed and polymerized before melting. Polymerized PMMA is inserted into the push-

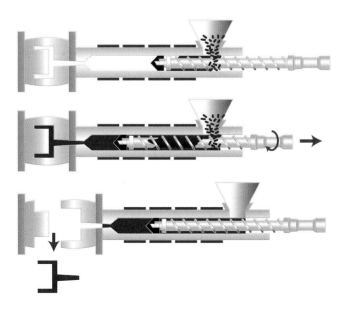

Figure 6.19 A schematic drawing showing the important part of the injection molding machine.

ing machine with heater where the PMMA is melted. Molten PMMA is guided into the mold under pressure. Optical patterns may be incorporated in the mold shape in order to make necessary optical structure on LGP. On the other hand, a stamper with appropriate optical patterns may be inserted in the mold. When the PMMA inside the mold is cooled down, gate cutting is done. The injection pressure, the injection time, and the cooling condition are important fabrication parameters to be optimized. Since the area of LGP that can be manufactured by using the injection molding method cannot become very large, this method is applied to LCDs of small and middle sizes.

The contraction of the polymer resin during cooling process should be considered in the design of the master mold, because maintenance of the LGP dimensions to the target specification is very important. In addition, the resin begins to cool instantly as it is input in the mold, which is the reason for the formation of flow mark on the LGP and should be avoided by appropriate process optimization. Another important point is to prevent the formation of warpage of the LGP by controlling the temperature distribution of the mold, in particular, the temperatures of the upper and the lower parts of the mold. Significant warpage of the LGP degrades the reliability of the BLU due to the scratch failure caused by the frictional contact between the optical film and the LGP or warpage failure.

6.3.3.2 Screen Printing

Figure 6.20 shows the extrusion machine for fabricating original plate. Polymer resin is inserted into the extruder where it is melted and extruded through a T-die having a certain cross-section. The extracted plate is cooled and then undergoes polishing, cutting, and drying processes, if necessary. The cut surface should be mirror-like because rough surface is the origin of the light leakage caused by light scattering. The area of the LGP fabricated by this method may become large, thus the extrusion method is more appropriate for large-size LCDs, such as large monitors and TVs. The extrusion molding process can be handled continuously and thus exhibits higher productivity compared to the injection molding method.

In case of extruding LGP having a periodic pattern such as a lenticular lens pattern, the periodic pattern is formed on the second roll, on which liquid resin is solidified resulting in the formation of the desired pattern. When the pattern is formed during the extrusion process of LGP, the hardening speed depends on the thickness of the LGP. This induces different replication quality depending on the thickness, which is mainly due to the contraction of the pattern. This point should be considered in the pattern design, and periodic patterns with gentle slopes are preferred and formed in this process in general.

The most conventional way to extract the light guided in the LGP toward the LCD panel is to form diffusing element on the back surface of LGP by using a screen printing method. Light-diffusing white ink is printed on this surface according to the optimized pattern design in order to satisfy the required brightness and uniformity on LGP. Figure 6.21 shows a schematic figure of this method, where the white ink is squeezed on the patterned mesh and then printed on the LGP. The pressure/velocity of the squeeze and the viscosity of the ink

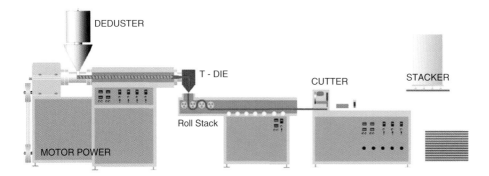

Figure 6.20 A schematic drawing showing the extrusion machine.

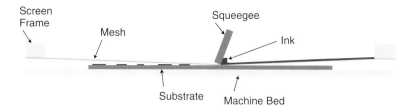

Figure 6.21 A schematic drawing showing the screen printing process.

are important parameters that should be optimized for exactly reproducing the designed patterns. In particular, it should be noted that the ink viscosity may change as the number of printing increases and more ink is replenished. This is due to the fact that the solution and the scattering agent may not pass through the printing screen at the same ratio and thus same amount. Therefore, control of the ink viscosity during printing is important for maintaining the same printing quality. This screen printing technique is the most conventional and reliable way for LGP fabrication, but the lifetime of the screen mesh is relatively short and should be replaced frequently.

From the viewpoint of printing procedure, circular patterns are desirable. The minimum distance between dot patterns should be approximately 100 μm in order to avoid bridges among the patterns. This means maximum pattern density may be smaller than those of other techniques. If other patterns are adopted such as square or diamond pattern, the maximum pattern density may increase compared to the circular pattern.

6.3.3.3 Other Methods

Recently roll stamping method has attracted attention due to its short fabrication process. Figure 6.22 shows a schematic diagram of the roll-stamping method for fabricating LGP. One of the two rolls is wrapped with a metal stamper film with optimized three-dimensional patterns being inscribed. As the pre-heated LGP passes through the two rolls, the stamper film forms the desired pattern on the LGP under appropriate temperature and pressure conditions. Various three-dimensional patterns can be tried by this technique, but replication quality may be worse for the case of complex patterns. The fabrication time can be decreased by using this technique. The transfer speed, the pressure/temperature conditions of and the distance between the roles are important process parameters that should be optimized. Flatness of the LGP is important in this process, because the depth of the inscribed pattern may change if the surface of the LGP is uneven. Non-uniform pattern depth may cause non-uniform optical properties. On the other hand, temperature control is also very

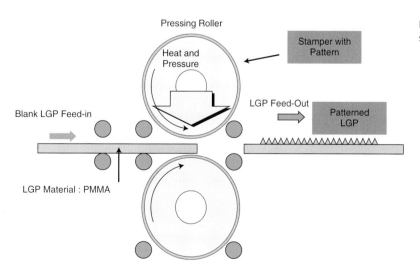

Figure 6.22 A schematic drawing showing the roll stamping process.

important during the stamping process, because the LGP resin will be pushed to deform the side of the LGP if the temperature is too high.

Other techniques have also been developed for LGP fabrication such as laser engraving and imprinting methods. The laser engraving method uses laser beam to partially melt the surface of the LGP resulting in a certain pattern. The dimensions of the pattern can be controlled by the amount of laser power incident on the LGP. The beam power, the location of the focal point, and the repetition number/duty ratio of the laser pulse are the main process parameters of this method. The luminance gain may increase by adopting this method, but the fabrication time becomes longer compared to the screen printing method.

In case of imprinting method, resin is coated on the soft mold with appropriate patterns being formed. The soft mold is then in contact with the LGP. The resin is cured on the LGP by UV light and the mold is detached. The UV curing conditions (uniformity, irradiation time, UV flux) and the physical properties of the resin are important process parameters. Complex three-dimensional patterns may be formed on a large area by using this technique, but the protruded patterns may be in contact with the reflection film beneath the LGP, which may cause scratch problem.

6.4 OPTICAL FILMS

There are several optical films on and under the LGP in BLU. These are diffuser, prism, and reflector films. Photographs of the surface and/or cross-sectional images of these films are shown in Figure 6.23. These films are usually fabricated by using a biaxially stretched PET substrate because it exhibits high thermal and mechanical stability. Especially PET film has a low thermal expansion coefficient of 20 to 60×10^{-6}/K. Diffusers have a random surface morphology, while prism films have one-dimensional prism grooves (Figure 6.24). On the other hand, reflector films have a multi-interface structure formed by two different media with different refractive indices.

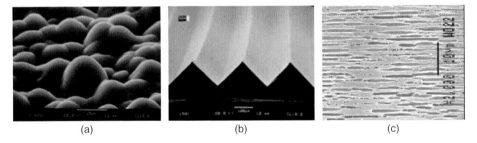

Figure 6.23 Optical films in BLU: (a) diffuser, (b) prism film and (c) reflector.

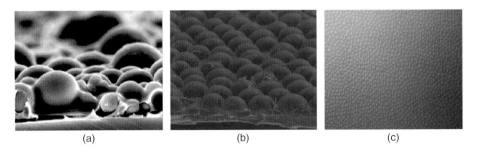

Figure 6.24 (a) Beads-coated diffuser and (b),(c) micro lens arrays.

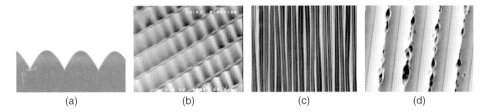

Figure 6.25 A few examples of modified prism films. (a) Rounded prism film, (b) waved side-plane prism film, (c) non-periodic prism film and (d) sand-blasted prism film.

To form a random surface morphology on the diffuser, spherical beads are mixed with binder resin and are coated on the PET substrate film (Figure 6.24(a)). When the beads and binder resin are mixed for coating, the amount of binder resin should be minimized in order to maximize the diffusing function of the film. However, beads may be easily detached due to low adhesion if the amount of the binder resin is too small. Therefore, there should be a compromise between the diffusing function and the amount of binder resin. To overcome this limitation, hemispherical lenses can be formed by using the imprinting method on a substrate (Figure 6.24(b)), which is called a micro-lens array (MLA) film. Even though the MLA film belongs to the category of diffuser, there is additional optical function, i.e., collimating effect of light due to the hemispherical shape of micro-lenses. In this imprinting technology it is important to make the master mold, because the three-dimentional MLA pattern is difficult to make by using micromachining. This point will be explained in the next sub-section. Besides, for MLA film it is important to realize a compact distribution of the MLA on the substrate. Figure 6.24(c) shows one example of the MLA film having a high density of hemispherical lenses [24].

The prism film with a sharp top edge is desirable from the viewpoint of optical performance but this structure is easily damageable easily during handling. Most prism films have a right apex angle, as was mentioned earlier. In some cases, the prism shape is modified and rugged as shown in Figure 6.25 by several methods. First one is a prism film having a rounded shape [25]. Figure 6.25((b) and (c)) shows that side planes of prism grooves are waved with a short period and a small amplitude [26] or with a long period and a large amplitude [27].These modifications relieve the periodicity of the prism pattern, which is favorable for the prevention of moire formation. In addition, weak damage on the top edge can be shaded. In some designs, the top edge is weakly damaged in a random manner on purpose by, for example, sand blasting as shown in Figure 6.25(d). Therefore, additional damages in the prism grooves are not perceived easily. All these methods reduce the brightness gain of the prism film by a few %, but we can eliminate the protect film on it. In general, very small beads are coated or matte pattern are formed under the prism film in order to prevent it from sticking to the film below it.

One prism film with horizontal grooves or two prism films with crossed prism grooves are adopted in BLU for enhancing its brightness. The prism film with periodic prism grooves and the periodically pixelated LCD panel may induce interference, which is called a "moire" pattern. To avoid this moire failure the pitch of prism grooves should be adjusted appropriately because the pitch of pixels on the LCD panel is determined by its resolution. However, it is difficult to produce numerous prism films with all necessary pitches. The pitch is modified by tilting the prism film slightly. Its haze also has important role in hiding this moire interference. All these factors are optimized when the prism film is assembled in BLU and then with the LCD panel.

Light is partly reflected at the interface where refractive indices of both sides are different from each other. The larger the difference of the refractive indices, the higher the reflectance at the interface between two materials. In general, refractive indices of polymeric materials are in the range of 1.4 to 1.7, so the Fresnel reflection at the interface between the polymeric material and air is not so small. Thus, if micro/nano-size voids or plate-like voids are formed in the polymeric film, it can play the role of a reflector film. Figure 6.26

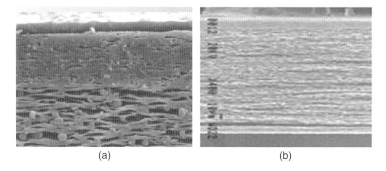

Figure 6.26 (a) Filler type reflector and (b) micro/nano-void type reflector.

shows cross-sections of these reflection films. These reflectors are diffusing ones which are effective against visual mura stains. In some cases reflection films with a mirror-like specular surface is effective in increasing brightness. For this purpose, Ag is evaporated on a substrate to form a thin metal layer for specular reflection. Another way for realizing specular reflection is to use a multi-layer structure with materials of different refractive indices. It will be explained in more detail in Section 6.4.4.

As the LCD module becomes slimmer, thicknesses of all BLU components need to be reduced. However, the light leakage through a thin reflection film cannot be avoidable completely, which becomes more substantial as the thickness decreases.

6.4.1 Diffuser

Diffusers as shown in Figure 6.24(a) are made by using a coating process. The beads are dispersed in binder resin without precipitation and flocculation. The beads are made from cross-linked PMMA and the binder resins are thermoset urethane acrylate in general. It is important to coat a monolayer consisting of beads because smothered beads are not suitable for making appropriate surface morphology. Usually, some volatile ingredient is included in the binder resin, which controls the viscosity of the resin. The volatile ingredient is removed after drying process and the volume of residual resin is minimized. There are other additives in the binder resin to control other properties such as adhesion. The general fabrication process of the beads-coated diffuser is shown in Figure 6.27.

In the case of the MLA film, the imprinting method is used as mentioned earlier. This method is illustrated by Figure 6.28(a). In mass production the film is made in terms of the roll-to-roll process with a soft mold. On the other hand, the prism film is produced by using a hard mold. The determination of the mold type between the soft and hard ones depends on the shape pattern to be formed on a substrate during fabrication process.

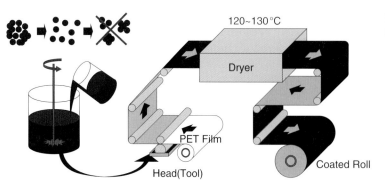

Figure 6.27 Process for producing a beads-coated diffuser.

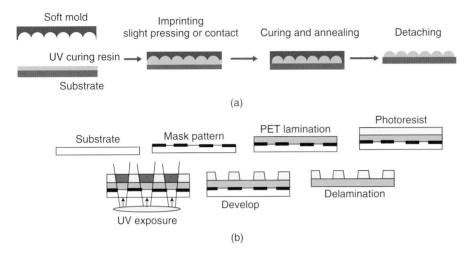

Figure 6.28 (a) A process flow of soft mold-based imprinting method and (b) a fabrication process of the soft mold with photolithography technique.

Figure 6.29 A diamond tip, a master roll and the process for producing a prism film.

The hard mold is more durable and robust, but it is shaped in terms of mechanical machining. It is difficult to machine three-dimensional patterns mechanically even though there are some suggested method for making hard mold [28]. Hemi-spherical lens patterns can be made by using several methods. For example, after coating a thin adhesive layer on a substrate, narrow-dispersive beads are attached on it, which can be used as a master mold. In this case, however, it is difficult to make very compact patterns. To make compact lens patterns photolithography patterning processes are used in general. Figure 6.28(b) shows one example. Using a photo-mask pattern we can accurately control the position and the shape of micro-lenses. In particular, the curvature of the lens can be controlled by several material and/or process parameters such as haze of photoresist, resin shrinkage in imprinting process, and so on. For soft mold it is important to make uniform lens shapes on a large area.

6.4.2 Prism Film

Prism patterns can be machined mechanically. A cylindrical roll is machined by using a diamond tip. When the roll rotates, the diamond tip is moved by a pitch per turn. To ruggedize this master roll Cr or Ni layer is electroplated on it. With this hard roll mold, prism films are produced by the imprinting method as shown in Figure 6.29. If the shape of tip is rounded, we can make rounded prism films without any sharp edges. Moreover, if the tip is vibrated a little bit during machining, prism films with wavy surfaces can be fabricated. After machining a master roll, prism surfaces can be modified by using a sand blasting process. Based on these modifications we can make various prism shapes as shown in Figure 6.25. It is important to minimize the optical gain losses caused by these modifications; on the other hand, handling problem may be reduced due to the robustness of these modified films against handling process.

Prism films can also be produced by using a soft mold, which is transferred from the hard mold in order to extend its lifetime. There are merits and demerits for using the soft mold, and cost competitiveness may be the most important point for the choice of the mold type.

6.4.3 Reflector

There are several methods to make voids in reflection films, but the most typical process is explained in this section. The resin is extruded by T-die and is expanded along the machine and transverse directions. In the PET resin, white pigment such as TiO_2 and specific filler of polyolefin, which has no miscibility with PET, are added. By drawing along both machine and transverse directions air voids can be formed around the fillers as shown in Figure 6.26(a). In the case of Ba_2SO_4 filler the film can be a good reflector without any voids since Ba_2SO_4 is itself a good reflective material in the visible wavelength region. Another way to create micro/nano air pores is to dissolve gas in the resin under high-temperature and high-pressure conditions. If the gas is dissolved during film extrusion, it creates voids in the film (Figure 6.26(b)). The process conditions have to be controlled carefully.

6.4.4 Other Films

Even though reflective polarizer is not a basic component of the BLU, it has a very important role in improving the optical efficiency of BLU. Reflective polarizer laminated on a bottom polarizer may be attached onto the LCD glass or it can be mounted as a separate film in BLU. The two linear polarizers attached on the LCD panel are absorptive type; they absorb the light having polarization direction orthogonal to their transmission axis and convert the absorbed light into heat. On the other hand, a reflective polarizer reflects the light whose polarization component is perpendicular to its transmission axis toward the BLU, which is partially reused via polarization recycling process. The luminance gain of BLU may be improved significantly by using the reflective polarizer.

The most typical reflective polarizer used in BLU is a multi-layer optical film (MOF), which consists of two different birefringent materials with ABAB⋯ periodic structure as shown in Figure 6.30. The refractive indexes of both layers are the same along the y direction, which means the light of y-polarization does not feel any interface and is transmitted throughout the film. On the other hand, there is a difference in the refractive index between the two layers for the light having polarization direction along the x-axis. Therefore, the light with x-polarization is partially reflected according to the Fresnel equations, similar to the reflection mechanism in a reflector film. This multi-layer type reflective polarizer has hundreds of layers with different thicknesses for reflecting light of all wavelengths in the visible light range.

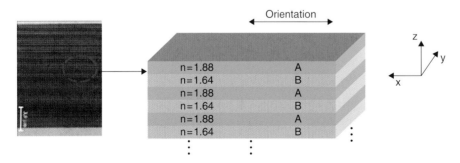

Figure 6.30 Structure of reflective polarizer based on multi-layer optical film.

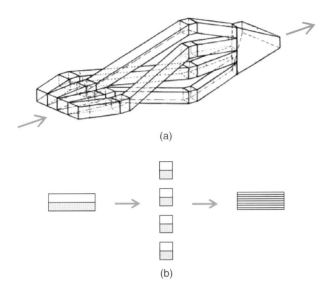

Figure 6.31 A schematic diagram of a four-channel interfacial surface generator (a) and its layer pattern resulting from a two-layer input (b). Connecting N units in series produces multilayer stack with $2*4^N$ layers [30].

If the refractive indexes for the two orthogonal polarization directions do not match between two materials, the incident light is reflected regardless its polarization. If the film has flat interfaces, its light reflection becomes specular. This is another way to make a specular reflector, which is mentioned in Section 6.4 [29].

The multi-layer film is made by using the co-extrusion process. Two different materials are merged in a feedblock. Then, they split into several channels horizontally and join into a channel vertically without mixing, as shown in Figure 6.31(a) and (b). In this way the number of layers increases exponentially according to the number of elements in the channel. For the multi-layer type reflective polarizer two different materials are divided into hundreds of branches and sandwiched between each other, forming ABAB··· structure with a gradient in the layer thickness [30]. The extruded film is stretched along a certain direction to make the designed layer thickness. When the film is stretched, refractive index of one layer material does not change, while that of the other layer material is changed along the stretching direction because the main chains of the polymeric material in the resin are aligned along that direction. The molecular design of two materials is important to fulfill the required difference in the refractive indexes of two materials for the light having the x-polarization direction.

Besides the MOF-type reflective polarizer, there are other types of reflective polarizing technologies such as the cholesteric liquid crystal film (CLCF) [31] and the wire grid polarizer (WGP) [32]. However, the MOF structure is the most fundamental structure in optics and shows the best performance among them.

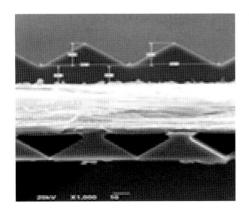

Figure 6.32 A photograph of the prism-on-prism film.

Another film to be introduced is a prism-on-prism film. As the thickness of LCD module is thinner, those of optical films also should be reduced. It may be difficult to maintain the flatness of a thin film, the substrate thickness of which is about or less than 100 μm, especially a prism film. To make rugged thin films, two crossed prism films are laminated, forming one film as shown in Figure 6.32. The prism grooves of normal prism films do not have enough area for attachment on top of it, thus the prism grooves of the lower prism film should be of the flat-topped shape. If the flat area of the prism lenses of the bottom film is larger, the adhesion between the two films becomes stronger but the optical gain is decreased. So the prism structure should be optimized by considering both adhesion and optical performance. A microlens-on-prism film can be made by the same process.

Recently a quantum dot (QD) film has received a great deal of attention. The emitting spectrum of ODs is relatively sharp compared to that of phosphor materials, which is a favorable condition for achieving high color gamut. Organic light emitting display (OLED) also has good color performance due to its sharp spectral components. To adopt a QD film in BLU is one way for LCD to compete with the OLED. QD particles are dispersed in the resin and sandwiched between barrier films. QD absorbs blue light from LED chips and emits light of longer wavelength with narrow half width. However, QD has very weak vapor-resistance, so the barrier property of outer layers of the QD film is important for long-term reliability [33, 34].

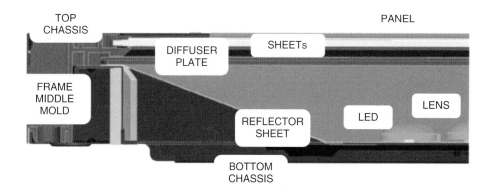

Figure 6.33 LED direct lighting backlight structure.

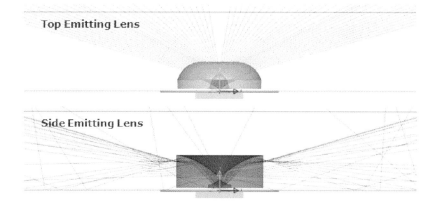

Figure 6.34 Top-emitting lens (top) and side-emitting lens (bottom).

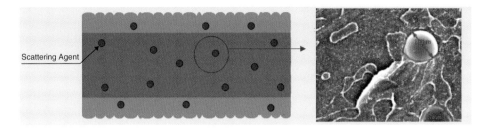

Figure 6.35 A cross-sectional view (left) and a photograph of the diffuser plate with optical patterns on both skin layers.

6.5 DIRECT-TYPE BLU

Up to now, we have described edge-lit type BLUs. For large-size display direct-lit BLU is adopted for cost reduction, even though it is thicker than an edge-lit type BLU. Instead of LGP, a diffuser plate is put over high-brightness LEDs (Figure 6.33) [35]. It supports optical films on it and homogenizes the light from LEDs thus hiding bright spots caused by individual LED. In the case of direct-type BLU it is important to reduce both its thickness and the number of LEDs simultaneously.

Therefore a special optical lens is applied on each LED to make wide angular output profile. There are two types of lens for this purpose. The direction of light from LED is converted into side-direction via the lens as shown in Figure 6.34 (bottom) [36]. Because a little amount of light is forwarded to the LCD panel directly, LED patterns can be shaded by BLU structure and films. In this case the assembly of LED and lens is relatively easy, but light efficiency is lower than the other type because of multiple reflections. The other top-emitting lens shown in Figure 6.34 (top) can be used to achieve a precise wide angular beam profile for uniform distribution. In this case the luminous uniformity is very sensitive to the variation of the LED-lens assembly condition. These variations should be considered at the lens design stage. Similar to the LGP pattern design in edge-lit BLU, the lens design in direct-type BLU is one of critical factors that determine the optical performances of BLU. These lenses are made by multi-cavity injection molding method. All lenses from individual cavities should be identical from the viewpoint of optical performance.

A light diffusing plate, which is called LDP, has the role of diffusing light from individual LEDs and making uniform luminous distribution. For this purpose scattering agents are added in the resin as shown in Figure 6.35. For higher diffusivity the bigger difference in refractive indexes between host resin and scattering agent is desirable. It is better to minimize the agent mixing ratio as possible as LED pattern can be hidden for higher transmittance of light. The size distribution of scattering agent is also important to have similar diffusivity with respect to R, G, B colors, since the diffusivity or scattering distribution of light depends on the size of the particle and the wavelength of light via Mie scattering mechanism. When the LDP is extruded, some surface morphology can be formed to split the image of light sources. Using the patterned extruding roll the pattern is transferred onto the LDP. In some cases a screen printing pattern can be printed on the bottom surface of LDP to compensate for non-uniformity of light from light sources.

6.6 SUMMARY

The optical characteristics of the components in BLU and their manufacturing processes were described in this chapter mainly based on the edge-lit backlight technology from the practical point of view. There are many other optical structures for certain optical functions and many other possible processes for fabricating certain components. Various attempts and competitions have been tried to improve the BLU technology.

Many of these methods and processes tried before could not be covered in this chapter except for a few representative ones.

In addition, innovative optical designs have been proposed such as air-guide BLU without any LGP, hybrid LGP structures with no or significantly reduced number of optical films, and even polarizing BLU which may increase the optical efficiency of LCD significantly. However, most of these proposed technologies suffer from high fabrication cost and/or insufficient reliability and thus could not be dealt with in this chapter.

Important characteristics of LCD devices, such as low power consumption, thin and slim design, wider color gamut, and optical performances, depend crucially on BLU technologies. So, the research and development in the BLU area have been done intensively and will be continued for the progress of LCD technology. Since LCD display technology has been matured and the TV and mobile device markets have been saturated, cost competition has been becoming increasingly intensified. Nevertheless, we expect the effort for developing new technologies to improve display performance will continue.

References

1. M. Anandan, LCD backlighting, Society for Informational Display Seminar Lecture Notes, pp. 169–250 (2002).
2. D. M. Brown, R. Dean, J. D. Brown, LED backlight: design, fabrication, and testing," Proc. SPIE, 3938, pp. 180–187 (2000).
3. http://www.kopti.re.kr/index.sko?menuCd=AB02001000000
4. Y.-H. Won, H. S. Jang, K. W. Cho, Y. S. Song, D. Y. Jeon, H. K. Kwon, Effect of phosphor geometry on the luminous efficiency of high-power white light-emitting diodes with excellent color rendering property, Optics Letters, 34(1), p. 1 (2009).
5. http://en.wikipedia.org/wiki/Corundum http://en.wikipedia.org/wiki/Wurtzite_crystal_structure
6. K. Hiramatsu, Epitaxial lateral overgrowth techniques used in group III-nitride epitaxy, J. Phys.: Condens. Matter, 13, p. 6961 (2001).
7. L. Jastrzebski, SOI by CVD: epitaxial lateral overgrowth (ELO) process – review, J. of Crystal Growth, 63, p. 493 (1983).
8. O. Manasreh (Ed.), III-nitride semiconductors: electrical structural and defects properties, Elsevier (2002).
9. A. Kitai, Principles of solar cells, leds and diodes; the role of PN junction, John Wiley and Sons, p. 237 (2011).
10. M. Sumiya, N. Ogusu, Y. Yotsuda, M. Itoh, S. Fuke, T. Nakamura, S. Mochizuki, T. Sano, S. Kamiyama, H. Amano, I. Akasaki, J. of Appl. Phys., 93, p. 1311 (2003).
11. C. Hemmingsson, G. Pozina, Optimization of low temperature GaN buffer layers for halide vapor phase epitaxy growth of bulk GaN, J. of Crystal Growth, 366, p. 61 (2013).
12. J. Ellis, GaN on silicon: A breakthrough technology for LED lighting, LEDs magazine, 11(2) (2014).
13. M. Seelmann-Eggebert, J. L. Weyher, H. Obloh, H. Zimmermann, A. Rar, S. Porowski, Polarity of (001) GaN epilayers grown on a (001) sapphire, Appl. Phys. Lett., 71, p. 2635 (1997).
14. Y. Muramoto, M. Kimura, S. Nouda, Development and future of ultraviolet light-emitting diodes: UV-LED will replace the UV lamp, Semicond. Sci. Technol., 29, p. 084004 (2014).
15. Cranking up the efficacy of Green LEDs, Compound Semiconductor, 19(7), p. 33 (2013).
16. M. R. Kramers, M. Ochiai-Holcomb, G. E. Hofler, C. Carter-Coman, E. I. Chen, I.-H. Tan, P. Grillot, N. F. Gardner, H. C. Chui, J.-W. Huang, S. A. Stockman, F. A. Kish, M. G. Craford, T. S. Tan, C. P. Kocot, M. Hueschen, J. Posselt, B. Loh, G. Sasser, D. Collins, High-power truncated-inverted pyramid $(Al_xGa_{1-x})_{0.5}In_{0.5}P$/GaP Light-emitting diodes exhibiting > 50% external quantum efficiency, Appl. Phys. Lett., 75(16), p. 2365 (2000).
17. J.-W. Pan, C.-S. Wang, Light extraction efficiency of GaN-based LED with pyramid texture by using ray path analysis, Opt. Express, 20(S5), p. A630 (2012).

18 J. Carey, New LED architectures and phosphor technologies lower cost and boost quality, LEDs magazine, 11(7) (2014).
19 Y. J. Heo, H. T. Kim, S. Nahm, J. Kim, Y. J. Yoon, J. Kim, Ceramic-metal package for high power LED lighting, Front. of Optoelectron., 5(2), p. 133 (2012).
20 G. Lee, J. H. Jeong, S.-J. Yoon, D.-H. Choi, Design optimization for optical patterns in a light-guide panel to improve illuminance and uniformity of the liquid-crystal display, Opt. Eng., 48(2), p. 024001 (2009).
21 B.-Y. Joo, J. J. Kang, J.-P. Hong, Analysis of the light-scattering power of patterned dot material printed on the light guide plate in liquid crystal display, Displays, 33, p. 178 (2012).
22 W. Y. Lee, T. K. Lim, Y. W. Lee, I. W. Lee, Fast ray-tracing methods for LCD backlight simulation using the characteristics of the pattern, Opt. Eng. 44(1), p. 014004 (2005).
23 T. Idé, H. Mizuta, H. Numata, Y. Taira, M. Suzuki, M. Noguchi, Y. Katsu, Dot pattern generation technique using molecular dynamics, J. Opt. Soc. Am. A., 20(2), p. 248 (2003).
24 A. Bastawros, J. Zhou, M. J. Davis, Z. Chen, W. Harsono, Making Displays Work for You: Diffuser Films and Optical Performance in LCDs, J. of the SID, 28(1), p. 20 (2012).
25 W. G. Lee, J. H. Jeong, J.-Y. Lee, K.-B. Nahm, J.-H. Ko, J. H. Kim, Light Output Characteristics of Rounded Prism Films in the Backlight Unit for Liquid Crystal Display, J. of Information Display, 7(4), p. 1 (2006).
26 http://products3.3m.com/catalog/us/en001/electronics_mfg/vikuiti/node_V6G78RBQ5Tbe/root_GST1T4S9TCgv/vroot_S6Q2FD9X0Jge/gvel_GD378D0HGJgl/theme_us_vikuiti_3_0/command_AbcPageHandler/output_html
27 M. Tjahjadi, G. Hay, D. J. Coyle, E. G. Olczak, Advances in LCD backlight film and plate technology, Information Display, 22(10) (2006).
28 E. G. Olczak, Optical structure and method of making, U.S. Patent No. 6862141 (2005).
29 F. Bastawros et al., Light collimating and diffusing film and system for making the film, U.S. Patent No. 7,889,427 (2011).
30 M. F. Weber, C. A. Stover, L. R. Gelbert, T. J. Nevitt, A. J. Ouderkirk, Giant Birefringent Optics in Multilayer Polymer Mirrors, Science, 287, p. 2451 (2000).
31 D. Joseph, Viscoelastic Flow Effect in Multilayer Polymer Coextrusion, Universiteitsdrukkerij TU Eindhoven, Eindhoven, the Netherlands, (2002).
32 J. S. Choi et al., Optical lens system, backlight assembly and display device, U.S. Patent No. 20060087863 A1 (2015).
33 J. S. Seo, T. E. Yeom, J.-H. Ko, Experimental and Simulation study of the optical performances of a wire grid polarizer as a luminance enhancement film for LCD backlight applications, J. of the Opt. Soc. of Korea, 16(2), p. 151 (2012).
34 K. Marrin, Quantum dot technology progresses, ships in LED-backlit LCD TVs, LEDs magazine, 10(12), (2013).
35 J. Chen, V. Hardev, J. Yurek, Quantum-dot displays: giving LCDs a competitive edge through color, Information Display, 29(1), p. 12 (2013).
36 D. DeAgazio, Design and manufacturing considerations for LED backlights, Information Display, 21(11) (2005).
37 R. S. West, H. Konijn, W. Sillevis-Smitt, S. Kuppens, N. Pfeffer, Y. Martynov, Y. Takaaki, S. Eberle, G. Harbers, T. W. Tan, C. E. Chan, high brightness direct LED backlight for LCD-TV, SID 03 Digest (2003).

7

TFT Backplane and Issues for OLED

Chiwoo Kim

Seoul National University, Daehak-dong, Gwanak-gu, Seoul, Korea

7.1 INTRODUCTION

Active matrix organic light emitting diode (AMOLED) is an electroluminescent device where the electric current passing through the diode produces electron–hole pair recombination. Because AMOLED is a self-emission display, it has several advantages over AMLCD such as thin and light form factor, wide viewing angle, and fast response time. Lower power consumption and cheaper production cost are also expected in the foreseeable future. Most of all, AMOLEDs are better suited for the next-generation flexible display since it does not need cell gap, color filter (for RGB separate patterning type AMOLED) and backlight system, which limit the curvature of the display. Based on these advantages, AMOLED has been adapted to many mobile devices and now it is growing explosively (Figure 7.1).

Display manufacturers are also focusing on the large-sized AMOLED TVs where more productive and lower-priced AMLCD dominate the market. AMOLED TVs can easily achieve a perfect black state, wide color gamut and slim form factor compared to AMLCD TVs. For TV application, from a backplane point of view, a-Si TFT was the first choice for AMOLED TV because its production cost is so low, but a-Si TFT's bias stress stability and low electron mobility issues have prevented it from being adapted to AMOLED applications [1]. a-Si based AMOLED had been studied by many groups, but failed to overcome the long-term stability issue [2].

Low-temperature polycrystalline silicon (LTPS) TFTs are found in small- and medium-sized mobile products. The mobile display applications demand a high resolution as well as a high display image quality for which current AMOLED manufactures have been using LTPS TFT backplane solutions. Since AMOLED is a current driven device, pixel TFTs are operated as an analog current driven mode. The output uniformity of drive TFTs is directly related to display image quality, and it is almost impossible to obtain a mura-free display with the random grain sized poly-Si alone. In addition to the precise control of the fabrication processes, LTPS TFT requires compensation circuit for each pixel. For mobile AMOLED applications, LTPS TFT's threshold voltage (V_{th}) and electron mobility compensation circuit is used to minimize the display mura associated with the non-uniform short-range TFT characteristics. A typical pixel compensation circuit is composed of five to seven transistors and one to two capacitors for each pixel.

Compared to a-Si based AMLCD, LTPS fabrication requires additional processes such as Excimer Laser Annealing (ELA), ion-doping and thermal annealing for dopant activation. High-resolution photolithography process is also necessary to accommodate the complicated compensation circuit. For RGB patterning, the

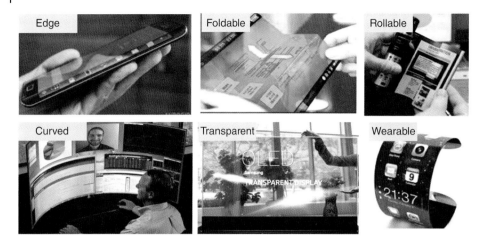

Figure 7.1 Application area of flexible AMOLED displays.

high accuracy fine metal mask (FMM) is used. RGB OLED shows high optical efficiency, deep color saturation, and does not need a color filter.

For large size TV application, oxide semiconductor (OS) TFT backplanes are being used [3–5]. Since the amorphous nature of the OS TFT leads to better short-range uniformity compared to laser crystallized poly-Si TFTs combined with the scalability of a-Si TFTs, OS TFT backplane technologies have been adopted for TV applications. In addition to OS TFT backplane, AMOLED TV further uses white OLED stacks because the OLED material evaporation process does not need the complicated FMM process. White OLED deposition uses an open shadow mask to shield the display border areas, which requires a coarser alignment accuracy of ~100μm. Further, a color filter array is used to convert the white emitted light into R, G, and B.

The LTPS and OS backplane technologies have pros and cons related to TFT mobility, short range V_{th} uniformity, productivity, and scalability. Most of the AMOLED manufacturing companies are developing both technologies. At the moment, LTPS with internal compensation is the technique for manufacturing high-resolution mobile displays, and oxide with external compensation is the technique for manufacturing large-sized displays.

7.2 LTPS TFT BACKPLANE FOR OLED FILMS

The ELA technique has been used for mass production of high-resolution mobile AMLCDs. The system consists of a UV lasers (308nm), beam delivery optics, and a precisely controlled positioning stage. The laser energy density can be controlled from 350mJ/cm^2 to 600mJ/cm^2. The typical pulse duration of the UV laser is 28 nanoseconds. For mobile AMOLED display applications, ELA LTPS TFT backplane technology is used primarily in high-volume manufacturing. LTPS fabricated from line beam ELA poly-Si has become a standard material for AMOLED backplanes of mobile devices because of its high device performance. However, the ELA system has several hurdles to overcome. ELA process requires multiple shot overlap (10 to 30 shots), high maintenance cost, and limited laser beam length. The fluctuation of the laser shot energy affect the grain size of the poly-Si and the TFT characteristics in one location is different from the other location. The horizontal and vertical line mura, that is directly related to the short-range TFT uniformity. By increasing the number of shots on the same area, ELA process can enable better uniformity of the TFT performance across the display area [1, 6, 7].

The ELA system can itself be improved for higher laser shot stability. In addition, by using laser beam mixing techniques where the laser beam is modulated along the long beam axis (see Figure 2.8), the vertical line mura can be reduced. Now the technical issues related to the use of ELA for OLED displays, such as the ELA induced shot mura, have been improved by implementing hardware solutions and adjusting the compensation circuit in AMOLED panel. The ELA process is now achieving reasonable production yield in volume manufacturing of OLED displays.

However, the existing ELA technique still requires a high cost of ownership with low process throughput. There are two reasons for those facts. First, the equipment cost is very high. This arises from the difficulties of manufacturing the large size excimer laser optics and the high power lasers. With the limited laser beam length (length of the long beam axis in Figure 2.8), it is difficult to handle large substrate (larger than Gen 6 size glass). The limited laser beam length comes from the constraints for laser power and optics size. Today, one scan ELA system is available for Gen 6 size glass but the equipment cost is too high. Even worse, the ELA system should be maintained and monitored daily, therefore the uptime is well below 80%.

Second, the laser beam is scanned with a slow speed and ~95% overlap depending on the display resolution and the pixel design. High-resolution displays with the pixel pitch of ~20μm or less require higher degree of poly-Si grain size uniformity. The translation distance between shots and the pixel pitch associated with the driving transistor's design are closely related. Higher resolution display requires shorter laser beam scan pitch. This leads to lower process throughput under mass production condition.

Manufacturing throughput of the conventional ELA can be increased by reducing the number of shots, or the selective crystallization of the TFT area [7, 8]. New laser crystallization techniques like sequential lateral solidification (SLS) [9–14], phase-modulated excimer laser anneal (PM-ELA) [15], selectively enlarging laser crystallization (SELAX) [16], continuous wave laser crystallization (CLC) [17], and single-crystal-like grain by CW laser crystallization with cylindrical micro-lens were also reported [17]. Among many high throughput laser crystallization techniques, the conventional line beam ELA is still mainly used for manufacturing AMOLED. However, new laser sources and optics as well as new concepts for improving the short-range uniformity are needed for higher productivity and better display quality.

For the AMOLED backplane process architecture, coplanar self-aligned P-channel TFT structure is widely used. P-channel TFTs are more stable than n-channel TFTs against bias stress. The advantage of using P-channel TFTs in AMOLED displays is that the pixel electrode can be the indium tin oxide (ITO) anode and the inverted OLED structure is not needed.

7.2.1 Advanced Excimer Laser Annealing (AELA) for Large-Sized AMOLED Displays

High equipment cost and the low throughput issues are the main hurdles that prohibit the use of ELA technique in AMOLED-TV application. There are several approaches to overcome the issues. For large area displays, where the pixel pitch is much larger than the ELA translation distance between shots, the conventional 95% overlap rule for mobile application can be changed. The advanced excimer laser annealing (AELA) process using conventional ELA equipment is developed. The new technique recrystallizes selected areas of the a-Si layer where TFTs will be fabricated using several bi-directional scans of the laser to achieve higher throughput and uniform structure is developed. The AELA process consists of three steps. First is the alignment to the key pattern on a-Si film layer, as shown in Figure 7.2. (1) Second step is the AELA crystallization. Final step is the active patterning on the poly-crystallized area via AELA.

Macroscopically, the AELA crystallization process results in two separated regions: one is the irradiated poly-Si region, and the other is the a-Si non-irradiated region. The following three key elements of the AELA process are different from the conventional ELA process.

First is the location controlled crystallization (LCC) with micro-meter precision. The basic purpose in the LCC is to re-crystallize the pixels' circuit area selectively with the excimer laser beam. The non-circuit regions in the pixel and border regions remain as a-Si.

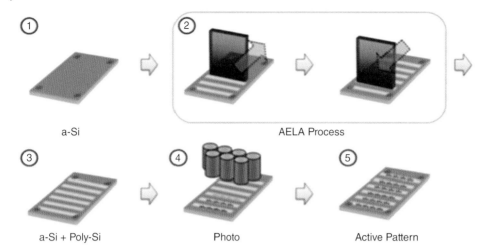

Figure 7.2 Schematic diagram shows AELA process with intermediary steps. Step 1 illustrates a-Si deposited substrate. Step 2 shows AELA process and the result of AELA process is shown at step 3, schematically. Step 4 is a general photolithography process, leading to the AELA processed substrate in step 5 [8, 9].

Second is the use of a constant speed scanning process. The circuit regions are crystallized by precisely positioning the beam to the circuit region and irradiating, then scanning to the next circuit region and irradiating, and so on until whole circuit of the substrate is covered. Since the irradiation pulse is only 28 nanosecond (much shorter than the scanning speed of the substrate) and the irradiation frequency is 300 Hz and fixed, the circuit positions are selected and irradiated simply by scanning the substrate with the constant speed associated with the periodic pixel circuit position.

Third is the minimization of the multiple scan. To achieve the required poly-Si TFT performance for operation of the AMOLED-TV, single scan AELA process is not sufficient. Again, we need multiple scan to obtain reasonably uniform microstructure of the poly-Si film. To obtain a satisfactory level of poly-Si crystallinity without sacrificing the throughput of the AELA process, a minimized multiple scan process is essential. We were able to find the minimized multiple scan process as follows (Figure 7.3).

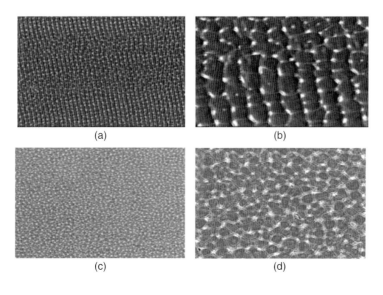

Figure 7.3 Microstructure of the conventional ELA at different magnifications (a, b), and the AELA processed poly-Si films at different magnifications (c, d) [8, 20].

A six-shot AELA process is evaluated, consisting of six irradiation scans of the substrate achieved by three forward and backward scans. Microstructures and TFT characteristics are analyzed. The six-shot process is selected not only for achieving uniform microstructure of the poly-Si film, but also for process reproducibility in manufacturing. Microscopically, there are four regions: a crystallized region, an edge region, a transition region between crystallized and edge regions, and a non-crystallized region. The crystallized region has been optimized to obtain uniform grain size. As expected, this region has very uniform TFT characteristics. The transition region is very narrow. The grain distribution in the transition region is not uniform due to the random formation of large and small grains. The edge region is the partially melting region that comes from the irradiation of the sloped profile laser beam. Grains in this region are much smaller than the grains in the crystallized region. The non-crystallized region is of course the a-Si film. The crystallized region is designated for pixel circuit. To obtain uniform TFT characteristics, the crystallized region must have a large size with uniform microstructure grains. Interconnections are fabricated in the edge region because the edge region can have high conductivity after a doping process. The non-crystallized region is intriguing for its potential applications, including light emitting window for the bottom emitting OLED display structures, photo-sensors with a-Si properties, and the a-Si TFTs for special purposes (Figure 7.4).

SEM images of conventional ELA and AELA-processed Si films are shown in Figure 7.3 [7, 8, 20]. Conventional ELA leads to a periodic grain structure that is well known and is shown in Figure 7.3b. The microstructure of the crystallized region for AELA is shown in Figure 7.3(c, d). Compared to the conventional ELA, the AELA microstructure shows spatial randomness: the grain boundaries are not aligned as much as the ELA grains are aligned. This is because of the reduced number of irradiation pulses in the AELA technique. The AELA uses a six-shot process whereas the conventional ELA process uses a twenty-shot process. In other words, more irradiation shots on Si films produce higher spatial order of grain distribution. TFT mobility between the non-crystallized region and the effective crystallized region is shown in Figure 7.4(b). In the non-crystallized region, near 0 μm, the TFTs perform similarly to a-Si TFTs, with a low mobility around 1 cm^2/V-sec. In the edge region from 0 um to 50um, the mobility increases according to the distance from the non-crystallized region. The crystallinity increases as the energy density and the number of shots of irradiation increase. In the crystallized region, between 50 μm and 200 μm in Figure 7.4(b), the P-MOS mobility is saturated at near 70 cm^2/V-sec and is uniform enough to be used for OLED displays.

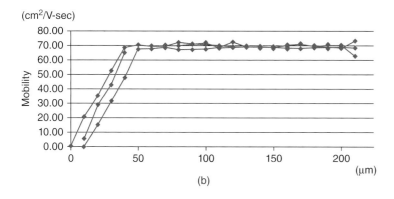

		V_{th} (V)	Mobility (cm^2/V-sec)	S-Factor (V/dec)
TFT	average	−2.07	73.81	0.40
	σ	0.44	1.55	0.03

(a)

(b)

Figure 7.4 (a) Summary of TFT characteristics in crystallized region. Conventional P-MOS process was used [18, 19]. (b) Graphs of grain location versus mobility with three different samples (see Figure 7.3a) [8].

7.2.2 Line-Scan Sequential Lateral Solidification Process for AMOLED Application

It was reported that the two-shot sequential lateral solidification (TS-SLS) process could be used to make TFT backplanes for large area AMOLED [1]. In addition, the TS-SLS process has wider process windows because it utilizes the controlled super lateral growth (C-SLG) phenomenon at the energy density of complete melting region. TS-SLS process requires just two shots (for 2D projection system with 25mm × 1.5mm beamlet) or less (for line scan system with 730mm long beam with 4um width) per unit area, resulting in a high productivity and a low operation cost [4]. In addition, TS-SLS process can be utilized for the large area substrates. Productivity of the SLS system is several times higher than the conventional ELA system. Alternative way of obtaining two-shot SLS materials is also possible.

The simple directional SLS process includes (1) inducing complete melting of predetermined regions, resulting in controlled super-lateral growth (C-SLG) and (2) micro-translating and re-irradiating the sample, leading to the extension of the C-SLG grains formed by the previous irradiation. Depending on the stepping distance (D_{step}) between laser pulses, the crystallized Si films have different microstructures. When D_{step} is less than the half of the length of C-SLG distance (lC-SLG), we can obtain directionally solidified materials in which the grains elongated along the crystallization direction without any protrusion. When D_{step} is between 1/2 × lC-SLG and lC-SLG, the polycrystalline films contain periodically arranged protrusion and the grains are located between the protrusions, which is called TS-SLS materials.

Another approach is the thin line beam approach developed by TCZ. ELA has a short-axis beam width of 400μm (see Figure 2.8). TCZ's thin line beam has an adjustable short-axis beam width of 4μm and long-axis length is 750mm. The short-axis beam width was chosen to avoid nucleation at the short-axis's center of the irradiated area along the entire long-axis beam length. The laser intensity was selected to minimize the laser related mura on the crystallized Si films. Based on the parametric studies on D_{step}, D_{step} = 1.5μm for directional SLS (D-SLS) and D_{step} = 3μm for TS-SLS materials were obtained (Figure 7.5). That is, by selecting the D_{step}, we can get either TS-SLS or D-SLS.

	Advantages	Disadvantages	Scan method and Poly-Si microstructure
ELA	Mass production proof	Narrow process margin Low productivity	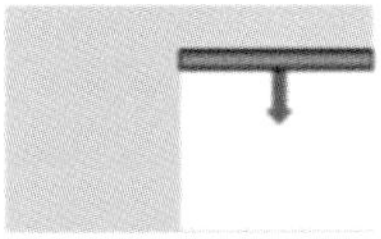 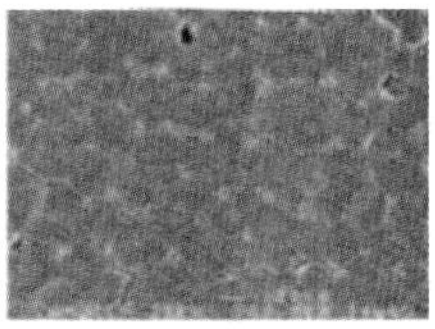
TS-SLS	Productivity Scalability	Anisotropic TFT	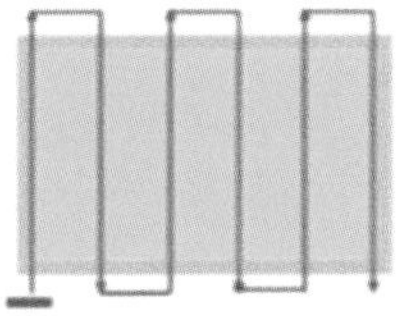 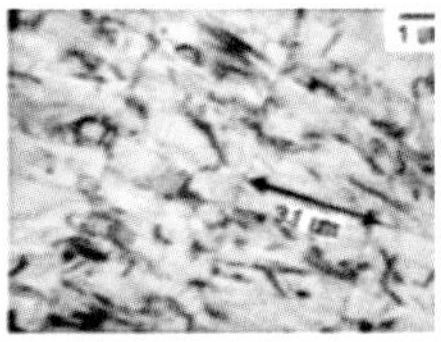
D-SLS	Scalability Mobility	Productivity	 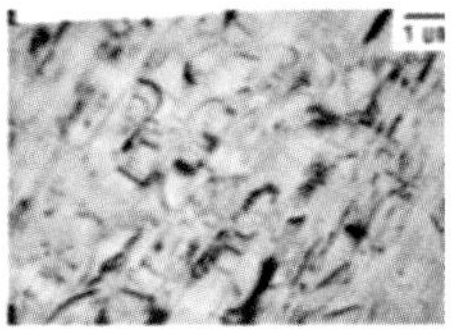

Figure 7.5 Comparison of ELA, TS-SLS, D-SLS.

TFT mobility can be further enhanced by using D-SLS process, which results in elongated grain poly-Si film. In the D-SLS process, the stepping distance (D_{step}) between the two laser pulses is less than half of the laser beam width so that the resulting poly-Si film does not have any protrusion. The protrusions, which are formed at the center of the beam width by the previous irradiation, are melted by the successive irradiation. The grain width, perpendicular to the scan direction, changes from 0.1 μm to 1 μm, irregularly. The electron backscattered diffraction (EBSD) analysis reveals that the grain width depends on the crystallographic orientation of the grains: the occlusion of the slowly growing grains, sub-grain boundary generation, or defect formation in the grains.

The lower mobility of TFTs with TS-SLS materials compared to TFTs with D-SLS materials is associated with the protrusions on the current path in the channel. When the channel direction is perpendicular to the grain growth direction, we call it perpendicular TFT. When the channel direction is parallel to the grain growth direction, we call it parallel TFT. The electron mobility of perpendicular TFTs have 30% to 40% lower than parallel TFTs, which is consistent with other reports [5]. The higher grain boundary density in the channel area of the perpendicular TFTs reduces their mobilities. On the contrary, the significant advantage of TS-SLS process over the D-SLS process is the productivity; since D_{step} is twice, the irradiation time reduces half. The optimum grain size is around 0.4μm for ELA processed materials and 2.5 to 3.5μm for TS-SLS processed materials. In Figure 7.5, the typical microstructures of ELA, two-shot SLS (TS-SLS), and directional SLS (D-SLS) processed poly-Si materials are summarized.

Since TFTs with SLS materials have different performances with respect to the channel directions, two types of TFTs were evaluated to optimize the image quality of AMOLEDs. For high-mobility TFTs such as switching circuit, the parallel TFT works better than the perpendicular TFT. For better uniformity TFTs such as OLED driving circuit, perpendicular TFT is the choice.

However, despite low mobility and poor productivity, the conventional ELA process is the choice for manufacturing AMOLED displays because ELA-based TFT can provide more uniform TFT performances than TS-SLS or D-SLS TFTs. Table 7.1 summarizes the TFT characteristics obtained by three different crystallization methods. ELA-based TFT has isotropic characteristics, while SLS-based TFT has bi-directional characteristics with respect to the laser scan direction.

Although the ELA process has been adapted for the mass production of mobile displays, the current large substrate for TV application require (1) longer laser beam, (2) better laser intensity uniformity along the beam length, and (3) enhanced productivity without the degradation of TFT performance. Compared to the existing ELA system with 750mm length beam, newly developed 1500mm length beam enables to make displays up to 65" diagonal sized TV [21]. Still, in this case the throughput remains an issue.

Finally, if a solution can be found to overcome the directionality characteristics of the TS-SLS, it will be possible to get more uniform TFTs with no laser shot mura in AMOLED display and to revolutionize the productivity of the laser crystallization process. TS-SLS materials can be obtained either by (1) multiple beamlet–based method or (2) by a thin line beam scan method. The advantages of multiple beamlet–based method in which a mask is utilized to define irradiated area are (1) wider crystallization process window, (2) improved productivity compared to the conventional ELA process and (3) scalability to large-sized substrates. On the contrary, the thin line beam scan SLS process, which has a laser system with higher frequency and 1500mm long beam, can be used for AMOLED TVs with single scan with improved

Table 7.1 TFT characteristics of ELA, TS-SLS, D-SLS (measured by conventional top-gate TFTs).

	Mobility (cm^2/Vs)	ΔV_{th} (V)	S.S. (V/dec)
ELA	95~105	0.37	0.27
TS-SLS	98~108	0.46	0.25
D-SLS	110~122	0.54	0.24

productivity. In addition, we can obtain the (1) directionally solidified materials, which can provide higher TFT mobility as well as the (2) TS-SLS materials, which can provide more uniform TFT performances by simply changing the stepping distance. This enables us to select optimum materials for each product; a product with highly integrated circuits, such as high-level de-multiplexing circuits, source drivers, and T-con, can use the directionally solidified materials and a product without high-speed switching devices can utilize the TS-SLS materials.

7.3 OXIDE SEMICONDUCTOR TFT FOR OLED

OLED TVs have better image qualities than LCDs, such as a high contrast ratio that is theoretically infinite (OLED displays can completely turn off sub-pixels), a wide color gamut and a wide viewing angle. They are suitable for motion images as they have much better GTG (gray-to-gray) response time, MPRT (motion picture response time), and much lower 3D crosstalk than LCDs. LCD TV panels use a-Si TFT backplanes whose manufacturing process is well established and low cost. There have been attempts to use a-Si TFTs for OLED displays, but the low mobility and the gate-bias-induced V_{th} (threshold voltage) shift make it impossible to use for OLED backplane. LTPS backplanes, on the other hand, have a high mobility and a high stability, but the production cost is relatively high and the OLED application is limited to the small- and medium-sized mobile displays.

Since the introduction of the amorphous Indium-Gallium-Zinc-Oxide (a-IGZO) thin-film transistor (TFT) [22], oxide semiconductor (OS) TFTs have been developed for various kinds of display backplanes [23–26]. Figure 7.6 shows the OS TFT's R&D history. The advantages of a-IGZO TFTs for AMOLED are high electron mobility ($>10\,\text{cm}^2/\text{Vs}$) and excellent short-range uniformity. OS backplanes using a-IGZO show a high performance and can exploit existing amorphous silicon process lines. To achieve mass production of OS AMOLED, various kinds of TFT structures are studied. With back channel etched (BCE) structure, the number of mask steps and parasitic capacitance can be reduced just like the a-Si TFT process. One major drawback of the BCE oxide TFTs is that device performance and bias-stability are not always reproducible. In BCE TFT structure, the TFT back channel is inevitably exposed to the process environment before it is covered by passivation layer. This means that the material properties such as chemical composition and carrier concentration of oxide semiconductors used for TFTs can be contaminated and changed between the runs on OS TFTs. Even worse, V_{th} is very sensitive to the partial pressure of O_2 in Ar/O_2 mixture during the sputtering process and it is very difficult to obtain the reproducibility.

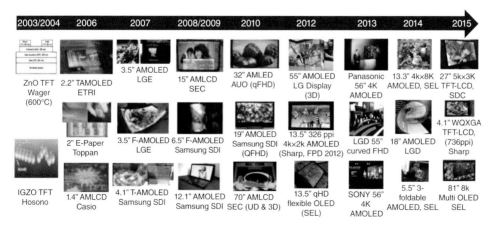

Figure 7.6 R&D history of oxide TFTs (*Source:* Prof. Sanghee Park of KAIST).

To overcome the BEC issues, two distinct structures of oxide TFTs are investigated: the etch stopper (ES) structure with a channel blocking layer and the coplanar structure in which the gate layer acts as an etching stopper.

7.3.1 Oxide TFT–Based OLED for Large-Sized TVs

The etch-stopper (ES) structure is the first choice for AMOLED-TV backplane because the ES can protect the oxide surface of the TFT channel area during the TFT process against plasma and wet chemical processes. The ES structure has a higher TFT reliability then the BCE, but it has larger parasitic capacitances because of the layer overlapping areas and it also needs more photomasks than the BCE structure. To drive a large-sized high-resolution OLED-TV, the unnecessary cross-over capacitances should be minimized and the coplanar structure has the advantage because the coplanar structure is inherently self-aligned non-overlap structure with respect to the gate line. In addition, optimized thermal process conditions are developed to improve the TFT reliability.

Besides coplanar structure, the bulk-accumulation (BA) a-IGZO TFTs (Figure 2.17(b)) are introduced [6, 27, 32–34]. BA TFTs can be achieved by shorting the top-gate (TG) and bottom-gate (BG) of the dual-gate (DG) a-IGZO TFTs with thin active layer. Thanks to the dual-gate sandwich structure, the channel accumulation layer extends the entire depth of the active layer, rather than being confined to the active-layer/gate-insulator interface. A big advantage of this device is that its on-currents can be five to seven times higher compared to that of single gate oxide TFTs. Owing to bulk accumulation/depletion, the TFT's turn on voltage (V_{th}) is always close to zero volt because the TFT's off-state and on-state are dominated by the values of V_{gs}, rather than interface states or bulk states that may affect the initial position of the Fermi level (EF). To reduce the parasitic capacitance, top gates with 2μm offsets at the source/drain electrodes are introduced in achieving high-frequency operation. Bulk-accumulation results in high TFT performance and stability, while keeping a simple manufacturing process.

Even though OLED-TV shows better display quality than LCD-TV, OLED-TV is still not the mainstream displays because of the high manufacturing cost. To reduce it, process cost of the OS TFT should be compatible to a-Si TFT, production yield should be improved to LCD level, and the materials and electronic parts costs should be reduced significantly [12]. With all these difficulties, OS TFT process is now considered the favorable material for OLED-TV because of its superior scalability and the low production cost compared to LTPS TFT [3–6]. To compete with the BCE type a-Si TFT used for LCD in view of the manufacturing cost, coplanar self-aligned top gate TFT structure [28–31] is compatible. A TFT backplane is fabricated through four photo-lithography steps. For the top emission OLED, three additional photo-lithography steps of PLN/anode/WIN (pixel define layer) are necessary after the TFT fabrication. Top gate 7 mask OS TFT with top emission OLED structure can be the standard process architecture of the OLED-TV. By merging source/drain (S/D) metal and anode metal, five-mask top gate TFT top emission structure is possible (Table 7.2). But, in this case, the choice of the S/D and the anode metal is limited to the same one. The key for this five-mask process is the choice of the S/D metal or multi-layer metals [5, 6]. It should be noted that conventional a-Si TFT for TN mode LCD uses five masks as well.

For large-sized OLED evaporation process, instead of the FMMs (fine metal masks) commonly used in small-sized RGB type OLED production, simple open masks are used for organic material deposition across the pixel area while shielding the panel border and fan-out pad area during evaporation. Open masks require WOLED (white OLED) technology with tandem color layers. Since WOLED doesn't need color separation in evaporation processes, it is free from misalignments and color mixing caused by the alignment error of FMM. But this WOLED has a disadvantage of reduced efficiency because WOLED needs color filter to convert the white light to RGB and only one-third of the emitted white light contribute to the display brightness. To compensate this issue, a white pixel is normally added (no color filter), but the overall efficiency remains much less than RGB side-by-side type OLEDs.

Table 7.2 Standard oxide TFT-LCD process flows and proposed oxide TFT-OLED five-mask flow [28].

Display mode	<LCD>	<OLED display>		
Structure	Bottom gate Back channel etch.	Bottom gate Etch. stopper	Top gate (Standard)	Top gate (Anode merge)
Process step (Photo-lithography) — TFT	Gate / Si / S/D / PSV / PIX	Gate / Oxide / Etch. stopper / S/D / PSV	Oxide / Gate / IL / S/D	Oxide / Gate / IL / S/D, Anode
Process step (Photo-lithography) — Anode		PLN / Anode / WIN	PLN / Anode / WIN	WIN
Channel material	a-Si	Oxide	Oxide	Oxide
Mask number (for TFT only)	5 (5)	8 (5)	7 (4)	5 (4)

There are two opposite emission types for OLED displays, namely top-emission and bottom-emission. For top-emission structure, the pixel emission layer is positioned on top of the pixel driving circuit. For that reason, the pixel aperture ratio for top-emission is much higher than bottom-emission. High aperture ratio means larger light emission area and the OLED lifetime is enhanced. Unfortunately, top-emission structure is not well established for large size OLED-TV because the semi-transparent low resistive cathode material is not available yet. In addition, the light must be extracted through cathode and the large area transparent encapsulation process is under development for manufacturing. In the long run, still the top-emission structure is the right approach for large-sized display just like in the case of mobile displays. On the contrary, the bottom-emission structure is being used for mass production of large-sized OLED-TVs. Even though part of the OLED emission area is blocked by pixel circuits, the thick low resistive cathode electrode structure gives a better display uniformity and mass productivity than the top-emission structure. The cathode can be a thick common metal layer, and the encapsulation is much easier than that of the top-emission because we do not need the transparent encapsulation. What is required is a water vapor transmission rate (WVTR) $< 10^{-6}$ g/m^2/day, which can be readily achieved with glass or barrier film encapsulation solutions.

The biggest issue for OS OLED-TV is the display mura. The output current fluctuation of the individual driving TFTs is caused by initial process induced non-uniformity and from the long-term degradation through differential aging under applied electrical stresses. Since OLED-TV is a current driven device, to reduce the display mura arising from the driving TFT's on-current fluctuation, it requires excellent OS TFTs uniformity and stability. Threshold voltage (V_{th}) compensation is also necessary to match the OS TFT's V_{th} [35]. OLED pixel with compensation circuit traditionally employs additional TFTs, capacitors, and signal lines. In general, a pixel compensation circuit requires three to seven transistors per pixel. To achieve large-sized high-resolution OLED-TV, a simple pixel structure is necessary to reduce defect density and improve aperture ratio. Minimizing the number of TFTs in a compensation circuit can not only reduce defects but also simplify driving signals. When the resolution increases, the signal delay becomes a major problem and the self-aligned top gate structure with organic inter-layer between signal line and gate line is proposed [3–5].

Because OS TFT's electrical stability is not yet good enough and the lifetime of OLED materials is relatively short, it is not yet possible to meet TV's lifetime requirement of one hundred thousand hours. This is the major huddle for the commercialization of the OS OLED-TV. In addition to the internal pixel compensation, the external real time individual pixel read-out compensation is the key technology for achieving the mura free display over the display lifetime. For further improvements of compensation quality, several new techniques other than V_{th} and mobility compensations are considered, including a luminance compensation that measures voltage-luminance characteristics at shipment, and degradation compensation.

7.4 BEST BACKPLANE SOLUTION FOR AMOLED

For the high-end mobile devices, high resolution, wide color gamut, thin and light, flexible, low-power consumption, and mura-free display technologies are essential. To meet these requirements, ELA-based LTPS TFT process becomes the standard back plane technology of AMLCDs, AMOLEDs, and flexible displays. LTPS TFTs have a high electron mobility of ~100cm^2/Vs and excellent bias stability. This enables implementing fine design rule TFTs of 1.5μm, various kinds of circuit integrations on glass, low power driving, and V_{th} compensation circuit for each pixel. While LTPS has the advantages to meet the requirements, further work is critical to improve its irregular poly silicon grain boundary issue that is related to the narrow laser process margin. Scalability and the productivity are also big issues to overcome. These are the reasons why high-performance LTPS process is limited to small- and medium-sized displays.

To compete with the highly efficient AMLCDs and have a proven productivity and scalability, an a-Si TFT compatible technology with higher electron mobility and reliability is necessary. The best candidate currently is the OS TFT. Typically, a-IGZO TFT has the electron mobility of 10cm^2/Vs. A-IGZO TFTs are applicable to large-sized AMOLED displays and relatively high-resolution AMLCDs. In addition, OS TFT backplane has an extremely low leakage current. This enables low refresh rate AMLCD and save the power consumption. The I-V characteristics of oxide, amorphous silicon (a-Si), and poly-silicon (LTPS) TFTs are shown in Figure 7.7. To address the known oxide semiconductor TFT reliability issues, a so-called CAAC-IGZO (c-axis-aligned crystal) TFT was introduced [24]. It has good reliability for bias temperature stress. Application of OS to non-volatile memory and x-ray sensors utilizing its high mobility and low leakage current are also explored to achieve higher sensitivity [36].

Table 7.3 compares the key manufacturing technologies of a-Si, LTPS, and OS TFT displays. The major issues for ELA LTPS TFT backplanes are the process cost, scalability, and the laser-induced mura. For OS TFT backplanes, the long-range uniformity, run-to-run reproducibility, and long-term TFT stability need to

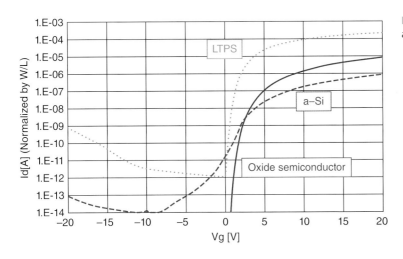

Figure 7.7 I-V characteristics of a-Si, LTPS and OS TFTs [24].

Table 7.3 Comparison of a-Si, LTPS, and oxide TFTs [6, 8, 39–41].

	a-Si TFT	LTPS TFT	OS TFT
Key Processes	Large process margin and scalability	Crystallization methods: ELA, SLS, AELA TFT channel grain control	Oxide sputtering: control of the composition ratio, post-treatments
TFT Photo Mask	4 to 5	7 (PMOS) to 9 (CMOS)	6 (BCE) to 8 (coplanar)
TFT Structure	BCE	Top gate self-aligned coplanar	BCE, ES, Coplanar
Process T. (C)	~300	450 (Furnace) ~ 550 (RTA)	RT ~ 350
Substrate Size	Gen 10	Gen 6	Gen 8
Mobility(cm^2/Vs)	0.5		~10
S Slope (V/dec)	0.4 to 0.5	0.2 to 0.3	0.09 to 0.2
I_{off} (A)	~10^{-13}	~10^{-12}	>10^{-13}
PBS / NBS	Bad	Very good	Not good
Light & T. Sensitivity	Medium Medium	Medium Low	High (NBTI) High
TFT Uniformity: Short Range	Good	Not good	Very good
Run to Run Variation	Excellent	Good	Issue
TFT Type Circuit Integration	NMOS gate driver	PMOS/NMOS/CMOS gate driver, pixel compensation, data mux,	NMOS gate driver, pixel compensation
Advantages	Large process margin and scalability	High mobility (~100 cm^2/Vs) excellent stability	a-Si like process and scalability
Disadvantages	Low mobility, poor Vth stability	Narrow laser process margin high fabrication cost	Relatively Poor Stability
Application	Dominant backplane technology of AMLCD	Advanced mobile AMLCD and AMOLED	Large-sized AMLCD and AMOLED

be improved. The V_{th} can be precisely controlled in LTPS TFTs by using ion doping process, but for oxide TFTs, it depends on the carrier concentration in oxide semiconductors. Oxygen vacancy (VO) is a donor in a-IGZO and thus the density of VO should be precisely controlled to achieve a uniform TFT threshold voltage distribution [37]. The OS material sputtering environment control and long-range uniformity are also crucial for OS TFTs [38].

Table 7.4 shows a comparison of the LCD, OLED-rigid, and OLED-flexible manufacturing processes. Flexible OLED has the most difficult processes including the PI coating, ELA-based LTPS backplane,

Table 7.4 Comparison of LCD, OLED-rigid, and OLED-flexible manufacturing processes.

	LCD	OLED-Rigid	OLED-Flexible
Back-plane	a-Si:H, 4–6M LTPS, OS 5–9M	LTPS, OS, 7–9M	PI w. carrier glass LTPS, oxide, 7–9
Display	Color filter 4M, LC	Mobile: EL 9–12 layers, FMM TV: white+ CF	Mobile: EL 9–12 layers, FMM TV: white+ CF
Cell TSP	Color Filter, LC separate, in/on cell	Frit + Glass Encap. separate, on cell	Multi-layer TFE, laser-cut separate, on cell
Module	Chip on glass (COG)	Chip on glass (COG)	Chip on plastic (COP)

RGB-type OLED patterning, carrier glass separation, and thin-film encapsulation. Rigid type LPTS fabrication line can be converted to flexible fabrication line by adding PI, TFE, and flexible cell processes. OS OLED-TV has been mass produced in Gen 8 size (2,200 × 2,500mm) fabrication line with white OLED technology. The existing a-Si fabrication line can be converted to OS by adding OS sputtering and annealing process. It is expected that the OS process can be scaled up to Gen 10 size or larger because of the a-Si process compatibility.

For the time being, high-resolution mobile displays use ELA-based LTPS backplane technology for both AMOLED as well as for AMLCD. To compete with a-Si or OS process, the scalability, grain size uniformity, and shot overlap of the ELA process need to be improved. Several attempts are being made to replace ELA technology. When the productivity and the scalability issues are cleared, LTPS process can be the solution for flexible OLED displays as well as for large-sized TV application. Presently, for OLED-TV, OS remains the primary choice because of its process compatibility with a-Si backplane infrastructure. OS AMOLED-TV requires both initial and ongoing external compensations for OS backplane's V_{th} shift and white OLED's differential aging issues. This external compensation and relatively low electron mobility prohibit OS to penetrate the mobile display applications. It is exciting to watch whether LTPS or OS technology will dominate the backplane technology of the next-generation display market.

References

1. C.-W. Kim, J.-G. Jung, J.-B. Choi, D.-H. Kim, C. Yi, H.-D. Kim, Y.-H. Choi, J. Im, Invited Paper: 59.1: LTPS backplane technologies for AMLCDs and AMOLEDs, Proc. SID 2011 (2011).
2. S. M. Choi, O. K. Kwon, H. K. Chung, An improved voltage programmed pixel structure for large size and high resolution AM-OLED displays, Proc. SID Symp. Dig., pp. 260–263 (2004).
3. H.-J. Shin, S. Takasugi, K.-M. Park, S.-H. Choi, Y.-S. Jeong, B.-C. Song, H.-S. Kim, C.-H. Oh, B.-C. Ahn. 7.1: Novel OLED display technologies for large-size UHD OLED TVs. Proc. SID 2015 (2015).
4. R. Tani, J.-S. Yoon, S.-I. Yun, W.-J. Nam, S. Takasugi, J.-M. Kim, J.-K. Park, S.-Y. Kwon, P.-Y. Kim, C.-H. Oh, B.-C. Ahn, 64.2: Panel and circuit designs for the world's first 65-inch UHD OLED TV, Proc. SID 2015 (2015).
5. C. Ha, H.-J. Lee, J.-W. Kwon, S.-Y. Seok, C.-I. Ryoo, K.-Y. Yun, B.-C. Kim, W.-S. Shin, S.-Y. Cha 69.2: High reliable a-IGZO TFTs with self-aligned coplanar structure for large-sized ultrahigh-definition OLED TV Proc. SID 2015 (2015).
6. M. Mativenga, D. Geng, J. Jang 3.1: Oxide versus LTPS TFTs for active-matrix displays, Proc. SID 2014 (2014).
7. M. Choi, Advanced ELA for large-sized AMOLED displays, ITC 2014 Abstract Books. p. 12 (2014).
8. M. Choi, S. Kim, J.-M. Huh, C. Kim, H. Nam. 3.4: Advanced ELA for large-sized AMOLED displays. Proc. SID 2014 (2014).
9. J. S. Im, MRS Symp. Proc. 1426, p. 239 (2012).
10. R. S. Sposili, J. S. Im, Sequential lateral solidification of thin silicon films on SiO2, Appl. Phys. Ltt. 69, p. 2864 (1996).
11. S. M. Choi, C. K. Kang, S. W. Chung, M. J. Kim, M. H. Kim, K. N. Kim, B. H. Kim, Proc. SID41, p. 798 (2010).
12. C. W. Kim, K. C. Moon, H. J. Kim, K. C. Park, C. H. Kim, I. G. Kim, C. M. Kim, S. Y. Joo, J. K. Kang, U. J. Chung, 21.4: Development of SLS-based system on glass display, Proc. SID 35, p. 868 (2004).
13. J. B. Choi, C. H. Park, I. D. Chung, K. H. Lee, H. K. Min, C. W. Kim, S. S. Kim, Proc. SID40, p. 88 (2009).
14. J. B. Choi, W. K. Lee, Y. J. Chang, J H. Oh, W. H. Jin, C. H. park, B. K. Choo, I. D. Cung, K. H. Lee, H. B. Hwang, Y. S. Lee, H. D. Kim, S. S. Kim, 53.2: 30" AMOLED based on sequential lateral solidification process. Proc. SID41, p. 794 (2010).
15. C. H. Oh, M. Ozawa, M. Matsumura, A novel phase- modulated excimer-laser crystallization method of silicon thin films, Jpn. J. Appl. Phys., 37, pp. L492–L495 (1998).
16. M. Hatano, T. Shiba, M. Ohkura, Selectively enlarging laser crystallization technology for high and uniform performance poly-Si TFTs, In SID Tech. Dig., pp. 158–161 (2002).

17. A. Hara, F. Tkeuchi, M. Takei, K. Yoshino, K. Suga, N. Sasaki, AM-LCD '01 Digest, p. 227 (2001).
18. S. D. Brotherton, D. J. McCulloch, J. P. Gowers, Jpn J Appl Phys 43, p. 8 (2004).
19. R. Paetzl, J. Brune, L. Herbst, F. Simon, B. A. Turk, Proceedings of 5th international TFT conference ITC' 09, 10.2, France (2009).
20. P. C. van der Wilt, 13.1: Excimer-laser annealing: microstructure evolution and a novel characterization technique, Proc. SID 2014 (2014).
21. M. Sobey, K. Schmidt, B. Turk, R. Paetzel. 8.2: Status and future promise of excimer laser annealing for LTPS on large glass substrates. Proc. SID 2014 (2014).
22. K. Nomura, H. Ohta, A. Takagi, T. Kamiya, M. Hirano, H. Hosono, Room-temperature fabrication of transparent flexible thin-film transistors using amorphous oxide semiconductors, Nature, 432, pp. 488–492 (2004).
23. S. Yamazaki 3.3: Future possibility of C-axis aligned crystalline oxide semiconductors comparison with low-temperature polysilicon, Proc. SID 2014 (2014).
24. T. Matsuo et al., 8.3: Advantages of IGZO oxide semiconductor, Proc. SID 2014 (2014).
25. Y. G. Mo, M. Kim, C. K. Kang, J. H. Jeong, Y. S. Park, C. G. Choi, H. D. Kim, S. S. Kim, Amorphous-oxide TFT backplane for large-sized AMOLED TVs, J SID 19(1), pp. 16–20 (2011).
26. Y. Kataoka, et al., IGZO technology for the innovative LCD, IDW 2013, pp. 12–15 (2013).
27. S. Steudela, et al., 29.4: Flexible AMOLED display with integrated gate driver operating at operation speed compatible with 4k2k, Proc. SID 2015 (2015).
28. T. Arai, 69.1: The advantages of the self-aligned top gate oxide TFT technology for AM-OLED display, Proc. SID 2015 (2015).
29. N. Morosawa, et al., SID '13 Digest, p. 85 (2013).
30. S. Hong, C. Jeon, S. Song, J. Kim, J. Lee, D. Kim, S. Jeong, H. Nam, J. Lee, W. Yang, S. Park, Y. Tak, J. Ryu, C. Kim, B. Ahn, S. Yeo, 25.4: Development of commercial flexible AMOLEDs, SID 14 Digest (2014).
31. F. M. Li, et al., Flexible barrier technology for enabling rollable AMOLED displays and upscaling flexible OLED Lighting, SID 13 Digest, pp. 199–202 (2013).
32. M. Mativenga, J. W. Choi, J. H. Hur, H. J. Kim, J. Jang, Highly stable amorphous indium–gallium–zinc-oxide thin-film transistor using an etch-stopper and a via-hole structure, J. Info. Dis., 12(1), pp. 47–50 (2011).
33. S. H. Ryu, Y. C. Park, M. Mativenga, D. H. Kang, J. Jang., Amorphous-InGaZnO4 thin-film transistors with damage-free back channel wet-etch process, ECS Solid State Let., 1(2) pp. Q17–Q19 (2012).
34. D. Geng, H. M. Kim, M. Mativenga, Y. F. Chem, J. Jang, 29.3: high resolution flexible AMOLED with integrated gate-driver using bulk-accumulation a-SGZO TFTs, Proc. SID 2015 (2015).
35. N.-H. Keum, K. Oh, S.-K. Hong, O.-K. Kwon, 7.2: A pixel structure using switching error reduction method for high image quality AMOLED displays, Proc. SID 2015 (2015).
36. T. Nishijima, S. Yoneda, T. Ohmaru, M. Endo, H. Denbo, M. Fujita, H. Kobayashi, K. Ohshima, Y. Shionoiri, K. Kato, Y. Maehashi, J. Koyama, S. Yamazaki, SID 43.1, pp. 583–586 (2012).
37. S. D. Brotherton, Introduction to thin film transistors, physics and technology of TFTs, pp. 302–305, Springer (2013).
38. T. Goto, S. Sugawa, T. Ohmi, 3.2: Application of rotation magnet sputtering technology to a-IGZO film depositions, Proc. SID 2014 (2014).
39. R. Chaji, A. Nathan, 13.2: LTPS vs oxide backplanes for AMOLED displays, Proc. SID 2014 (2014).
40. H. Oshima, 8.1: Value of LTPS: present and future, Proc. SID 2014 (2014).
41. S. Idojiri, M. Ohno, K. Takeshima, S. Yasumoto, M. Sato, N. Sakamoto, K. Okazaki, K. Yokoyama, S. Eguchi, Y. Hirakata, S. Yamazaki, 4.1: Apparatus for manufacturing flexible OLED displays: adoption of transfer technology, Proc. SID 2015 (2015).

8A

OLED Manufacturing Process for Mobile Application

Jang Hyuk Kwon and Raju Lampande

Department of Information Display, Kyung Hee University, Hoegi-dong, Dongdaemun-gu, Seoul, 130-701, Korea

8A.1 INTRODUCTION

Organic light-emitting diode (OLED) is a widely researched technology due to its potential to become a cost-effective and energy-efficient next-generation display. Because of its many advantages such as simple device structure, light weight compared to its liquid crystal display (LCD) counterpart, wide viewing angle and others, intensive research work has been done to improve the performance of this promising technology [1–4]. In addition, OLED displays also provide high-contrast ratio and brightness, much faster response time, higher color saturation as compared to those of conventional LCDs [3]. As a consequence, progress in OLED materials, device technology, and operational lifetime is almost no longer an issue for mobile display application.

Generally, active matrix OLED (AMOLED) displays are mass-produced on glass as well as on flexible substrates by thermal evaporation of small molecule materials through a shadow mask. Recently, top-emitting OLEDs are used for the production of mobile displays. Top-emitting architecture is much more efficient and enables higher aperture ratio and higher resolution compared to that of bottom emission OLED structure [5–7]. This device structure includes a strong optical micro-cavity effect between two reflective electrodes. Proper control of the optical cavity modes leads to enhanced light efficiency and color purity in OLED devices. Though LCD is a more popular technology, it requires a backlight and color filters to generate red, green, and blue (RGB) pixels [8]. OLED displays are free from backlighting due to their self-emissive nature. The individual RGB subpixels can be formed by thermally evaporating small molecule emissive materials through shadow masks or by filtering the white OLED light through RGB color filters [9]. Additionally, an OLED lifetime is very sensitive to impurities, film quality, and environmental conditions [10, 11]. Several companies succeeded in developing high-efficiency materials for better performances and high production yield of the devices. But due to the sensitivity of the organic materials to oxygen and moisture, encapsulation technology plays a crucial role to improve the lifetime of OLED display. Indeed, to prevent image quality and luminance degradation, proper encapsulation technique is required in OLED panel fabrication.

In this chapter, we discuss the details of current OLED manufacturing process for mobile display application including top-emission micro-cavity device, fine metal mask (FMM) technology for RGB patterning, and encapsulation technologies for OLED displays. The present status of AMOLED technology in mobile display, FMM for subpixel patterning, and encapsulation techniques are covered in the second, third, and fourth

Flat Panel Display Manufacturing, First Edition. Edited by Jun Souk, Shinji Morozumi, Fang-Chen Luo, and Ion Bita.
© 2018 John Wiley & Sons Ltd. Published 2018 by John Wiley & Sons Ltd.

sections of this chapter. Finally, an OLED manufacturing process on glass as well as on the flexible substrates for mobile display application is illustrated in the last section of this chapter.

8A.2 CURRENT STATUS OF AMOLED FOR MOBILE DISPLAY

AMOLED displays for smartphones, cellular phones, and personal multimedia players have already been in the market for several years. AMOLEDs have exceptional display merits for instance high-resolution, excellent contrast ratio, good image quality, and low power consumption [12–15]. Other benefits of this display are fast response to changing images and excellent color quality provided by OLEDs [12, 16]. Moreover, AMOLED displays are particularly appropriate for mobile application because they are very thin, can function over a wide temperature range, and have a wide viewing angle. The main technologies for fabricating mobile AMOLEDs consist of vacuum evaporation of organic materials, high yield shadow mask technique for pixel patterning, efficient top-emitting stack structure and Low-temperature polycrystalline silicon (LTPS).

The previous substrate size for manufacturing of AMOLED display was up to $750 \times 650\,\text{mm}^2$ (which represents one-fourth of a Gen 5.5 size substrate of $1500 \times 1300\,\text{mm}^2$). Current OLED technology is using Gen 6 size ($1500 \times 1800\,\text{mm}^2$) glass substrates due to the availability of LTPS equipment size. After the fabrication of LTPS TFT (thin-film transistor) backplane substrate with this size, it is divided into two separate parts ($750 \times 1800\,\text{mm}^2$) for the OLED process.

In AMOLEDs, high-cost components such as color filter (CF) and the backlight unit are not required. To make a cost-effective structure of AMOLED, a large area TFT backplane technology with a low mask count and an appropriate pixel forming technique for large size substrates are presently main focus of intensive research.

8A.2.1 Top Emission Technology

Initially, a significant amount of research has been dedicated to bottom emission OLED configurations, where, light is emitted toward the viewer through the transparent anode (ITO) and the glass substrate. The benefits of a bottom emission configuration are its simple and efficient device configuration. But, the emission aperture has to be shared with the pixel electronics required to drive the OLED device, thus restricting the pixel size in case of high resolution displays. This could be prevented by using top-emission architecture, where the emitted light pass through a semitransparent cathode and encapsulation system of device. Normally, active matrix organic display with top-emission OLEDs (TEOLEDs) offers an excellent aperture ratio and saturated color purity compared to those of bottom-emission OLEDs. Additionally, top-emission device induce following advantages: (i) large pixel size, (ii) high efficiency at the front viewing angle, (iii) low power consumption, and (iv) better lifetime.

Recently, TEOLED architecture has proved to be a main contender for active matrix display application because of its several merits over bottom emission device. The standard architecture of TEOLED is shown in Figure 8A.1. Typically, TEOLED comprises a reflective anode on glass substrate, organic light emitting layer, charge injection and transport layers, semitransparent cathode, and a high refractive index capping layer for reducing waveguide losses. Usually, high performance p- and n-doped charge injection and transport materials are included in the OLED stack to maintain proper charge balance and to achieve high efficiency. Strong micro-cavity effects can be generated in TEOLED because of the interaction between the reflective anode and semitransparent cathode layers. As a result, these strong micro-cavity effects can induce significant impact on the out-coupled light and viewing angle characteristics of OLED [17]. Furthermore, by considering micro-cavity structure, saturated colors, and improved color gamut can be attained via controlling spectral characteristics in OLED. Therefore, micro-cavity based OLED structure is an essential and widely used candidate for the active matrix display application [5, 7]. Similarly, high refractive index material as capping layer on top of

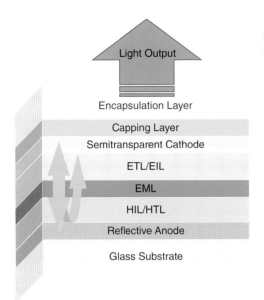

Figure 8A.1 Schematic of top-emission OLED structure. It is composed of organic layers such as hole injection and transport layers (HIL and HTL), emissive (EML), and electron injection and transport layer (EIL and ETL) between reflective anode and thin semitransparent cathode.

the semitransparent cathode is helpful to decrease the wave guiding losses. As a result, significant improvement in the luminous efficiency can be attained. Typically, the efficiency improvement factor is two times for green and three times for red micro-cavity OLEDs; however, no substantial enhancement is attained in blue micro-cavity devices. The existing AMOLED display with top-emission architecture for mobile phone application have nearly 100% color gamut.

Numerous pros and cons of the top-emitting configuration have been described in the literatures. The optical outcoupling efficacy, color gamut and color quality can be enhanced substantially because of strong micro-cavity effects generated by multi-reflection interference between the reflective anode and semitransparent cathode [5, 17–22]. The luminance characteristic of micro-cavity OLED is primarily affected by the optical properties of the semitransparent electrode and the cavity length [6, 7, 23, 24]. Therefore, proper control of the layer thicknesses in the optical cavity and optimizing the optical properties of semitransparent cathode (low absorption and high reflectivity) are essential for high performance of OLED. Usually, the radiance intensity of OLED device depends on the thickness condition of both hole transporting and electron transporting layers (HTL and ETL).

Based on the separation between the reflective anode and semitransparent cathode electrodes as well as the emission wavelength of emitted light, we can quantify the exact cavity-modes such as first-order, second-order, third-order, and so on [6, 7]. Higher radiance intensity is detected for the first-order cavity mode; however, total thickness condition of organic layers is too thin (under 1000 Å) to fabricate efficient devices and avoid issues such as short and proper charge balance, and so on [6]. On the other hand, total thickness of organic layers in second-order cavity mode TEOLED is adequate and the device shows high efficiency and good color purity. Because of the high performance and appropriate total thickness (about 2000 Å), the second-order micro-cavity TEOLED is currently considered for the mass production of AMOLED mobile displays. Similarly, TEOLED with third-order cavity mode can achieve high production yield because of very thick condition of organic layers (~3000 Å). Especially, third-order micro-cavity devices may be suitable for large-size TV display, where high production yield is extremely desirable. Normally, in the case of second-order cavity mode, two probable locations of emissive layer (EML) can be achieved (i) near the anode side and (ii) near the cathode side. Similarly, third-order cavity mode has three possible positions of EML, mainly near the anode side, in the middle of device, and near the cathode side, respectively. The micro-cavity based red, green, and blue TEOLED device architectures for

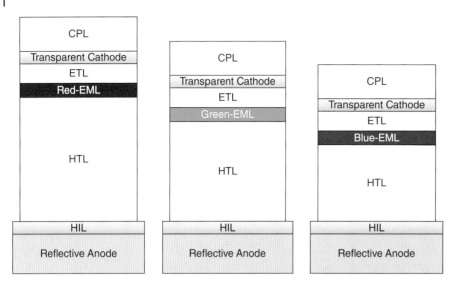

Figure 8A.2 Top-emission structure of red, green, and blue OLED for thick HTL condition. CPL is a capping layer.

thick HTL condition are shown in Figure 8A.2. Recently, it is observed that the position of emissive layer in micro-cavity devices, particularly second- and third-order cavity mode can have a significant impact on the current efficiency, driving voltage, color coordinates, and the viewing angle properties of TEOLED [5–7]. The third-order cavity mode with thick HTL can have significant variations of the emission color coordinates with respect to viewing angle. However, it has been reported that the center EML third-order TEOLED can provide a color coordinates similar to second-order thick HTL device and color shift is also very small with viewing angle. Such variation in the color coordinates could be due to the smaller optical path length between emitted and reflected light in the micro-cavity TEOLED device [6].

For a basic understanding of the micro-cavity effects, theoretical optical evaluation of the micro-cavity based TEOLEDs can be explained by considering the below equations [7]. Herein, the EL spectrum of device with micro-cavity effect consists of Fabry-Perot multiple beam interference and two-beam interference, which is expressed as Δ_{FP}, and Δ_{TBI}. This shows that the spectrum shape and peak position are related to the refractive index and extinction coefficient of the electrodes as well as the reflectance, transmittance, absorbance, and phase of the electrodes.

$$\left|I_{ext}(\lambda)\right|^2 = \frac{T_{top}\sum_j\left[1+R_{bottom}+2\sqrt{R_{bottom}}\cos(\Delta_{TBI})\right]}{1+R_{top}R_{bottom}-2\sqrt{R_{top}R_{bottom}}\sin^2\left(\frac{\Delta_{FP}}{2}\right)}\left|I_{int}(\lambda)\right|^2 \tag{8A.1}$$

$$\Delta_{FP} = \phi_{top}+\phi_{bottom}-\frac{4\pi\sum_i n_i d_i \cos\theta}{\lambda} \tag{8A.2}$$

$$\Delta_{TBI} = \phi_{bottom}-\frac{4\pi\sum_j n_j z_j \cos\theta}{\lambda} \tag{8A.3}$$

Whereas R_{top} and R_{bottom} denote the reflectance at the interface of top and bottom electrodes, T_{top} is the transmittance of semitransparent electrode. Δ_{FP} and Δ_{TBI} represents the phase terms of Fabry-Perot multiple beam interference and two-beam interference. Here, λ, n, d, and z represent the resonance wavelength, refractive index, physical thickness of each cavity layers, physical thickness from the reflective bottom electrode to the emitting dipole, respectively. The $\varnothing_{top}$ and $\varnothing_{bottom}$ are phase change of reflection at the interfaces of top and bottom electrodes in TEOLEDs, θ=angle in cavity layers, $|I_{int}(\lambda)|^2$ = original spectrum in free space, respectively. By considering this equation, the relation between cavity length and spectrum shape such as a full width half medium (FWHM) and the spectral peak position can be easily evaluated. Furthermore, spectral narrowing can be attained for higher cavity length as well as greater reflectivity of bottom and top electrodes.

The main drawbacks of the top-emission architecture are the difficulty in fabrication, color coordinate deviation, highly transparent encapsulation requirement, and efficiency fluctuation. The optical output from the micro-cavity changes significantly with total organic layer thickness. The following equation defines micro-cavity effects in relation to the optical output. The total optical thickness of the cavity, L, is given by [25]

$$L(\lambda) \approx \frac{\lambda}{2}\left(\frac{n}{\Delta n}\right) + \sum_j n_j L_j + \left|\frac{\varphi_m}{4\pi}\lambda\right| \qquad (8A.4)$$

The first term in Eq. 8A.4 denotes the penetration depth of electromagnetic field into the mirror dielectric stack where, λ is center wavelength, n is average refractive index, and Δn is index difference between the layers. The next term indicates the sum of optical thicknesses of organic layers between two mirrors. Final term is the effective penetration depth into the top metal mirror, where φ_m is the phase shift at the metal reflector. This equation signifies that the exact control of the organic layer thickness is essential to attain uniform optical properties across a large area substrate. The film thickness variation in real device should be less than 2% or lower can be considered. The larger change in thickness uniformity may effects on cavity length, which leads to a change in emission color. Indeed, in a real production, it is challenging to achieve thickness uniformity on a large size substrate. However, an exact process control of organic layer thickness in a real production has been already rooted on half-size Gen 6 substrates (750 × 1800 mm^2).

8A.3 FINE METAL MASK TECHNOLOGY (SHADOW MASK TECHNOLOGY)

OLED manufacturing requires precise process control, which is critical to make a high-resolution RGB color pattern. Different techniques are available to deposit organic materials such as inkjet, laser, and thermal evaporation [8]. Currently, the active matrix organic displays with OLED subpixels for mobile application are manufactured by evaporation of small molecule light emitting materials under vacuum pressure of 10^{-7} to 10^{-10} torr through a less than 50 μm thick metal shadow mask [12]. When one emissive material is being deposited via thermal deposition, the other color zones are blocked by shadow mask. The schematic demonstration of red, blue, and green pixel patterning through shadow mask with a point source is shown in Figure 8A.3.

The organic materials for RGB subpixels are thermally evaporated through FMM with a point source or with a linear source. This is a conventional process used for development and mass production of commercial AMOLED mobile displays. FMM is made by patterning invar steel using chemical etching or electroforming or laser process [26, 27]. A light-sensitive photoresist layer is coated onto the invar sheet, and the pixel design is recorded through light exposure. Suitable mask pattern is formed on the invar steel sheet by developing the resist and chemical etching. To smoothen the sharp edges of the mask, two-step etching from top and bottom surfaces is commonly used. Figure 8A.4 illustrates the cross section of a FMM after etching process.

There are few issues related to FMM, which need to be considered for achieving high-quality subpixel patterning. Mainly, FMM can exhibit a sagging problem due to its size, thickness, and the material properties. However, this issue can be overcome by using a mask tensioning process. The openings formed in FMM by chemical etching can have a variation of ±10 μm because of etching non-uniformity. This dimensional

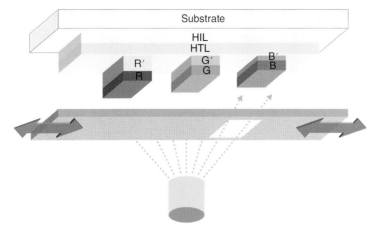

Figure 8A.3 RGB subpixel patterning process using shadow mask and a point source.

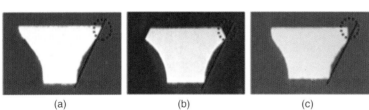

Figure 8A.4 Cross-sectional view of the opening area of shadow mask after a) one, b) two, and c) one etching steps. Red circle shows the etched portion of the mask.

variation is very high compared with other dimensions in the active matrix display technologies. Furthermore, other issues affecting the patterning precision are shadow effects related to mask thickness, total pitch deviation caused by mask tension, and the alignment process margin between the mask and substrate panel. The shadow effect, total pitch deviation, and alignment margin can cause dead space in the pixel structure. As a result, the aperture ratio for the RGB subpixels can be decreased in display area. In addition, further deviation of the pixel opening size may arise due to accumulation of material after repeated depositions; therefore, the mask needs regular cleaning.

Shadow mask technology is a mature technology for the production of AMOLED, however, the limitations in the fabrication of high quality shadow masks directly affect the development of high-resolution AMOLED displays. As an alternative, accounting for the FMM limitations, new RGB subpixel arrangements have been developed for increasing the resolution of AMOLED. Normally, RGB stripe and RGB dot-type pixel arrangements have been employed in the previously commercialized AMOLED mobile displays [28]. These arrangements of subpixels are shown in Figure 8A.5. Recently, Pentile RGB stripe and Pentile RGB diamond-type pixel arrangements have been preferred in mobile AMOLED displays due to their "effective"

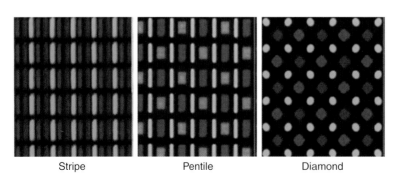

Figure 8A.5 Different arrangements of RGB subpixels (stripe, pentile, and diamond-type pentile arrangement) to achieve an "effective" high-resolution AMOLED display.

high-resolution and low-power consumption benefits. Typically, Pentile pixel arrangements use RGBG configuration with smaller size and higher density of green subpixels because of the higher sensitivity of human eye to green color. Such arrangements also provide higher brightness as well as improved lifetime yield for the blue subpixel. Similarly, Pentile diamond-type pixels also have two small green dot subpixels and diamond-shaped blue and red subpixel. This arrangement is favorable to make high-resolution display because of the long distance between individual color subpixels. In order to make a full-color OLED with high resolution, shadow mask should have an opening area of less than 20 µm. However, it is hard to create pitching area of less than 20 µm with a high yield due to the limitations in shadow mask fabrication, for example, thickness, distance between adjacent opening area, and other mask effects.

8A.4 ENCAPSULATION TECHNIQUES FOR OLEDS

To maintain a long lifetime and high image quality in AMOLED display, organic layers need protection from water vapor and oxygen. Device encapsulation is thus an important part of OLED display development. Therefore, to prevent image quality degradation, proper encapsulation techniques need to be effectively adapted for OLED display.

The encapsulation barrier should be appropriate to decrease the O_2 and H_2O permeation and assure longer lifetimes for OLED devices. Normally, diffusion of O_2 and H_2O can be evaluated using standard commercial Mocon instruments [29]. However, this instrument has a resolution of about 5×10^{-3} g/day m^2. This permeation rate limit is not enough to satisfy the measurement criteria for OLED. Normally, it is considered that to achieve 50,000 hours of operational lifetime in OLEDs, water vapor transmission rate (WVTR) should be lower than 10^{-6} g/day m^2 [10, 11]. Generally, WVTR is evaluated using well known calcium (Ca) test [30–33]. This test can be performed in two ways either electrically or optically [32]. In the optical measurements, thermally evaporated thick Ca layer is encapsulated using thin-film technique or glass technique and their transmittance measurements are performed as a function of storage time under oxygen and water exposure. The original metallic Ca layer upon exposure to oxygen and water gets converted into transparent Ca salts. The transmittance of this Ca layer is measured and plotted with respect to the storage time. The approximate WVTR rate through the encapsulation can be obtained from the slope of measured transmittance change. In case of electrical measurements, thermally evaporated Ca film is deposited between two metal electrodes. Herein, conductance of the Ca film is evaluated under the constant voltage bias. Indeed, conductance can be determined from the resistance of film under test, which is inversely proportional to its thickness. This measurement consents to define the amount of decomposition of Ca film. Presently, several encapsulation techniques are available for OLED application such as metal can, glass lid sealing, UV curable epoxy, frit sealing, and thin-film encapsulation. Among them, frit sealing and thin-film encapsulation techniques are generally used in the mobile OLED display application [34–41].

8A.4.1 Frit Sealing

For the manufacturing of rigid TEOLED displays, frit sealing encapsulation is an effective approach to prevent organic materials from the penetration of oxygen and water vapor. Melted frit glass-bonded OLED and encapsulation glass substrates can prevent any penetration of oxygen and water vapor because of the perfect barrier properties of glass. However, frit is too brittle to be safe from external impact. As shown in Figure 8A.6, the frit pattern can be made with screen printing, and this cover glass with patterned frit material is bonded to the OLED substrate. Then, a laser beam with near infrared wavelength (810-940 nm) tuned to the absorption of frit material is incident on the frit seal edge through the glass cover. Due to the high moving speed of laser, heat generation effect is minimized during frit sealing. The main function of the laser is to melt the frit because it delivers localized energy, preventing other assembly damage. After absorbing the laser energy, the frit material melts and forms strong bonds between the glass cover and OLED

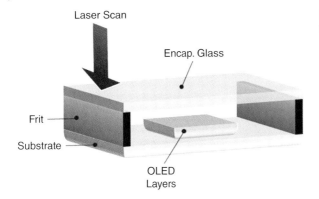

Figure 8A.6 Schematic of widely used laser based frit sealing method for rigid OLED encapsulation.

substrate. The complete assembly after frit sealing process for the OLED encapsulation is demonstrated in Figure 8A.6.

8A.4.2 Thin-Film Encapsulation

The glass encapsulation method is not suitable for flexible OLED display application because of its stiffness. The thin-film encapsulation is lighter and highly flexible, which is preferred for flexible AMOLED display devices. The thin-film encapsulation contains alternating organic and inorganic thin layers coated on the OLED device using thermal evaporation and PECVD (plasma enhanced chemical vapor deposition) or sputtering techniques.

A WVTR rate below 10^{-5} g/day m^2 and 10^{-6} g/day m^2 is considerable for long life OLED operation: anything higher than this could degrade the image quality and performances of OLED display. Several years ago, the Vitex encapsulation technology is reported for flexible OLED application, which used alternating polyacrylate and Al_2O_3 deposited using flash evaporation and sputtering, respectively [29]. The thin-film encapsulation process with Vitex technology is illustrated in Figure 8A.7. In theory, thick inorganic layers can protect moisture and oxygen permeation perfectly, while thin-film inorganic layers have a lot of defects such as particles or pinholes. The moisture and oxygen can permeate through this defect. When we stack multiple organic/inorganic layers, this multiple layers can delay water oxygen permeation time, and it can effectively protect water and oxygen penetration. To consider the particle contamination and complete coverage including sidewall of any layers, thick polymer insulating layers of several microns are used for surface planarization. Indeed, micron-sized particle contaminations may exist on any surface. Currently, organic and inorganic multiple layers, such as four to six pairs, have been used for thin-film encapsulation. Recently, silicon nitride by PECVD or Al_2O_3 by atomic layer deposition shows higher barrier performance, which could reduce the

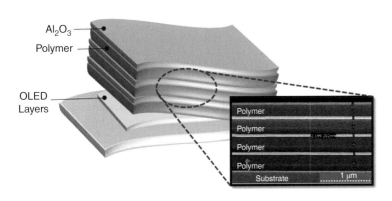

Figure 8A.7 Thin-film encapsulation process for flexible OLED display with multilayers of alternate organic and inorganic films. The marked area with dotted line shows the well-known Vitex encapsulation technology.

number of organic and inorganic layer pairs (e.g., two or three pairs). The flash evaporation process is not sufficient to coat thick organic layers rapidly without any issues. To achieve rapid coating of organic layers, the inkjet printing process has been adopted recently.

8A.5 FLEXIBLE OLED TECHNOLOGY

Recently, flexible AMOLED displays attracted significant attention due to their mechanical flexibility, light weight, and wide variety of applications [42–44]. Particularly, OLED on plastic substrate has already been commercialized for smartphone application. These displays are thin, bendable and, most importantly unbreakable. Flexible OLED displays can be fabricated on plastic substrates such as polyethylene naphthalate (PEN), polyethylene terephthalate (PET), polyethersulphone (PES), polyimide (PI), or metal foils. Among the plastic substrates, polyimide is the only that can endure high temperature over 450 °C required for LTPS backplane processing. It has yellowish color and its transparency is not sufficient. However, top-emission OLED structure can avoid this issue. Metal foil substrates have very high processing temperature capability, good chemical resistance, and excellent moisture barrier properties, but suffers from bad surface roughness. Such high roughness requires coating of an additional polymer layer for surface planarization. Therefore, metal foil substrates are not ideal for AMOLED displays.

Typically, the fabrication of flexible AMOLED display is carried out in four different steps: (i) polyimide solution is coated on the glass substrates to form thin polyimide layer, (ii) fabrication of TFT and OLED using LTPS and evaporation techniques, (iii) thin-film encapsulation, and (iv) polyimide detachment using laser treatment from the glass surface. Figure 8A.8 describes a typical fabrication process flow for flexible AMOLED display. To make a good barrier layer under the TFT pixel array, a multi-layer configuration has been adapted. This multilayer structure is composed of two alternate organic layer and silicon oxide or nitride layer followed by thick polyimide layer using printing method.

8A.6 AMOLED MANUFACTURING PROCESS

A block-wise representation of the complete AMOLED manufacturing process is shown in Figure 8A.9. Typically, two TFTs are required to drive each pixel, where the switching TFT store the data signal in storage capacitor to provide a voltage without time loss and then driving TFT delivers current modulated by the signal voltage to the

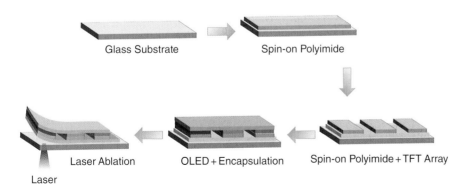

Figure 8A.8 Stepwise representation of substrate handling process in flexible AMOLED display manufacturing. This process is composed of formation of polyimide layer on the glass substrate followed by TFT array/OLED layer deposition and encapsulation. Finally, use laser treatment to detach the completed device on polyimide from the glass substrates.

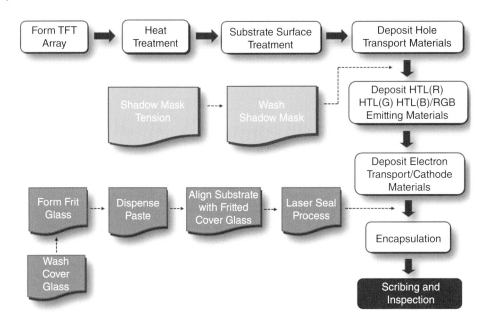

Figure 8A.9 Complete process flow of AMOLED manufacturing.

pixel for light emission. Presently, most companies use polycrystalline silicon TFT for the manufacturing of mobile device AMOLED [45].

As explained in the previous section of this chapter, the second-order micro-cavity based top-emission OLED structure is currently used for mobile AMOLED manufacturing due to its several advantages including high aperture ratio, excellent color purity, high efficiency, and so on. The fabrication of full-color OLED comprises several steps, such as substrate surface treatment, thermal evaporation of organic molecules, thin-metal cathode deposition, and coating of encapsulation layers. Currently, all display companies are using thermal evaporation under high vacuum for deposition of small molecules, electroluminescent layer and shadow mask technique for the patterning of RGB pixels. For the manufacturing of full-color OLED, RGB side-by-side patterning is done using FMM. Herein, the mask is positioned close to the glass surface to prevent shadow effect. Figure 8A.10 shows an example of shadow mask deposition with a linear source where the charge injection and transport layers are thermally deposited sequentially through an open mask under high vacuum conditions, and then R,G,B side-by-side emitter layers are formed through FMM with stripe shaped open area.

When shadow masks are used repeatedly during continuous manufacturing of full-color OLED, evaporated materials accumulate in the fine open areas of the metal mask. Such accumulation of materials on the pitch of metal mask may reduce the size of the open area. Similarly, there is also a chance of thermal expansion of the shadow mask due to the heat from the evaporation source. Therefore, proper cleaning of shadow mask and alignment adjustment between shadow mask and substrate is required after every few deposition runs to remove any unwanted effects. The distance and alignment adjustment between substrate and shadow mask are crucial in terms of high-quality resolution and device efficiency.

Encapsulation is the final step in full-color OLED manufacturing. After the deposition of the thin cathode and the thick out-coupling layer, the device needs to be encapsulated to protect from oxygen and moisture. As explained in the previous section, laser frit sealing is used in the manufacturing of rigid OLED display. Therefore, before applying this technique glass cover slide needed to be cleaned properly to prevent device from contamination. Similarly, proper alignment between OLED device substrate and fritted cover glass is

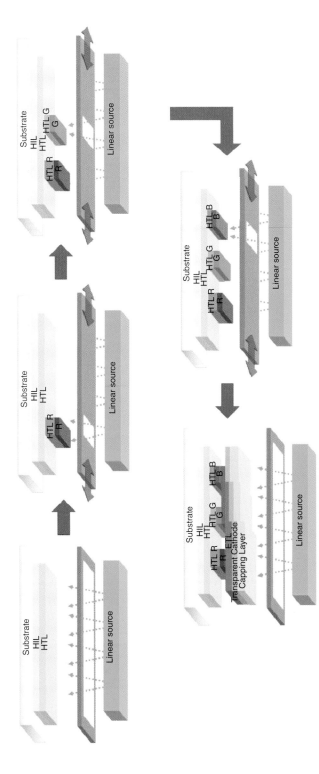

Figure 8A.10 Schematic illustration of the fabrication process of full-color OLED using metal shadow mask. The charge injection and transport layers are thermally deposited through open mask condition under high vacuum condition, and side by side emitter layer of red, green, and blue formed through metal mask with stripe shape open area.

also essential. Lastly, laser process is applied on the frit seal to join the device substrate and the glass encapsulation cover together. In contrast, the thin-film encapsulation technique is employed for flexible OLED devices, with several periods of organic/inorganic material layers. Until now, this method is known to provide the best protection compatible with the high volume production of flexible AMOLED.

8A.7 SUMMARY

AMOLED displays are extremely promising because of their outstanding performance, simple device structure, and cost-effective manufacturing process potential. In addition, thin and flexible form factors are compatible with the strong driving force in the growth of future advanced mobile and wearable devices.

References

1. J. K. Mahon, History and status of organic light emitting device (OLED) technology for vehicular applications, SID Symposium Digest of Technical Papers, 32, pp. 22–25 (2001).
2. H. Sasabe, J. Kido, Multifunctional materials in high performance OLEDs: challenges for solid state lighting, Chem. Mater., 23, pp. 621–630 (2011).
3. A. M. Bagher, OLED fabrication for use in display system and comparison with LCD and plasma, Greener Journal of Electronics and Communications, 1, pp. 001–012 (2014).
4. W. C. H. Choy, C. Y. Ho, Improving the viewing angle properties of microcavity OLEDs by using dispersive gratings, Optics Express, 15, p. 13288 (2007).
5. Y. H. Son, M. J. Park, R. Pode, J. H. Kwon, High efficiency top-emission organic light emitting diodes with second and third-order micro-cavity structure, ECS J. Solid State Sci. Technol., 5, pp. R3131–R3137 (2016).
6. M. J. Park, Y. H. Son, G. H. Kim, R. Lampande, H. W. Bae, R. Pode, Y. K. Lee, W. J. Song, J. H. Kwon, Device performance of third order micro-cavity green top-emission organic light emitting diodes, Org. Electron., 26, pp. 458–463 (2015).
7. M. J. Park, G. H. Kim, Y. H. Son, H. W. Bae, J. H. Kong, J. H. Kwon, High efficiency red top-emitting micro-cavity organic light emitting diodes, Optics Express, 22, pp. 19919–19929 (2014).
8. J. H. Kwon, RGB color patterning for AMOLED TVs, Information Display, 2/13, pp. 12–15 (2013).
9. C. W. Han, K. M. Kim, S. J. Bae, H. S. Choi, J. M. Lee, T. S. Kim, Y. H. Tak, S. Y. Cha, B. C. Ahn, 55-inch FHD OLED TV employing new tandem WOLEDs, SID Symposium Digest of Technical Papers, 43, pp. 279–281 (2012).
10. B. I. Choi, S. B. Woo, J. C. Kim, Measurement of ultra-low moisture permeation through epoxy sealing in OLED encapsulation, XX IMEKO World Congress, Metrology for Green Growth, September 9-14, Busan, Republic of Korea (2012).
11. R. S. Kumar, M. Auch, E. Ou, G. Ewald, C. S. Jin, Low moisture permeation measurement through polymer substrates fororganic light emitting devices, Thin Solid Films, 417, pp. 120–126 (2002).
12. J. J. Lih, C. L. Chao, C. C. Lee, The challenge of high resolution to active matrix OLED, SID Symposium Digest of Technical Papers, 37, pp. 1459–1462 (2006).
13. M. Hack, Richard Hewitt, J. Brown, J. W. Choi, J.H. Cheon, S. H. Kim, J Jang, Analysis of low power consumption AMOLED displays on flexible stainless steel substrates, SID Symposium Digest of Technical Papers, 38, pp. 210–213 (2007).
14. R. Q. Ma, K. Rajan, M. Hack, J. J. Brown, J. H. Cheon, S. H. Kim, M. H. Kang, W. G. Lee, J. Jang, Highly flexible low power consumption AMOLED displays on ultra thin stainless steel substrates, SID Symposium Digest of Technical Papers, 39, pp. 425–428 (2008).
15. N. H. Keum, O. K. Kwon, High resolution AMOLED pixel using negative feedback structure for improving image quality, SID Symposium Digest of Technical Papers, 44, pp. 461–464 (2013).

16. M. Y. Chang, Y. K. Han, C. C. Wang, S. C. Lin, Y. J. Tsai, W. Y. Huang, High color purity organic light emitting diodes incorporating a cyanocoumarin derived red dopant material, J. Electrochemical Soc., 155 (12), pp. J365–J370 (2008).
17. S. Hofmann, M. Thomschke, B. Lüssem, K. Leo, Top-emitting organic light emitting diodes, Opt. Express, 19(S6), pp. A1250–A1264 (2011).
18. C. J. Lee, Y. I. Park, J. H. Kwon, J. W. Park, Microcavity effect of top-emission organic light emitting diodes using aluminium cathode and anode, Bull. Korean Chem. Soc., 26, pp. 1344–1346 (2005).
19. C. J. Yang, S. H. Liu, H. H. Hsieh, C. C. Liu, T. Y. Cho, C. C. Wu, Microcavity top-emitting organic light emitting devices integrated with microlens arrays: simultaneous enhancement of quantum efficiency, Cd/A efficiency, color performances, and image resolution, Appl. Phys. Lett., 91, p. 253508 (2007).
20. H. Peng, J. Sun, X. Zhu, X. Yu, M. Wong, H. S. Kwok, High-efficiency microcavity top-emitting organic light-emitting diodes using silver anode, Appl. Phys. Lett., 2006, 88, p. 073517 (2006).
21. S. F. Hsu, S. W. Hwang, C. H. Chen, Highly efficiency top-emitting white organic electroluminescent devices, SID Symposium Digest of Technical Papers, 36, pp. 32–35 (2005).
22. C. Xiang, W. Koo, F. So, H. Sasabe, J. kido, A systematic study on efficiency enhancement in phosphorescent green, red and blue microcavity organic light emitting devices, Light Sci. Appl., 2, p. e74 (2013).
23. S. Hofmann, M. Thomschke, P. Freitag, M. Furno, B. Lüssem, K. Leo, Top-emission organic light-emitting diodes: influence of cavity design, Appl. Phys. Lett., 97, p. 253308 (2010).
24. A. W. Lu, J. Chan, A. D. Rakić, A. M. Ching, A. B. Djurišić, Optimization of microcavity OLED by varying the thickness of multi-layered mirror, Optical and Quantum Electronics, 2006, 38, pp. 1091–1099 (2006).
25. Application of organic and printed electronics (Chapter 3), Springer Publications, pp. 57–81 (2013).
26. S. N. Kumar, R. John, S. Lauer, W. Little, B. Daul, Electroforming technology for manufacturing thin metal mask with very small apertures for OLED display manufacturing, SID Symposium Digest of Technical Papers, 46, pp. 2111–2214 (2015).
27. J. Heo, H. Min, M. Lee, Laser micromachining of permalloy for fine metal mask, International Journal of Precision Engineering and Manufacturing-Green Technology, 2, pp. 225–230 (2015).
28. S. Chen, H. S. Kwok, Color filter pixel arrangement for improving the color gamut of AMOLED microdisplay, SID Symposium Digest of Technical Papers, 43, pp. 1484–1487 (2012).
29. J.S Park, H. Chae, H.K. Chung, S.I. Lee, Thin film encapsulation for flexible AM-OLED: a review, Semicond. Sci. Technol., 26 p. 034001 (2011).
30. S. Forrest, P. Burrows, M. Thompson, IEEE Spectrum 37, pp. 29-34 (2000).
31. A. Nissen, The low-temperature oxidation of calcium by water vapor, Oxidation of Metals, 11 pp. 241–261 (1977).
32. R. Paetzold, A. Winnacker, D. Henseler, V. Cesari, K. Heuser, Permeation rate measurements by electrical analysis of calcium corrosion, Rev. Sci. Instrum. 74 pp. 5147–5150 (2003).
33. G. Nisato, M. Kuilder, P. Bouten, L. Moro, O. Philips, N. Rutherford, Thin film encapsulation for OLEDs: encapsulation of multilayer barriers using the Ca test, SID Symposium Digest of Technical Papers, 34, pp. 550–553 (2003).
34. D. Yu, Y. Q. Yang, Z. Chen, Y. Tao, Y. F. Liu, Recent progress on thin-film encapsulation technologies for organic electronic devices, Optics Communications, 362, pp. 43–49 (2016).
35. J. S. Park, H. Y. Chae, H. K. Chung, S. I. Lee, Thin film encapsulation for flexible AM-OLED: a review, Semicond. Sci. Technol., 26, p. 034001 (2011).
36. C. Y. Li, B. Wei, Z. K. Hua, H. Zhang, X. F. Li, J. H. Zhang, Thin film encapsulation of OLED displays with organic-inorganic composite film, Electronic Components and Technology Conference, pp. 1819–1824 (2008).
37. S. P. Subbarao, M. E. Bahlke, I. Kymissis, Laboratory thin-film encapsulation of air-sensitive organic semiconductor devices, IEEE Transactions on Electron Devices, 57, pp. 153–156 (2010).
38. H. N. Lee, H. J. Kim, Y. M. Yoon, Thin film barriers using transparent conductive oxides for organic light emitting diodes, J. SID, 17/9, pp. 39–744 (2009).

39. S. W. Seo, H. Y. Chae, S. J. Seo, H. K. Chung, S. M. Cho, Extremely bendable thin-film encapsulation of organic light-emitting diodes, Appl. Phys. Lett., 102, p. 161908 (2013).
40. L. Zhang, S. Logunov, K. Becken, M. Donovan, B. Vaddi, Impacts of glass substrate and frit properties on sealing for OLED lighting, SID Symposium Digest of Technical Papers, 41, pp. 1890–1893 (2010).
41. S. Logunov, S. Marjanovic, J. Balakrishnan, Laser assisted frit sealing for high thermal expansion glasses, J. Laser Micrp/Nanoeng., 7, pp. 326–333 (2012).
42. D. Jiin, S. An, H. Kim, H. Koo, T. Kim, Y. Kim, H. Min, S. Kim, Materials and components for flexible AMOLED Display, SID symposium Digest of Technical Papers, 42, pp. 492–493 (2011).
43. S. Hong, J. Yoo, C. Jeon, C. Kang, J. Lee, J. Ryu, B. Ahn, Sangdeog Yeo, Technologies for flexible AMOLEDs, Information Display, 1/15, pp. 6–11 (2015).
44. J. Y. Yan, J. C. Ho, J. Chen, Foldable AMOLED display development: progress and challenges", Information Display, 1/15, pp. 12–16 (2015).
45. J. J. Lih, C. F. Sung, C. H. Li, T. H. Hsiao, H. H. Lee, Comparison of a-Si and poly-Si for AMOLED displays, J. SID, 12/4, pp. 367–371 (2004).

8B

OLED Manufacturing Process for TV Application

Chang Wook Han and Yoon Heung Tak

LG Display, E2 Block, LG Science Park 30, Magokjungang 10-ro, Gangseo-gu, Seoul, 07796, Korea

8B.1 INTRODUCTION

OLED (organic light-emitting diode) displays are rising as the next-generation display device, surpassing LCD in its color reproduction capability. In terms of performance, OLED demonstrates (1) an excellent contrast ratio as it can realize perfect black state as it's a self-light-emitting device, (2) a superb color expression characteristic as it is capable of reproducing colors accurately without distorting the image quality, (3) a wide viewing angle due to the viewing angle independent discoloration or deterioration, and (4) a crisp clear motion picture quality without delay thanks to its fast response. In terms of design, OLED offers multiple features differentiated from those of LCD. In particular, it can access relatively easy to the commercialization of curved, wall-paper type, transparent, double-sided, wave-type displays with OLED technology, as shown in Figure 8B.1.

To turn the concept of large-screen OLED TV into a tangible reality, the development of some key technologies such as the TFT backplane, OLED pixel fabrication process, and encapsulation is essential. Unlike the displays for mobile phones, OLED TV requires a technology that enables pixel generation on a large glass substrate. In the case of LCD, the technology has grown fast since the successful commercial mass production from the Gen 1 size panel ($300 \times 350\,mm^2$) until the commercial mass production of the Gen 8 size glass panels ($2500 \times 2200\,mm^2$) that can be divided into six 55-inch panels. The successful mass production using the eighth-generation oversized glass substrates increased the productivity and lowered the unit price significantly, thereby laying the groundwork for the fast growth of the OLED TV market.

The OLED pixel generation technology is divided into the RGB side-by-side method, the white OLED-color filter method [1], and the blue OLED+CCM method [2, 3], as shown in Figure 8B.2.

The RGB side-by-side method has the advantage of excellent light utilization efficiency because three primary colors (RGB) are generated by each pixel. The RGB side-by-side method is again divided into the vacuum evaporation method [4], which generates RGB pixels using the fine metal mask (FMM); the soluble method [5, 6], which generates RGB pixels by printing a liquid-type material on the panel; the laser transfer method [7, 8], which generates RGB pixels by forming a thin layer on the donor board via metallization or printing followed by laser beam patterning onto the layer; and organic vapor jet printing (OVJP) [9], which jets material onto the RGB pixels via a heated nozzle.

The white OLED+color filter (CF) method emits lights by filtering the white-light spectrum with the RGB color filter. Its overall performance relies significantly on the efficiency of the white OLED and its spectrum.

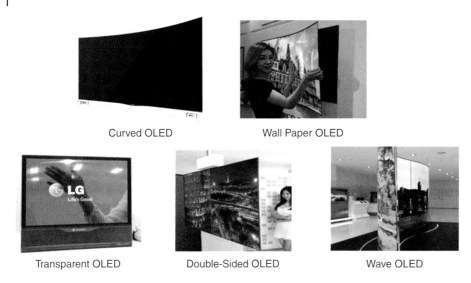

Figure 8B.1 Different types of OLED displays.

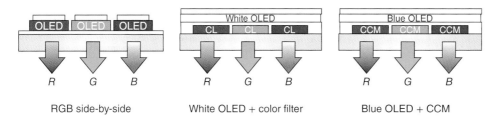

Figure 8B.2 Different OLED pixel generation technologies.

The blue OLED+color change medium (CCM) method turns the short-wavelength blue spectrum into long-wavelength red and green light via CCM [2, 3].

In this chapter, the FMM and white-OLED technology, the current two methods applied to the OLED TVs, will be introduced.

8B.2 FINE METAL MASK (FMM)

In the FMM method, hole transport layer (HTL), electron transport layer (ETL), and cathode will be deposited as common layers by using an open mask, as shown in Figure 8B.3, but the RGB light-emitting array layer will be generated by depositing red, green, and blue on the desired pixels by attaching the FMM close to the panel. This method has been used with Gen. 6 half size glass panels for producing portable device displays, such as smartphones and smart watches, but it is yet to be applied to the manufacturing of OLED TVs.

As explained earlier, the production technology capable of generating pixels with the Gen 8 size line is essential for the production of OLED TVs with competitive price points. However, the weight of the metal mask increases with the growing size of the glass substrate, thereby creating a gap between the panel and the metal mask called "sagging." The gap between the panel and the metal mask should be kept less than 5 μm to flawlessly generate red, green, and blue pixels. In fact, the gap between the $1210 \times 400\,mm^2$ panel and the metal mask was measured 440 μm of serious sagging, which would lead to intermixing of the red, green, and blue colors (see Figure 8B.4). If the size of the panel grows even larger to a Gen 8 size line, this

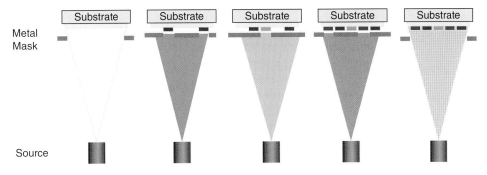

Figure 8B.3 Process flow of the RGB patterning by FMM method.

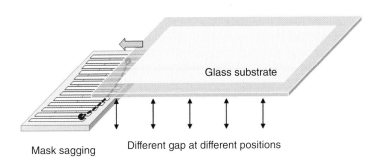

Figure 8B.4 Mask sagging observed in the FMM method.

problem will become even more serious and practically limit the production of 4K and 8K high-resolution displays.

In the 2013 SID Forum, a team from AUO proposed a 65-inch panel fabrication method that relies on the FMM technology [4]. A Gen 6 size full-size glass panel ($1500 \times 1800\,\text{mm}^2$) capable of yielding two 65-inch FHD displays was used for pilot production. Figure 8B.5 shows an OLED process with FMM in which 65-inch displays were produced, with the 65-inch prototype having up to 200nit brightness and FHD resolution (1920×1080) shown in Figure 8B.6.

Small-mask scanning (SMS), also an FMM method, was developed by Samsung Display Co. for OLED TV fabrication purpose [10]. The SMS method is capable of generating independent RGB pixels with performance level equivalent to OLEDs produced via the FMM method, while preventing the metal mask sagging problem with the use of minimized mask size. Thereby SMS is theoretically an ideal technology for the production of OLED TVs. In the SMS method, the emitting layer is patterned by fixing the evaporation source and the mask in place while the panel is moving as shown in Figure 8B.7, whereas in the FMM method, the emitting layer is patterned with the panel and the metal mask fixed in place.

Some prototype 55-inch FHD displays were made via the SMS method, however, it is known that the SMS method did not make successful production scale.

Figure 8B.8 shows an OLED TV produced via the SMS method. It is a 55-inch TV with its full-white brightness measured up to $150\,\text{cd/m}^2$ and 1920×1080 FHD resolution.

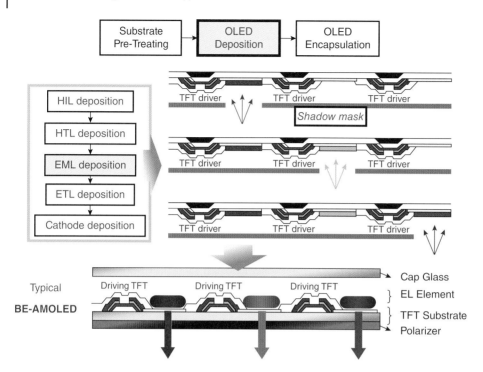

Figure 8B.5 Process flow of the FMM method for large-sized OLED TV fabrication [4].

Figure 8B.6 65-inch FHD OLED TV prototype produced via the FMM method [4].

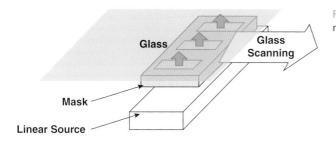

Figure 8B.7 RGB patterning process based on the SMS method [10].

Figure 8B.8 55-inch FHD OLED TV produced via the SMS method.

8B.3 MANUFACTURING PROCESS FOR WHITE OLED AND COLOR FILTER METHODS

As the white OLED and CF methods do not use FMM for emitter deposition process but adopt the open mask and avoid potential mask sagging issues, this White OLED+CF method is regarded as a suitable technology for producing large OLED TV panels on Gen 8 size mother glass. LG Display Co. realized OLED TV production employing the white OLED and CF methods. This chapter will introduce white OLED-based OLED TV panels manufacturing process.

Figure 8B.9 shows a cross-sectional view of the panel that incorporated a TFT backplane, a color layer, white OLED layer, and encapsulation layer. The panel structure is based on the bottom-emission type, in which the light of the OLED is emitted toward the TFT layer.

The process flow of the panel production before OLED deposition starts with the fabrication of oxide TFT in the Gen 8 line, as shown in Figure 8B.10, which is then followed by the formation of an RGB color filter layer via photo-lithography. If a red color layer as a light shield layer is formed on the TFT, it can prevent degradation of TFTs by the lights emitted from the OLED. Next, the overcoat layer is formed using a high-molecular polymer material. The overcoat layer is formed for three purposes: (1) it prevents any solvent ingredient left on the color layer from diffusing into the OLED layer and deteriorating it during operation, (2) this layer planarizes the surfaces of the red, green, and blue pixels, (3) it can work as a white subpixel in WRGB pixel structures. As there is no color filter layer in the white pixels, the empty space is filled with

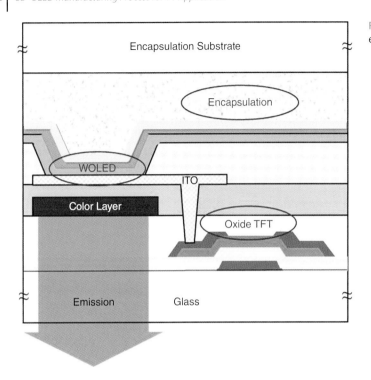

Figure 8B.9 Cross-sectional view of the bottom-emission type OLED TV panel.

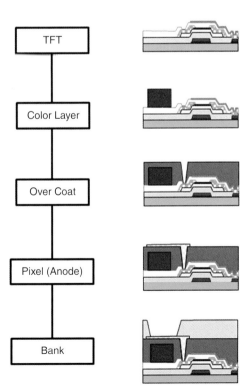

Figure 8B.10 Production flow of TFT/CF/OC/anode/bank processes.

overcoat material, making it a white subpixel, for this reason, the overcoat layer should be thicker than the color layers. The white OLED improves the light efficiency by using four WRGB subpixels and thus improve the overall display power efficiency. In general, the WRGB four-pixel system consumes 40% less energy than the RGB three-pixel system. Next, an organic insulator film is used to set apart different colors with a blank layer, and to prevent short circuit between the electrodes. The organic insulator film should be selected among those materials that would not absorb moisture and that show minimum outgassing [11].

Next, the white OLED manufacturing process based on the bottom-emission structure will be introduced. The panels produced via the production process shown in Figure 8B.10 are put into the WOLED deposition/encapsulation processes after a cleaning process. The process flow is shown in Figure 8B.11. As a first step in the deposition system, plasma treatment is applied to the panel with the anode pattern formed. For effective surface cleaning, wet treatment [12], plasma treatment [13], and UV ozone treatment [14] have been developed to lower the hole injection barrier by removing the contaminants on the ITO surface and by increasing the work functions of the ITO. Among such treatments, plasma treatment (e.g., oxygen, nitrogen, and argon plasma treatment) is known to be effective in lowering the hole injection barrier. Next step is depositing the organic materials and cathode metallic layer using a thermal evaporator. In the case of white OLED, deposition is performed using the open mask instead of FMM. To prevent the potential deformation of FMM due to the high heat generated by the organic material evaporation, the panel and the evaporation source should be separated with a sufficient distance. In this case, the organic materials evaporated from the source are deposited not only on the FMM but also on the vacuum chamber wall, thereby increasing the material loss and lowering the material utilization efficiency. The white OLED using the open mask, however, can narrow the distance between the panel and the evaporation source, thereby pushing up the material utilization efficiency. Further, Al (aluminium) is often used as a cathode layer because it has an excellent electron injection property. OLED needs an encapsulation layer because its properties deteriorate upon exposure to air or moisture. An encapsulation layer, a glass or metal panel, is formed on top of the cathode after filling the gap using high-molecular polymer materials.

8B.3.1 One-Stacked White OLED Device

Emission from more than one molecular species is typically required to create a white spectrum. Two methods can be employed to realize the excitation. First, the host and the dopant can be deposited on the same layer where energy is transferred from the host to the dopant, or it can be so configured as to make charge trapping

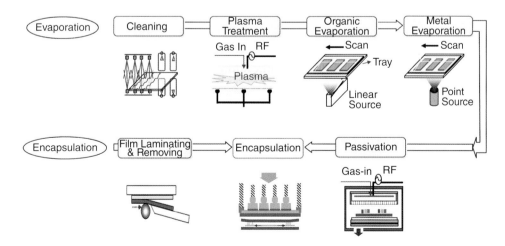

Figure 8B.11 Production flow of white OLED deposition/encapsulation processes.

occur in the dopant. Second, the location of the excitation recombination zone may be adjusted using the carrier blocking layer [15]. Figure 8B.12 shows the one-stacked white OLED applied with BCP interlayer and DCM2 dopant. The white spectrum was realized on this OLED by adjusting the intensity of the blue and red colors depending on the thickness of the BCP, and by changing the lighting wavelength according to the concentration of DCM2 [15]. The OLED demonstrated the following performance data: up to 11.5V, up to 1.23cd/A current efficiency, and up to 0.31lm/W power efficiency when the brightness was rated at 100cd/m^2.

Figure 8B.13 shows a white OLED that adopted the four-color RGBY emitter system. The OLED demonstrated the following performance data: up to 3.03V and up to 15.1cd/A current efficiency [16].

Figure 8B.14 shows a white OLED device where fluorescent blue was positioned under the phosphorescence green and red mixing layers and an interlayer placed between the two light-emitter layers [17]. The device demonstrated up to 18.7cd/A efficiency and (0.31, 0.35) color coordinates when the red doping was 0.4%, and up to 20cd/A efficiency level with (0.29, 0.37) color coordinates when the red doping was 0.2%.

Figure 8B.15 shows the spectra of the white OLED and of the color filters, as well as the emission spectra and efficiency for the blue, green, and red subpixels through color filter.

Figure 8B.16 shows a white OLED used in medical displays. Phosphorescent yellow and fluorescent blue were applied to the OLED. The intensity of the yellow and blue unit can be controlled by changing the thickness of the HTL. The OLED demonstrated an up to 16.7cd/A efficiency level with (0.305, 0.317) color

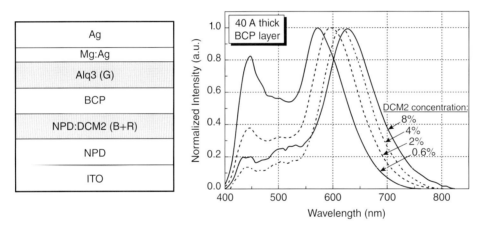

Figure 8B.12 One-stacked white OLED structure and effect of the DCM2 concentration in the NPD on the device spectrum [15].

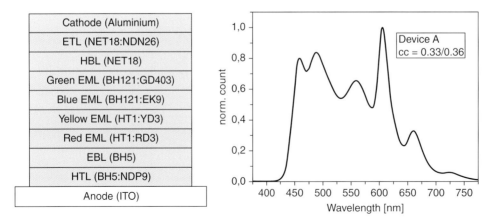

Figure 8B.13 Fluorescent white PIN white OLED and EL spectrum [16].

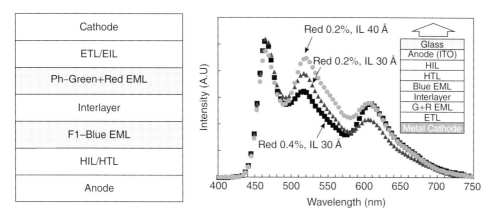

Figure 8B.14 A white OLED with phosphorescent green, red, and fluorescent blue emitters. EL spectrum dependence on red doping and the thickness of the interlayer [18].

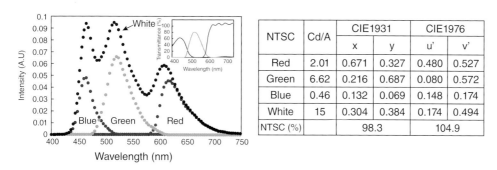

NTSC	Cd/A	CIE1931		CIE1976	
		x	y	u'	v'
Red	2.01	0.671	0.327	0.480	0.527
Green	6.62	0.216	0.687	0.080	0.572
Blue	0.46	0.132	0.069	0.148	0.174
White	15	0.304	0.384	0.174	0.494
NTSC (%)		98.3		104.9	

Figure 8B.15 EL spectrum and color filter spectrum of the white OLED [18].

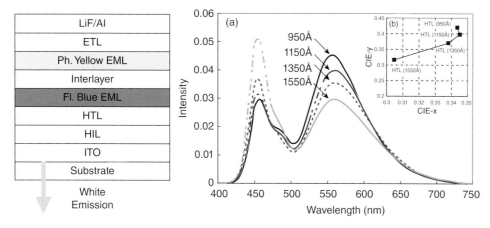

Figure 8B.16 A one-stacked white OLED to which phosphorescent yellow and fluorescent blue were applied. EL spectrum dependence on the thickness of the HTL [18].

coordinates when the thickness of HTL was 1550 Å, thereby making it possible to produce medical displays with a 7080 K color temperature.

The one-stacked white OLED has a simple structure that helps lower the material and equipment investment cost, but due to its low energy efficiency, its brightness level is low and corresponding power consumption high. Moreover, as the color temperature of the white OLED is about 6000 K, it demands more current to meet the required over 10,000 K OLED TV color temperature.

8B.3.2 Two-Stacked White OLED Device

Due to its low efficiency, the one-stacked white OLED has low brightness and consumes more power. To address these problems, two-stacked white OLED, which consists of two OLEDs in series connection, was devised. Figure 8B.17 shows the schematic structure of the two-stacked white OLED where the fluorescent blue unit and the phosphorescent red and green unit are interconnected via the charge generation layer (CGL) [19]. The OLED demonstrated the following performance data: an up to 20.7 cd/A efficiency, an up to 14.1 V voltage level, with (0.26, 0.34) color coordinates at 1,000 cd/m^2.

CGL is composed of the p-type and n-type, which play the role of the anode and cathode, respectively, and injects electric charges, making it easy to inject opposite electric charges to the adjacent sub-OLED units. In general, CGL should be transparent and capable of generating a low-voltage junction with the adjacent OLED unit. Figure 8B.18 shows the schematic structure of the two-stacked white OLED [20] that adopted Li-doped p-n connector and Li-free p-n connector CGLs. Their efficiency was 33 cd/A in both cases, but their driving voltages were 6.1 and 6.7 V, respectively, at 1,000 cd/m^2. The voltage of the organic Li-free doped CGL was about 0.6 V higher than that of the other, and the Li-free p-n connector CGL needs an additional buffer layer to prevent the interlayer diffusion of the dopant.

Figure 8B.19 shows the schematic structure of the two-stacked white OLED consisting of two units: a fluorescent blue unit and a phosphorescent green and red mixing unit [21]. Figure 8B.20 shows their respective EL spectrum and operating efficiency. The fluorescent blue unit demonstrated up to 8.1 cd/A efficiency, the phosphorescent green and red mixing unit each showed an up to 44.8 cd/A efficiency, and the white unit

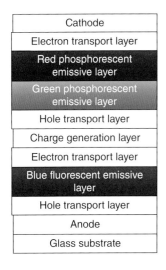
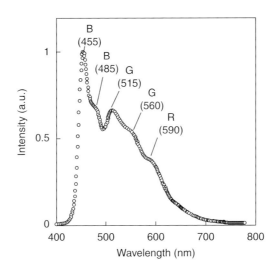

Figure 8B.17 Two-stacked white OLED to which phosphorescent green, red, and fluorescent blue were applied, and EL spectrum [19].

8B.3 Manufacturing Process for White OLED and Color Filter Methods

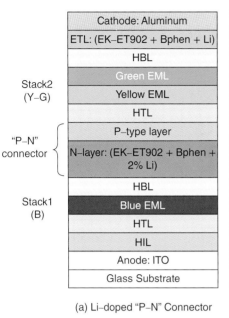

Figure 8B.18 Two-stacked white OLED with (a) Li-based and (b) Li-free CGLs [20].

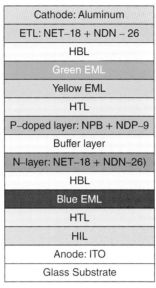

Figure 8B.19 Schematic structure of the (a) blue, (b) R+G, and (c) two-stacked white OLED [21].

demonstrated an up to 7.5 V performance level with up to 56.7 cd/A efficiency at a 10 mA/cm^2 current density.

Figure 8B.21 shows a spectrum of the color filter well matching the EL spectrum of the two-stacked white OLED so that it could realize an excellent color reproduction. As the excitation energy is transmitted from the green dopant to the red dopant when red and green are simultaneously doped on the EML, the color may change easily according to a miniscule change in the red-dopant content. Typically, green dopant 10% and red dopant 0.2% are doped with the color coordinate of the R+G stack (0.445, 0.532). However, if the red dopant is doped by 0.1% the color coordinates shift to (0.411, 0.562) with perceived color becoming greenish-yellow. Alternatively, if the red doping is at the 0.3% level the color coordinates shift to (0.501, 0.482) with the perceived color becoming reddish-yellow. Therefore, the R+G EML type is not applicable to commercial OLED TVs because its color uniformity is not satisfactory in large OLED panels.

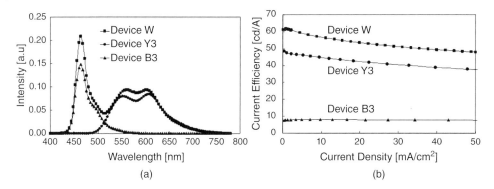

Figure 8B.20 Comparison of the operational characteristics of the three devices W, B3, and Y3: (a) EL spectra; and (b) current density versus current efficiency [21].

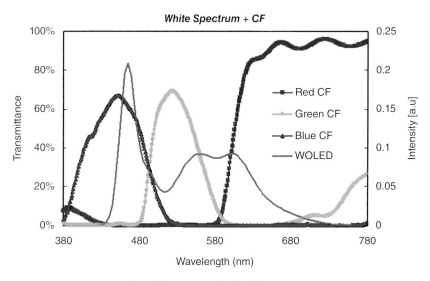

Figure 8B.21 Spectrum of the white OLED and color filter [21].

Cathode (Al)	
ETL	
Ph. YG EML	
HTL	
CGL	
ETL	
FL. Blue EML	
HTL	
HIL	
Anode (ITO)	
Substrate	

Voltage	7.1 (V)
Efficiency	78.7 (cd/A)
EQE	32.5 (%)
CIE (X,Y)	0.318, 0.331

Figure 8B.22 Schematic structure and electrical characteristics of the two-stacked white OLED consisting of fluorescent blue and phosphorescent yellow green stacks [22].

Figure 8B.23 Photograph of the 55-inch FHD OLED TV [23].

Table 8B.1 Specifications of the 55-Inch OLED TV [23].

Item	Content	Unit
Panel size	55	inch
Resolution	1920(RGBW) × 1080	-
Brightness	>400	Cd/m^2
Contrast ratio	>100,000: 1	-
White Color	0.285, 0.294	
Color Gamut (ATSC)	118	%
Thickess	4	mm

Figure 8B.22 shows a schematic structure of the two-stacked white OLED consisting of fluorescent blue and phosphorescent yellow green stacks, which was proposed to raise color uniformity, with its performance indicators being 7.1 V and an up to 78.7 cd/A efficiency at $10\,mA/cm^2$ current density [22]. The yellow green stack is capable of realizing the red and green colors with color filter, respectively, because its peak wavelength is 560 nm. It can also attain excellent color uniformity in the Gen 8 size glass panels.

Figure 8B.23 and Table 8B.1 show a 55-inch OLED TV model and its specifications, which was made with a two-stacked white OLED device composed of fluorescent blue and phosphorescent yellow green units.

As the two-stacked white-OLED device can raise its efficiency over threefold despite the fact that it doubles the number of layers, the device can be applied to OLED TVs with $100\,cd/m^2$ brightness at full white and $400\,cd/m^2$ at peak white.

8B.3.3 Three-Stacked White-OLED Device

Figure 8B.24 shows a schematic structure of the three-stacked white-OLED device composed of two fluorescent blue units and one phosphorescent yellow green unit with the following performance data: $81.2\,cd/A$ current efficiency at $10\,mA/cm^2$. The two-stacked white-OLED device composed of fluorescent blue and phosphorescent yellow green stacks demonstrates relatively lower efficiency in the blue stack, thereby creating a bottleneck in raising the brightness of the OLED TV. Therefore, the 77-inch ultra-high-definition OLED TV (Table 8B.2) with 150 nit full-white brightness and 450 nit peak brightness, as shown in Figure 8B.25, becomes feasible if the three-stacked white OLED composed of two fluorescent blue units is adopted.

As the three-stacked white OLED uses two blue stacks, it is feasible to produce OLED units with $150\,cd/m^2$ brightness at full white and $450\,cd/m^2$ at peak white by improving the efficiency of the blue units. Furthermore,

Cathode
EIL
ETL
B EML
HTL
CGL
ETL
YG EML
HTL
CGL
ETL
B EML
HTL
HIL
Anode
Glass

Figure 8B.24 Schematic structure of a three-stacked white OLED consisting of two fluorescent blue units and one phosphorescent yellow green unit.

Table 8B.2 Specifications of the 77-Inch OLED TV [24].

Item	Content	Unit
Display type	RGBW OLED	
Panel size	1714 × 978	mm
Resolution	3840 × 2160 (UHD)	
Brightness	150 (full) / 450 (peak)	cd/m^2
Color gamut	118 (BT. 709)	%
Curve	R = 5000	mm

Figure 8B.25 Photograph of the 77-inch UHD OLED TV [24].

the device also allows the extension of the blue unit as it can lower the current density to the level required by the blue unit. However, further studies are required to lower the driving voltage because its driving voltage is too high.

References

1. J. P. Spindler, T. K. Hatwar, M. E. Miller, A. D. Arnold, Lifetime- and power enhanced RGBW displays based on White OLEDs", SID Symposium Digest, Vol. 36, pp. 36-39 (2005).
2. A. R. Duggal, J. J. Shiang, C. M. Heller, D. F. Foust, Organic light-emitting devices for illumination quality white light, Applied Physics Letter, 80, pp. 3470–3472 (2002).
3. A. P. Ghosh, W. E. Howard, I. Sokolik, R. Zhang, V. M. Shershukov, A. V. Tolmachev, N. I. Voronkina, V. A. Dudkin, Color changing materials for OLED microdisplays, SID Symposium Digest, 31, pp. 983–985 (2000).
4. C.-Y. Chen, L.-F. Lin, J.-Y. Lee, W.-H. Wu, S.-C. Wang, Y. M. Chiang, Y.-H. Chen, C.-C. Chen, Y.-H. Chen, C.-L. Chen, T.-H. Shih, C.-H. Liu, H.-C. Ting, H.-H. Lu, L. T., H.-S. Lin, L.-H. Chang, Y.-H. Lin, A 65-inch smorphous oxide thin film transistors active-matrix organic light-emitting diode television using side by side and fine metal mask technology, SID Symposium Digest, 44, pp. 247–250 (2013).
5. M. O'Regan, Solution processed OLED sisplays: advances in performance, resolution, lifetime and appearance, SID Symposium Digest, 40, pp. 600–602 (2009).
6. Y. Iizumi, Y. Kobayashi, T. Tachikawa, H. Kishimoto, K. Itoh, H. Kobayashi, N. Itoh, S. Handa, D. Aoki, T. Miyake, A novel hole-injection layer with the ability of forming hydrophobic-hydrophilic patterns, SID Symposium Digest, 36, pp. 1660–1663 (2005).
7. S. T. Lee, M. C. Suh, T. M. Kang, Y. G. Kwon, J. H. Lee, H. D. Kim, H. K. Chung, LITI (Laser Induced Thermal Imaging) Technology for High-Resolution and Large-Sized AMOLED, SID Symposium Digest, 38, pp. 1588–1591 (2007).
8. T. Hirano, K. Matsuo, K. Kohinata, K. Hanawa, T. Matsumi, E. Matsuda, R. Matsuura, T. Ishibashi, A. Yoshida, T. Sasaoka, Novel laser transfer technology for manufacturing large-sized OLED Displays, SID Symposium Digest, 38, pp. 1592–1595 (2007).
9. M. Shtein, P. Peumans, J. B. Benziger, S. R. Forrest, Micropatterning of small molecular weight organic semiconductor thin films using organic vapor phase deposition, Applied Physics Letter, 93, pp. 4005–4016 (2003).
10. J. H. Kwon, RGB color patterning for AMOLED TVs, Information Display, 29, pp. 12–15 (2013).
11. H. Shindo, T. Tsutsumi, T. Sakurai, M. Hanmura, A. Honma, Transmissive low outgassing organic insulator suitable for various OLED displays, SID Symposium Digest, 43, pp. 1538–1541 (2012).
12. F. Li, H. Tang, J. Shinar, O. Resto, S. Z. Weisz, Effects of aquaregia treatment of indium–tin–oxide substrates on the behavior of double layered organic light-emitting diodes, Applied Physics Letter, 70, pp. 2741–2743 (1997).
13. C. C. Wu, C. I. Wu, J. C. Sturm, A. Kahn, Surface modification of indium tin oxide by plasma treatment: An effective method to improve the efficiency, brightness, and reliability of organic light emitting devices, Applied Physics Letter, 70, pp. 1348–1350 (1997).
14. W. Song, S. K. So, D. Wang, Y. Qiu, L. Cao, Angle dependent X-ray photoemission study on UV-ozone treatment of indium tin oxide, Applied Surface Science, 177, pp. 158–164 (2001).
15. R. S. Deshpande, V. Bulović, S. R. Forrest White-light-emitting organic electroluminescent devices based on interlayer sequential energy transfer, Applied Physics Letter, 75, pp. 888–890 (1999).
16. S. Murano, E. Kucur, G. He, J. Blochwitz-Nimoth, T. K. Hatwar, J. Spindler, S. Van Slyke, White fluorescent PIN OLED with high efficiency and lifetime for display applications, SID Symposium Digest, 40, pp. 417–419 (2007).
17. H.-S. Choi, H. K. Kim, H. S. Pang, S. H. Pieh, C. J. Sung, M.-S. Kim, C.-W. Han, Y. H. Tak, White OLED panel with RGBW color filters based on dual-plate OLED display (DOD) structure, SID Symposium Digest, 42, pp. 1748–1751 (2009).

18 C. J. Sung, J. J. Kim, J. M. Lee, H.-S. Choi, Fabrication of simple white OLED with high color temperature for medical display applications, IMID Digest, pp. 489–492 (2009).
19 S. Ishihara, K. Masuda, Y. Sakaki, H. Kotaki, S. Aratani, High-efficiency white organic light-emitting diodes with a two-stack multi-photon emission structure, SID Symposium Digest, 40, pp. 1501–1503 (2007).
20 T. K. Hatwar, J. P. Spindler, W. J. Begley, D. J. Giesen, D. Y. Kondakov, S. Van Slyke, S. Murano, E. Kucur, G. He, J. Blochwitz-Nimoth, High-performance tandem white OLEDs using a li-free "P-N" connector, SID Symposium Digest, 42, pp. 499–502 (2009).
21 C.-W. Han, Y.-H. Tak, B.-C. Ahn, 15-in. RGBW panel using two-stacked white OLED and color filters for large-sized display applications, Journal of the SID, 19, pp. 190–195 (2011).
22 C.-W. Han, J.-S. Park, H.-S. Choi, T.-S. Kim, Y.-H. Shin, H.-J. Shin, M.-J. Lim, B.-C. Kim, H.-S. Kim, B.-S. Kim, Y.-H. Tak, C.-H. Oh, S.-Y. Cha, B.-C. Ahn, Advanced technologies for UHD curved OLED TV, Journal of the SID, 22, pp. 552–563 (2015).
23 C.-H. Oh, H.-J. Shin, W.-J. Nam, B.-C. Ahn, S.-Y. Cha, S.-D. Yeo, Technological progress and commercialization of OLED TV, SID Symposium Digest, 46, pp. 239–242 (2013).
24 H.-J. Shin, S. Takasugi, K.-M. Park, S.-H. Choi, Y.-S. Jeong, H.-S. Kim, C.-H. Oh, B.-C. Ahn, Technological progress of panel design and compensation methods for large-size UHD OLED TVs, SID Symposium Digest, 47, pp 720–723 (2014).

9

OLED Encapsulation Technology

Young-Hoon Shin

OLED Technology Development Division, LG Display, Co., Ltd. 245, LG-ro, Wollong-myeon, Paju-si, Gyeonggi-do, 413-779, Korea

9.1 INTRODUCTION

It is a well-known fact that OLED is very vulnerable to moisture. Moisture triggers the formation of so-called "dark spots" or non-luminous areas by hydrolysing the electrode materials or the electron injection layer [1, 2]. Sometimes, OLED displays would show pixel shrinkage along the moisture infiltration path on the panel on which the pixels are arranged. The key objective for the development of OLED encapsulation technologies is to enable moisture-proof properties in OLED devices by using a deep understanding of the mechanisms causing their extreme moisture sensitivity.

The first commercialized OLED encapsulation technology was developed by Tohoku Pioneer in Japan, which had already been mass-producing passive-matrix OLEDs. The company came up with an OLED structure in which a tape-type desiccant is put inside a metal can whose misfits have been leveled by a pressing machine, and sealant is dispense on the outskirts of the metal can before it is compressed onto the TFT panel. The most frequently adopted encapsulation technology as of 2015 is the frit seal, which is applied to smartphone displays, and the face seal, which is applied to large OLED TVs. Both technologies are perceived by the OLED panel manufacturers as their key competitive advantage that would guarantee them a head start in the race to mass-produce OLED panels. The encapsulation technology has been continuously developed and perfected as the base technology that would enable the mass production of lighter and thinner displays. Lately, R&D efforts have been focused on developing new manufacturing methods that can help increase the size and flexibility of OLED displays.

In this chapter, definition and classification of the encapsulation technology shall be discussed first, which is then followed by in-depth discussion on the detailed structures and manufacturing processes of various technologies as well as a brief introduction of the future direction of the encapsulation technology.

9.2 PRINCIPLES OF OLED ENCAPSULATION

As briefly mentioned in the introduction, OLED devices have been known to be extremely susceptible to moisture. The cathode layer included in OLED devices would easily corrode when exposed to moisture and oxygen, while an electron transport layer also forms oxide that would hinder the free transport of electric charges and hence prevent OLED devices from emitting light. The susceptibility to moisture of OLED devices is shown in Figure 9.1.

Flat Panel Display Manufacturing, First Edition. Edited by Jun Souk, Shinji Morozumi, Fang-Chen Luo, and Ion Bita.
© 2018 John Wiley & Sons Ltd. Published 2018 by John Wiley & Sons Ltd.

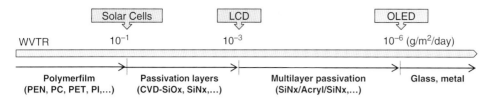

Figure 9.1 Comparison of water vapor transmission rate (WVTR) by application.

Among optoelectronic devices, it is known that while a packaging technology with about 10^{-1} g/m^2/day WVTR level is required for solar cells, for LCDs a WVTR of up to 10^{-2}~10^{-3} g/m^2/day is required to prevent contamination of the liquid crystal and malfunction of the TFT array. In the case of OLED, however, it has been reported that a moisture proof technology with an unprecedented WVTR level of up to 10^{-6} g/m^2/day is required in theory due to the unique properties of OLED devices. What is notable here is that the WVTR level of up to 10^{-6} g/m^2/day is such an extreme target that even MOCON, one of the most broadly adopted commercial WVTR measuring instruments, with its measuring limit of up to 5×10^{-4} g/m^2/day, would fail to measure it. Nevertheless, many companies eventually produced OLED panels by coming up with a variety of creative ideas to overcome this challenge, and by applying the optimal encapsulation technology to OLED devices.

Some of the technologies that have been developed and applied to the panels for this purpose are the following: (1) absorbing or eliminating the moisture that infiltrated the cavity of the OLED, using desiccant; (2) delaying the moisture infiltration of the device by filling the inner cavity to prevent the further spread of the moisture; and (3) realizing near-perfect sealing to achieve an up to 10^{-6} g/m^2/day WVTR level.

As such, it should be emphasized that a high-performance encapsulation technology that is an order-of-magnitude higher than what is required for conventional electronic devices is required for OLED panel manufacturing.

9.2.1 Effect of H$_2$O

Figure 9.2 shows actual dark spots formed in OLED due to moisture infiltration. The luminescent samples in the pictures below were exposed to the atmosphere right after the bottom electrodes, organic luminescent layer, and top electrodes were deposited on the panel, without going through the encapsulation process. The exposure time was 30 min, 1 h, and 2 h for (a), (b), and (c), respectively. The dark spots started to form just a few minutes after they were exposed to the atmosphere, and grew large enough to be visually detectable throughout the front side. In short, the experiment proved that OLED will start to develop a non-luminous area on the panel due to the airborne moisture within just a few hours after

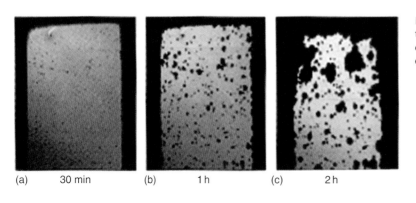

Figure 9.2 Change of the dark spots in the OLED device without encapsulation over time due to exposure to the atmosphere.

it is exposed to the atmosphere, and what actual form the dark sport will eventually take. It can be seen in the pictures below that the non-luminous area grew within 2 h to occupy near 30% of the entire panel area.

The formation and growth pattern of such dark spots may vary depending on the moisture infiltration path and on the panel structure. For instance, dark spots with random shapes and locations, as shown in Figure 9.4(a), will be formed in the edge seal structure where the bottom panel metalized with OLED and the top panel for encapsulation will be combined via the dispensing of sealant onto the edge of the panel, and whose inside is filled with inert gases as the moisture will spread all across the panel within the inner cavity once the external moisture infiltrates the sealant (Figure 9.3(a)).

On the contrary, in the dam and fill or face seal structure, where no cavity will be left between the bottom and top substrates as a result of the application of stuffing resin (filling material) or of an adhesive sheet, the external moisture will permeate the panel via the hardened solid-type inner stuffing material (Figure 9.3(b)). In such case, the non-luminous area will grow from the four sides facing the outside atmosphere at a constant rate in the same direction as that in which the moisture infiltrated the panel (Figure 9.4(b)).

Furthermore, in the case where a passivation layer is formed, as shown in Figure 9.3(c), the moisture will spread via the stuffing material, thereby forming a non-luminous area around the panel edge. The non-luminous area, however, will grow in a circular form at the defects in the passivation layer near the panel edge as shown in Figure 9.4(c), because the moisture should pass through the defect site in the passivation layer given that the direct contact of the moisture with the OLED will be blocked by the passivation layer.

Figure 9.5 shows microscopic images of an actual shrinkage caused by the moisture that infiltrated a display panel with fine arrays of pixels. Images (a) and (b) in Figure 9.5, respectively, correspond to images (b) and (c) in Figure 9.4.

To recap, it should be emphasized that OLED devices are extremely susceptible to moisture, so much so that a miniscule amount of moisture infiltration can cause deficiencies in product reliability, such as dark spots or pixel shrinkage. The physical forms of such defects are closely associated with the encapsulation method and the active moisture infiltration mechanism.

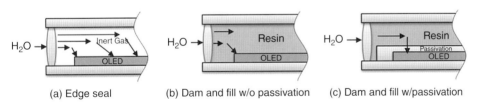

Figure 9.3 Moisture infiltration path depending on the encapsulation structure (cross-sectional view).

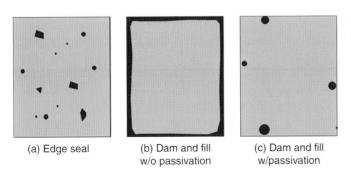

Figure 9.4 Formation of dark spots or contraction depending on the encapsulation structure.

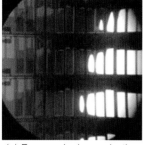

(a) Face seal w/o passivation   (b) Face seal w/passivation

Figure 9.5 Pixel shrinkage pattern depending on the presence or absence of a passivation layer.

9.3 CLASSIFICATION OF ENCAPSULATION TECHNOLOGIES

The diverse encapsulation technologies that have been developed or applied for mass production can be classified in multiple ways depending on the criteria used. Figure 9.6 shows a general classification of the key encapsulation technologies applied for mass production or the present condition in the R&D phase.

First, an encapsulation technology can be classified mainly as a cavity structure or a non-cavity structure, depending on the presence of an empty space between the encapsulation substrate and the TFT substrate. The encapsulation types with a cavity structure include the UV edge seal and the frit seal. These types have been applied in the majority of the small-flat PM OLED panels that came into existence beginning from the birth of the OLED industry, down to the present day. Such types did not need a passivation layer for blocking moisture because either moisture absorbing desiccant was added inside the cavity of panel with UV edge seal, or a hermetic sealant was used. Nevertheless, they demonstrate excellent stability in terms of reliability. Still their critical drawback is that it is difficult to make them flexible or larger.

On the other hand, no-cavity structure whose inner space is filled with resin to lessen the deficiencies of the cavity structure, a separate moisture passivation layer is usually added to the OLED device. If the vacuum processes such as Sputter, PECVD, or ALD are added to deposit passivation layer, the no-cavity structure incurs cost overrun related to equipment investment and increased process complexity. Despite such shortcomings, the no-cavity structure is considered indispensible for producing large and flexible displays because it allows the producers to prevent the scratches caused by direct contact of the two panels (e.g. Cover glass and TFT glass), as well as the resulting faulty pixels. For this reason, people across the industry are working hard to develop a no-cavity-structure encapsulation technology that is cost-effective, highly reliable for the mass production of flexible and large-screen displays.

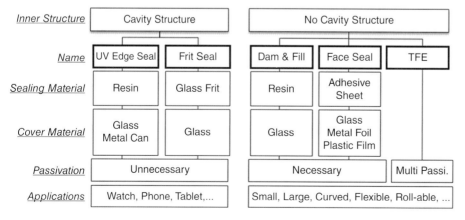

Figure 9.6 Classification of encapsulation technologies.

In the succeeding sections, the materials, processes, and applications associated with the key encapsulation technologies classified earlier will be discussed in further detail.

9.3.1 Edge Seal

As mentioned previously in the introduction, in 1997, Tohoku Pioneer in Japan started the first mass production of OLED panels for use as a car audio display screen. The model, which was used to display various icons based on the passive-matrix driving method, was significant in that it was the first such display based on the encapsulation technology with sufficient reliability and with productivity good enough for mass production. The edge-sealed panel structure that was adopted by Tohoku Pioneer is shown in Figure 9.7(a).

Figure 9.7(a) shows an encapsulation structure in which specific-sized depressions were formed by pressing the metal cap, and desiccant is fixed inside the freshly formed depressions, thereby preventing them from physical contact with the OLED. The method, which used the metal cap processed as such, was called "Metal Can Encapsulation (MCE)". Applied to the most of early OLED devices, it was used mainly in producing low-resolution PMOLED panels adopted by numerous mobile handsets. The structure shown in Figure 9.7(b) belonged to the Glass Edge Seal, which was thinner and simpler than the metal MCE. This full-glass structure made OLED panels thinner and improved their productivity compared to the metal can method, by using the chemical etching process to form depression in the encapsulation glass.

Regardless of whether a metal can or glass cap is used as the capping material, they are sealed with the TFT glass using moisture-proof sealant dispensed onto the perimetric edges. But, no matter how the sealant's moisture-proof property is excellent, external moisture can infiltrate the panel only through the edge sealant (not metal can nor glass cap).

Epoxy resins are typically used as the edge sealant material due to their excellent chemical resistance and storage stability. Another benefit of epoxy resins is that it does not shrink significantly or generate volatile substances. The WVTR of the epoxy sealant is about several to values of $g/m^2/day$, although it may vary by maker. If the sealant is dispensed onto 1 to 2 mm wide area, the panel will form a non-luminous area with the shape shown in Figure 9.2. Because the organic luminous layer will deteriorate due to the moisture infiltration through the sealant within six months if the completed panel remains at room temperature. Therefore, in the edge seal method, an adhesive-tape-type desiccant is put inside the panels to remove the moisture that infiltrated through the sealant. For the desiccant, a material with excellent moisture adsorption capability, such as $CaCl_2$ or CaO, is to be used. The formula for the desiccation with CaO is shown below.

$$CaO + H_2O \rightarrow Ca(OH)_2$$

OLED panel manufacturers ensure reliability against moisture by calculating the volume of desiccant required to interact with the moisture expected based on the estimated WVTR and the required lifetime in the application that the panel manufacturers wish to produce. Such methods can be related to those considered in the packaging of food or of electronic parts that are known to be susceptible to moisture. For instance, in many food packages, vinyl package multicoated with aluminium foil and acryl protective film will block moisture first while the miniscule amount of moisture that infiltrated via the pinhole of the package will be removed by the chemical reaction of the silica gel pack that was put inside the package along with the product.

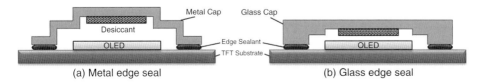

Figure 9.7 Typical structures of the edge-sealed OLED panel.

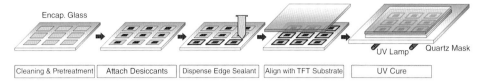

Figure 9.8 Typical glass UV sealing process.

The advantage of the UV edge seal method is that an additional moisture passivation layer is not required because the moisture adsorption performance of an included desiccant demonstrates a foreseeable level of reliability. Figure 9.8 shows a conceptual schematic of the UV edge seal process using a glass cap.

First, the encapsulation cap glass is cleaned and prepared to remove foreign substances and contaminants, thereby improving the adhesion performance of the sealant. In the pre-processing stage, the glass surface is physically or chemically etched using argon plasma, or is subjected to chemical activation in which the groups on the surface will be excited. After pre-processing, the glass is then laminated with the desiccant tape along its central axis. Next, sealant shall be dispensed onto the panel using a dispenser nozzle tracing along the cell edges. The viscosity of the sealant dispensed at this stage, which is decided based on the width and height of the seal, shall be set at hundreds of thousands cP in the pre-manufacturing stage. The encapsulation glass prepared as such is then loaded onto and aligned precisely with the panel formed with TFT and EL layers before the two are fixed together in an atmosphere that is semi-vacuum and moisture-free but is filled with inert N_2 gas. The bonding quality of the completed unit will be improved when the negative pressure in the vacuum chamber is increased such that the pressure difference between the external and internal (package) atmospheres will narrow the gap between the glass substrates due to deformation, however, the desiccant may contact and scratch the EL unit or induce Newton ring defects, a concentric stain from optical interference caused by the irregular glass surface. Therefore, it is important to adjust the atmospheric pressure appropriately in the bonding phase by considering the target productivity and quality.

When the top and bottom glass substrates have been fixed together, the sealant will be flooded with UV lights for the hardening process. As the wavelength of the most frequently used UV lamp is 365 nm, the sealant designed to harden at this wavelength is used in general, and to prevent the potential damaging of the TFT unit by the UV light, a quartz mask is to be utilized so as to selectively transmit the UV light only to the edges dispensed with the sealant. After the sealant hardening, the panel is cut into cell units.

With its excellent moisture elimination performance thanks to the desiccant, the edge seal method contributed greatly to the successful realization and commercialization of the early OLEDs. The edge seal method, however, has since been replaced gradually by other types of new encapsulation technologies because it failed to overcome one critical limitation: the method could not meet the requirements for thinner and flexible OLED displays because it hindered the adoption of a plastic film or of metal foil due to dependency on space for desiccant, and that illumination defects may occur if the top and bottom substrates contact each other due to the external pressure.

9.3.2 Frit Seal

Frit seal (also called "laser sealing"), an encapsulation technology developed early by Corning Inc. in the United States [3, 4], is applied in accordance with the following steps: (1) a paste-type frit shall be produced by mixing an organic binder and solvent with powder-type glass/metallic oxides; and (2) the seal produced as such shall be printed or dispensed onto the glass substrates, and the latter shall be pre-heated and flooded with IR laser before being sealed together (Figure 9.9).

After the pre-heating and laser sealing, the binder and solvent shall be removed from the seal, leaving only the glass ingredient, thereby acquiring an excellent moisture resistance performance (WVTR<10^{-8} g/m^2/day) equaling that of glass to the seal, and hence negating the need for a separate moisture adsorbent or a passivation layer. Therefore, it became feasible to develop the "zero gap" technology, in which the organic spacer

Figure 9.9 Typical structure of the frit-sealed OLED panel.

formed by applying photolithography on the TFT panel would be in contact with the glass cap, as it became possible to apply a flat glass cap without any separate etched area that was to be laminated with the desiccant tape, as shown in Figure 9.9. The zero gap technology not only allowed producers to decrease the incidences of Newton ring, which is caused by the pressure difference between the internal and external atmospheres, but also contributed to the thinning of the OLED panels by preventing potential scratches due to the contact between the top and bottom substrates.

Additionally, the advancement in the material and process technologies enables a narrow bezel structure by reducing the width of the frit seal to a dramatic extent. The narrow bezel is fast becoming the most critical differentiated point in the industry in terms of aesthetic design.

Thanks to such diverse benefits, the majority of the existing OLED panels are known to be produced using the frit seal method, with the prominent exception of some curved panels produced by Samsung Mobile Display, which dominates the OLED panel market for smartphones.

Figure 9.10 shows one typical frit-sealing process. The glass frit mixed in the form of paste shall be printed by screen-printing process onto the cleaned and pre-processed panel; as such, its seal area should be patterned. Next, any organic content or solvent in the mixture shall be vaporized by heating the glass, thereby pre-sintering the frit. It will then be aligned with the TFT substrate before the glass frit is flooded with IR laser to fuse together the substrates of the top and bottom substrates. Once the two substrates are fused together as such, the completed panel shall then be cut into cell units.

The following are some of the serious drawbacks of the frit seal method: (1) other encapsulation materials, such as metal foil and plastic film, cannot be applied due to the narrow selection of fusion materials, which relies on the fusion bonding of the glass panels by firing IR laser; (2) some fine cracks may occur in the seal part if extreme care is not taken when handling the panels during the printing of frit onto the glass substrates in the fusion fixing, or during transport; and (3) the sealed part can be easily broken even after the completion of the panels, due to its brittleness. In particular, the method cannot be applied to the manufacturing process of large or flexible OLEDs due to its critical deficiency: with increased size of the panel, the seal will easily crack when it is subjected to any stress arising from the irregular curvature caused by the potentially slight torsion of the panel. In fact, the method has never been applied to any OLED larger than 20 inches; nor has any frit seal method with a markedly improved property sufficient to overcome such limitation been known to exist in the industry. The method succeeded in realizing an unprecedented level of reliability by adopting glass, which has a higher barrier than any other material, but ironically, the highly brittle nature of glass caused the method to be overtaken by competing methods.

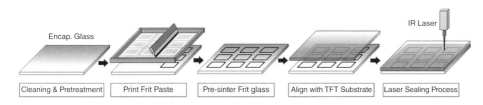

Figure 9.10 Typical frit-sealing process.

9.3.3 Dam and Fill

The dam and fill method was designed to overcome the size limitations of the edge sealing methods which have vacant space between cover and TFT substrates as noted previously. The method was developed and actively honed by Japanese panel and encapsulation materials manufacturers, including Sony, in the early days of the OLED industry (Figure 9.11).

The method realized the no-cavity encapsulation structure by dispensing the panel edge with highly viscous resin (as in the UV edge seal) while filling the inner cavity with less viscous resin before the hardening process, thereby leaving no empty space throughout the panel (Figure 9.11). The method, however, needs an additional measure to prevent moisture infiltration through the resin because it is not equipped with a separate desiccant. One of the most-well-known moisture prevention methods is forming a separate moisture-protection layer such as a passivation layer on the EL unit.

Among the inorganic films that are used in semiconductors, Si or Al oxide films are usually adopted as a passivation layer, because they form an amorphous structure with no grain boundary and enable high moisture resistance, excellent electric insulation, and a high light transmittance. In addition, the method protects the device from any physical damage that may be inflicted on it during the manufacturing process, thanks to its hardness, and allows the utilization of sputter or PECVD, equipment frequently used by the panel manufacturers in the TFT panel manufacturing process (Figure 9.12).

Figure 9.12 shows the schematic flow of the dam and fill process. The cover glass finished with cleaning and pre-processing will be dispensed onto with highly viscous resin similar to one used in the UV edge seal. Next, resin with low viscosity will be dropped as a fill material to fill the empty space inside the panel. The fill material is the material that is added to the hardening agent to promote the complete hardening of the material during the curing process, and is designed to minimize the amount of outgassing so that the various gases generated during the hardening process would not influence the OLED device.

The cover glass dispensed with a dam and fill material will be aligned and laminated with the TFT panel in the vacuum chamber. When the vacuum chamber is vented to the atmosphere, external pressure is exerted on the panel due to the pressure difference between inside and outside of the panel. In the process, the fill material with low viscosity will spread across the panel, thereby filling the void between the highly viscous dam material and the top/bottom glasses. As the empty space not yet stuffed with the fill material will incur a defect in the form of a void with an increasing panel size, precision tuning of specific process conditions, such as the dispensing pattern or the dispensing interval of the fill material, is required. Once the lamination is finished, the encapsulation process will be completed after the dam and fill material goes through the UV or thermal curing process, whereas the completed panel shall be cut into smaller cell units.

Figure 9.11 Typical structure of dam and fill encapsulation.

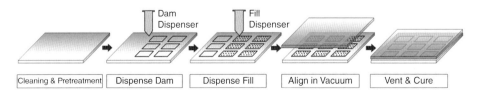

Figure 9.12 Typical dam and fill encapsulation processes.

The dam and fill method has been considered the favored OLED panel passivation technology since the early days because it is effective in filling the cavity inside the panel and is a proven technology with no significant issue in the manufacturing process, but it has yet to overcome some issues associated with its shelf lifetime reliability. As the resin used as the dam and fill material has a WVTR value of over $10\,g/m^2$day given the unique property of the polymer material, even though it is one of the most moisture-resistant materials available, the diffusion of moisture into the inner cavity of the panel is inevitable. The infiltrating moisture is therefore being blocked by the passivation layer, but with one caveat: the moisture will eventually infiltrate the panel due to the diverse particles added in the various manufacturing processes, the potential cracks in the passivation layer arising from surface unevenness, or pinholes in the passivation layer.

It has been known that a passivation layer thicker than 5 μm is being applied to some OLED panels that adopted the dam and fill method to ensure mass production reliability by complimenting some deficiencies in the passivation layer. This is to capitalize on the phenomenon that the deficiency density will drop and the coverage of the particles will get better with the increasing thickness of the passivation layer. The use of the standard vacuum deposition equipment to form a passivation layer thicker than 5 μm, however, is detrimental in terms of productivity, thereby dealing a fatal blow to the process compared to other encapsulation methods. For this method to be adopted as the mainstream technology suitable for mass production by overcoming such deficiency, it is essential that some ground-breaking passivation technology with a marked improvement in terms of the step coverage be applied, or that a new type of dam and fill material with an extremely low WVTR be developed.

9.3.4 Face Seal

Sometimes called "film seal method" or "hybrid encapsulation method" as compared to the "thin-film encapsulation method" described in section 9.2.2.5, but usually called "face seal," this method uses film-type rather than liquid-type seal materials among the no-cavity encapsulation methods. LG display succeeded in the first commercial application of the face seal method, which was named "SPE (solid-phase encapsulation)" for its use of metal foil and adhesive, to the world's first 55-inch OLED TV [5]. Below are the advantages of the film-type seal materials compared to the liquid-type seal materials (Figure 9.13).

(1) Warpage is minimized when laminating two different substrates thanks to its low shrinkage during hardening as it remains in the semi-solid state.
(2) No rupture of the seal material has occurred, and it is easy to adjust its thickness and uniformity.
(3) There is no limitation in the selection of the substrates (e.g., glass, metal, or polymer film).
(4) The roll-to-roll process will be applicable to the process in the future.

As shown in Figure 9.13, the face seal method demonstrates the simplest structure among all the encapsulation structures that have been developed so far. The sealing adhesive cements the top and bottom boards across the panel, while inside the panel only the passivation layer covers the EL layer. For a brief account of how reliability is assured in this method, the diffusion speed of the external moisture is being slowed down primarily by the adhesive seal material and secondarily by the passivation layer's blocking of its contact with the EL device. To sum up, similar to the dam and fill method, its shelf life is determined by the function of the moisture resistance property of the adhesive as well as by the performance and thickness of the passivation layer.

Figure 9.13 Typical structure of the face-sealed OLED panel.

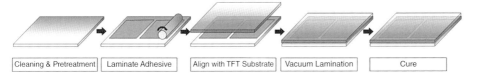

Figure 9.14 Typical process of face seal encapsulation.

The passivation layer in the SPE method developed by LG display is very thin (5 μm), which implies that it has a structure optimized for mass production compared to the dam and fill type OLED panels, and that the moisture resistance property of the adhesive is exceedingly superior to compared to the existing resin materials for sealing.

The process flow of the face seal method is shown in Figure 9.14. The cleaned and pre-processed cover material is laminated by roll-laminating the film seal produced in the form of a double-sided adhesive sheet. Next, it is aligned with the TFT panel coated with the passivation layer on the EL layers in the vacuum chamber before proceeding with the vacuum lamination process. As the adhesive sheet is made with thermal cure-type material to prevent any UV damage that may be inflicted on the TFT unit as well as to ensure process convenience, it is made to go through the curing process before completion.

The face seal may look very simple in terms of both its structure and process, but advanced material performance and a mass production technology are indispensible to ensure reliability in mass production compared to other methods. Also, as mentioned earlier, the face seal is drawing attention from the industry as the only encapsulation technology that has ever been applied to the commercial production of large OLED TVs. In particular, its structural simplicity guarantees no vulnerability in terms of mechanical strength. The other exceptional advantages of the method compared to other encapsulation methods are the following: (1) metal foil, plastic film, and other various materials, besides the obvious glass choice, can be used as the encapsulation material, thereby ensuring an entire array of applications, including curved, flexible, or rollable displays; and (2) it will be possible to manufacture it with the roll-to-roll process in the end. Therefore, the encapsulation technology is expected to bring about further technological advancements in the future.

9.3.5 Thin-Film Encapsulation (TFE)

Continuous efforts have been made by both the industry and the academy for an extended period to apply the multi-layered organic/inorganic thin-film encapsulation technology, which is frequently being used in the semiconductor industry to form a protective coating or to seal the integrated circuits [7, 8]. The principal aim of TFE is to eliminate the cover materials, such as glass or metal, by maximized moisture barrier property of the multi-layered organic/inorganic passivation layers. Thus, this technology is being highlighted as the most advantageous and ultimate encapsulation structure in light of the goals being pursued by the OLED industry: light, thin, flexible screen.

Figure 9.15 shows a cross-sectional structure of the OLED panel to which the TFE method is applied. No separate cover material was used; instead, a multilayer passivation was fabricated on the EL layer. It should be noted, however, that the borderline will often be unclear as the multilayer passivation is similar in structure to the face seal as an additional protective film or polarization film may be laminated onto the multilayer passivation to prevent scratches during the manufacturing process or while being used by the users.

In the TFE structure, one pair of inorganic/organic layers in the multilayer passivation stack is called a "dyad." For instance, the multilayer passivation in Figure 9.15 is composed of 2.5 dyads. The inorganic layer is composed of the same material used in the passivation layer adopted by other methods classified as no cavity structure in Fig. 9.6, and will serve as a direct barrier against infiltrating moisture, whereas the organic layer will complement the various defects of the inorganic layer, or will planarize it. It is inevitable for the thin inorganic layer to have some moisture infiltration paths as it is more likely to have pinholes or various other particles. It has been proven

Figure 9.15 Typical structure of TFE.

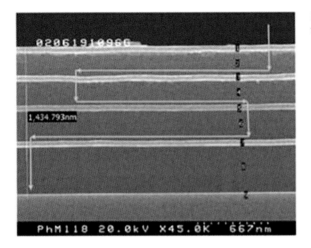

Figure 9.16 Extremely long "effective" diffusion path due to the large spacing between the defects [8].

time and time again by various researchers that the moisture infiltration path will become much longer (also called "tortuous path," Figure 9.16) when the inorganic layer is deposited after the surface is planarized with the organic layer rather than simply increasing the inorganic layer thickness, thereby resulting in a significantly increased WVTR performance [6–8].

The process flow of the TFE method is shown in Figure 9.17. First, the first inorganic layer is deposited onto the EL devices. Sputter, PECVD, or ALD are frequently used for this purpose, but the particular film deposition method is selected considering comprehensive factors such as the generation of particles during the process, the step coverage, and the throughput. Next, the organic layer is either printed or coated onto the panel.

In coating the organic film, which is just a few μm thick, one method is selected from among flash evaporation, thermal evaporation, PECVD, printing, slit coating, and so on. The development of a matching organic material appropriate for each method should also be realized, while other factors such as the coverage characteristics, outgassing, shrinkage ratio, productivity, and ease of patterning, should be considered depending on the combination of the material and equipment. As the second inorganic layer will be deposited only after the organic layer is hardened by thermal or UV light, the inorganic deposition-organic curing process will be repeated. After the desired number of dyads has been created, the

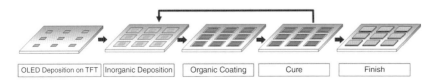

Figure 9.17 Typical TFE process flow.

encapsulation process will be completed without any additional processing before the panel is transferred to the next process.

The biggest hurdle in the mass production of OLED with the application of the TFE method is the total number of layers. If the number of layer increases, the density of various defects will be lowered, and the allowable size of the particles that can be covered up will increase, thereby improving the moisture barrier property. If the share of the process in the entire production line becomes disproportionately larger, however, it will not be acceptable in terms of investment cost or productivity. Vitex in the United States, the company known to have pioneered the TFE method, coated the TFE by combining the sputter and flash evaporation methods, and named it "Barix^{TM}." The company explains that four to five dyads are necessary for the encapsulation of the OLED when applying this method [8].

Many panel manufacturers and those engaged in the development of panels for flexible displays are striving to apply TFE or similar technologies in their respective manufacturing processes, but in reality, it is not easy to decrease the number of layers without lowering the level of reliability in covering all the product defects. For this reason, they are still struggling in terms of productivity or of the enlargement of the display size, thereby limiting the number of cases applied to actual mass production to a negligible number considering the long history of TFE development as well as associated efforts. It is obvious, however, that there is huge possibility of applying the aforementioned technology to various high-value-added products by developing or improving various other methods, such as combining the technology with the face seal or dam and fill method or adding a barrier film, simply because of its unsurpassed advantage in producing lighter, thinner, and flexible displays.

9.4 SUMMARY

Various methods of encapsulating OLED have been discussed. The early OLEDs were mass-produced by installing a desiccant inside the panel through the invention of the sealing technology inspired by the packaging of foods and electronic devices. Diverse encapsulation technologies with matching properties, such as lightness, thinness, and flexibility, are currently being actively developed to realize various commercial applications that will maximize the benefits of OLED. Now that curved, flexible, and roll-able displays are being commercialized, it should be emphasized, in particular, that the competitive advantage of a panel manufacturer will largely depend on its development of a matching encapsulation technology befitting the unique requirements of different applications.

References

1 Aziz et al. Degradation processes at the cathode/organic interface in organic light emitting devices with Mg:Ag cathodes, Applied Physics Letters, 72(21), p. 2642 (1988).
2 Liew et al. Investigation of the sites of dark spots in organic light emitting devices, Applied Physics Letters, 77(17), p. 2650 (2000).
3 B. G. Aitken., J. P. Carberry, S. E. DeMartino, H. E. Hagy, L. A. Lamberson, R. J. Miller, R. Morena, J. E. Schroeder, A. Streltsov, S. Widjaja, Glass package that is hermetically sealed with a frit and method of fabrication, US Patent 6,998,776.
4 R. M. Morena, L. A. Lamberson, S. Widjaja, S. L. Logunov, Frit-sealing at high heating and cooling rate, 6th Pacific Rim Conference on Ceramic and Glass Technology, September 11–16, Hawaii USA (2005).
5 C.-W. Han, J.-S. Park, H.-S. Choi, T.-S. Kim, Y.-H. Shin, H.-J. Shin, M.-J. Lim, B.-C. Kim, H.-S. Kim, B.-S. Kim, Y.-H. Tak, C.-H. Oh, S.-Y. Cha, B.-C. Ahn, Advanced technologies for UHD curved OLED TV, Journal of the SID, 22, pp. 552–563 (2015).

6 M. S. Weaver, et al. Organic light-emitting devices with extended operating lifetimes on plastic substrates. Applied Physics Letters, 81(16), p. 2929 (2002).
7 G. Nisato, et al. Evaluating high performance diffusion barriers: the calcium test, 21st Annual Asia Display, 8th International Display Workshop, Nagoya, Japan, p. 1435 (2001).
8 R. J. Visser, Barix ,ultilayers: a water and oxygen barrier for flexible organic electronics, Vitex Systems, Google, p. 13 (2015).

10

Flexible OLED Manufacturing

Woojae Lee[1] and Jun Souk[2]

[1] E&F Technology, 14 Tapsil-ro 35 beon-gil, Yongin-si, Gyeonggi-do, Korea, 17084
[2] Department of Electronic Engineering, Hanyang University, South Korea

10.1 INTRODUCTION

Flexible displays have been a topic that attracted a high level of interest and attention for more than 10 years. They offer very desirable features for product design related to thinness, lightweight models as well as free-form factor models. Thus, there have been sustained research efforts across the display industry to achieve flexible displays in different types of display technologies, such as flexible ebook [1], cholesteric LCD [2], flexible thin-film transistor-liquid crystal display (TFT-LCD) [3], and flexible OLED display [4].

In recent years, the flexible OLED technology in particular has finally reached commercial production stage due to the progress in materials and process technology, enabling an ideal combination of display performance and form factor flexibility. In this chapter, we focus on the manufacturing of flexible OLED displays for small- and medium-sized panels for smartphone and smartwatch devices. These applications are the first generation in the volume production of flexible OLED displays.

This chapter focuses on the materials and fabrication process of small- and medium-sized flexible OLEDs that are currently used in display industry. There are several different approaches to achieve flexible OLED, by selecting from different types of TFT backplane, flexible substrate, and barrier films as shown in Figure 10.1. However, the definition of flexible OLED, at least in this chapter, is the OLED display fabricated on a thin flexible plastic film as currently used in production. Therefore, a broad meaning of flexible OLED, such as OLED on metal foil or on an ultra-thin flexible glass (below 0.1 mm thickness) is not the scope in this chapter.

Today's commercially available flexible OLEDs are found in edge type Samsung Galaxy phones [5] and smartwatches [6] as shown in Figure 10.2.

The OLED devices shown in Figure 10.2 are regarded as the first-generation flexible OLED products. The flexibility of these displays is limited to shaping with a large radius of curvature of about 40 mm and without requirement for repeated bending capability. There have been many technical hurdles for flexible OLED to reach even this first level, mainly due to the poor plastic substrate materials, poor substrate handling methods, and a lack of proper flexible encapsulation methods.

Flat Panel Display Manufacturing, First Edition. Edited by Jun Souk, Shinji Morozumi, Fang-Chen Luo, and Ion Bita.
© 2018 John Wiley & Sons Ltd. Published 2018 by John Wiley & Sons Ltd.

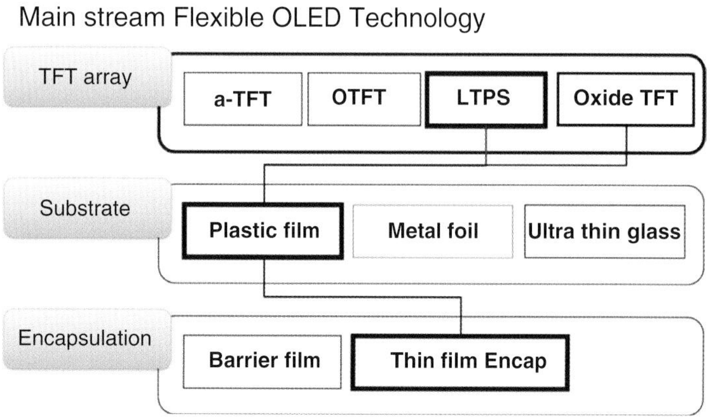

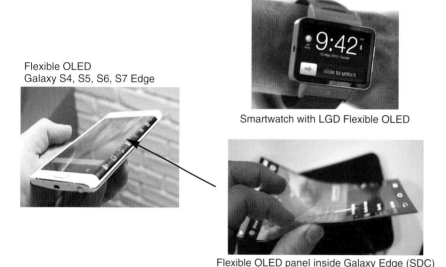

Figure 10.1 Various approaches to achieve flexible OLED. Boxes in bold line indicate the mainstream technology used in current products.

Figure 10.2 Flexible OLEDs and their current applications in smartphones and smartwatches.

10.2 CRITICAL TECHNOLOGIES IN FLEXIBLE OLED DISPLAY

The major differences between the manufacturing of rigid and of flexible OLEDs are related to substrate handling and encapsulation technologies. In order to fabricate flexible OLED displays, (1) the plastic substrates materials must withstand high process temperature of LTPS TFT, (2) the encapsulation layer must be transparent and flexible, (3) the overall process should be performed in equipment designed to handle glass substrates since all of the existing process equipment is designed to handle glass substrates only. Therefore, the current flexible OLED process is performed on flexible films attached on a carrier glass substrate. The flexible substrate film is then detached from the carrier glass after all the processes are completed [7].

The fabrication of flexible OLED displays requires critical processes in three areas that differ from the rigid OLED process: (1) substrate preparation process, coating and curing a thin, high temperature resistant polyimide (PI) polymer on the carrier glass, (2) flexible and transparent multi-layer encapsulation process, and (3) delamination process, which is lifting off the TFT/OLED fabricated on PI film from the carrier glass.

Table 10.1 List of critical areas in flexible OLED process.

1. Spin/Slit coating Polyimide varnish on the carrier glass.
2. Polyimide curing process.
3. Encapsulation process after OLED layer deposition.
4. Laser lift off process.
5. Touch sensor attach or pattern on TFE.

These areas also require additional processes in order to handle plastic film. Table 10.1 lists these critical processes.

10.2.1 High-Temperature PI Film

As a substrate material for flexible OLED, polyimide (PI) varnish is a well-known material [8]. The process temperature of LTPS TFT backplane for OLED is usually above 450 °C. In order to use PI as a film, an adhesive with high temperature stability would be needed to attach the film on the carrier glass. However, a proper adhesive which withstands TFT process conditions is not available yet. Thus, PI varnish material spin or slit-coated on the carrier glass is the method currently used for production of flexible OLED panels. Samsung display and LG display use this method in commercialization of flexible OLED panels used in edge type smartphones and smartwatches.

The PI varnish is an amber-colored solution. Small- and medium-sized OLED displays for mobile applications such as smartphones and tablets have a top emission structure (see Ch 8A.2, Figure 8A.1). Therefore, the transparency of the PI substrate is not an issue. The physical properties of PI can be controlled by its molecular structure and functional groups. For flexible OLED, it endures near 500 °C process temperatures [9]. Since the coefficient of thermal expansion (CTE) of the carrier glass is about 3 ppm/°C [10]. PI material with low CTE value is required to minimize bending by thermal expansion mismatch during the various temperature processes. The properties and molecular structure of high temperature PI are shown in Table 10.2 and Figure 10.3.

Today, only a few chemical companies are able to manufacture high-temperature PI varnish compatible to LTPS process. Today's large-sized flexible OLED displays such as OLED TV or transparent OLED use bottom emission structure. Development of transparent high-temperature PI varnish is required in order to fabricate future large-sized flexible OLEDs. Several companies are also developing transparent high-temperature PI varnish material.

Table 10.2 Properties of high-temperature PI film.

- Polyimide Film by Solution Coating
- Coating Thickness ~10 μm
- Properties:
  - Optical: Amber color
  - Thermal: CTE < 7 ppm/°C
    (TFT glass CTE ~ 3 ppm/°C)
  - Tg: 340 °C

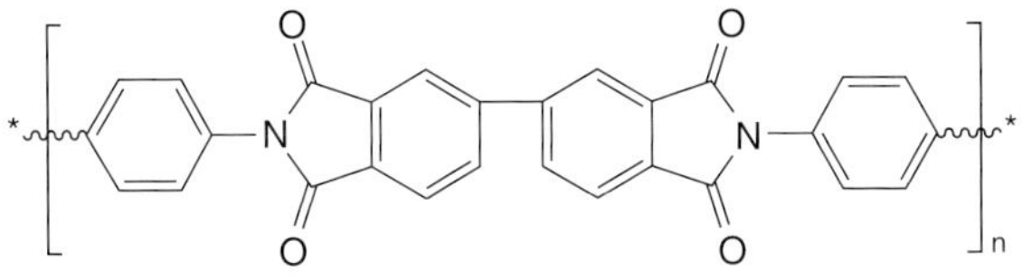

Figure 10.3 Molecular structure of high-temperature PI.

10.2.2 Encapsulation Layer

OLED panel encapsulation, or barrier layer performance, plays an important role in the device lifetime and reliability [11]. Since OLED and cathode materials are very sensitive to water vapor and oxygen, a very high-quality barrier coating is essential with typically below 10^{-6} g/m^2/day level of water vapor transmission rate (WVTR). For comparison purpose, WVTR requirements for OLED and other electronic application are shown in Figure 10.4 [12].

Almost all of the small-and-medium OLEDs use top-emission structure to enhance the emission efficiency of high-resolution OLED. Therefore, a transparent encapsulation solution was required and was developed as the thin-film encapsulation composed of multiple organic/inorganic layers to protect the OLED layers from permeating water vapor and oxygen. This thin-film encapsulation layer replaces the encapsulation glass that was used on rigid glass-based OLED.

Encapsulation technology for flexible OLED can be divided into two categories: TFE (thin-film encapsulation) [13, 14] and hybrid [15]. TFE method is very effective for barrier layer and dominantly used in flexible OLED production. Hybrid method is also used in current products; not only for flexible OLED but also for rigid OLED devices.

10.2.2.1 Thin-Film Encapsulation (TFE) Method

The TFE method was originally developed by Vitex and commercialized by Samsung Display Company (SDC) [16]. Since SDC occupies more than 95% of current OLED market share, TFE has become a dominant encapsulation method from a commercial basis. The core technology involves depositing organic and inorganic layers in a multilayer stack (Figure 10.5). Even if defects such as pinholes exist in the inorganic barrier layer, the adjacent organic layer and use of additional inorganic layers allows delaying significantly the moisture/oxygen diffusion due to the long diffusion path required to penetrate the entire stack. The Vitex technology is composed of multiple pairs of organic and inorganic layers called a dyad. The original method used a combination of sputter deposition of aluminium oxide and vapor deposition of organic material followed by UV

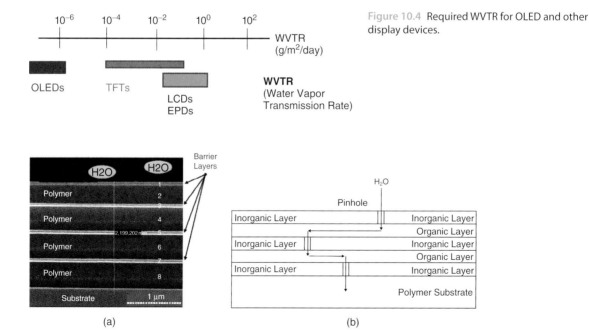

Figure 10.4 Required WVTR for OLED and other display devices.

Figure 10.5 (a) Barix barrier by Vitex (b) Tortuous moisture diffusion path in multilayers system.

curing process to make four layers of dyad [17]. Increasing the number of layers make the encapsulation more expensive and time consuming [18, 19]. The original Vitex TFE method progressed to a simpler structure and faster process, such as using the combination of ink-jet printed organic layer and inorganic insulator layers. The combination of organic/inorganic layer structure is not a fixed process yet and still evolving. ALD (atomic layer deposition) method is under investigation to form a dense inorganic layer replacing PECVD and sputtered films.

There are several advantages for TFE method. The barrier quality is higher than other encapsulation methods used for flexible OLED displays. TFE is also thinner than the hybrid method, and could be used in foldable displays, which requires much higher level of flexibility. The material cost is also lower since it does not use high-quality barrier film substrate and adhesive, although the initial equipment investment cost is higher for the TFE case. In addition, the edge seal is narrower which enables super narrow bezel display.

10.2.2.2 Hyrid Encapsulation Method

The hybrid encapsulation method consists of a combination of thin-film inorganic layers and a barrier film attached with pressure-sensitive adhesive (PSA). The advantage of this technology is a lower initial investment cost and short process time. Even though the material cost is higher than TFE, pre-inspection of barrier film is possible, which can lead to a higher production yield. In addition, hybrid method can be applicable to larger OLED panels. The hybrid encapsulation technology is shown in Figure 10.6 [20].

There are two ways to build the hybrid encapsulation: one is to use dam and fill type with a liquid resin-type adhesive and the other is to laminate barrier film with PSA as shown in Figure 10.7 [21]. A dam structure is built to prohibit moisture penetration from the sidewall of first inorganic barrier layer (Figure 10.8) [22].

A remaining challenge for the hybrid method is to reduce material cost. Additionally, having a transparent barrier polymer film is very critical to allow use in top-emission OLED [23]. The base polymer film for this purpose is intrinsically expensive due to the required quality. To minimize the defect and particle, a highly pure and clean resin must be used for the film fabrication under highly clean environment. Additional planarization layer is required to enhance the surface smoothness.

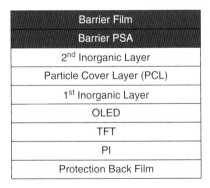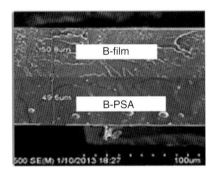

Figure 10.6 Vertical stack structure of hybrid encapsulation and SEM image of top two layers.

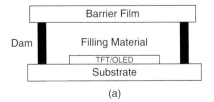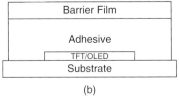

Figure 10.7 Two types of hybrid encapsulation method: (a) Dam and fill (b) Lamination barrier film with PSA.

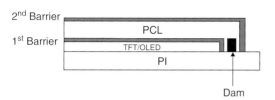
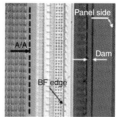

Figure 10.8 LG display's unique dam structure located at edge side of first inorganic barrier layer.

10.2.2.3 Other Encapsulation Methods

Many R&D efforts were made to achieve high barrier properties in simpler structures with low cost. Symmorphix Inc. reported in 2005 that they succeeded in making a single barrier layer performing under 6×10^{-6} g/m^2/day WVTR value (Figure 10.9) using physical vapor deposited aluminosilicate (AlSi$_x$O$_y$) with less than 200 nm thickness. Around the same time, other single barrier layer approaches based on different materials and deposition methods such as ALD (atomic layer deposition) aluminium oxide by University of Colorado [24] and PECVD (plasma enhanced chemical vapor deposition) silicon oxycarbide by Dow Corning have been reported but did not reach the Symmorphix result.

Even though using a single barrier layer performance is not good enough for encapsulation, it still is eligible for use in the barrier film for hybrid encapsulation. Thus, several companies commercialized films not only for OLED encapsulation but also e-paper application. PET film with alumina (Al$_2$O$_3$) or silicon oxide (SiO$_2$) barrier formed via e-beam evaporation method is available by Toppan Printing. Its WVTR is known to be 2.10×10^{-4} g/m^2/day (40°C / 90% RH). Mitsubishi's X-Barrier™ is in the market with 10^{-4} g/m^2/day WVTR value [25]. Silicon oxide by vacuum deposition on polyester film made the barrier performance to this level. A Korean company, i-Components made i-Barrier® film with 10^{-4} g/m^2/day WVTR value by sputtered silicon nitride (SiN$_x$) [26]. Its application expands from flexible OLED hybrid barrier to quantum dot barrier film.

Besides, many companies and research institutes focus on developing simpler and inexpensive way than multiple barrier layers of Vitex' Barrix. 3M has been developing technology to make a WVTR of 5×10^{-4} g/m^2/day with a dyad on PET and multiple coatings to improve it to 10^{-6} g/m^2/day for OLED application. For OLED devices, they use PC (polycarbonate) or COP (cyclo olefin polymer) film instead of PET, which has high birefringence [27]. The multiple layer barrier mentioned above is normally consisted of organic and inorganic layer on top of each other. There are several different approaches. One example is a graded ultra-high barrier technology from GE Global Research [28]. By using a PECVD method, silicon oxynitride (SiO$_x$N$_y$)

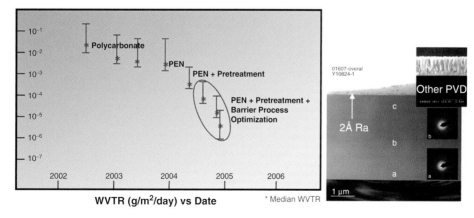

Figure 10.9 Symmorphix's single layer barrier with WVTR 10^{-6} g/m^2/day level.

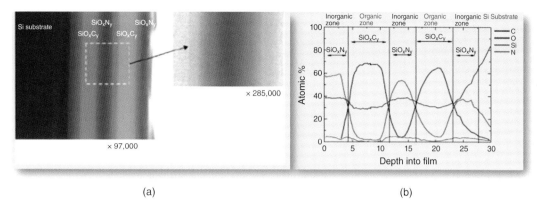

Figure 10.10 Graded ultra-high barrier by GE [29]. (a) Cross-sectional TEM image, (b) XPS spectrum of graded ultra-high barrier layer.

and silicon oxycarbide (SiO_xC_y) are continuously deposited with changing composition of the materials achieving 5×10^{-6} g/m^2/day WVTR with 500 nm thickness (Figure 10.10).

The other approach is to use an all inorganic layers stack to make barrier. Like in the case of graded ultra-high barrier, roll to roll physical vapor deposition is done in a single chamber by Vitriflex [30]. The performance reaches 10^{-6} g/m^2/day WVTR level.

Another method is replacing the organic layer in multilayer stack barriers with nanoparticles of traditional oxide inorganic layer (Figure 10.11). Terra Barrier succeeded in achieving under 10^{-6} g/m^2/day WVTR level [31] since nanoparticles do not only seal the defects but also react with moisture and oxygen and seize them within the layer.

10.2.2.4 Measurement of Barrier Performance

The measurement of barrier layer performance can be quantified directly by several ways: using MOCON test equipment under isostatic condition, using radioactive tritium tracer tester, or using mass spectrometer for water molecules. WVTR value above 10^{-3} g/m^2/day is normally obtained with MOCON WVTR measuring equipment. MOCON is a standardized test method not only in electronics industry but also all in other major industries such as the food industry and film industry. Lately, MOCON released AQUATRAN II model for measuring WVTR up to 10^{-5} g/m^2/day level but it takes more than 100 days for steady state measurement

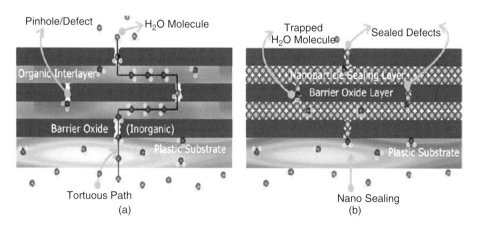

Figure 10.11 Terra Barrier concept. (a) Conventional multilayer barrier, (b) Nanoparticle sealed barrier Terra Barrier.

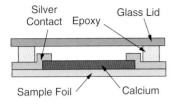

Figure 10.12 Calcium test cell.

Table 10.3 Features for various WVTR measurement methods.

	MOCON (AQUATRAN II)	Tritium	Mass Spectrometry	Calcium cell test
Detection Limit (g/m²/day)	5×10^{-5}	2.4×10^{-7}	1×10^{-6}	1×10^{-6}
Detection Species	Water Oxygen	Water	Multiple molecule	Water
Limited condition (Temperature/Humidity)	Yes	Yes	Yes	No

below 40 °C [32]. The other method commonly used for WVTR data is the calcium test (Figure 10.12) [33]. Calcium is sealed in the cell and during the test, moisture-absorbed calcium becomes transparent. The WVTR value can be calculated based on the time and area of degraded. The advantage of the calcium test is that the acceleration test is possible, like 85 °C and 85 % relative humidity condition test. The detection limitation and other properties in measurement methods are summarized in Table 10.3.

10.2.3 Laser Lift-Off

After the flexible OLED (F-OLED) fabrication process is completed on the PI/carrier glass, including TFT layers, OLED layers, and TFE layers, while the delamination of the PI film carrying the device from the carrier glass follows. Since the delamination process requires a complete lift-off of the PI film stack without residues and damage to the OLED device, it is regarded as a critical step that dominates the product yield.

In the laser lift-off process, when the laser beam scans across the carrier glass about 70 % to 80 % of the short wavelength (308 nm) beam energy is absorbed in the amber-colored PI film and generating heat localized near the interface with glass causing delamination of the PI film. This process is called LLO (laser lift-off) [34] and the wavelength (308 nm), power density, and beam scan speed of the excimer laser are critical parameters to avoid device damage (Figure 10.13) [35, 36]. Longer wavelength laser could damage the device. Since the excimer laser equipment is expensive and requires high operation cost, solid state laser (wavelength at 343 nm) is also used for production because of its low maintenance cost. Various non-laser delamination approaches are attempted without success in production scale. Additionally, in the earlier days a sacrificial layer of thin a-Si was used under the PI film for easy PI delamination, however, the sacrificial layer was eliminated afterward.

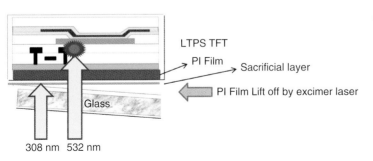

Figure 10.13 Laser lift-off process.

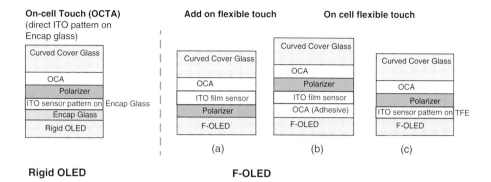

Figure 10.14 Difference in touch sensor structure between rigid and F-OLED: (a) Touch sensor pattern on the encapsulation glass, (b) Touch sensor add on the TFE, (c) Touch sensor pattern on the TFE.

In the case of using a transparent PI film, the sacrificial layer may be used again for the absorption of laser energy.

10.2.4 Touch Sensor on F-OLED

Unlike the rigid OLED case, where the touch sensor layer is directly patterned on the encapsulation glass (on-cell touch structure) as shown in Figure 10.14, integrating a touch sensor layer in flexible OLED has limited options because the top layer is typically a soft TFE layer instead of encapsulation glass. It is more difficult to pattern ITO electrode pattern on the thin-film encapsulation layer because TFE is easily attacked by the process chemicals. Three options are available in this case as shown in Figure 10.14: (1) ITO film sensor attach on the polarizer, (2) ITO film sensor attach on the TFE, and (3) ITO direct pattern on TFE. Recently, ITO direct patterning process on TFE has been successful in production and is regarded as a most advanced flexible touch solution.

10.3 PROCESS FLOW OF F-OLED

The F-OLED processes typically consist of the following steps: (1) PI layer coating and curing process on carrier glass substrate, (2) a standard LTPS TFT backplane process, (3) a standard OLED stack deposition, (4) thin-film encapsulation process, (5) laser lift off the finished PI film from the carrier glass,(6) scribe to the individual panel size, (7) attach a touch sensor film, (8) attach a circular polarizer and (9) module assembly with driver-IC attach. The overview of process is shown in Figure 10.15.

10.3.1 PI Film Coating and Curing

As a first step for the flexible film substrate preparation, PI varnish is slit coated on the carrier glass as shown in Figure 10.16 [37, 38].

After slit coating of PI varnish, a thermal curing process follows. The curing process requires a long thermal process time to make a fully cured solid PI film, typically over 2 hours at high temperature between 400 and 450°C (Figure 10.17).

An example of the typical PI curing process equipment is shown in Figure 10.18.

The thickness of the cured final PI film layer ranges from 10 to 20 μm. Generally, around 10 μm thick PI is obtained for a one-step coating and curing processes. Before the PI varnish coating process, a sacrificial layer can be deposited to allow easier delamination of the PI film by laser after completing the device fabrication,

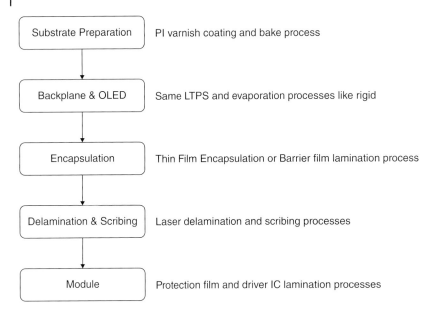

Figure 10.15 Overview of F-OLED manufacturing process flow.

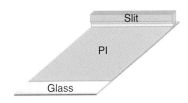

Figure 10.16 Slit coating of PI varnish to prepare flexible OLED substrate.

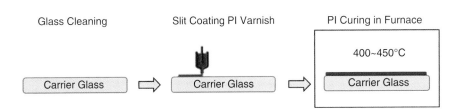

Figure 10.17 PI varnish coating on a cleaned glass and subsequent curing of film.

Figure 10.18 Photograph of PI curing equipment.

as described in section 10.2.3. Also, a silicon nitride or silicon oxide film is deposited on the cured PI film to provide a barrier for the subsequent LTPS structure.

10.3.2 LTPS TFT Backplane Process

Commercial grade small-sized OLED panels for mobile applications use LTPS (low temperature poly silicon) TFT, while large-sized rigid and curved OLED panels for TV application adopt metal oxide TFT such as IZGO for their backplane [39]. LTPS TFT is solely used for flexible OLED backplane technology today.

From the material and process standpoint, a standard LTPS process almost identical to rigid glass-based OLED is used for flexible OLED (Figure 10.19). The only different step is dehydrogenation process that is performed somewhat lower temperature than rigid OLED process to protect the PI film [40]. The dehydrogenation process is required in order to outgas hydrogen from the amorphous silicon film deposited by PECVD before laser annealing process, which is usually done at about 500 °C for glass substrate OLED [41]. One additional consideration point in handling PI film is related to particle contamination, since plastic material is usually easy to attract particles compared to glass substrate due to its static electricity. Therefore, more caution and anti-static treatment are necessary for flexible backplane processes.

10.3.3 OLED Deposition Process

A standard OLED deposition process with FMM follows after completing LTPS backplane process and subsequent TFT panel inspection steps. The TFT mother glass is scribed into smaller sizes before loading into the OLED deposition chamber (Figure 10.20).

The flexible OLED deposition process is identical to rigid OLED process except for the encapsulation portion. The equipment is divided into three clusters as shown in Figures 10.21 [42] and 10.22, with each cluster being composed of six vacuum process chambers. The sequence for OLED layer deposition is summarized in Table 10.4.

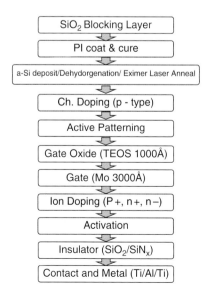

Figure 10.19 LTPS process flow on PI film.

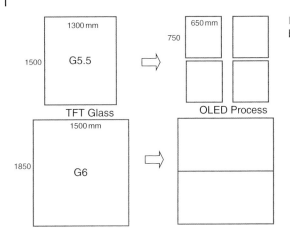

Figure 10.20 TFT processed panel is scribe into two or four pieces before OLED deposition.

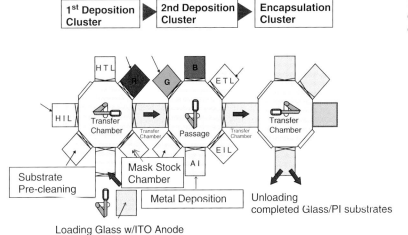

Figure 10.21 A schematic layout of OLED process chambers (*Source:* Canon-Tokki[42]).

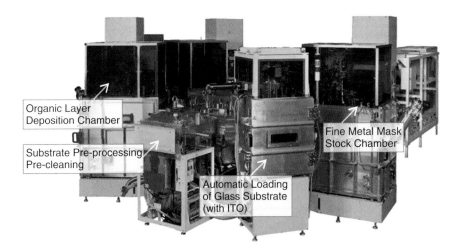

Figure 10.22 Photograph of OLED deposition equipment (*Source:* Canon-Tokki website).

Table 10.4 Process sequence of OLED layer deposition and thin-film encapsulation.

- Film Deposition Cluster #1
  1. A robot transfers glass substrates (with Patterned ITO) to the cluster.
  2. Pre-treatment cleaning of substrates.
  3. Evaporation of HIL/HTL/R organic layers.
  4. Transfer of substrates to Deposition Cluster #2.
  5. Loading chamber is vented for tray exchange
- Film Deposition Cluster #2
  6. Evaporation of G/B/ETL/EIL organic layers.
  7. Metal cathode deposition.
- Encapsulation Chamber
  8. Transfer substrates to the encapsulation cluster.
  9. Substrate UV cleaning - Deposit Organic layer UV cure → deposit PECVD SiN_x → second organic layer deposition → UV cure → second PECVD SiN_x (repeat the layers)
  10. Unload the substrate

10.3.4 Thin-Film Encapsulation

After deposition of the cathode electrode in Cluster #2, the substrate is transferred to the encapsulation cluster without breaking vacuum. The encapsulation process has been described in detail earlier in section 10.2.2. Recently, the original Vitex TFE method composed of multilayer structure progressed to a simpler structure for production, using the combination of SiN_x layers, and inkjet printed organic layer. Figure 10.23 illustrates the schematic layout of TFE equipment.

10.3.5 Laser Lift-Off

After TFE processing, the finished OLED can be unloaded to regular atmosphere environment for proceeding with detaching the finished PI film from the carrier glass. There are two methods in delamination of PI film: one method is to detach the film at mother glass size followed by scribing into individual cell size. The other method is to scribe into cell size followed by detaching the film in a cell size. The former method has less particle issue that can happen during the scribing the mother glass. On the other hand, the latter method can use smaller delamination equipment and lower investment and operation cost.

As described in section 10.2.3, the wavelength (308 nm), power density and beam scan speed of the excimer laser beam are critical parameters to avoid device damage during the laser lift-off process. Table 10.5 indicates the typical laser beam parameters for lift-off process.

10.3.6 Lamination of Backing Plastic Film and Cut to Cell Size

The lamination of a protection film on the bottom of PI of OLED cell is the next process after delamination, since the 10–20 μm thickness of the PI film would be a challenge for transfer and driver IC bonding. A thicker

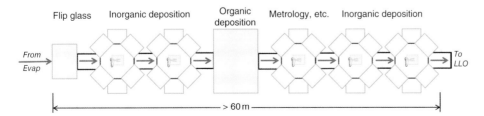

Figure 10.23 Schematic layout of TFE equipment for two SiN_x layers with ink jet organic layer interlayer.

Table 10.5 Process parameters in Gen 5.5 size PI lift-off.

Beam width :	750 mm
Wavelength :	308 nm
No. of scan :	2
LLO power :	235 mJ/cm^2
Throughput:	~70 panels/hour

supporting plastic film is needed for cell handling purpose. The second purpose of laminating thicker plastic film is as a mechanical stress absorbing layer and better reliability of the flexible OLED, by shifting the neutral zone to reduce layer stress and avoid cracks in the TFT and barrier layers [43] during flexing. The laminated PI/plastic film with total thickness of 80–100 μm is then cut into the individual cell size.

10.3.7 Touch Sensor Attach

As described earlier in section 10.2.4, the touch sensor for rigid OLED uses on-cell touch structure where the patterned ITO sensor layer is directly fabricated on the encapsulation glass. Since for flexible OLED structure this structure would require patterning the ITO touch sensor directly on top of TFE and risking damage to underneath layers, initially the touch sensor used a transparent film solution attached on the TFE surface by OCA (optically clear adhesive) [44]. However, in recent products, direct patterning of touch sensor on TFE has been adopted successfully.

10.3.8 Circular Polarizer Attach

The next step is attaching a circular polarizer film on top of the touch sensor layer by OCA lamination. The circular polarizer is required to enhance the image quality of OLED display by eliminating the mirror-like reflected image from the top aluminium electrode array.

10.3.9 Module Assembly (Bonding Drive IC)

Bonding the display controller IC to the flexible OLED panel requires a new structure called chip on film (COF) structure because the bonding substrate is changed from glass to PI film and the smartphone design trend requires slim and narrow bezel feature. Therefore, chip on glass (COG) and chip on flexible PCB methods are no longer valid for flexible OLED. The drive IC and the controller IC have been integrated into a single-chip solution a long time ago. Furthermore, recently the touch controller IC has also been integrated with the display controller IC forming truly a one chip solution. The single chip is mounted on the pre-wired film that forms COF structure, and then the conductive pads in the COF are ACF bonded to the pads on the flexible OLED to deliver the electrical signal to drive the display [45, 46]. Then the COF is bended to the backside of panel forming a slim feature as shown in Figure 10.24.

Up to this point, a mainstream flexible OLED fabrication process currently used for today's commercial products was described. Flexible OLEDs available today, such as in the Galaxy edge type and smartwatches, are regarded as first-generation flexible OLED. These panels are bendable but the bending radius is relatively large. The cost of today's flexible OLED is relatively higher than rigid OLED due to the longer process time and lower product yield. In particular, TFE is a costly process that needs to be improved.

10.4 FOLDABLE OLED

The foldable smartphone is considered as a next generation product expected to be introduced in the market within 1 or 2 years. The essence of foldable smartphones is the use of a foldable OLED display, which is

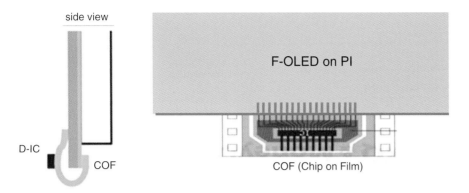

Figure 10.24 COF package solution for bonding one chip drive IC.

an extension of flexible OLED technologies. The display in a folded position is a typical smartphone screen, and when unfolded it becomes a larger screen as shown in Figure 10.25. Although foldable OLED draws a high level of attention because it can possibly provide breakthrough applications, many technical hurdles exist today that should be solved before launching the foldable product.

Stringent material properties are required in every layer of the foldable OLED to allow withstanding over 200,000 times repeated folding and unfolding motions [47, 48]. The layers that make up the foldable OLED may include the foldable OLED, touch sensor, polarizer, and cover plastic as shown in Figure 10.26. Each layer is bonded together by folding endurable OCA-type adhesives. The cover plastic replaces the cover glass that was used in rigid OLED and first-generation flexible OLEDs [49].

Figure 10.25 Photograph of a foldable smartphone with foldable OLED display (*Source:* Samsung Electronics).

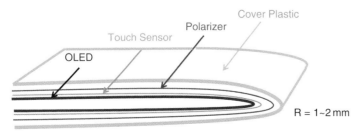

Figure 10.26 Layer structure in foldable OLED.

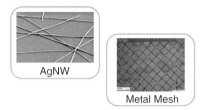

Figure 10.27 Examples of touch sensor candidates for foldable display.

Almost all layers in the foldable OLED structure require new and very challenging material criteria compared to the first-generation bendable type flexible OLED. Selected list of major material properties required in foldable OLED application are as follows:

- For cover plastic window: hardness requirement >7 H (current hardness of cover glass is 7–8 H), should maintain optical transparency after folding test at 1–2 mm radius of curvature.
- For polarizer: the current polarizer with 100 μm thickness cannot withstand the repeated folding motion. Ultra-thin flexible circular polarizer or other methods replacing polarizer film should be searched.
- For touch sensor: traditional ITO-based electrode cannot be used because of its brittleness. Metal mesh or AgNW (silver nanowire) on a polymer film can be candidates for touch sensor material in foldable touch sensor application (Figure 10.27).

Besides the above listed areas, TFT and TFE still require further material improvement for mechanical reliability during the folding action. In general, the material requirement criteria in foldable needs many new composite materials that never been used before and the hurdles are very high.

10.5 SUMMARY

The flexible OLED process described in this chapter reflects what is being used in the current products, regarded as the first generation flexible OLED. The overall fabrication flow for current F-OLED final module is summarized in Figure 10.28.

The foldable OLED is regarded as the next wave of flexible OLED products and is under intensive R&D to be introduced to the market within 1 to 2 years. Therefore, the foldable OLED technology is at its near final stage. The families of flexible OLEDs can be divided into bendable, foldable, rollable, and stretchable and so on. The bending radius for each of the flexible group is different and requires different degree of mechanical strength. The content in this chapter does not represent the whole scope of these flexible OLED family. The process and new materials for next-generation flexible OLED are still evolving. As the new process and new

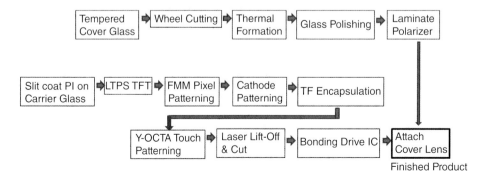

Figure 10.28 Current process flow of flexible OLED module for smartphone.

materials evolve, the expansion of flexible OLEDs to the new areas in conjunction with the new product design concept will continue.

References

1 Y. Chen, J. Au, P. Kazlas, A. Ritenour, H. Gates, M. McCreary, Electronic paper: Flexible active-matrix electronic ink display, Nature, 423, p. 136 (2003).
2 W.-S. Choi, M.-K. Kim, Effect of small liquid crystal molecules on the driving voltage of cholesteric liquid crystal, Displays, 25(5), pp. 195–199 (2004).
3 W. Lee, M. Hong, T. Hwang, S. Kim, W. S. Hong, S. U. Lee, H. I. Jeon, S. I. Kim, S. J. Baek, M. Kim, I. Nikulin, H. I. Jeon, S. I. Kim, S. J. Baek, M. Kim, I. Nikulin, Transmissive 7" VGA a-Si TFT plastic LCD using low temperature process and holding spacer, SID Symposium Digest of Technical Papers, 37(1), pp. 1362–1364 (2006).
4 D.-U. Jin, T.-W. Kim, H.-W. Koo, D. Stryakhilev, H.-S. Kim, S.-J. Seo, M.-J. Kim, H.-K. Min, H.-K. Chung, S.-S. Kim, Highly robust flexible AMOLED display on plastic substrate with new structure, SID Symposium Digest of Technical Papers, 41(1), pp. 703–705 (2010).
5 http://www.samsung.com/ca/consumer/mobile-devices/smartphones/galaxy-s.
6 http://www.apple.com/watch/.
7 J. H. Souk, W. j. Lee, A practical approach to processing flexible displays, SID Symposium Digest of Technical Papers, 18(4), pp. 258–265 (2010).
8 H. Lifka, C. Tanase, D. McCulloch, P. Van de Weijer, I. French, Ultra-thin flexible OLED device, SID Symposium Digest of Technical Papers, 38(1), pp. 1599–1602 (2007).
9 D. Jin, S. An, H. Kim, Y. Kim, H. Koo, T. Kim, Y. Kim, H. Min, S. Kim, Materials and components for flexible AMOLED display, SID Symposium Digest of Technical Papers, 42(1), pp. 492–493 (2011).
10 A. Pecora, L. Maiolo, M. Cuscunà, D. Simeone, A. Minotti, L. Mariucci, G. Fortunato, Low-temperature polysilicon thin film transistors on polyimide substrates for electronics on plastic, Solid Stat. Elect. 52, p. 348 (2008).
11 Z. D. Popovic, H. Aziz, Reliability and degradation of small molecule-based organic light-emitting devices (OLEDs), IEEE Journal of Selected Topics in Quantum Electronics, 8(2), pp. 362–371 (2002).
12 J. Lewis, Material challenge for flexible organic devices, Materials Today, 9(4), pp. 38–45 (2006).
13 P. E. Burrows, V. Bulovic, S. R. Forrest, L. S. Sapochak, D. M. McCarty, M. E. Thompson, Reliability and degradation of organic light emitting devices, Appl. Phys. Lett. 65, 2922 (1994).
14 A. B. Chwang, M. A. Rothman, Sokhanno Y. Mao, Richard H. Hewitt, Michael S. Weaver, Jeff A. Silvernail, Kamala Rajan, Michael Hack, Julie J. Brown, Xi Chu, Lorenza Moro, Todd Krajewski, Nicole Rutherford, Thin film encapsulated flexible organic electroluminescent displays, Appl. Phys. Lett. 83, p. 413 (2003).
15 J. S. Lewis, M. S. Weaver, Thin-film permeation-barrier technology for flexible organic light-emitting devices, IEEE Journal of Selected Topics in Quantum Electronics, 10(1) (2004).
16 A. B. Chwang, M. A. Rothman, S. Y. Mao, R. H. Hewitt, M. S. Weaver, J. A. Silvernail, K. Rajan, M. Hack, J. J. Brown, X. Chu, L. Moro, T. Krajewski, N. Rutherford, Thin film encapsulated flexible organic electroluminescent displays, Appl. Phys. Lett. 83, p. 413 (2003).
17 M. S. Weaver, L. A. Michalski, K. Rajan, M. A. Rothman, J. A. Silvernail, J. J. Brown, P. E. Burrows, G. L. Graff, M. E. Gross, P. M. Martin, M. Hall, E. Mast, C. Bonham, W. Bennett, M. Zumhoff, Organic light-emitting devices with extended operating lifetimes on plastic substrates, Appl. Phys. Lett. 81, p. 2929 (2002).
18 J. Greener, K. C. Ng, K. M. Vaeth, T. M. Smith, Moisture permeability through multilayered barrier films as applied to flexible OLED display, Applied Polymer, 106(55) (2007).
19 J. D. Affinito, M. E. Gross, C. A. Coronado, G. L. Graff, I. N. Greenwell, P. M. Martin, "A new method for fabricating transparent barrier layers, Thin Solid Films, 290–291(15), pp. 63–67 (December 15, 1996).

20 S. Hong, C. Jeon, S. Song, J. Kim, J. Lee, D. Kim, S. Jeong, H. Nam, J. Lee, W. Yang, S. Park, Y. Tak, J. Ryu, C. Kim, B. Ahn, S. Yeo, Development of commercial flexible AMOLEDs, SID Symposium Digest of Technical Papers, 45(1), Article first published online: 7 JUL (2014).

21 D. Herr, L. Strine, PSLA adhesive material options for flexible OLED packaging, Global LEDs/OLEDs, 2(2), pp. 18–23 (2012).

22 S. Hong, J. Yoo, C. Jeon, C. Kang, J. Lee, J. Ryu, B. Ahn, S. Yeo, Technologies for flexible AMOLEDs, Information Display, 31(1), pp. 6–11 (2015).

23 R. S. Kumar, M. Auch, E. Ou, G. Ewald, C. S. Jin, Low moisture permeation measurement through polymer substrates for organic light emitting devices, Thin Solid Films, 417, pp. 120–126 (2002).

24 M. D. Groner, S. M. George, R. S. McLean, P. F. Carcia, Gas diffusion barriers on polymers using Al_2O_3 atomic layer deposition, Appl. Phys. Lett. 88, p. 051907 (2006).

25 http://www.mpi.co.jp/english/news/200811140450.html.

26 http://www.i-components.co.kr/product/product-000300.html.

27 W. A. MacDonald, M. K. Looney, D. MacKerron, R. Eveson, R. Adam, K. Hashimoto, K. Rakos. Latest advances in substrates for flexible electronics, Journal of the Society for Information Display, 15, pp. 1075–1083 (2007).

28 M. Schaepkens, T. W. Kim, A. G. Erlat, M. Yan, K. W. Flanagan, C. M. Heller, Ultrahigh barrier coating deposition on polycarbonate substrates, J. Vac. Sci. Technol. A 22, p. 1716 (2004).

29 Y. Min, K. T. Won, E. A. Gun, M. Pellow., D. F. Foust, J. Liu, M. Schaepkens, C. M. Heller, P. A. McConnelee, T. P. Feist, A. R. Duggal, A transparent, high barrier, and high heat substrate for organic electronics, Proceedings of the IEEE, 93, pp. 1468–1477 (2005).

30 R. Prasad, D. R. Hollars, Inorganic multilayer stack and methods and compositions relating thereto, US patent, US 20140060648 A1 (2014).

31 http://www.tera-barrier.com/technology.html.

32 http://www.mocon.com/instruments/aquatran-model-2.html.

33 J.A. Bertrand, S. M. George, Atomic layer deposition on polymers for ultralow water vapor transmission rates: the Ca test, 54 th Annual Technical Conference Proceedings, Chicago, IL, pp. 492–496 (April 16–21, 2011).

34 R. Delmdah, The excimer laser: precision engineering, Nature Photonics, 4, p. 86 (2010).

35 R. Delmdahl, M. Fricke, B. Fechner, Laser lift-off systems for flexible-display production, Journal of Information Display, 15, p. 1–4 (2014).

36 C. H. Lee, S. J. Kim, Y. Oh, M. Y. Kim, Y.-J. Yoon, H.-S. Lee, Use of laser lift-off for flexible device applications, J. Appl. Phys. 108, p. 102814 (2010).

37 I. French, D. George, T. Kretz, F. Templier, H. Lifka, Flexible displays and electronics made in AM-LCD facilities by the EPLaR™ Process, SID Symposium Digest of Technical Papers, 38(1), pp. 1680–1683 (2007).

38 J.-C. Ho, Y.-Y. Chang, C.-M. Leu, G. Chen, C.-P. Kung, H.-C. Cheng, J.-Y. Yan, S.-T. Yeh, L.-Y. Jiang, Y.-H. Chien, H.-L. Pan, C.-C. Lee, A novel flexible AMOLED with touch based on flexible universal plane for display technology, SID Symposium Digest of Technical Papers, 42(1), pp. 625–628 (2011).

39 http://informationdisplay.org/IDArchive/2016/JanuaryFebruary/FrontlineTechnologyOxideTFTsReport.aspx.

40 H.-H. Hsieh, C.-H. Tsai, C.-S. Yang, C.-J. Liu, J.-H. Lin, Y.-Y. Wu, C.-H. Fang, C.-S. Chuang, Flexible AMOLEDs with low-temperature-processed TFT backplane technologies, SID Symposium Digest of Technical Papers, 21(8,) pp. 326–332 (2013).

41 S. An, J. Lee, Y. Kim, T. Kim, D. Jin, H. Min, H. Chung, S. S. Kim, 2.8-inch WQVGA flexible AMOLED using high performance low temperature polysilicon TFT on plastic substrates, SID Symposium Digest of Technical Papers, 41(1), 2010, pp. 706–709 (2010).

42 http://www.canon-tokki.co.jp/eng/product/el/mass.html.

43 H.-J. Kwon, H. S. Shim, S. Kim, W. Choi, Y. Chun, I. S. Kee, S. Y. Lee, Mechanically and optically reliable folding structure with a hyperelastic material for seamless foldable displays Appl. Phys. Lett. 98, p. 151904 (2011).

44 S. Kim, W. Choi, W. Rim, Y. Chun, H. Shim, H. Kwon, J. Kim, I. Kee, S. Kim, S.Y. Lee, J. Park, A highly sensitive capacitive touch sensor integrated on a thin-film-encapsulated active-matrix OLED for ultrathin displays, IEEE Transactions On Electron Devices, 58(10), pp. 3609–3615 (2011).
45 S.-y. Jung, K.-H. Lee, S.-Y. Song, Y.-s. Choi, O.-j. Kwon, J.-h. Ryu, Y.-C. Joo, E.-a. Kim, Display device, organic light emitting diode display and manufacturing method of the same, US20090167171 A1 (2016).
46 C.-S. Ko, J.-D. Lee, Organic light emitting display and method of manufacturing the same, US8436529 B2 (2015).
47 H.S. Shim, I.S. Kee, S.K. Kim, Y.T. Chun, H.J. Kwon, Y.W. Jin, S.Y. Lee, D.W. Han, J.H. Kwack, D.H. Kang, H.K. Seo, M.S. Song, M.H. Lee, S. C. Kim, A new seamless foldable OLED display composed of multi display panels, SID Symposium Digest of Technical Papers, 1, pp. 257–260 (2010).
48 S. Kim, H.-J. Kwon, S. Lee, H. Shim, Y. Chun, W. Choi, J. Kwack, D. Han, M.S. Song, S. Kim, S. Mohammadi, I.S. Kee, S. Y. Lee, Low-power flexible organic light-emitting diode display device Advanced Materials, 23(31) (2011).
49 S. Kim, H.-J. Kwon, S. Lee, H. Shim, Y. Chun, W. Choi, J. Kwack, D. Han, M.S. Song, S. Kim, S. Mohammadi, I.S. Kee, S. Y. Lee, An 8.67-in. Foldable OLED display with an in-cell touch sensor, SID Symposium Digest of Technical Papers, 46(1), pp. 246–249 (2015).

11A

Metal Lines and ITO PVD

Hyun Eok Shin, Chang Oh Jeong, and Junho Song

Samsung Display Co. Ltd, Youngin City, Gyeonggi-Do, 446-711, Korea

11A.1 INTRODUCTION

The TFT array structures used for display include TFTs, capacitors, pixel electrodes, and interconnect wires. Metal electrodes constituting the above structures have an important role in determining the properties, quality, and yield of the display device.

The metal electrode has progressed together with the simplification of the TFT backplane process architecture, from the earlier Cr and Al electrodes processed in an 8-mask flow, to present Cu electrodes with 4-mask TFT flow. Representative process simplifications were combining photolithography with contact and passivation process through top pixel process architecture, and integrating the process of active-layer and source-drain (see Chapter 2.3). The simplification of process architecture is also accompanied by changes in the metal electrode. For example, in the case of combining the contact and the passivation process, a metal electrode with larger selectivity to the insulating film and durability against dry etching process is required. The metal lines applied to the early stage TFT-LCDs when screen size was relatively small and low resolution, used refractory metals such as Cr, Ta, and Mo for their mechanical strength, thereby helping COF (chip on FPC) package between bus lines and bonding pads. The reasons for selecting these metals during the early stage a-Si TFT production can be summarized as follows: (1) process simplification, (2) excellent chemical resistance and mechanical durability that lead to the high product yield, (3) possible to reinforce gate insulator by anodization.

As the demand for lower resistivity metals gradually increased, two-layered metal stacks such as Al/Cr, Mo/Al, and so on, were investigated to combine the mechanical and chemical resistance to etching process of the clad layer, with the Al layer ohmic characteristics. In these days, Cu based multi-layer metals like Ti/Cu, MoTi-alloy/Cu are commonly used for larger and high resolution panels.

11A.1.1 Basic Requirements of Metallization for Display

Resistivity is the first criterion for selecting the electrode metal in backplane of display such as TFT-LCD or OLED. The resistance of a uniform slab of conducting layer is expressed as in Figure 11A.1. As an intrinsic property of metal, resistance is proportional to resistivity ρ.

$$R = \frac{\rho l}{tw} = R_s \frac{l}{w}$$

where R_s: sheet resistance

Figure 11A.1 Resistance of a rectangular of slab with resistivity ρ.

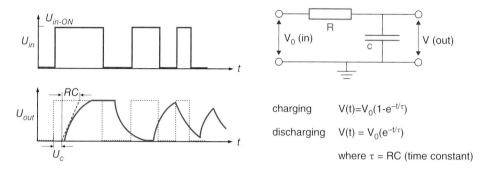

charging $\quad V(t) = V_0(1 - e^{-t/\tau})$

discharging $\quad V(t) = V_0(e^{-t/\tau})$

where $\tau = RC$ (time constant)

Figure 11A.2 RC-delay phenomena of a signal.

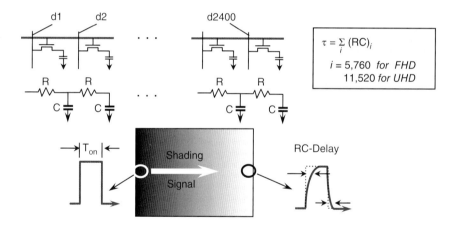

$$\tau = \sum_i (RC)_i$$

$i = 5{,}760$ for FHD
$11{,}520$ for UHD

Figure 11A.3 RC-delay of a gate signal and its shading effect upon propagation across the display panel.

A general issue for transmitting signals through metal wires is the time-delay between input and output in the structure. This delay is generally known as the time delay or time constant (τ) of the RC circuit, and it is the time response of the circuit when a step voltage signal is applied (Figure 11A.2).

High resistance of electrode in RC circuit causes a higher time constant and results in signal distortion in display area as shown in Figure 11A.3. Therefore, lower resistivity metal lines are required to improve the time delay caused by panel size increase, high-resolution, and high-frequency driving.

The development of metal electrodes progressed to a simplified TFT backplane process architecture, from 8-mask-step Cr electrode to the current 4-mask-step Cu electrode. Table 11A.1 shows the metals used in display panels and we will discuss the details of physical properties of each metal.

Cu and Al have very low electrical resistivity and can be processed with low material cost, therefore they are widely used as main electrode metals in displays and semiconductor devices today. Silver has the lowest resistivity among all metals but it is a high cost material (Figures 11A.4 and 11A.5).

Table 11A.1 Physical properties of basic conductive materials used in display panel manufacturing.

Metal	Ag	Cu	Al (Al alloy)		Mo	Ti	*ITO	*TiN
			Pure Al	Al alloy				
ρ ($\mu\Omega$-cm) BULK	1.59	1.68	2.82	-	5.6	42	-	25
ρ ($\mu\Omega$-cm) Thin Film (After annealing at 350 °C)	1.8	2.1	3.0	3.2~4.3	14	70	180~250	100~1000
Hardness (Mohs)	2.5	3.0	2.8	>3.0	5.5	6.0	>5	9
Density (g/cm3)	10.5	9.0	2.7	2.7	10.3	4.5	7.2	5.2
Young's Modulus	83	120	70	>100	329	116	190	600
Adhesion to dielectric (SiO2 or SiNx)	Not bad	Bad	Good	Good	Not bad	Excellent	Good	Excellent
Compound	Ag_2O Ag_2S	Cu_2O Cu_2S	Al_2O_3	Al_2O_3	MoO_3	TiO_2		
Material cost	Very high	Low	Low	Low	High	Low	Very high	Low

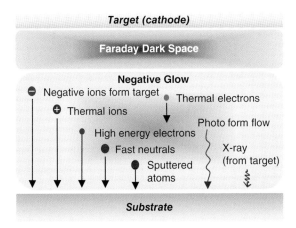

Figure 11A.4 The basic processes occurring at the surface of the substrate.

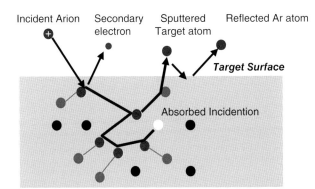

Figure 11A.5 Mechanism of linear collision cascade [1].

11A.1.2 Thin-Film Deposition by Sputtering

The physical mechanism of the sputtering process is momentum transfer between the colliding ions in the gas phase and the atoms in the target material. Colliding ions are fast moving inert argon (Ar) ions from plasma. Due to momentum transfer, atoms can be ejected from the target material [1]. As shown in Figures 11A.4 and

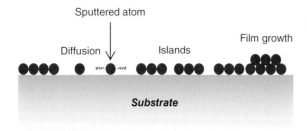

Figure 11A.6 Arrived species on the substrate in a sputtering chamber.

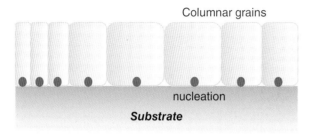

Figure 11A.7 Nucleation and growth of thin film.

11A.5, ejected atoms from the target materials have sufficient energies to reach the substrate where the film is being deposited. Sputtered atoms have energy of 10–30eV with velocity $3-6\times10^3$ m/sec initially, but undergo collisions that reduce their energy to 1–10eV but still high enough to make a dense thin film. For comparison, the energy of evaporated atoms is only 0.1eV in the thermal evaporation method.

The basic processes occurring at the surface of the substrate are shown in Figures 11A.6 and 11A.7. The mobility of the incident atoms arriving at the substrate is highly dependent upon the sputtering parameters (pressure and power), the substrate temperature, the distance between target and substrate and the surface conditions.

There are several factors that determine the growth and quality of thin film. These factors are as following:

(1) Incidence energy of sputtered atom → High incidence velocity hamper surface diffusion and results in small grain size, which is controlled by Ar gas pressure and DC power of sputter system. Figure 11A.4 shows the particles in sputtering chamber.
(2) Surface diffusion effect → High surface temperature makes high diffusion of atoms on the substrate.
(3) Seed of thin film → Source of grains of thin film depends on defects, polarity, and surface roughness, high seed density reduce grain size and makes small grain size.
(4) Negative Ions → Formed near cathode by secondary electrons.
(5) Impurity gas effect → Usually oxygen and nitrogen in the process chamber.

When the sputtered atoms arrive at the substrate, a thin film is formed. The morphology, microstructure, and crystallographic orientation of the thin films depend on the growth conditions, which is sometimes depicted in a so-called structure zone model. Structure zone diagrams of Figure 11A.8[2] shows that the microstructural evolution as a function of homologous temperature (Ts/Tm) (shown on left figure), and as a function of the energy of the impinging atoms on the substrate (shown on right figure). Films deposited at substrate temperatures far from their melting temperature (Ts/Tm < 0.2) usually consist of fibrous columnar grains with voided grain boundaries while films deposited at higher temperatures (0.5 < Ts/Tm < 0.8) consist of large columnar grains with dense grain boundaries. At temperatures close to the melting point (Ts/Tm > 0.8), the deposited films consist of large equiaxed grains formed due to recrystallization and segregation of impurities to the surface and grain boundaries. These growth regimes are called Zone I, II, and III in the order

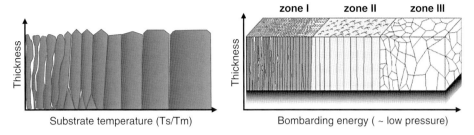

Figure 11A.8 Structure zone model by Thorton [2].

of rising deposition temperature. Increasing the surface diffusivity, by higher deposition temperature and/or energetic bombardment of the growing film leads to gradual changes in the film structure: slightly larger grains, smoother grain surface, and denser grain boundaries. Low energy bombardment only affects the first few atomic layers leading mainly to enhance surface diffusion. Higher energy bombardment (few tens of eV depending on the material) leads to defect creation in the first few atomic layers. Refractory metals such as Mo or Ti having high melting point film can be controlled not by substrate temperature but generally by DC power or Ar pressure.

The resistivity of thin film metal is quite different from bulk metal. When film thickness decreases, film's bulk characteristics disappear and size effect changes the electric properties dramatically as shown in Figure 11A.9. Grain boundary and electron-scattering phenomena in thin metal film are main reasons for the higher resistivity than bulk metal. Fuchs-Sondheimer model (FS model) [3] of an ideal continuous metallic thin film represents the following conductivity equation of thin film with the condition of scattering process by two plane surfaces and assuming the Boltzmann equation for the distribution function. Besides geometrical limitations that enhance surface scattering, there are other sources of resistivity:

- Volume defects and impurities.
- Lattice vibrations.
- Strain and discontinuities including grain boundary.

The presence of grain boundaries may reduce significantly the conductivity of metallic materials. Mayadas and Schatzkes (MS model) [3] were the first to provide a model that takes into account both the scattering at external surfaces (size effect) and scattering due to grain boundaries. They showed how a reduction of the grain size can induce a significant increase of resistivity as shown in Figure 11A.10. On the other hand, voids may also play a role in determining the electrical behavior of ultra-thin metal films. Thickness of metallic films is comparable to the electron mean free path and one-dimensional effect is occurred in the plane of the film.

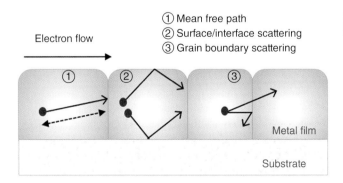

Figure 11A.9 Schematic diagram showing the grain boundary size effects on the electron flow.

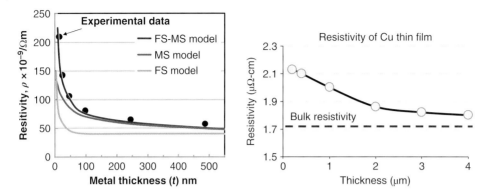

Figure 11A.10 Comparison of the resistivity of Cu/Ta multi-layers measured experimentally and predicted in terms of the F-S, M-S, and the new FS-MS combined models [3].

Resistivity of thin film by Fuchs-Sondheimer (FS) model is given by

$$\rho = \rho_o \left(1 + \frac{3\lambda}{8t}(1-p)\right) \tag{11A.1}$$

and resistivity by Mayadas and Schatzkes (MS) model is given by

$$\rho = \rho_o \left[1 - \frac{3\alpha}{2} + 3\alpha^2 - 3\alpha^3 \ln\left(1 - \frac{3\lambda}{8t}\right)\right] \tag{11A.2}$$

where $\alpha = \frac{\lambda}{d}\frac{R}{1-R}$

ρ_o : resistivity of bulk metal
p : surface reflection coefficient
λ : mean free path of conducting electrons
R : grain boundary scattering coefficient
d : grain size of thin film

11A.2 METAL LINE EVOLUTION IN PAST YEARS OF TFT-LCD

The electrode lines in the early stage TFT-LCDs, when screen size was relatively small and low resolution, used metals such as Cr, Ta, and Mo due to their mechanical strength, thereby helping COF (chip on FPC) package between bus lines and bonding pads. The reasons for selecting these metals at early stage can be summarized as:

(1) for the process simplification purpose by using same metal in metal lines and bonding pads
(2) excellent chemical resistance and mechanical durability that lead to the high product yield
(3) possibility to reinforce gate insulator by anodization.

As the demand for the lower resistivity of metals gradually increased, double-layered metal clads like Al/Cr, Mo/Al, and so on, were studied. There was no hillock phenomenon observed in Al/Al-Nd alloy layers. Toshiba applied MoW alloy that has relatively low resistivity. Additionally, triple-layer structures such as Mo/Al/Mo and Ti/Al/Ti capping layer structures were used for source/drain electrodes since they endure etching process and satisfy the ohmic characteristics as compared to using Al single layer, which has weak chemical and mechanical resistances. Presently, Cu-based multilayer metals such as Ti/Cu, MoTi/Cu are commonly used for larger and high-resolution panels.

11A.2.1 Gate Line Metals

11A.2.1.1 Al and Al Alloy Electrode

The selection of metal material is limited by the physical properties and the etching process. Table 11A.2 shows the properties of some metals that are used in LCD panel manufacturing. For the gate electrode metal, low resistivity, thermal resistance, and easy etching properties are necessary. Aluminum electrode has low resistivity and is etched easily using a PAN etch solution (mixture of phosphoric-acetic-nitric acids).

The compressive residual stress in the Al layer resulting from the thermal process history can be relieved by the extrusion of a hillock, which sometimes causes cracking of the insulating dielectric overlayer and results in an electrical short between gate and source drain electrodes. Figure 11A.11 shows the mechanism of hillock formation in pure Al thin films and a representative surface morphology.

Capped double-layer structures such as Mo/Al or Ti/Al are a good gate electrode structure, require only one photomask and one wet etching process, and have low material cost. A low temperature ILD (inter layer dielectric) process was developed to minimize Al hillock formation. Refractory metals such as Mo, Ti, W with

Table 11A.2 Metals and their properties used in TFT-LCDs.

	Ag	Cu	Al	AlNd	Mo	Cr
Resistivity (μΩ-cm)	2.1	2.2	3.1	4.5	14	22
N+ contact	Ohmic	Non-ohmic	Non-ohmic	Non-ohmic	Ohmic	Ohmic
ITO contact	~1 E04 Ω	~1 E04 Ω	~1 E09 Ω	~1 E09 Ω	~1 E04 Ω	~1 E05 Ω
Adhesion	X	X	O	O	O	O
Dry etchable	X	X	O	X	O	X
Application	Reflector	Gate or SD Line	Gate or SD Line	Gate or SD Line	Barrier	Barrier

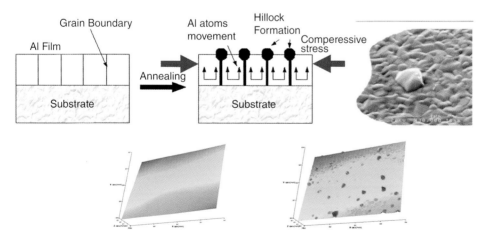

Figure 11A.11 Mechanism of hillock formation in pure Al thin film (top figure). The CTE difference between Al and glass (23.9 versus 4.5 ppm/K, respectively) results in compressive stress, driving hillock formation in Al for stress relaxation. Surface morphology of Al-alloy (bottom left figure) and hillock formation in pure Al film after annealing at 350 °C (bottom right figure).

Table 11A.3 CTE (Coefficient of thermal expansion) of selected materials.

Materials	CTE ($10^{-6} K^{-1}$)	Materials	CTE ($10^{-6} K^{-1}$)
Al	23.9	Glass	4.5
Mo	4.8	IZO	6.95
Cu	16.5	ITO	8.1
Ti	8.6	SiO_2	0.5
a-Si	0.5	Si_3N_4	0.8

high Young's modulus are used as capping metals. Thermally resistive Al alloys were also developed to suppress hillock formation up to 400 °C annealing temperature. Table 11A.3 shows the coefficient of thermal expansion of some materials.

To prevent Al hillock in the early days of LCD panel manufacturing, Al anodizing process was performed and the top surface of Al was changed into Al_2O_3 by oxidation process. Top layer of Al film was converted into Al_2O_3 layer which can prevent Al hillock formation. But this anode oxidation process required additional photomask to protect gate pad area by photoresist to prevent formation of insulator Al_2O_3 layer. Capping metal with high Young's modulus can efficiently prevent Al hillock formation as shown in Figure 11A.12 and Table 11A.4.

To reinforce the thermal stability of Al-based electrodes, a variety of Al alloy were developed. Al-Nd is one such example—while pure Al has hillock formation at 230 °C annealing temperatures, Al-Nd alloy can suppress hillock formation at annealing temperatures higher than 350 °C due to the formation of intermetallic Al_4Nd as shown Figure 11A.13. On the other hand, the resistivity of Al-Nd is around 4.5 μΩ-cm, which is 45% higher than that of pure Al. In the early days of LCDs, a bilayer gate metal structure with Cr/AlNd (bottom/top) was introduced but the Cr/AlNd structure needed a separate etching step for each layer. Later, an AlNd/Mo structure was developed where the AlNd/Mo layer was batch etched in one step with PAN wet etchant. By introducing bilayer gate metal, it was possible to reduce photo mask number from seven to five, including the modified contact process. Reducing the mask steps in the process architecture is key to achieve low cost and higher throughput in TFT-LCD mass production.

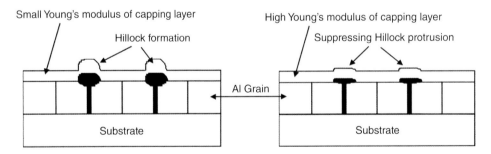

Figure 11A.12 Capping materials with higher Young's modulus suppressed Al hillock formation.

Table 11A.4 Young's modulus of metal or capping layer (*1E+10 Nm^{-2}).

Mo	Cr	Cu	Ag	Al	Ta	Ti	AlNx	MoNx	TiNx	AlOx
33	15.7	12.3	7.9	7.1	18.2	11	≤33	≤45	≤60	≤35

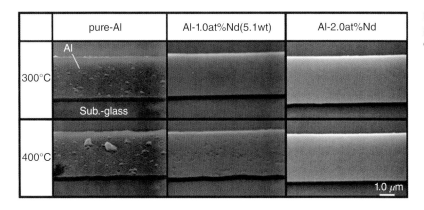

Figure 11A.13 Thermal resistance of pure Al versus Al alloy (AlNd, for example).

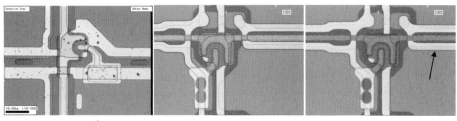

Pure Al only (2500Å)
→ G/D Short defect by Al hillock

AlNd/Mo (2500/500Å)

Al/Mo (2500/500Å)

Figure 11A.14 Al and Al-based bilayer gate structure.

As the demand increased for larger size and higher resolution LCD panels, low-resistance electrode materials become much more important. Although Al-Nd alloy is a thermally strong metal, resistivity is much higher than pure Al. Alternatively, bi-layer structures with pure Al and Mo capping layer were developed to suppress hillock formation within the process temperature of TFT-LCD flow. As shown in Figure 11A.14, there was no difference observed between AlNd/Mo and Al/Mo cases after TFT process, and thus AlNd was replaced by pure Al in bilayer structures.

11A.2.1.2 Cu Electrode

The resistivity of Cu is 30% lower than that of Al (Table 11A.5), which means that either the thickness of the electrode can be reduced 30% or the line width can be 30% narrower. A narrower metal line width allows reducing the metal area in the pixel and increasing the LCD panel transmittance. Another benefit for using Cu is related to its thermal stability and absence of hillock formation issues, and thus can be used as a gate metal without capping layer. However, Cu has inherently low adhesion characteristic to glass or to dielectric

Table 11A.5 Comparison of Al and Cu.

	Al	Cu
Resistivity ($\mu\Omega$-cm)	3.1	2.1
Gate thickness (Å)	4000	2700
S/D thickness (Å)	6000	4000

layers such as SiO_2 or SiN_x and thus requires adding a bottom adhesion layer. Additionally, using Cu metal for source/drain electrodes requires adding a diffusion barrier to separate Cu from directly contacting a-Si layer. Mo, Mo-alloys, or Ti are generally used as a barrier layer of Cu metal electrodes.

Cu-based source/drain electrodes have additional advantages for reducing the number of metal layers. The triple layer structure of Mo/Al/Mo can be reduced to Mo/Cu or Ti/Cu double layers. This simpler structure enables reducing the process time and the material cost.

11A.2.2 Data line (Source/Drain) Metals

Data line metals are required to have a low resistance and no diffusion into a-Si layer. Al and Cu have very low resistance characteristics, but they have strong inter-diffusion with Si layer (Table 11A.6). Thus, the use of Al and Cu metals for data line used bilayer or triple layer structures with Mo, Mo-alloy, or Ti as barrier layers.

The structure and selection of data line metals have progressed with the process architecture. In the early days, Cr 5-Mask, Cr 4-Mask, Cr/AlNd 5-Mask, Mo 4-Mask processes were used. Presently, Mo/Al/Mo 5-Mask, Mo/Al/Mo 4-Mask and Mo/Cu or MoTi alloy/Cu, Ti/Cu 4-Mask structure are used.

11A.2.2.1 Data Al Metal

Al has been used as the main material for gate or data lines (or source/drain) electrodes for a long time. Al has strong inter-diffusion with a-Si and a high contact resistance with ITO pixel layer, therefore Al is used with a barrier layer such as Mo, Ti, and Cr and in triple layer structure such as Mo/Al/Mo, Ti/Al/Ti, and Cr/Al/Cr as shown in Figure 11A.15. The bottom barrier layer is necessary to prevent inter-diffusion between Al and

Table 11A.6 Reactive temperature with Si for selected metals as source/drain line.

Metal	Resistivity (μΩcm)	Melting point (°C)	Coefficient of thermal expansion (ppm/°C)	Reactive temperature with Si (°C)
Al	3.1	660	23.9	~250 (junction spike)
Cu	2.3	1083	17	Less than 100
Mo	11.5	2610	5	400 ~ 700
Ti	86.5	1668	8.5	400 ~ 1000
Cr	21	1907	4.9	400 ~ 450

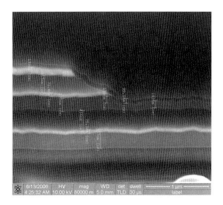
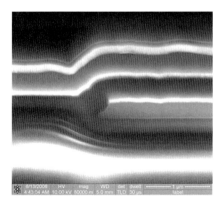

Figure 11A.15 Cross-section of Gate Al/Mo and SD Mo/Al/Mo structures.

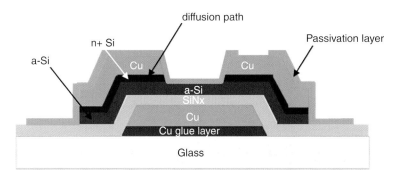

Figure 11A.16 Cu-based TFT structure for TFT-LCD.

n+ a-Si. The top barrier layer is used for preventing Al hillock and reducing contact resistance with pixel layer, which is deposited above the data line metal.

11A.2.2.2 Data Cu Metal
Use of Cu material for source and drain (S/D) electrodes has the advantage of low material cost and high electrical performance. But, Cu atoms can easily diffuse into a-Si during thermal processing and also Cu tends to have poor adhesion properties with dielectric under layer. Thus, adding a barrier layer for Cu is essential in Cu metal structures (Figure 11A.16).

Barrier metals that can prevent Cu atom drift into the semiconductor active layer (a-Si, p-Si, or IGZO) are: Ta, Ti, TiN, Mo, Mo alloy, MoN, and so on. Without such a barrier, TFT devices will typically develop a reliability problem (tested by bias temperature stress) as Cu atoms diffuse into active Si layer. Considering process integration and wet etching, a proper selection of the barrier material is very important to allow etching with batch process. Mo or Mo alloy metals are widely used, employing peroxide-based chemical etching, as well as Ti barrier metals with persulfate-based wet etching process.

11A.2.2.3 Data Chromium (Cr) Metal
Cr with thin film resistivity of ~21 μΩcm was used as data line metal for TFT-LCD in the early days of notebook PC applications. Cr has excellent chemical resistance and is corrosion proof even against aggressive ITO etchants such as aqua regia solution. However, its high resistivity and high tensile stress properties limited the film thickness and resulted overall in limited applications in LCD panel manufacturing. The bilayer structure of Cr/AlNd was used, but due to the environment issues of Cr, replacement with Mo or other metals was developed.

11A.2.2.4 Molybdenum (Mo) Metal
Mo also has relatively high resistivity of 13 to 16 μΩcm and limited use as single layer for gate or S/D electrode applications. However, it has been used as metal electrode for limited panel sizes of less than 20″ such as notebook PC, desktop monitors, or mobile OLED panel applications.

Figure 11A.17 shows the cross-section of S/D electrode stack with Mo layer. Mo has high tensile stress at high-pressure sputtering condition and has possibility of micro-cracking problem. To prevent this issue, low-pressure sputtering is necessary to release strong tensile stress. Figure 11A.17 shows the stress-induced crack of Mo film deposited in a high sputtering pressure condition. As a refractory metal, Mo is relatively easily controlled by process pressure or DC power more than the temperature of substrate. Mo also has excellent direct contact resistance properties with ITO or IZO pixel electrodes, and can be used as a-Si diffusion barrier or capping layer for Al or Cu electrodes.

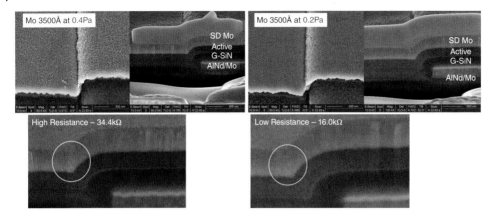

Figure 11A.17 Crack defect of S/D Mo layer as a function of process pressure (FIB image).

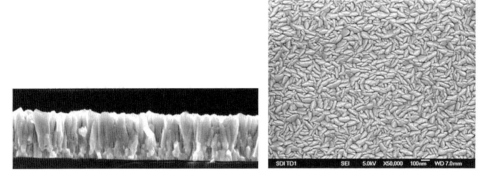

Figure 11A.18 Columnar microstructure of Mo thin film.

Intermetallic Mo-silicide compounds, $MoSi_2$ and $MoSi_3$, form excellent barrier layers between Al or Cu and active layers such as a-Si, p-Si, or IGZO layer preventing diffusion of Al or Cu atoms into the active layer. Mo has a columnar microstructure as shown in Figure 11A.18.

11A.2.2.5 Titanium (Ti) Metal

Ti is a corrosion-resistant transition metal with very high melting point, and excellent adhesion properties that allow it to be used as glue layer for Al or Cu metals. Ti_3Al intermetallic compound is formed at Al interface and Cu_4Ti intermetallic layer is formed at Cu interface. Ti can form intermetallic Ti-silicide compounds, such as TiSi, $TiSi_2$, and Ti_5Si_4, allowing it to be an excellent barrier layer between Al or Cu and active layers such as a-Si, p-Si, or IGZO (Figure 11A.19).

Its oxide, TiO_2, has insulating characteristics, and thus the Ti/ITO interface has a relatively higher contact resistance than Mo/ITO. Titanium nitride is very hard and conductive layer and could be used as a good conducting barrier layer, however, titanium nitride is hard to etch by general chemical solutions. Ti/Al/Ti structures are generally etched by dry etching process and used for high-resolution array panels. A fluorine chemical source is necessary to etch Ti, which requires careful process integration since fluorine sources also etch dielectric oxides and glass.

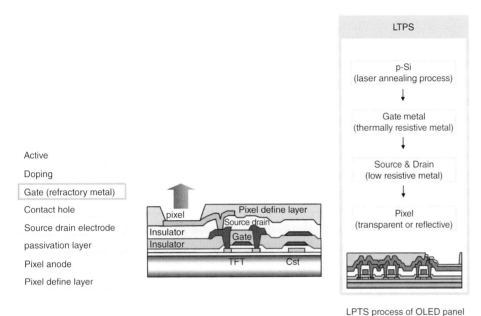

Figure 11A.19 Typical top gate LPTS structure and process for OLED panel (example of top emission structure).

11A.3 METALLIZATION FOR OLED DISPLAY

11A.3.1 Gate Line Metals

OLED devices typically use LTPS or oxide semiconductor TFTs, and thus their substrates are exposed to a high thermal process during fabrication of up to 500 °C, or 350 °C, respectively. The gate metal has to endure very high thermal stresses in the LTPS process and thus refractory metals such as Mo or hillock free Cu or Al alloy need to be used as gate metal electrodes.

For small-sized OLED panels, Mo can be used for gate electrode. Since Mo is patterned with dry etch processes, Mo-based gate electrode are advantageous to make high-resolution TFT arrays. Mo has resistivity of 13~16 μΩ-cm, which, although is much higher than Al or Cu, is sufficient for applications in panels less than 10″. However, due to the high intrinsic residual stress properties of Mo, the thickness of gate Mo layer structure is limited.

11A.3.2 Source/Drain Metals

In conventional TFT manufacturing processes, the data electrode metals for OLED experience thermal processes limited by the polyimide curing temperature of 230 to 300 °C. Cu/Mo, Cu/Mo-alloy, and Cu/Ti bilayer structures are generally patterned with wet etch processes in large OLED panel fabrication. Direct contact between Cu and pixel electrode is possible, therefore, Cu bi-layer structure is possible in OLED process architecture but top TCO layer is necessary on the Cu/barrier structure for pad layer.

The function of the bottom barrier layer is to prevent Cu atoms diffuse into the active layer. Refractory metals such as Mo, Mo alloys, or Ti are used as diffusion barrier and as glue layers on dielectric materials such as SiO_2 or SiN_x

Pure Al can also be used as source drain (or data line) metals. Triple layer Ti/Al/Ti structures are necessary for the Al-based S/D electrode, with the top capping Ti necessary for direct contact with anode layer of ITO or IZO and the bottom Ti as good barrier metal for the active layer (LTPS or IGZO). The Ti/Al/Ti structure can be dry etched in one patterning step, and thus this trilayer electrode structure is suitable for high-resolution patterning with widely implementation for high resolution OLED S/D metal. Stable intermediate phases, Si_2Ti, $SiTi$, Si_4Ti_5, Si_3Ti_5, and $SiTi_3$, can form at the interface and prevent Al diffusion into the Si layer. Mo/Al/Mo structures can also be used for OLED S/D electrodes, but the usage is limited typically to large-sized OLED panel with wet etching process.

11A.3.3 Pixel Anode

There are two types of OLED structure based on their light emission direction: bottom emission, and forward (or top) emission structure as shown in Figure 11A.20.

The anode layer needs conductive and high work function characteristics to allow efficient hole injection into the organic emission layers shown in Figure 11A.21. Indium oxide based transparent electrode materials such as ITO or IZO are generally used for anode layers with high work function of 4.6 to 4.8eV. Other metals, such as Au, Ni, and Pt, and oxides, such as molybdenum oxide, can also have high work function and be used as anode in OLED devices.

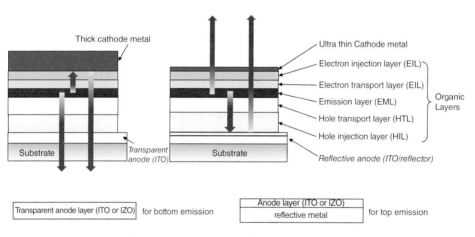

Figure 11A.20 Bottom- and top-emission structure of OLED.

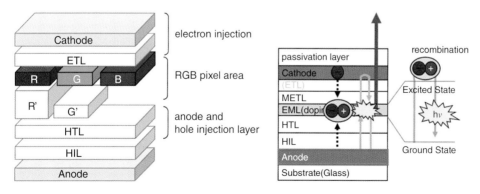

Figure 11A.21 Basic EL structure of OLED and emitting mechanism.

The bottom-emission OLED structure requires a transparent conductive anode, and thus ITO or IZO are typical choices. The top-emission OLED structure needs a reflective anode, typically a metal such as Ag or Al that has both high reflectivity and high work function properties.

11A.4 TRANSPARENT ELECTRODE

Transparent electrode materials are typically used for the pixel electrode, and thus should have a high light transmittance (>80–90%) across the visible wavelength range and relatively low resistivity (<1 × 10^{-3} Ω·cm) characteristics. Thus, the characteristics of transparent electrodes can be obtained by using wide band gap materials ($E_g > 3$ eV) with appropriate dopants to create electric carriers. Indium oxide is the main component in transparent conductive oxide (TCO) materials in combination with tin or zinc dopants, that is, indium tin oxide (ITO) and indium zinc oxide (IZO) respectively.

Table 11A.7 shows selected properties of ITO and IZO films. Polycrystalline ITO is used as a common transparent electrode due to its high transparency and low resistivity. Amorphous ITO, a-ITO, or IZO are also widely used as pixel electrodes since they can be more easily etched, and patterned a-ITO can be converted into polycrystalline ITO by subsequent thermal annealing processes. An example ITO composition is 74% In, 18% O_2, and 8% Sn by weight %.

The resistivity of a transparent electrode depends on the carrier mobility, which can be affected by the deposition process temperature, annealing process, grain size, and so on. In general, the low resistivity materials have good crystallinity and lower defects density.

Weak acids are preferred to be used for pixel electrode etching in order to minimize risk for damage to TFT structure due to the etchant penetrating through the pinholes or cracks in the passivation layer. IZO films have excellent patterning properties due to their fully amorphous film structure. In the case of ITO, after repeated cycles of sputtering processing, often nodules are formed on the surface of the ITO target, which become a source of particles and potential defects, one of the main reasons for yield loss in the OLED pixel process. IZO PVD targets generate nearly no nodule, as shown Figure 11A.22. IZO films have a refractive index of around 2.0 and result in lower transmittance compared to ITO films which have a lower refractive index around 1.9.

To further improve the wet etching rate of amorphous ITO (a-ITO) films, H_2O or H_2 gas can be introduced in the ITO PVD process chamber at room temperature. The electrical and optical characteristics of ITO films are heavily dependent on the O_2 partial pressure during deposition process. Furthermore, a-ITO can be changed to polycrystalline ITO after annealing above 150 °C and achieves lower resistivity around 180 to 240 uΩ·cm.

Table 11A.7 Selected properties of ITO and IZO films.

	ITO		IZO (amorphous)
	a-ITO	p-ITO	
Composition (wt%)	90 In_2O_3: 10 SnO	90 In_2O_3: 10 SnO	90 In_2O_3: 10 ZnO
Resistivity	500~600 μΩ·cm	180~200 μΩ·cm	300~350 μΩ·cm
Transparent	60%–80%	≥90%	≥85%
Hardness	100 (ref.)	~110	90–100
Target Nodule	Yes	Yes	No
Etch Rate	~600 Å/min (with oxalic or sulfuric acid)	~600 Å/min (with aqua regia)	~1,800 Å/min (with oxalic or sulfuric acid)

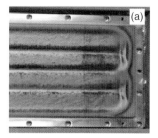
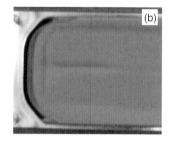
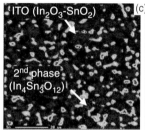
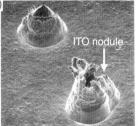

Figure 11A.22 (a) ITO target surface after extensive usage showing nodule formation, (b) no nodule on IZO targets, and (c) ITO target surface composition and SEM images of cone-shaped nodules [4].

References

1 M. Fluss, J. Marian, J. Marian, Charged-particle irradiation for neutron radiation damage studies, characterization of materials, edited by E. N. Kaufmann, John Wiley & Sons, Inc., SID 96 Digest, p. 2113 (2011).
2 F. Vollertsen, Micro Metal Forming, pp. 177–199, https://www.researchgate.net/publication/278653688_Size_Effect-Enabled_Methods (2013).
3 M. Wang, B. Zhang, G. P. Zhang, Effects of interface and grain boundary on the electrical resistivity of Cu/Ta multilayers, *J. Mater. Sci. Technol.*, **25**(5) (2009).
4 M. Schlott, M. Kutzner, Nodule formation on indium-oxide tin-oxide, SID 96 Digest (2015).

11B

Thin-Film PVD: Materials, Processes, and Equipment
Tetsuhiro Ohno

FPD PV Division, ULVAC, Inc., 2500 Hagisono, Chigasaki, Kanagawa, Japan, 253-8543

11B.1 INTRODUCTION

In 1993 when the FPD (flat panel display) market was booming, ULVAC developed and released the first cluster-type sputtering PVD (physical vapor deposition) equipment for use in the TFT (thin-film transistor) array manufacturing process. Since then ULVAC provided sputtering equipment as a standard for deposition processes in TFT array fabrication and continued to timely evolve the equipment concept to meet the industry needs. The major evolution in equipment has been the plate size enlargement, matching the corresponding panel size expansion. Presently the largest equipment currently available in production is for Gen 10 size (2880×3130 mm) glass substrates for TV applications (Figure 11B.1), and the next-generation is already in progress (Figure 11B.2).

For Gen 2 size and Gen 3 size, cluster tools with a simple multi-chamber architecture were employed, and then evolved to X-type for Gen 4 size to Gen 6 size, and vertical cluster type for Gen 8 size and Gen 10 size. The glass size had grown by 52 times from Gen 2 size to Gen 10 size, however, the foot print increase of the sputtering tool has only increased about 18 times.

Recently the LTPS (low-temperature poly silicon) technology for small- and medium-sized high-definition displays in the smartphones and tablets has also expanded to using Gen 6 size equipment. Also, it is expected that Gen 6 size fabs will soon release new TFT devices to the market for sophisticated high-definition panels and new product applications (e.g., flexible display and on-vehicle display). Under these circumstances, ULVAC is accelerating the sputtering equipment development for transparent oxide semiconductor (TOS) film deposition, since this film is very promising for TFTs for electro-luminescent (EL) devices as well as for LCDs.

Sputtered Al (aluminum) film has generally been used as the wiring material (metallization) in TFT devices from the beginning stages of TFT-LCD production. However, when the display panel sizes increased and requirements expanded to higher-definition specifications, the problems related to hillock and electrical contact for Al have become more critical. Although high melting point metals or aluminum alloys can be solutions against those problems, the process of those metals is more complex and increases the risk of abnormal discharge during the sputtering process. Additionally, another critical issue in metallization has been requirement for lower resistance.

For these reasons, copper (Cu) as a new wiring material has been actively developed, and is now being used in some of the devices. Its simple structure is also useful to reduce the processing cost.

Flat Panel Display Manufacturing, First Edition. Edited by Jun Souk, Shinji Morozumi, Fang-Chen Luo, and Ion Bita.
© 2018 John Wiley & Sons Ltd. Published 2018 by John Wiley & Sons Ltd.

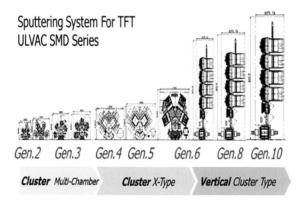

Figure 11B.1 Sputtering equipment configurations according to glass substrate generation (ULVAC).

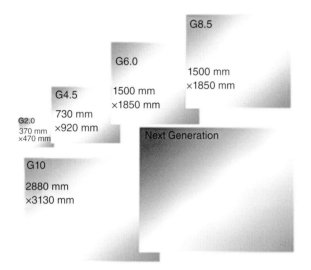

Figure 11B.2 Transition of substrate size for TFT-LCD.

In this paper, the evolution of ULVAC sputtering equipment used across all the TFT manufacturing generations is presented. Then, the new TFT material used in the TOS represented by IGZO (indium gallium zinc oxide) is discussed, including suggestion of the optimal cathode type as well as the sputtering technology.

11B.2 SPUTTERING METHOD

Sputtering is a phenomenon where the surface atoms or molecules of a solid target are ejected when high energy particles collide against its surface. In the typical sputtering process, argon gas is ionized by DC or RF discharge and collides into an electrode called "target" (a metal or oxide made of the base material to be deposited), so that the sputtered material accumulates on the substrate. The planar magnetron sputtering method with a transversal magnetic field was invented by J. S. Chapin in 1974, and since then this PVD method has been used widely in the industry though the slower deposition speed compared to vacuum deposition is a problem.

Figures 11B.3 and 11B.4 describe the electron motion in the planar magnetron sputtering method [1]. The magnetic circuit composed of, for example, an S-pole magnet at the center and an N-pole magnet on the outer

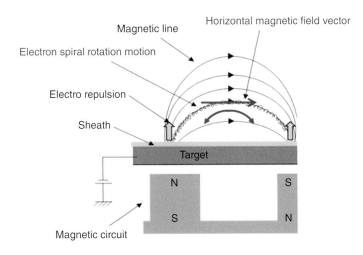

Figure 11B.3 Diagram of electron motion in planar magnetron sputtering method.

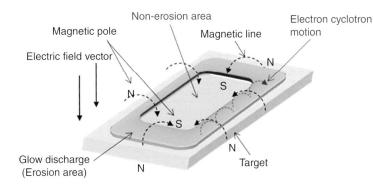

Figure 11B.4 Discharge and erosion in planar magnetron sputtering method.

periphery placed in a N-S-N magnetic arrangement on the rear of the electrode (target) create warped magnetic lines.

Secondary electrons are generated as gas ion atoms collide against the target surface, and they move in a spiral rotary motion due to the magnetic field lines. When secondary electrons enter the electric field on the target surface, they move in a reciprocating motion due to repulsion. As the number of collisions with gas atoms increases through this motion, the ionization rate of gas atoms increases leading to formation of high-density plasma. The density of the formed plasma is highest at the point where the magnetic field has the strongest vector horizontal to the target (horizontal magnetic field). Therefore, in the magnetron sputtering method, the plasma density distribution depends on the strength of the horizontal magnetic field. This causes uneven sputtering of the target affecting the uniformity of the film deposited on the substrate and degrading the target usage efficiency. Control using both the magnetic field and electric field is important for improvement in such event. In addition, it is well known that the quality of the sputtered thin film greatly depends on the absolute value of the plasma density.

Another type of magnetron sputtering method has the magnetic circuit located inside a cylindrical target and the target is rotated during film deposition (rotary cathode). In response to panel manufacturer requirements for sputtering equipment in TFT array process, ULVAC have been using a planar target based on its original process technology (cathode selection) and on CoO (cost of ownership) estimations (details are described later).

11B.3 EVOLUTION OF SPUTTERING EQUIPMENT FOR FPD DEVICES

11B.3.1 Cluster Tool for Gen 2 Size

Before the rise of the FPD market, pass-by type inline sputtering equipment had been regularly used for production that needs higher productivity (i.e., the substrate moves parallel to the sputter target surface, as shown in Figure 11B.6). However, this method had two major problems in FPD production: inconvenience in multi-layer film deposition and particle generation caused by film peel-off from the substrate transfer carrier.

In order to solve those problems, in 1993 ULVAC developed a cluster-type sputtering system that made sideways static deposition possible (the substrate and the target are positioned upright, parallel with each other as shown in Figure 11B.7), with an early Gen 2 size tool example shown in Figure 11B.5. That cluster-type sputtering equipment had earned a good reputation from panel manufacturers and the sideways static deposition method had become standard for the sputtering process in TFT array manufacturing.

Introducing this cluster-type sputtering tool configuration, such as ULVAC SMD-450 G2 system in Figure 11B.5, required the following changes and technologies:

(1) Unnecessity of carrier (tray)
(2) Large vacuum transfer robot
(3) Horizontal transfer and sideways deposition
(4) Multistage heating mechanism
(5) Magnetron cathode (magnet-moving type cathode)
(6) Long life deposition-preventive shield (to prevent film peeling)

Figure 11B.5 Gen 2 size sputtering equipment (ULVAC).

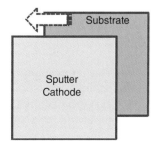

Figure 11B.6 Concept of pass-by sputtering.

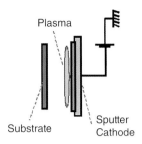

Figure 11B.7 Method of sideways sputtering.

11B.3.2 Cluster Tool for Gen 4.5 to Gen 7 Size

In the late 1990s, in response to industry needs ULVAC had made further significant advancement in sputtering tool design and released Gen 5 size sputtering systems (X-type) featuring:

– integrated substrate pre-heating mechanism into the load/unload chamber
– double platen mechanism to hold a substrate in the sputtering chamber,
– mounting capability of the multiple cathode (up to three cathodes) to address the number reduction of the sputtering chambers
– independent sputtering power supply to each sputtering chamber

The conventional cluster-type equipment is composed of multiple modules where one function (e.g., deposition processing) is assigned to one equipment module (sputtering chamber). Therefore, there were restrictions in equipment performance:

(1) Footprint
   The equipment size increased according to the expansion of the substrate size (advancement of generation)
(2) Processing time:
   Processing time including sputtering rate could not be improved despite the expansion of substrate size

In order to address those limitations, the X-type equipment configuration with multifunctional chambers was developed and enabled the following upgrades:

(a) Reduction in footprint
(b) Reduction in equipment price
(c) Improvement of processing speed (particularly when processing multilayered films)

Recently LTPS TFT has become of high interest for high-resolution panel applications with higher sensitivity to particle defects. A tray-less cluster PVD system was developed to reduce particle generation

11B.3.3 Vertical Cluster Tool for Gen 8 Size

The multi-chamber architecture was essential for developing sputtering equipment suitable for TFT array fabrication. However, as the glass substrate size increased in response to demand from LCD TV applications, a number of critical problems resulted from the related increase in PVD system footprint and total weight:

(1) Technical problems:
   Sputtering uniformity for the large substrate, target usage efficiency, handling of thin glass substrates
(2) Logistics:
   Transportation restrictions on roads and by ship, delivery and installation to the clean room upper floor
(3) Cost:
   Substantial cost increase in development, plant facilities, processing, and materials

Although the transportation problem was solved by dividing the core chamber properly into smaller pieces, many of those problems cannot be solved if the same equipment architecture as that used until G7 was maintained. That is the background for the development of new architecture for Gen 8 size featuring "the vertical cluster-type sputtering equipment that uses the sideways static deposition method."

The pass-by vertical sputtering system, though it is not a cluster tool, has been used for the CF (color filter) deposition since it offers high productivity with lower equipment cost. The new tool had taken advantage in vertical deposition and implemented it into the cluster tool. A lot of new elemental technologies have been employed for that new tool, maintaining the merits of the cluster-type system and incidental film deposition technique established in the past years. Here processing modules are disposed in line without using the transfer chambers, which are commonly used in Gen 2 size up to Gen 7 size tools.

The design concept in development of Gen 8 size tool was "simple, space saving, and high productivity." In achievement of such new concept, the following issues have been considered and investigated.

(1) Maintenance and improvement of film performance: Combination of conventional sideways static deposition method and new AC cathode (the AC cathode will be described later)
(2) Stable operation: Simple mechanism (robot-less substrate transferring system)
(3) Reduction of footprint: Elimination of substrate transfer chambers from conventional cluster-type equipment, replacement by an original simple substrate transfer system for each chamber
(4) Improvement of productivity: Same throughput regardless of single-layer or multilayered film
(5) Substrate scale-up: Use of substrate carrier, vertical chamber structure (to minimize dead space)

Figure 11B.8 shows a vertical sputtering equipment configuration that uses the sideways static deposition method based on the above concept. Although the area of Gen 8.5 size substrates is almost double from that of Gen 6 size, the system footprint was significantly reduced using the vertical architecture (Figure 11B.9). This reduction of the footprint required in the cleanroom facility has made processing and distribution easier. A "positioning system" that changes the substrate position from horizontal to vertical to load onto the carrier is included in such systems. Additionally, instead of the transfer chambers typically used in the previous

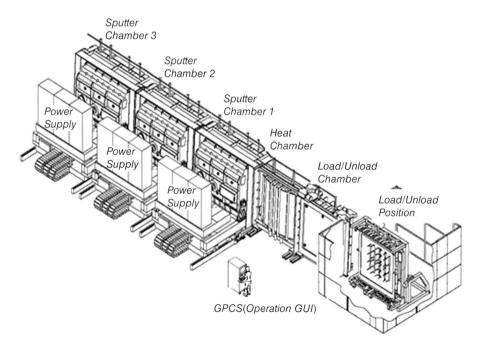

Figure 11B.8 Gen 8.5 Size ULVAC sputtering equipment.

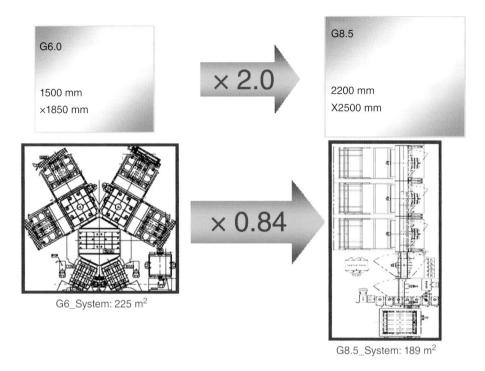

Figure 11B.9 Footprint and glass size comparison between Gen 6 size and Gen 8.5 size.

cluster-type equipment, a simple carrier transfer mechanism is used. The new architecture allows the substrate to be static relative to the carrier, to be covered with a deposition-preventive shield during film deposition. Carrier transfer lanes are installed on both the deposition and transfer sides of each chamber, without a partition. The original Traverse System is installed in sputtering chambers so that substrate carriers can quickly change lanes after deposition completion. This allows the carrier of the processed substrate to return to the load/unload chamber without contacting the carrier of the unprocessed substrate enabling a clean film deposition processed in a continuous vacuum environment, similar to the previous-generation cluster-type equipment. Besides this vertical-type cluster tool, a new special sputtering cathode has also been developed.

The vertical platform for sputtering equipment employing sideways static deposition method has become mainstream in production on substrates over Gen 8 size for large TV panel applications. Also in order to meet the requirement of the latest large high-definition TV panel production, further improvement in maintenance performance (e.g., time reduction), low-particle transfer mechanism, vent airstream control, processing speed enhancement, driving system precision control, and energy saving were developed.

11B.4 EVOLUTION OF SPUTTERING CATHODE

11B.4.1 Cathode Structure Evolution

Figure 11B.10 shows the evolution of ULVAC's cathode designs optimized for TFT device fabrication. In the case of the primary single magnet cathode used with small substrates, the plasma for one cathode (target with one backing plate) was controlled by a single magnetic circuit. To optimize the film quality distribution, the magnetic field was corrected by changing the shape of the magnetic circuit. To achieve uniform film thickness and film quality using the single magnet cathode, while preventing the target being sputtered non-uniformly

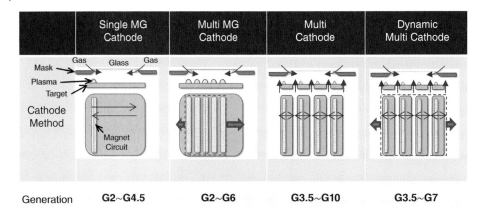

Figure 11B.10 Evolution of sputtering cathode designs (ULVAC).

(eroded), optimal control is necessary so that the single magnetic circuit oscillates over the target. In addition, the sputtering rate is highest directly above the circuit area due to the characteristics of the magnetic circuit. Therefore, to reduce the deposition time and improve productivity, it was necessary to increase the in-plane scanning oscillation speed of the magnetic circuit within the target and to increase the sputtering power (applied power). However, as the upper limits for controlling the speed of the magnetic circuit and for sputtering power density (definition: power applied to target / magnetic circuit area = W/cm^2) were determined for each target material, limitations were found in further reducing the deposition time.

To address such limitations, ULVAC developed a "multi-magnet cathode" design based on an arrangement of multiple magnetic circuits. As the plasma is simultaneously controlled by multiple magnetic circuits, this design can provide a superior film quality control compared to the single magnet cathode. In addition, as the total area of magnetic circuits is larger, the power to be applied to the target can be increased, leading to an improved deposition rate for the entire substrate in-plane.

However, as the substrate size has increased since then, it became necessary to produce a high-purity large backing plate or target material for the single-cathode-type sputtering equipment with single or multiple magnet arrangements. Fabrication of ever larger target becomes virtually unfeasible since material manufacturers had to modify their equipment, and technical problems remained unsolved. In addition, obtaining the ideal film thickness and film quality while guaranteeing good enough uniformity (a significant property of process film) was difficult due to the expansion of the cathode area. This is because, based on the principle of parallel planar DC sputtering, the plasma tends to become non-uniform as the substrate size increases. This is attributable to the fact that the anode exists only around the outer periphery of the substrate, and also to the fact that the difference between the anode area and the cathode area is too large, causing the plasma to concentrate around the outer periphery, hampering the uniform distribution on the plane of the substrate. Use of a rod anode bar between the cathode and substrate can be one solution, however it lowered the aperture ratio of the cathode area to substrate, and caused thermal distortion and particle problem due to film peeling-off. Furthermore, because the discharge space is enclosed by one target, substrate, and deposition-preventive shield, the gas must be supplied from outer periphery of the enclosure in a non-uniform distribution at the target center.

As a fundamental solution for the insufficient anode (disadvantage of single- and multi-magnet cathodes), a "multi-cathode" design was introduced for Gen 5 size and larger substrates (and currently available from Gen. 3.5) where multiple cathodes (targets) are arranged over the substrate to create a uniform arrangement of anodes. As anodes are arranged around each cathode (target) on a one-to-one basis, independent plasma control is possible. This made it possible to satisfy the process requirements of the increase in substrate size for the immediate future, and expand scalability for small substrates.

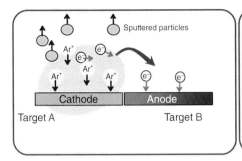
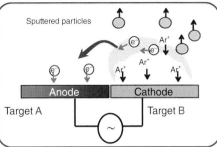

Figure 11B.11 Concept of AC sputtering.

To take advantage of "multi cathode" design, an AC power supply for driving the cathode was introduced to achieve further improvements in deposition uniformity. Figure 11B.11 shows the principle of discharge in AC sputtering based on using a pair of two multi-cathodes where one is anodic and the other one cathodic, and then switching polarity. To secure the same deposition rate as with DC power supply, the anode and cathode are switched alternately in the frequency range up to 100 kHz. Uniform discharge between the pair was achieved, ensuring uniform deposition distribution for larger substrates. Furthermore, this AC cathode can be used for sputtering dielectric materials, which was difficult with DC control. Therefore, using this cathode design with additional elements, we entered the stage of practical use for periphery insulation film production equipment compatible with the next generation oxide semiconductor device.

11B.4.2 Dynamic Multi Cathode for LTPS

After the introduction of the multi-cathode design, the independent control type multi cathode still had an issue to be eliminated. There was a concern that the film quality may fluctuate on the substrate directly under the cathode and anode electrodes (a problem generally called "mura") due to a variation in quality or thickness due to the cathode shape, making process optimization difficult. The degree of film quality distribution differs depending on the film type used for each device, film control value, or process conditions.

Higher film uniformity became further important as the demand for higher resolution panels increased. In order to improve uniformity, a new mechanism that oscillates the entire cathode between left and right for the distance up until the "mura" disappears (generally called "dynamic multi-cathode") was introduced. It can be used in the standard cluster-type sputtering equipment for LTPS. An oscillation control method is optimized for the film type in each device, and the oscillation mechanism was simplified for easy manufacturing. For large vertical equipment, the quality distribution problem was quickly solved by incorporating dynamic multi-cathode so that the substrate carrier is oscillated through precision control relative to the static cathode during film deposition.

11B.4.3 Cathode Selection Strategy

Figure 11B.12 shows the recommended strategy for cathode design selection. The cathode shall be determined considering the following two points:

(1) Achievement of the required device specifications
(2) CoO (cost of ownership) including maintenance performance, cleaning of deposition-preventive shields, and target material cost

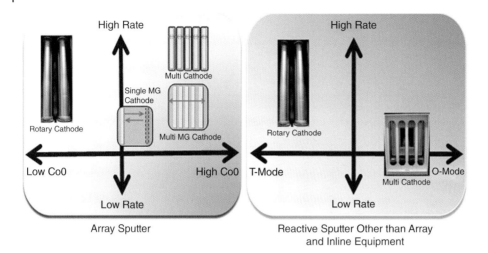

Figure 11B.12 ULVAC sputtering cathode selection.

Two types of the cathodes, planar and rotary, are available. So far, the planar type mainly had been used for TFT array, on the other hand, a rotary cathode is selected in the following cases:

(1) Inline equipment for non-TFT-array processes, special sputtering of oxide film (e.g., SiO_x, NbO_x, and AlO_x), or touch panel application (e.g., ITO deposition)
(2) Transition process mode is required

Since the initial phase of TFT-LCD manufacturing industry, the utilization efficiency of a rotary target has been superior to that of a planar target. However, the rotary target still has some challenges compared to a planar target. For example, target fabrication cost is still high, and in case of the target exchange maintenance, the entire rotary cathode (including target and magnetic circuit) must be dismounted. In parallel, the utilization of planar targets has also continued to improve reducing the differences between the planar and rotary target configurations.

11B.5 TRANSPARENT OXIDE SEMICONDUCTOR (TOS) THIN-FILM DEPOSITION TECHNOLOGY

11B.5.1 Deposition Equipment for TOS-TFT

Currently, a new semiconductor material for TFTs enabling displays with higher resolution and lower power consumption is under technical investigation. Sputtered IGZO (In, Ga, Zn, Oxide) films can replace conventional silicon-based semiconductor thin film formed by conventional plasma CVD method. IGZO, discovered by Professor Hosono's Group at the Tokyo Institute of Technology, is attracting considerable attention. It is an oxide compound of three elements (indium, gallium, and zinc) that can be formed by sputtering [2, 3], and has >10× higher mobility than the amorphous silicon (a-Si), and allows less off-current than high-mobility low-temperature poly-silicon TFT (LTPS). As the production process can also be simplified, IGZO is anticipated to enable high-definition and lower-power consumption panel manufacturing at low costs.

The characteristics of IGZO greatly depend on the oxygen partial pressure during film deposition and post-heat-treatment temperatures [4]. Based on this information, we have checked the process margin in mass-production equipment with the existing sputtering cathode. We have tested the variation in TFT characteristics during the life of the target using SMD950 sputtering equipment (applicable to Gen 4.5 size substrates) with multi-cathode and AC sputtering power supply, and confirmed that there was no variation in the deposition rate and TFT characteristics over time [5]. We have also checked the in-plane TFT characteristics distribution on the Gen 8 size substrate in order to know whether IGZO film deposition technology by sputtering can be applied to large equipment, and confirmed that the distribution was good enough [6].

11B.5.2 New Cathode Structure for TOS-TFT

In order to use the IGZO film in a current-controlled TFT device design for OLED application, it is necessary to further improve the reliability and quality of IGZO film over large areas. The conventional pass-by sputtering method is available for improvement in film thickness uniformity. However, this method is not feasible since the film is also deposited to the substrate carrier and such film may easily peel off from the carrier, generating particles in the deposition chamber. Alternatively, we developed a solution by using "moving cathode" design, maintaining the advantages of the conventional cluster deposition tool. In this configuration, it was possible to eliminate the particles from start to finish of deposition, and produce IGZO films with highly uniform quality and reliability when used in TFT arrays.

The concept and performance of the newly developed "moving cathode" are outlined as follows:

(1) Cathode moves in a chamber
   a. ⇒Improvement of in-plane uniformity with low volume of particle
(2) Use of pre-sputtering position
   a. ⇒Improvement of film uniformity in quality and thickness
(3) Use of control plate on the cathode side
   a. ⇒Incident component control enhanced to further improve film quality

Figure 11B.13 compares the performance of the conventional cathode and new "moving cathode" configurations. Film thickness non-uniformity is observed in conventional the film deposition method with fixed both substrate and cathode (substrate fixed type), since the sputtering rates are different for the section directly above and between the targets.

As the moving cathode uses the pass-by film deposition method where the substrate is static and the cathode moves, good in-plane film thickness distribution property is obtained. In addition, as the substrate tray (in the case of vertical equipment) or the substrate (in the case of cluster-type equipment) is masked during film deposition, particles are suppressed.

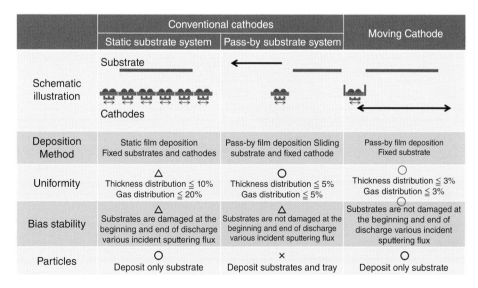

Figure 11B.13 Comparison of conventional cathodes and the moving cathode.

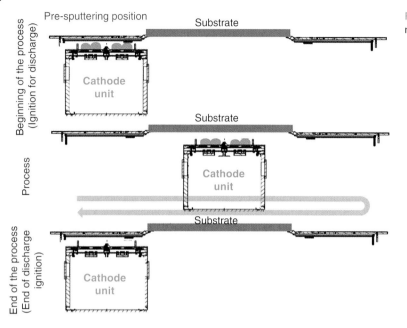

Figure 11B.14 Oscillation pattern of moving cathode.

In order to improve the film quality itself, two concepts are incorporated in the moving cathode. One is the use of a pre-sputtering position to prevent the inclusion of sputtering particles at the start (when plasma is ignited) and at the end of discharge. Figure 11B.14 shows the cathode oscillation pattern during pre-sputtering and film deposition. Uniform film quality can be obtained from the first to the last layer. Another concept is the use of a control plate to adjust the incident angle of particles sputtered from the target reaching the substrate.

When a control plate is not used, a layered structure is created by particles sputtered at a low incident angle (oblique component) and particles sputtered at a high incident angle (vertical component) as the cathode moves. When a control plate is used, particles sputtered at a low incident angle are cut off by the plate, allowing film deposition by only the particles sputtered at a higher incident angle. Here sputtered particles are confined vertically, so only those passing through the space under high density plasma condense on the substrate [7, 8], and it was confirmed that that TFT reliability was improved.

As shown in Figure 11B.15, this effect is presumably coming from the change of IGZO single layer quality by extracting only good sputtered particles for deposit thin-film deposition with limiting incident angle.

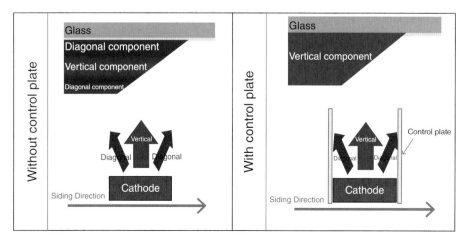

Figure 11B.15 Film deposition image with/without control plate.

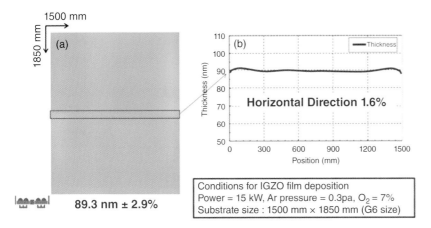

Figure 11B.16 IGZO film thickness distribution on Gen 6 size substrate (a) in-plane, and (b) horizontal directions.

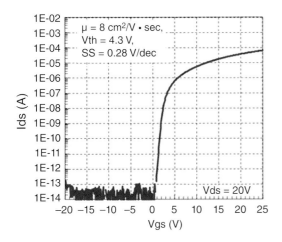

Figure 11B.17 Initial characteristics of IGZO-TFT at 25 points in-plane of Gen 6 size substrate.

In the test using a Gen 6 size substrate (1,850 mm × 1,500 mm) under this composition, a film thickness uniformity of ±2.9% (±1.6% in horizontal direction) was obtained (Figure 11B.16).

Figures 11B.16 and 11B.17 show the measurement results of the etch stopper type IGZO-TFT film formed by the moving cathode. Initial TFT characteristics (μ = 8 cm^2/V·sec, Vth = 4.3 V, SS = 0.28 V/dec) were obtained, and it has also shown a stability as good as ΔVth = 0.4 V in the reliability testing.

11B.6 METALLIZATION MATERIALS AND DEPOSITION TECHNOLOGY

There are two types of TFT structures: the top gate and bottom gate type (Figure 11B.18).

In the top gate type, which is used for LTPS and possibly TOS, the gate electrode is placed over the source/drain (S/D) electrode on the glass substrate, whereas the bottom gate is usually used in a-Si TFTs. In all the types, the gate electrode, source/drain (metal wiring film), and pixel electrode (transparent conductive film) are generally formed by sputtering, and the key requirements are listed below:

(1) Gate and source/drain electrode (metal wiring film)
- Low resistance
- Low cost
- Etching performance

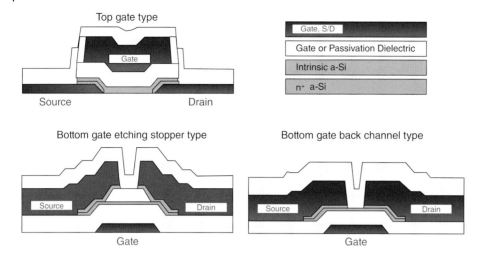

Figure 11B.18 Structure of thin-film transistor (TFT).

- Anti-hillock performance
- Contact performance with oxide-based transparent conductive film and semiconductor layer

(2) Pixel electrode (transparent conductive film)
- Low resistance
- High transparency
- Low cost
- Etching performance

Table 11B.1 shows the resistance values of the typical metal films using for metallization layers. Conventional high-melting point metals, such as chromium, molybdenum, titanium and tantalum, had been used, but they were replaced by aluminum-based films because of the low resistance requirements for the larger and higher-resolution panels. However, the hillocks in the post-heating process and insufficient contact between an oxide-based transparent conductive film (e.g., indium thin oxide) and semiconductor layer

Table 11B.1 Resistance values of typical metallization film.

Material	Resistance	Features
Ag	1.7 μΩcm	Low resistance material. Adhesion is poor.
Al	3.0 μΩcm	Anti-hillock performance is poor.
AlCe (2 at%)	4.5 μΩcm*	Arcing and anti-hillock performance are good.
AlNd (2 at%)	4.5 μΩcm*	Anti-hillock performance is good. Cost is high.
Mo	10 to 15 μΩcm	Corrosion resistance is poor. Cost is high.
MoW (35 wt%)	12 to 15 μΩcm	Corrosion resistance is good. Cost is high.
Cr	17 to 20 μΩcm	Stress is high. Adhesion is good.
Ta (α: body-centered cubic)	25 to 30 μΩcm	Base film is required.
Ti (4N)	60 μΩcm	Target purity is dependent on specific resistance.
Ta (β: Tetragonal crystal)	200 μΩcm	–

*After annealing (for 5 minutes at 300°C)

remain challenges. Introducing a high-melting point metal as a barrier layer, or using an aluminum alloy can help, however, those approaches make the process more complicated and may frequently cause abnormal plasma discharge.

Currently, copper is a new electrode metal under development that can substitute aluminum-based electrode materials to achieve lower metallization resistance in larger-sized panels. However, copper shows some challenges related to adhesion to glass substrate or base film degradation. As a countermeasure, a use of molybdenum or titanium-based barrier metal layer can be a solution, although these materials cannot satisfy both etching performance and low-cost requirements.

In 2008, ULVAC announced a low-resistance copper metallization technique at low cost. The base film is formed by oxygen-mixture sputtering using a Cu-based alloy target to improve adhesion to the glass substrate and base layer without barrier layer [9, 10]. However, in recent TFT processes, there are cases where hydrogen plasma treatment is used after S/D formation. That caused a problem that optimal adhesion cannot be obtained because oxide film formed on the boundary surface is deoxidized by the hydrogen plasma. The Cu-alloy (Cu-Mg-based copper alloy material) developed by ULVAC can achieve a low-resistance metallization film featured by better barrier performance and adhesion property with the glass substrate and base layer than those of existing metal films. That approach is cost-effective because Mo or Ti-based barrier metal layer is not necessary, and can be extended to use in IGZO-TFT as well as Si-base TFTs.

As a by-product of IGZO-TFT technology, dielectric film deposition by sputtering is a new option. Hydrogen contained in SiO_x passivation films deposited by CVD from SiH_4 precursors may affect the IGZO TFT characteristics because it is oxide semiconductor. It is reported that AlO_x prepared by sputtering is effective as a passivation film for the IGZO-TFT [11]. Since this material can also be applied as the encapsulation film for OLED components to allow a better waterproof performance, it can be a significant key technology in the future. A deposition technology to form the dielectric film by high-speed reactive sputtering equipment and related target is now ready for production.

References

1 B. N. Chapman, Glow discharge processes sputtering and plasma etching, John Wiley and Sons (1980).
2 K. Nomura, et al., Thin-film transistor fabricated in single-crystalline transparent oxide semiconductor, Science, 300(5623) pp. 1269–1272 (2003).
3 K. Nomura, et al., Room-temperature fabrication of transparent flexible thin-film transistor using amorphous oxide semiconductors, Nature, 432, pp. 488–492 (2004).
4 M. Takei, et al., Dependence of IGZO TFT characteristics on annealing temperature and O_2 partial pressure, Japan Society of Applied Physics, The 70th Meeting of The Japan Society of Applied Physics and Related Societies, pp. 8a-H-7 (in Japanese) (Autumn 2009).
5 T. Kurata, et al., IGZO TFT characteristics in large-scale substrate, Japan Society of Applied Physics, The 7th Meeting of The Japan Society of Applied Physics and Related Societies, pp. 17a-TL-5 (in Japanese) (Spring 2010).
6 T. Yukawa, et al., Development of sputtering process for IGZO TFT on large substrate, IDW'10, FMC 2-2 (2010).
7 J. Sakamoto, et al., Japan Society of Applied Physics, The 74th Academic Lecture, p. 17p-B4-14 (in Japanese) (Autumn 2013).
8 D. Kobayashi, et al., IGZO film characteristic uniformity and TFT reliability for large sputtering cathode, Japan Society of Applied Physics, The 61st Academic Lecture, 18p-E10-16 (in Japanese) (Spring 2014).

9 S. Takasawa, et al., Cu wiring process for TFTs – enhancement of adhesion and barrier characteristics achieved with an oxygen-mixed sputtering, ULVAC Technical Journal (in English), 69, pp. 8–13 (2008).
10 M. Shirai, et al., Cu wiring process for TFTs – improved hydrogen plasma resistance with a new Cu alloy, ULVAC Technical Journal (in English), 71, p. 24 (2009).
11 T. Kurata, et al., IGZO TFT characteristics in large-scale substrate, Japan Society of Applied Physics, The 7th Meeting of The Japan Society of Applied Physics and Related Societies, pp. 17a-TL-5 (in Japanese) (Spring 2010).

11C

Thin-Film PVD (Rotary Target)

Marcus Bender

Applied Materials GmbH & Co. KG, Alzenau, Germany

11C.1 INTRODUCTION

Magnetron sputtering deposition is one of the major technologies used for depositing thin films for many applications. Particularly, the production of flat panel displays (FPD) relies on sputtering technology for multiple process steps including backplane metallization, pixel electrodes as well as optical enhancement coatings, and touchscreen panel layers. Recently, the introduction of metal oxide active channel layers for high mobility TFTs raised a lot of interest of the industry.

There are multiple approaches to deposit materials by sputtering. For industrial applications, both planar and rotary cathodes are used in combination with either DC or AC powering schemes. AC powering is possible by sine-wave power supplies as well as using bi-polar pulsed units providing additional process freedom to the developers and users of these deposition tools. All of these items will be described in more detail in this chapter. radio-frequency sputtering (r.f.) is not commonly used in industrial applications, however, is very often found in research environments. The same holds for the high-power impulse magnetron sputtering (HiPIMS) techniques, which are being developed by several researchers.

For industrial use of magnetron sputtering, two generic tool architectures are commonly used: the dynamic (inline) or the static (inline or cluster) layouts. In a dynamic tool, the substrate is passing with constant velocity in front of one or more linear deposition sources. By keeping the distance between the substrates as small as possible the utilization of the deposition sources can be maintained at a very high level. Uniformity levels in the moving and perpendicular directions can be optimized separately giving rise to usually good overall uniformity and repeatability. The Applied Materials New Aristo™ system, which is available in different sizes, is an example of an inline dynamic sputter deposition platform. The modular layout of the tool enables adaptations to many applications for the layer materials, layer stacks, and productivity requirements. Figure 11C.1 shows a photograph of such a system equipped for ITO deposition.

On the other hand, a static system is operated by placing a substrate inside a process chamber and keeping it motionless while the deposition takes place. One static deposition system may contain several process chambers depending on the layers or layer stacks to be produced within the system. The deposition source in this case needs to be able to coat the whole area of the substrate, which can be realized by using a monolithic large area planar cathode or an array of planar or rotary cathodes. Depending on the arrangement of chambers the system may have a cluster or inline geometry or may contain elements of both architectures. Figure 11C.2 shows an Applied Materials PiVot™ static deposition system with inline architecture featuring

Figure 11C.1 Applied Materials New Aristo™ dynamic deposition system (*Source:* Photo: Applied Materials).

Figure 11C.2 Applied Materials PiVot™ static deposition system (*Source:* Photo: Applied Materials).

an array of rotary cathodes in the process modules. Both cluster and hybrid architectures can be realized with this platform.

11C.2 SOURCE TECHNOLOGY

11C.2.1 Planar Cathodes

The discovery of sputtering is commonly attributed to W.R. Grove [1, 2], who observed deposition taking place from a wire toward a silver surface while studying a glow discharge. From there toward the planar magnetrons still widely in use today it was a long way. Major milestones on this way were the invention of the magnetron by Penning and others in the late 1930s [3] and then the integration into a planar magnetron source by Chapin in 1974 [4]. The principle of a magnetron cathode is the superposition of electrical and magnetic fields, which are confining the electrons within a closed loop path near the target surface, which is operated as a cathode. This gives rise to an increased ionization probability close to the sputter target, providing enhanced plasma density and sputter efficiency, namely deposition rates [5].

The processes taking place in glow discharges can be very complex and have been described in detail, for example, by Behrisch et al. [6] and by Chapman [7]. In short, a glow discharge is ignited by establishing a sputter atmosphere typically consisting of noble gas like Argon at a pressure of not more than a few Pa into the evacuated chamber and by applying a voltage of several hundred of volt to the cathode. In the glow discharge, positively charged Argon ions are accelerated toward the cathode surface by the electrical field. When high-energy Argon ions hit the target surface, material is removed from the target and travels through vacuum until reaching a surface where it condenses—that is, not only building a layer on the substrate placed inside but also on the walls of the vacuum chamber. Also, as a consequence of the ion bombardment, secondary electrons are generated at the target (cathode) surface, which are repelled from it and end up ionizing more Argon atoms and thus keeping the discharge running. An anode arrangement placed in the vicinity of the cathode closes the electric circuit driving the glow discharge. Very often the walls of the vacuum chamber are serving as anodes eliminating the need to supply dedicated anodes.

In the presence of a magnetron the electrons are trapped and forced into a cycloidal movement pattern above the target surface. The density of charged particles in the closed loop created by the magnetron is increasing dramatically and the discharge voltage of this setup is decreased from several kV to a few hundred volts depending on target material, processing parameters and geometry. Sputtering occurs in the racetrack defined by the magnetic field lines where the electron confinement is taking place.

In principle, there are two different operation modes of sputtering cathodes—non-reactive and reactive. In a non-reactive sputter process, the sputter atmosphere consists of inert gases only, for example, Argon. The composition of the deposited layer is not identical but is correlated to the target material composition. Depending on specific sputter rates, the target composition at the surface will reach a steady state reflecting bulk composition and specific sputter yield. When sputter yields for the constituents of the target are very different, this effect may cause low total rates due to agglomeration of the low-yield component at the surface. This operation mode is widely used during metal deposition. In a reactive process, it is desired to stimulate a chemical reaction between a metallic target material and a reactive gas present, like oxygen or nitrogen, on the surface of the growing film in order to grow an oxide or nitride layer. This operation mode is commonly referred to as partially reactive sputtering. A particular topic to be addressed in reactive processing is the target surface coverage by reactive species which can lead to local charge build-up as well as sputter rate changes. An intermediate way to operate a sputter process is to use a target material consisting of the compound (e.g., oxide or nitride) material and adding some reactive gas in order to fine-tune the layer properties or balance property requirements and deposition rates relating to manufacturing productivity (Figure 11C.3).

Since its inauguration, Planar Magnetron Cathode Technology has been the workhorse for both industry and academia for many years. Various shapes and sizes have been realized. The most common geometries

Figure 11C.3 Planar magnetron cathode showing the electron racetrack marked by deep erosion of the target material (*Source:* Photo: Applied Materials).

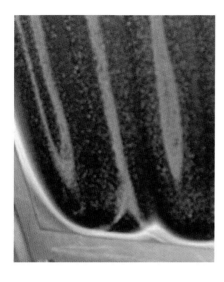

Figure 11C.4 Planar cathode equipped with moving magnetron. Two racetracks can be seen. The target corners exhibit re-deposition zones, also the target surface is contaminated with nodules growing during partially reactive sputtering of ITO (*Source:* Photo: Applied Materials).

include circular and rectangular planar cathodes. In the display industry, rectangular planar cathodes with a length of up to approximately 3 m have been used in order to coat large area substrates. In inline coating systems, elongated rectangular cathodes are used serving as linear deposition sources. For volume production several cathodes are arranged one after the other in order to achieve the necessary film thickness with a given system throughput. Multiple layer stacks consisting of different materials can be produced by using different target materials, or changing the sputter atmospheres locally in reactive processes. For static deposition, both large-area monolithic cathodes as well as arrays of elongated rectangular cathodes bonded onto a metal backing plate have been used.

Planar magnetron cathodes exhibit a couple of inherent architectural disadvantages. One disadvantage is caused by the fact that only at the racetrack position material from the target is removed during the process. This effect limits target material utilization and may also affect layer uniformities. Furthermore, sputtered material that is scattered back toward the target surface can accumulate in areas where no sputtering occurs. The re-deposited material gives rise to particle generation at a later stage of target usage and to process instabilities due to local charge build-up, particularly in the case of reactive processes when the sputtered material possesses a very low electrical conductivity due to chemical reactions with gases like oxygen or nitrogen during the process. The usage of magnetrons that are moving during deposition helps reducing these effects without being able to completely suppressing them (Figure. 11C.4).

Static sputtering arrangements with planar cathodes typically suffer from non-uniform anode availability. When anodes are arranged at the outer circumference of a static planar cathode array or a large-area planar cathode, the central area of the cathode suffers from build-up of space charges limiting the sputter efficiency there. This results in poor film uniformity, which can be overcome by placing anodes in-between the single cathodes of a planar cathode array or in-front-of a large-area planar cathode. In both cases, re-deposition collected at the anode surfaces increases particle generation after long operation time. Furthermore, during reactive processes the highly resistive coating on the anode induces process shifts difficult to overcome.

11C.2.2 Rotary Cathodes

Several years after the invention and implementation of planar magnetron cathodes for coating processes, the idea of cylindrical rotating magnetrons was born. The first to file this idea was McKelvey in 1981, at the time working at Shatterproof, a glass coating company in the United States [8]. At approximately the same time, in Eastern Germany a similar concept was described [9]. Within the last 30 years the concept found wide acceptance in various applications, starting with architectural glass and photovoltaic coatings and during the last decade also for production of flat panel displays [10].

The basic operation principle of a rotary magnetron is similar to the planar magnetron, however, the geometry of the cathode and target is cylindrical with the magnetron being placed inside the tube (see Figure 11C.5). Depending on the material to be sputtered, the target consists of either a backing tube with bonded cylinders of the material to be deposited or of a monolithic tube of the sputter material. The first mentioned assembly is usually realized for ceramic materials like indium tin oxide (ITO) whereas monolithic targets can be made of various metals like aluminium or copper. Although nowadays all relevant materials for flat panel display production are available in a rotary form factor, the production of rotary targets is more complex compared to planar targets and therefore some development effort of target manufacturers may be necessary to provide new materials. During operation the tube is rotated around its longitudinal axis making sure that each part of the target is being sputtered periodically.

The architecture of a cylindrical rotary magnetron cathode bears some inherent differences to the well-known and established planar architecture. The geometry allows both realizing a higher packing density of target material inside the deposition chamber and utilizing a higher percentage of the target material enabling extended operation times before the targets need to be exchanged. During process, the rotation of the tube provides a stable environment over a lifetime even in reactive processes as all surface areas are constantly being sputtered and the agglomeration of re-deposition can be effectively suppressed. Also, this feature increases cooling efficiency since the heat generated during sputtering is more efficiently and evenly distributed over the circumference of the cylinder. On the other hand, integration of a rotary target assembly into a sputtering system needs a rotary cathode drive, connecting the cathode to the outside world, providing cooling water and power as well as the rotation drive for the target itself. This increases complexity of deposition systems to a certain extent. Also, the space available for the magnet yoke is limited by the inner diameter of the target or backing tube, which may create restrictions with regards to magnet geometry.

Many implementations of rotary magnetron cathodes into inline deposition systems are known, not only in the display industry but also in other large area coating applications like architectural glass, roll-to-roll

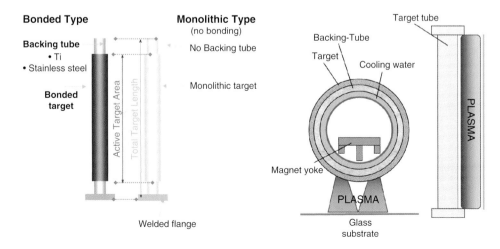

Figure 11C.5 Layout of rotary targets (left) and scheme of a rotary magnetron during operation (right).

coating or photovoltaics. In these cases the rotary cathode is implemented similarly to planar cathodes as described in section 11C.2.1. The cylindrical targets are rotated during operation. So there is no difference between back side and front sides of the cylindrical target in terms of particulate accumulation—all areas along the circumferences are sputtered when they move through the magnet race track.

11C.2.3 Rotary Cathode Array

The arrangement of a rotary cathode array for static coating of large area substrates was introduced about 10 years ago by Applied Materials and is available in its static sputter coater PiVot™ system (Figure 11C.6). The array consists of multiple rotary cathodes arranged side-by-side and opposite to the substrate to be coated. For a Gen 8.5 size substrate with $2.2 \times 2.5\,m^2$ area the rotary cathode array consists of 12 cathodes, Gen 6 size substrates with $1.8 \times 1.5\,m^2$ size are coated with 8 cathodes. The spacing between the cathodes and the distance between cathodes and substrates are optimized for coating uniformity and highest material usage efficiency.

Layer properties and uniformities can be controlled additionally by the unique feature of tilting the magnets during deposition [11] (Figure 11C.7). This feature is realized by an additional motor allowing control of target rotation and magnet movement independently. This so-called magnet-wobble technique can be used in various ways and allows efficient coating of extremely thin films as well as sensitive fully-reactive coatings by suppressing the plasma instabilities at the very beginning of discharge ignition. Several modes have been developed and are in use in industrial scale. The most important deposition modes are the so-called perfect wobble and the split-sputter-mode. Perfect wobble describes a deposition mode where the magnets are swept completely between two predefined angles during a predefined deposition time. The magnets could move one or several times the whole way from one position to the other depending on available time and angular velocity of the magnets. On the contrary for the split-sputter-mode (SSM) the magnets are positioned at one angle before the discharge is started and maintained for one part of the desired total deposition time. Then, the sputter power is switched off or reduced while the magnets are moved to a different position. Subsequently, the remaining deposition time is sputtered again with the desired sputter power. The magnet angles for all modes are optimized to match the sputter characteristics of the material to be deposited. In addition, it is possible to independently adapt angular positions for each rotary cathode of the rotary cathode

Figure11C.6 Array of copper rotary cathodes used for flat panel display backplane metallization in an Applied Materials PiVot™ coating system (*Source:* Photo: Applied Materials).

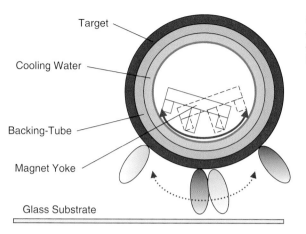

Figure 11C.7 Rotating the magnetrons in the rotary array during deposition enables very efficient uniformity optimization and tailoring layer properties [11] (see Figure 11C.1).

array. It is immediately obvious that magnet wobble opens a wide variety of options for optimizing layer thickness uniformities. Additionally, the technique allows tailoring of electronic, optical, or morphological properties and uniformity of the growing layer since the direction of the atoms arriving at the substrate and forming the layer can be controlled as well as the influx of charged particles (e.g., negative oxygen ions in reactive processes), which are following the electrical fields and are propagating perpendicular from the racetrack area toward the substrate [12]. The architecture of the rotary cathode array containing open space in-between individual cathodes allows arranging gas inlets not only along the outer frame of the cathodes but also inside the area covered by the array. A very uniform distribution of the sputter atmosphere is the consequence. This is particularly a benefit for reactive processes since the partial pressure and uniformity of the reactive gas in the deposition chamber is crucial for good layer homogeneity.

A particularly important feature to consider in an area deposition sources is the layout and availability of the anode. For the rotary target array the anodes provided are metal bars located in between the cathodes. This layout is possible due to the open nature of the array with cathodes at a certain spacing separating each other. It also ensures evenly distribution of effective anodes symmetrically placed on both sides for each cathode within the array: Each cathode is provided with two anodes, one located left of the cathode and one located right of the cathode. At the same time, it avoided anode bars in the space between the targets and the substrate, which would result in heavy anode coating and subsequent process instabilities and particle generation from the anode surfaces.

In order to make sure that anode symmetry is perfect, all other chamber surfaces are being electrically disconnected from the anode by a pull-up resistor, which is built-in between anodes and ground potential [13] (Figure 11C.8). The chamber walls are kept at ground potential and are not acting as additional anode surfaces due to the resistor. This electrical setup allows defining anode geometry and electrical effectiveness very precisely counteracting uniformity degradations that otherwise may be observed in less well balanced architectures with anodes positioned only at the outer edge of the cathode array.

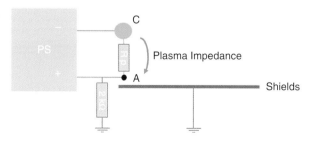

Figure 11C.8 Electrical connection scheme of cathode, anode, and ground in the rotary array system. C represents the cathode, while A represents the anode. Anode and ground potential are separated by a pull-up resistor. Shields and chamber walls are kept at ground potential.

11C.3 MATERIALS, PROCESSES, AND CHARACTERIZATION

11C.3.1 Introduction

The cross-section of a typical inverted staggered TFT used in the display industry is shown as an example in Figure 11C.9. Besides the shown example of indium gallium zinc oxide (IGZO), amorphous silicon or polycrystalline silicon are widespread as active layer materials. In general, the exact number and arrangement of functional layers depends strongly on the TFT architecture and the integration scheme. In any case, the TFT needs gate, source and drain (S/D) metals, often referred to as metal 1 and metal 2, both deposited by PVD processes. Additionally, there is a sputtered pixel electrode, for example, ITO or another transparent conducting material (not shown in Figure 11C.9), which is addressed by the TFT to drive the liquid crystal switching individual pixels in LC displays.

The most important layer properties are film thickness, electrical conductivity and, in several cases, the optical transmittance or reflectance. The electrical conductivity of thin layers can be measured with the four-point method or using contactless methods based on high-frequency electric fields [14]. Photo-spectrometers are used in transmittance or reflectance mode in order to determine the optical properties. Both methods have been integrated into automatic scanning measurement tables able to generate measurement patterns on large area substrates used for flat panel display production. Film thicknesses can be determined from either optical or electrical measurement results under some assumptions on specific optical constants or resistivity respectively. Direct thickness measurements are possible by step profilometers or cross section SEM/TEM, for example, however, it is very difficult to achieve the same degree of automation and measurement reliability on large-scale substrates with these methods.

11C.3.2 Backplane Metallization

The most common material for metal 1 and metal 2 has been aluminium for a long time, due to its good balance between material cost and layer properties. The bulk-resistivity of aluminium is ~2.8 µΩcm, which allows for producing sufficiently conductive bus lines for gate and source-drain electrodes in flat panel display applications. One major issue with aluminium is the known tendency to form hillocks during subsequent high-temperature steps, such as the gate insulator deposition. The aluminium hillocks decrease production yield by causing shorts in the device, therefore, hillock control is very important for the usage of aluminium metallization in flat panel displays. Two major methods have been discussed to avoid this phenomenon; one is using Al-alloys with higher stress-relaxation temperature and the other being the usage of a cap-layer like molybdenum or others [15]. The disadvantage of typical Al-alloys under consideration is, however, the strong increase in specific resistivity. For this reason the capping process has been the approach of choice for most industrial applications.

Another known issue with aluminium sputtering is the occasional occurrence of splashes. Splashes are micron-sized particulates generated at the sputter target by melting of target material during abnormal

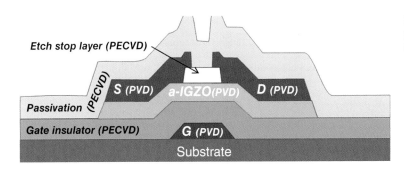

Figure 11C.9 Cross-section of a bottom-gate TFT with metal-oxide channel-layer and etch stop layer.

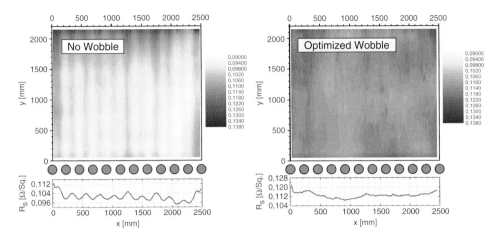

Figure 11C.10 Sheet resistance patterns of an aluminium coating coated on Gen 8.5 size substrates with a static rotary cathode array. The usage of magnet wobble allows for significant uniformity improvements.

discharges and subsequent deposition of aluminium droplets on the substrate. The generation rate of splashes can be influenced by several reasons such as target purity and morphology, vacuum condition of the deposition system, and process stability. Process stability can be improved with proper target conditioning assuring removal or stabilization of its surface oxide layer, which is easily built up even in high vacuum due to the high reactivity of aluminium. This implies that sputtering the whole surface of the target and avoiding any re-deposition zones on the target is beneficial for suppressing aluminium splashes. In Figure 11C.10 a typical Gen 8.5 size uniformity map of an aluminium layer coated in a static deposition system is shown illustrating the effect of process optimizations with a rotary cathode array.

The trend for displays to larger size, higher pixel resolutions, and faster refresh rates is demanding bus lines with high conductance to support low resistive-capacitive delays, for which even thick aluminium layers were found not to be sufficient. Thus, in the recent years, copper metallization was introduced and is used more and more for production of ultra large-area, high-resolution devices. The advantage of copper layers is a specific resistivity which is approximately 30% lower than that of aluminium ($\approx 1.7\,\mu\Omega$cm). Additionally, it is more stable to corrosion. Disadvantages that had to be solved before transferring to copper metallization included the diffusion of copper into silicon, which may cause contaminations of the channel layer. Moreover, the adhesion of copper needs to be improved. Similar to aluminium metallization, two solutions have been evaluated: usage of copper alloys as well as an intermittent barrier and adhesion layer. In most industrial applications, barrier layers made from molybdenum, titanium, or alloys thereof are in use.

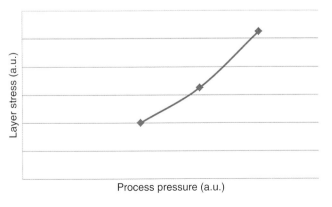

Figure 11C.11 Stress of a thin molybdenum layer versus process pressure.

For LTPS applications, molybdenum or molybdenum-tungsten alloys are used frequently as gate metal due to their low resistance, good etchability, high temperature stability and controllability of layer stress (Figure 11C.11). The molybdenum gate layer after patterning may also serve as self-aligned mask for S/D-doping by ion implant.

The sputter deposition processes of all metal layers are normally DC driven within a pure Argon gas atmosphere. For thicker layers and adhesion improvement layer stress control may be important. The total pressure during sputtering is an important control parameter for that purpose as the growth kinetics of thin layers is heavily dependent on the flux and energy distribution of incoming atoms and ions, which are in turn influenced by the mean free path directly related to the sputtering pressure [16].

11C.3.3 Layers for Metal-Oxide TFTs

The advent of ultra-high resolution devices with high refresh rates triggered not only the migration from aluminium toward copper as metallization material, but also the replacement of the a:Si channel layer in the TFT. One promising material family, which has been studied and evaluated for more than 20 years are amorphous oxide semiconductors [17], mostly represented by indium-gallium-zinc-oxide (IGZO). Today, academia and some display panel makers have succeeded in improving integration schemes in order to suppress device issues like uniformity and stress stability and allow transfer of the technology into volume production.

As shown in Figure 11C.9, an additional layer compared to the standard a-Si TFT can be used for suppressing backchannel damages during the S/D etching step. Although this approach requires one additional masking step it is used frequently in volume production. Adaptation of etching steps as well as modification of the channel layer are being investigated by many groups in order to omit this additional coating and patterning step. Typically, the TFT mobility that can be achieved with IGZO channel layers is in the order of $10\,\text{cm}^2/\text{Vs}$.

For production purposes, the uniformity of TFT properties over the substrate are of utmost importance. It is not surprising that the TFT uniformity is strongly related to the uniformity of the electronic properties of the IGZO film [18]. The microwave photoconductivity decay (μPCD) measurement technique has been established as a method for assessing the electronic properties and the uniformity of semiconductors. In μPCD measurements the semiconducting film is excited by a laser pulse and the reflectivity in the μ-wave region is measured in a time resolved manner. This method allows conclusions on number of excited charge carriers as well as on the recombination mechanisms by the decay time of the signal (Figure 11C.12, bottom side). The optimization of layer uniformity as determined by μPCD relies on control of influx of charged particles, for example, negatively charged oxygen ions. This can be done by optimizing the magnet orientation during static coating (Figure 11C.12, bottom side).

Several groups have been working to overcome the mobility limit of $\sim 10\,\text{cm}^2/\text{Vs}$ for IGZO-based metal-oxide TFTs. Approaches include the switch to other channel materials or using different channel architectures. Among alternative materials, a lot of attention was drawn to the use of ZnO as an active layer. A field effect mobility close to $100\,\text{cm}^2/\text{Vs}$ has been reported for ZnON-based TFTs with good stress stability [19]. Unfortunately, it was found to be very difficult to produce ZnON devices with threshold voltages close to 0V, which is important for practical applications in display backplanes. One alternative approach is to use a dual active layer consisting of a very thin high mobility layer as channel covered by a thicker low mobility layer serving as backchannel. With this method, a mobility of approximately $100\,\text{cm}^2/\text{Vs}$ was achieved with good stability and threshold voltages close to 0 V [20]. Backplanes exhibiting good uniformity are being produced with this method. Very important is achieving an excellent thickness control particularly for the thin high mobility layer, since the TFT properties are strongly dependent its thickness [21]. Figure 11C.13 shows a horizontal cross-section of the thickness of a thin ITO layer coated on a Gen 8.5 size substrate with a horizontal layer uniformity better than ± 3.5% achieved by using the split-sputter-mode (SSM) of a rotary cathode array system.

A lot of effort has been put into barrier and passivation layers for metal oxide devices since those are known to be very sensitive to environmental influences. The usage of Al_2O_3 coatings as passivation layer has been

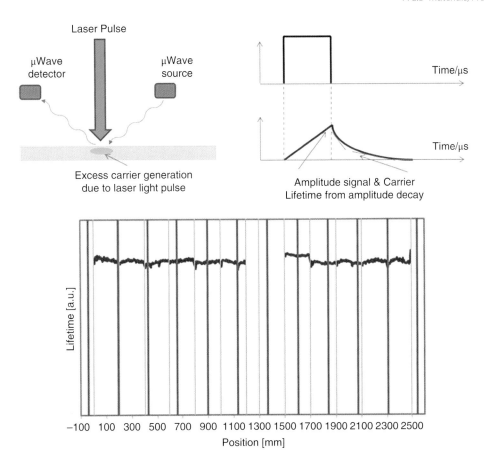

Figure 11C.12 Top: Principle of μ-wave photoconductivity decay measurements. Bottom: μPCD distribution of IGZO layer deposited with rotary cathode array and optimized process parameters.

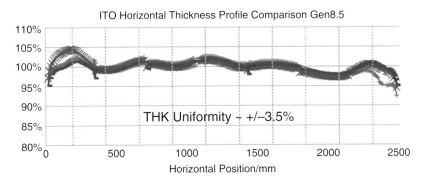

Figure 11C.13 Horizontal thickness profile of thin ITO layer coated with a rotary cathode array using the split-sputter-mode (SSM) (*Source:* deposition and measurement by Applied Materials).

reported to be beneficial for the bias stress behavior of IGZO TFTs [22]. For this application, processes for fully reactively sputtered alumina layers have been developed [23]. Due to the fully reactive nature of the process combined with the high reactivity of aluminium strong hysteresis phenomena, as described in

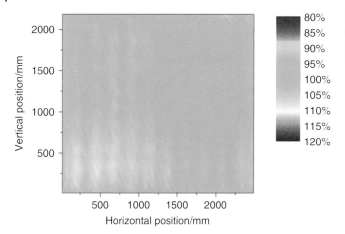

Figure 11C.14 Thickness uniformity of ±9% of an alumina layer deposited on Gen 8.5 size with the PiVot static array coater (*Source:* deposition and measurement by Applied Materials)

Section 10.2.4 are observed. Thus, the dual magnetron technology was used enabling a stable deposition process and good layer and uniformity control of ±9%, as shown in Figure 11C.14.

11C.3.4 Transparent Electrodes

Transparent conductors have widespread applications in flat panel displays. For LCD devices, two transparent electrodes are necessary in order to modulate the liquid crystal by the electric fields. For other devices, such as OLED, at least one electrode needs to be transparent to allow light outcoupling. In general, the idea of a transparent and at the same time-conductive material sounds contradicting since transparency in the visible spectral region requires an electronic bandgap larger than 3eV, which in turn limits the ability to provide a sufficient number of free charge carriers. Some metal-oxides offer the possibility to enhance conductivity while maintaining optical transmittance through control of stoichiometry and doping. The material with the highest conductivity among the family of metal oxides is indium-tin-oxide (ITO), for example [24]. For this reason it has been used widely as transparent electrode in flat panel display applications. A resistivity as low as 130 μΩcm has been reported after optimizing the charge carrier density by generating oxygen vacancies and by doping with tin atoms providing one excess electron compared to indium [25].

The material performance depends strongly on process parameters, with the substrate temperature playing a dominating role on the resistance of the growing ITO film, which undergoes a recrystallization at ~150 °C [26]. For this reason, films that were exposed to temperatures above this recrystallization point usually show a much lower resistivity than films deposited and kept at lower temperatures or ambient conditions. Another important aspect is the oxygen content of the layer, since controlling oxygen vacancies is

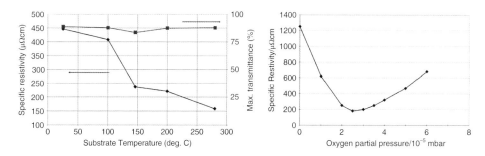

Figure 11C.15 ITO resistivity versus oxygen partial pressure and substrate temperature during deposition (*Source:* deposition and measurement by Applied Materials).

crucial for optimizing the electrical properties of the film. Figure 11C.15 (left) shows the temperature dependence of the ITO transmittance and resistivity with deposition at various substrate temperatures. Figure 11C.15 (right) shows the typical resistance minimum behavior as function of oxygen partial pressure, which underlines the importance of controlling oxygen vacancies of the film.

One issue of a crystallized ITO layer is the poor etchability compared to amorphous phase, affecting applications such as the pixel electrode of an LCD. Etching of amorphous ITO can be carried out with oxalic acid based etchants, which are mild and minimize risk for attacking underlying TFT layers. This approach provides for sufficiently large etch rates and high pattern resolution, however, the oxalic acid is not able to dissolve ITO crystal grains. This leads to process reliability challenges that even when the bulk layer is amorphous, since crystal seeds may remain as residues after etching. The effect becomes more severe at increased layer thicknesses due to spontaneous crystallization phenomena in the bulk. For this reason work has been done to improve ITO film etchability while maintaining good layer properties. A well-established way to suppress crystallinity is the use of hydrogen as an additional process gas during deposition [27]. Hydrogen can be added either as H_2 gas or as H_2O vapor, which dissociates in the sputter gas discharge. The functionality of both additives is similar, however, for practical applications some considerations have to be followed: introducing water vapor into a vacuum chamber can be easily managed by stacking a water evaporator and a special water vapor mass flow controller. Since the reaction time of the setup and the adsorption/desorption ratio of water vapor at the surfaces inside the chamber needs to be balanced for stable volume production, timing and stability of water vapor flow, substrate flow and sputter processes need to be controlled very well. On the other hand, from a technical point of view, it is straightforward to introduce hydrogen gas into the same vacuum chamber. However, since optimizing ITO layer properties requires at the same time controlling the partial pressure of oxygen, the safety aspects related to mixing O_2 and H_2 have to be considered when designing such a sputter system. Both process schemes are known to be used in production by various panel makers. By carefully controlling the process parameters and the gas atmosphere an optimum process window providing low resistivity and low level of residual ITO crystals can be obtained (see Figure 11C.16).

As an alternative material to ITO, indium-zinc-oxide (IZO), has been studied and used by several groups. IZO exhibits a conductivity about twice that of ITO, however, it has been found that the layer stress as well as

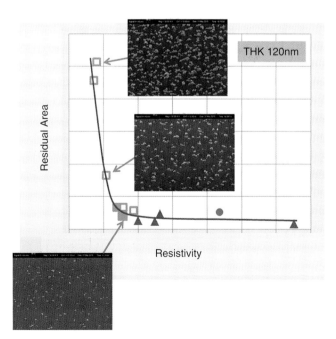

Figure 11C.16 Residue area versus layer resistivity for ITO deposited with various hydrogen incorporations. Carefully controlling the deposition parameters and reactive gas partial pressures enables process tuning for both, good etchability and resistivity (*Source:* depositions and measurement by Applied Materials).

the film crystallinity at moderate temperatures are significantly different [28]. The lower crystallinity of the layers enables production of fine structures with controlled taper angles and low residue levels necessary for high-resolution displays.

11C.3.5 Adding Touch Functionality and Improving End-User Experience

In recent years with the rise of smartphones and tablet computers, touch screen panels have become ubiquitous. Adding touch functionality to a display device can be done either by adding a dedicated touch device on top of the display providing the necessary layers and control logics, or alternatively the touch functionality can be fully integrated into the display panel. In either case, additional electric layers are necessary to detect and transfer the signal produced by the interaction of the user with the mobile device.

The most common touch screen architecture is the so-called projected capacitive touch technology, where a grid of transparent conductive electrodes is detecting local changes of the electric field impacted by the addition of a conductive object such as the finger of the user. The conductive electrodes in this application are usually thin ITO layers, which are patterned as usual. After ITO patterning, the optical appearance of areas with and without ITO layers may be different, which disturbs the user experience. Adding an index- or optical-matching layer-stack (often referred to as invisible ITO) can help suppressing these optical differences.

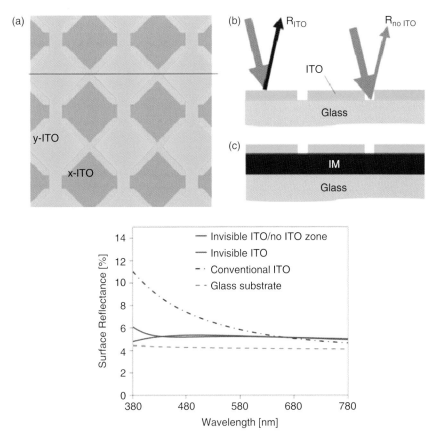

Figure 11C.17 Touch sensor patterning on two sides of a TSP internal glass substrate: top view (a), cross-section of conventional ITO (b) an example of invisible ITO stack concept (c). The graph at the bottom shows the reflectance of ITO with and without the Invisible ITO stack (*Source:* Data from Applied Materials).

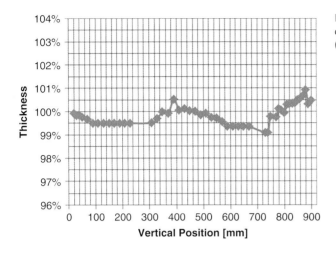

Figure 11C.18 Vertical thickness uniformity of SiO$_2$ layer of better than ±1.0% shown on Gen 4.5 substrate size (*Source:* Data from Applied Materials).

Similarly the contrast ratio of mobile displays and therefore the sunlight readability can be improved significantly by adding an anti-reflective coating. The contrast enhancement is realized by reducing the visible spectrum reflectance from the inner and top surfaces of the display. Furthermore, the optical appearance of anti-reflective coatings is optimized by tuning the design of the stack to produce a color-neutral reflection and avoid impacting the color perception from the display. In some cases, a slight bluish hue is tolerated and may help identifying a layer stack, which provides color stability with regards to viewing angle as well as a good robustness against production variations.

Both the invisible ITO and the anti-reflective coating applications use absorption-free dielectric materials to form optical thin-film stacks that control light reflections through interference effects resulting from the reflections at the interfaces of alternating layers with different refractive indices [29, 30]. A typical structure is shown in Figure 11C.17. Typically, combinations of Silica and Niobia (or Titania respectively) are used as low- and high-refractive index materials to form optical stacks. SiO$_2$ can be sputtered in a fully reactive deposition process with a doped silicon target using the dual magnetron approach. For niobium oxide, it is possible to make use of NbO$_x$ sputter targets and rely on a partially reactive process. Due to the non-conductive nature of the resulting coating also for this layer a dual magnetron setup is most commonly found.

For optical enhancement coatings, a particular attention must be given to the uniformity and long-term stability of the deposition process and equipment. Since the layer stack is forming an optical interference structure, small differences of either of the layer thickness or of the refractive index (resulting from the material composition) may cause significant visually detectable nonuniformities of the final product. A layer uniformity better than ± 2% is required for sufficient color fastness. To achieve this high performance and long-term production stability, besides optimizing the cathode layout and environment, it is important to add in-process monitoring with active feedback looping into the control of reactive gas flows and the process power. Figure 11C.18 shows the single layer uniformity of a SiO$_2$ film deposited with the dual rotary magnetron technique in a dynamic inline coating system.

References

1 D. M. Mattox, The foundations of vacuum coating technology, Noyes Publications, Norwich (2003).
2 W.bR. Grove, On the electrochemical polarity of gases, Phil. Trans. Royal. Soc. (London), B142, p. 87 (1852).
3 F. M. Penning, Coating by cathode disintegration, U.S. Patent 4,356,073, filed December 1935 (October 26, 1939).
4 J. S. Chapin, Sputtering Process and Apparatus, U.S. Patent 4,166,018, filed January 1974 (August 28, 1979).

5 H. Frey, G. Kienel, Dünnschichttechnologie, VDI Verlag, Düsseldorf (1987).
6 R. Behrisch (ed.) Sputtering by particle bombardment I, Springer Verlag, Berlin (1981).
7 B. Chapman, Glow discharge processes, Wiley, New York (1980).
8 H. bE. McKelvey, Magnetron cathode sputtering apparatus, U.S. Patent 4,356,073, filed February 1981.
9 (October 26, 1982).
10 W. Erbkamm, et al., Einrichtung zum Hochratezerstäuben nach dem Plasmatronprinzip, GDR-Patent DD 217964 A3, filed October 1981 (October 26, 1982).
11 R. De Gryse, J. Haemers, W. P. Leroy, D. Depla, Thin Solid Films, (2012).
12 F. Pieralisi, M. Hanika, E. Scheer, M. Bender, Proc. 17th International Display Workshops, December 1–3, Fukuoka, Japan, p. 1865ff (2010).
13 N. Ito, N. Oka, Y. Sato, Y. Shigesato, Japanese Journal of Applied Physicsm 49, p. 071103 (2010).
14 M. Hanika, T. Stolley, Apparatus for treating a substrate, U.S. Patent 8,083,911, filed February 2008 (December 27, 2011).
15 J. Krupka, Meas. Sci. Technol., 24, p. 062001 (2013).
16 Y. W. Ko, D. H. Choi, C. H. Lee, J. C. Lee, Journal of the Korean Physical Society, 33, p. S415 (1998).
17 Y. G. Shen, Materials Science and Engineering, A359, p. 158 (2003).
18 T. Kamiya, K. Nomura, H. Hosono, Sci. Technol. Adv. Mater., 11, 044305, p. 1ff (2010).
19 S. Yasuno, T. Kugimiya, S. Morita, A. Miki, F. Ojima, S. Sumie, Applied Physics Letters, 98, p. 102107 (2011).
20 M. Ryu, et al., Proc. IEEE Int. Electr. Dev. Meeting 2012, San Francisco, USA, pp. 5.6.1–5.6.3 (December 10–13, 2012).
21 S. I. Kim, et al., Proc. IEEE Int. Electr. Dev. Meeting 2008, San Francisco, USA, p. 1–4 (December 15–17, 2008).
22 H. C. Park, E. Scheer, K. Witting, M. Hanika, M. Bender, H. C. Hsu, D. K. Yim, Appl. Phys. A 121(2), p. 535 (2015).
23 T. Arai, et al., SID Symp. Digest of Techn. Papers, p. 1033 (2010).
24 A. Klöppel, J. Liu, E. Scheer, Proc. Int. Display Workshops 2012, Kyoto, Japan, p. 163 (December 4–7, 2012).
25 I. Hamberg, C. G. Granqvist, J. Appl. Phys. 60, p. R123 (1986).
26 C. Daube, et al., Proc. Electronic Displays, Chemnitz, Germany (April 9–10, 1997).
27 O. Tuna, et al., J. Phys. D: Appl. Phys. 43, p. 055402 (7pp) (2010).
28 S. Ishibashi, et al., J. Vac. Sci. Technol. A8(3), p. 1399 (1990).
29 D. S. Ginley (ed.) Handbook of transparent semiconductors, Springer, New York, Heidelberg, Dordrecht, London, p. 161ff (2010).
30 H. A. MacLeod, Thin-film optical filters, 3rd ed., IoP Publishing, Bristol and Philadelphia (2001).

12A

Thin-Film PECVD (AKT)

Tae Kyung Won, Soo Young Choi, and John M. White

Applied Materials, Santa Clara, CA 95054 USA

12A.1 INTRODUCTION

Plasma-enhanced chemical vapor deposition (PECVD) method is widely used for thin-film deposition of insulating and semiconducting layers in thin-film transistors (TFT) for AMLCD and AMOLED displays. This is because, among all thin-film deposition technologies, PECVD provides the most suitable film quality as well as production scalability as it can provide excellent uniformity at relatively high deposition rates while at the same time providing a very low level of defects in the films produced. PECVD has been an essential enabler of FPD manufacturing since the industry's inception in the early 1990s and through its many stages of growth from laptop screens, to desktop monitors and large-screen TVs and recently going through the smartphone and tablet revolution. In recent years, as the requirements for high-resolution and high-performance devices have increased, PECVD systems have also became more advanced and sophisticated in order to meet demands for higher film quality and uniformity, all while the size of the PECVD systems have become very large. The latest generation of PECVD systems can process substrates measuring up to $9\,m^2$. After holding such a long-favored position in the manufacturing of TFT backplanes, PECVD has recently expanded dramatically into the front plane as the preferred means of providing thin-film encapsulation (TFE) of OLED displays.

Most common PECVD systems are capable of simultaneously processing multiple substrates in a single cluster tool. Adoption of the cluster tool architecture enabled FPD manufacturing to enter into high volume mass production. A typical cluster tool system utilizes a central vacuum transfer chamber, which accommodates a stack of single-slot load lock chambers for substrate entry and exit from/to atmospheric environment, and up to as many as 7 PECVD chambers for thin-film deposition. The PECVD chambers can be outfitted to deposit amorphous Si (a-Si), doped a-Si, microcrystalline Si (μc-Si), doped μc-Si, SiN_x, SiON, and SiO_x films for either TFT or TFE applications. Figure 12A.1(a) shows a typical Gen 10 size PECVD system with a PECVD chamber on the left and a triple single-slot load lock stack on the right in the view. Figure 12A.1(b) shows a high-angle view of a full Gen 10 size system, having five PECVD chambers [1–4].

Figure 12A.2 shows a schematic illustration of a typical PECVD process chamber and shows the mechanisms of thin-film deposition and in situ chamber dry cleaning. The hardware configuration is basically a parallel plate capacitively coupled plasma (PP-CCP) reactor. Controllable process parameters available to tune thin-film properties include precursor gas flow mixture and flow rate, temperature of the substrate and the chamber walls, chamber pressure, electrode spacing, and RF power. One very key element of every

Figure 12A.1 Gen 10 size PECVD system (a) single chamber view, and (b) perspective of a five-chamber system.

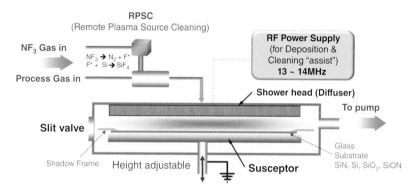

Figure 12A.2 PECVD process chamber illustration.

PECVD chamber is the upper electrode showerhead or "diffuser," which ideally features a very large array of hollow cathode cavities that face the substrate. These hollow cathodes can significantly enhance gas dissociation efficiency, particularly in the range of pressure around 1 Torr where PECVD typically operates most effectively. The hollow cathode effect results from the relatively large electrode surface area per unit volume of plasma inside each hollow cathode. Electron density may be increased within the hollow cathode volume by an order of magnitude over a standard non-HCE showerhead, resulting in substantially higher gas dissociation efficiency. The bottom electrode (also called a susceptor), where the substrate rests during deposition, ideally provides uniform heating and/or cooling, as required, to the glass substrate during processing. RF power, commonly in the range of 13 to 14 MHz, is delivered to the showerhead. All modern PECVD chambers are also equipped with in situ dry cleaning capability in the form of a remote plasma source cleaning (RPSC) unit. These RPSC units are commercially available, relatively small separate plasma sources, which are positioned upstream from the PECVD process chamber and effectively provide dissociated fluorine containing gases, which can then flow downstream to the process chamber. The RPSC unit is used periodically to etch away the unwanted film residues left in the chamber by the PECVD process. Nitrogen trifluoride (NF_3) or fluorine F_2 is used for cleaning because both species can relatively easily be dissociated to nearly 100% and the chamber effluent can then be easily abated, thereby avoiding emission of global-warming-potential gases [1–3].

While the PECVD mechanisms and chemical reactions are complicated and depend on various deposition parameters, Figure 12A.3 shows the basic deposition concepts that generally consist of dissociation, atomic ionization, molecular ionization, atomic excitation, molecular excitation, and surface adsorption. An illustration of the amorphous silicon deposition process is shown as an example (right).

Almost all commercially successful thin-film deposition in the FPD industry is carried out in parallel-plate capacitively coupled plasma reactors operating in the range of 13 to 14 MHz. It is well known that higher frequency excitation can produce higher density plasma with potential benefits of higher gas dissociation/deposition rates and lower ion energy (in fact, 27 MHz has been used successfully up to Gen 5 size). However, for Gen 6 and larger substrate sizes, deposition at frequencies above 13 to 14 MHz tend to suffer excessively

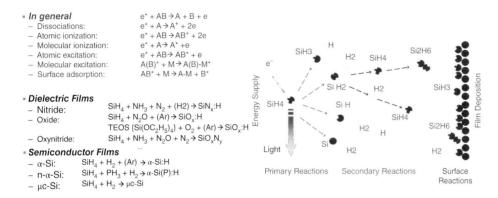

Figure 12A.3 PECVD deposition chemical reactions.

from the non-uniform effects of RF surface standing waves in the plasma chamber. A few alternative high-density plasma source technologies, such as inductively coupled plasma (ICP) and microwave plasma sources have been explored, again in efforts to increase plasma density, that is, increase deposition rate, and lower ion bombardment energy, as well as to lower processing temperature; but up to this point, none have been commercialized.

12A.2 PROCESS CHAMBER TECHNOLOGY

12A.2.1 Electrode Design

12A.2.1.1 Hollow Cathode Effect and Hollow Cathode Gradient

Figure 12A.4(a) schematically describes the hollow cathode effect (HCE), showing secondary electron generation and enhanced electron density inside and right below hollow cathode cavities, or showerhead holes. The electrons oscillated and accelerated by electric field across the plasma sheaths enhance the dissociation by about one order of magnitude [5]. Figure 12A.4(b) is a cross-section view of a pair of axisymmetric 2-D

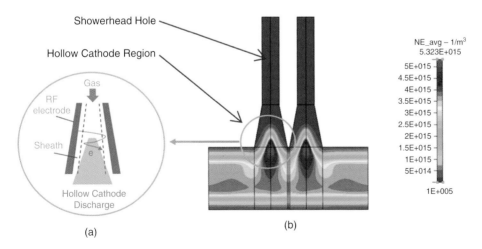

Figure 12A.4 Hollow cathode effect.

Table 12A.1 Silicon nitride film properties deposited by hollow cathode.

SiNx film	DR A/min	RI	Stress MPa	WER (6:1 BOE) A/min	FT-IR N-H	FT-IR Si-H
HCE	1944	1.89	80	1063	19.0%	6.4%
Non HCE	1641	1.86	500	5281	18.6%	20.4%

numerical models of the electron density distribution inside what would be two adjacent hollow cathode showerhead holes. Table 12A.1 shows a comparison of silicon nitride (SiNx) film properties deposited by a HCE diffuser and a non-HCE diffuser, and otherwise identical conditions. The non-HCE film shows significantly lower quality than the HCE film, such as lower refractive index (RI), higher tensile stress, higher wet etch rate (WER), and higher Si-H chemical bonding percentage. Si-H bonding percentage is one of the most important SiNx film qualities. Electrical quality of SiNx film can be characterized using FTIR measurements, which directly correlate to critical film characteristics, such as energy band gap, dielectric constant, and defect density. These points will be described in greater detail in a subsequent section on applications.

The basic guideline for scaling up PECVD processes generation to generation has been to maintain the same intensive deposition parameters, such as substrate temperature, deposition pressure, and electrode spacing, while somewhat proportionally increasing the extensive deposition parameters such as RF power and gas flow rate. These guidelines worked adequately through the Gen 5 size. However, as substrate size increased to Gen 6 size and beyond, some processes did not simply scale up according to the "usual rules." It was found through study and many experiments that film non-uniformity issues were being caused by non-uniformity of RF power density between the two parallel plate electrodes. As a solution to this problem, a hollow cathode gradient (HCG) diffuser was developed to compensate/control plasma uniformity by properly shaping the voltage distribution across the full array of hollow cathodes in the showerhead. The specific design is made to exactly compensate for the surface standing wave effect. By varying the hollow cathode cavity depth, angle, and/or cavity volume, the gradient in hollow cathode effect can be tailored as needed across the showerhead. The basic principles behind the HCG concept may be extended to any size of reactor. Figure 12A.5 shows typical deposition rate uniformity profiles which have been achieved over generations of products, with careful tuning of the HCE [1, 6].

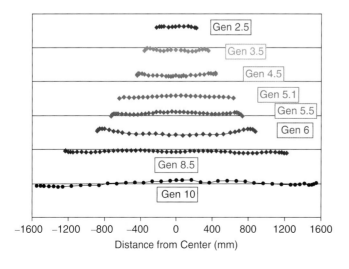

Figure 12A.5 Deposition rate uniformity over generations.

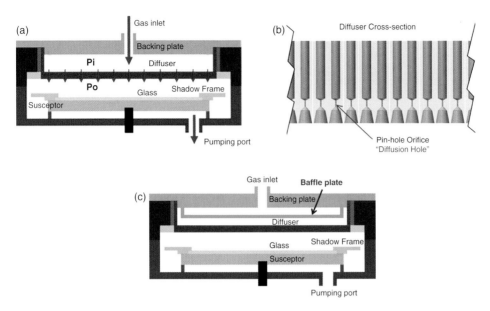

Figure 12A.6 Gas flow uniformity control.

12A.2.1.2 Gas Flow Control

The showerhead, or diffuser, in a parallel-plate CCP PECVD reactor has multiple functions: not only plasma voltage distribution control, as a powered electrode, but also gas flow uniformity control. While hollow cathode cavities enhance plasma dissociation efficiency, gas flow through each of those cavities is controlled by a pin-hole orifice as shown in Figure 12A.6. While SiN_x film uniformity is more sensitive to the surface standing wave effect in the plasma, SiO_x film uniformity is strongly affected by gas flow control. Certain processes, for example, the TEOS (tetraethylorthosilicate)-based SiO_x process may require an additional flow control mechanism beyond carefully controlling the diffusion hole size. In such cases, a baffle plate located between the backing plate and diffuser plate, as shown in Figure 12A.6(c), is provided. The baffle plate comprises its own unique pattern of gas distribution holes, which provides a unique desirable pressure distribution of Pi to the upstream side of the showerhead diffusion holes.

12A.2.1.3 Susceptor

The susceptor, or bottom electrode, also has multiple functions in a PP-CCP PECVD reactor. First, the susceptor is the substrate supporting surface during processing. It is critically important that this surface is flat and makes good thermal and electrical contact with the backside of the substrate, because the distance between the showerhead and the substrate at every location (i.e., the process spacing) is a critical control parameter of the plasma discharge and, in turn, a critical parameter for the process recipe. A flatness < 1 mm is typically required. Amorphous silicon, for example, is very sensitive to electrode spacing parallelism. Too much variation will degrade film quality and deposition rate. Because the substrate must be exchanged after every layer or layer stack is deposited and because the PP-CCP process spacing is by necessity too narrow to allow the exchange to take place between a stationary set of electrodes, the susceptor must move away (down) from the process spacing to allow room for a robot (blade) to execute a substrate exchange. As a consequence of having to move the susceptor to do substrate exchange and because the susceptor is the opposing (grounded) RF electrode to the showerhead, it is necessary to use flexible, electrically-conductive grounding straps around the perimeter of the susceptor, as shown in Figure 12A.7. These straps are essential in providing proper RF grounding and thus proper plasma confinement in the space between the showerhead and susceptor, which is the key to achieving high deposition rates, uniform high-quality films, as well as efficient use of RF power. Plasma confinement will be discussed in detail in a later section.

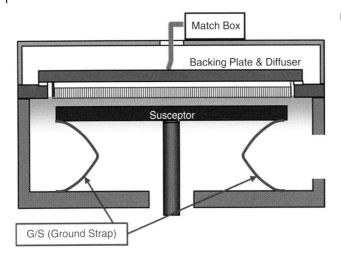

Figure 12A.7 Susceptor grounding.

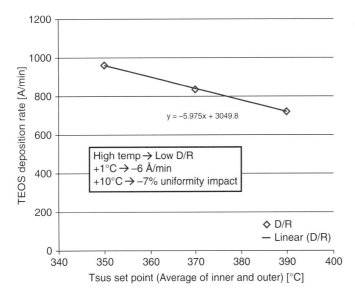

Figure 12A.8 Temperature sensitivity of TEOS-based SiO$_x$ process.

Second, the susceptor provides thermal energy during plasma processing. A susceptor is designed to have multizone electrical heaters for high-temperature processing in the range of 200 to 450 °C. Additionally, if the process requires that heat from the plasma is to be removed from the substrate to maintain a desired (lower) substrate temperature, the susceptor is designed to provide a circulating heat transfer fluid for cooling the substrate (as well as heating purposes). A low substrate temperature is required for CVD processes such as TFE in OLED manufacturing; where the deposition temperature must be maintained <100 °C. Temperature uniformity within a substrate is important and is usually controlled within a 5 °C range. TEOS-based SiOx processes are particularly sensitive to temperature. The deposition rate temperature coefficient of TEOS is 0.7%/°C, as shown in Figure 12A.8.

12A.2.2 Chamber Cleaning

Every PECVD process chamber inevitably accumulates unwanted deposits and must be cleaned otherwise particulates will at some point cause product yield losses. In situ plasma cleaning was a productivity revolution in the FPD manufacturing industry when first introduced in Gen 2 size PECVD systems because it turned many hours per week of downtime required for manual chamber cleaning into production time, as well as dramatically reducing the panel yield losses due to particulates.

Presently, in situ plasma cleaning has been replaced by remote plasma source cleaning (RPSC) [7, 8]. RPSC technology created a second productivity revolution in mass production PECVD tools due to further improvements in particles and yield performance, as well as enabling a longer lifetime of process chamber components such as diffusers and susceptors. The advantage of RPSC is that a very high cleaning rate can be achieved while there is virtually no plasma damage to the chamber hardware from the cleaning process, thereby reducing cost of ownership of the tools [3]. A typical RPSC reactor is an inductively coupled plasma source, which can efficiently dissociate fluorine-containing molecules into reactive fluorine radicals. NF_3 is widely used for dry cleaning because it is commercially viable and has less negative environmental impact as compared to other fluorine-containing gases such as SF_6 or carbon fluoride (C_xF_y) compounds. However, molecular fluorine (F_2) shows the least environmental impact, as it shows zero global warming potential (GWP), compared to NF_3 with a GWP_{100} of about 10,800, or SF_6 with GWP_{100} of about 22,200. GWP is a relative measure of how much heat a greenhouse gas traps in the atmosphere. It compares the amount of heat trapped by a certain mass of the gas in question to the amount of heat trapped by the same mass of carbon dioxide. A GWP value is established by carrying out the calculations over a specific stated time interval, commonly 20, 100, or 500 years. GWP is expressed as multiple of carbon dioxide, which has a GWP that is standardized to 1. Due to its weaker bonding energy, F_2 dissociates much easier than NF_3, reducing the energy required from the system. Unfortunately, F_2 has not been widely used as a cleaning gas because its toxicity is 10 times higher than NF_3, it is much more reactive/corrosive and it requires a large initial capital investment.

Detecting the end point of the cleaning process is most commonly done by fingerprinting the chamber pressure curve as the partial pressures of the various chemical species inside the chamber reactor change upon the cleaning process reaching completion, as described by the following chemical reactions:

$$NF_3 + \text{remote plasma}(RPS) \rightarrow N_2 + F^*$$
$$F^* + Si/SiN_x/SiO_x \rightarrow SiF_4 + \ldots$$

Although NF_3 or F_2 dissociation efficiency can be nearly 100%, it is very critical to properly design the gas flow conduit to deliver reactive radicals from the RPSC unit to the PECVD chamber without losing their reactivity. Once dissociated, fluorine radicals tend to recombine into molecular fluorine (F_2) by spontaneous exothermic reaction, in particular under high pressure conditions, although in situ RF power may be applied to break down the recombined F_2 species. Nonetheless, it is important to consider the gas flow conductance, materials and methods of construction, heat dissipation, and more in the chamber design to preserve and deliver >90% of the fluorine activated in the RPSC unit into the process chamber.

Whenever the substrate throughput of a PECVD system is limited by the processing rate (rather than the mechanical throughput) the cleaning rate performance directly impacts throughput. Figure 12A.9 clearly

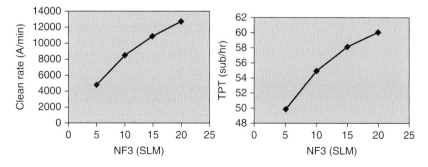

Figure 12A.9 NF_3 flow impact on clean rate and system throughput (TPT) in an example of Gen 6 size PECVD.

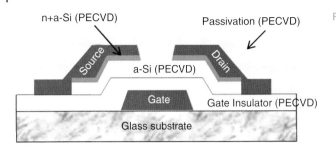

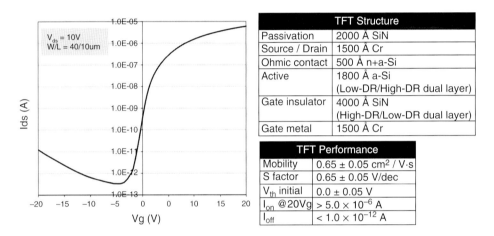

Figure 12A.10 a-Si TFT structure (BCE type).

Figure 12A.11 Typical I-V characteristics of a-Si TFT and its structure.

shows the effect of NF_3 flow on clean rate and system throughput for a certain film application on a certain Gen 6 size PECVD system.

12A.3 THIN-FILM MATERIAL, PROCESS, AND CHARACTERIZATION

12A.3.1 Amorphous Si (a-Si) TFT

Figure 12A.10 shows a typical back-channel-etch (BCE) type a-Si TFT structure. Figure 12A.11 shows its layer thicknesses and electrical characteristics. As described in the structure table, a dual-layer concept (low deposition rate (DR) process for interface layer and high DR process for bulk layer) has been used in production for the a-Si film and for the gate insulator SiN film to improve system throughput and TFT characteristics [9, 10].

12A.3.1.1 Silicon Nitride (SiN)

It is widely recognized that PECVD SiN is superior to other dielectric materials as gate insulator and passivation layers in a-Si TFTs. The dual layer structure (separate interfacial and bulk dielectric layers) is often used in large area TFT array manufacturing for improving productivity and achieving improved transistor characteristics. Table 12A.2 shows typical PECVD SiN_x film property ranges for a-Si TFT application.

PECVD process parameters for SiN_x film include gas flow rates for SiH_4, NH_3, and N_2, pressure, RF power, electrode spacing, and substrate temperature. Typical single parameter process trends of 600×720 mm size

Table 12A.2 Typical PECVD SiN$_x$ properties used for a-Si TFT devices.

Film property	Typical value	Measurement method
Deposition rate	50–300 nm/min	Thickness
Refractive index	1.85–1.95	Ellipsometer
Film stress	−1000–500 MPa	Stress tool
N-H bond	13–20%	FT-IR
Si-H bond	0.1–10%	FT-IR
H content	~25%	HFS/RBS
Wet etch rate	20–100 nm/min	6:1 BHF solution
Dielectric constant	6–7	Mercury probe
Optical band gap	5.0–5.5 eV	UV-Vis spectrometer

substrates are shown in Figures 12A.12 through 17, with the following film property trends. As the SiH$_4$ flow increases, the SiN$_x$ film becomes of lower quality (higher Si-H content, more tensile stress, higher wet etch rate (WER)). Pressure and electrode spacing increases show the same trends as for SiH$_4$ flow, however, the RF power and N$_2$ flow rate have opposite trends. Increasing RF power or N$_2$ flow produces a better quality SiN$_x$ film (lower Si-H content, compressive stress, lower WER). Therefore, SiN$_x$ film properties can be adjusted by controlling each process parameter depending upon the film requirements for target application, for example, gate insulator or passivation.

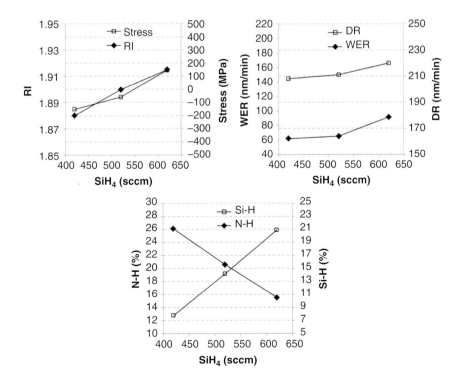

Figure 12A.12 SiN$_x$ film properties change as a function of SiH$_4$ flow.

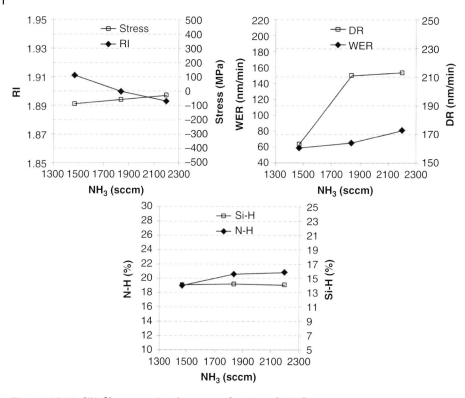

Figure 12A.13 SiN$_x$ film properties change as a function of NH$_3$ flow.

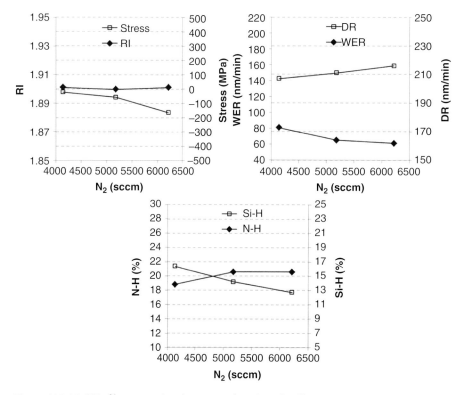

Figure 12A.14 SiNx film properties change as a function of N$_2$ flow.

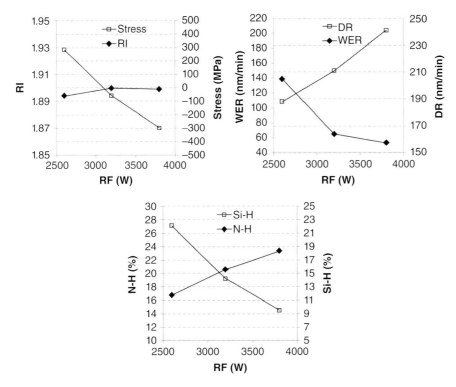

Figure 12A.15 SiN$_x$ film properties change as a function of RF power.

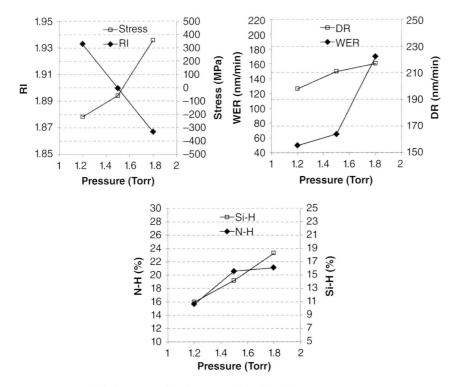

Figure 12A.16 SiN$_x$ film properties change as a function of pressure.

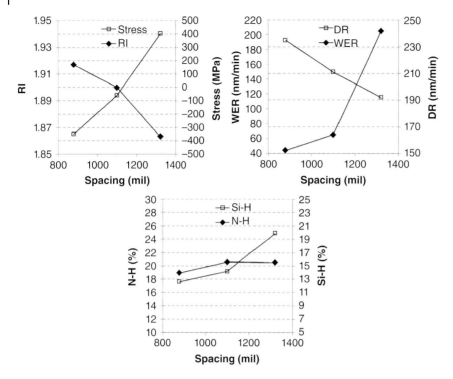

Figure 12A.17 SiN$_x$ film property change as a function of spacing.

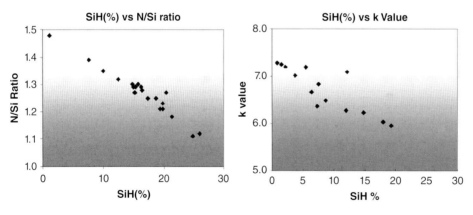

Figure 12A.18 Relationship between Si-H content and N/Si ratio and k value.

In general, a high-quality SiN$_x$ film is characterized by a low Si-H content, as shown in Figure 12A.18, and it can be obtained by increasing RF power, reducing pressure or reducing electrode spacing as shown in the trend charts.

The relationship between a-Si TFT characteristics and SiN$_x$ film properties is very complicated. Physical and chemical properties of the film are all critical to the TFT performance. As discussed, a wide range of SiN$_x$ film quality can be prepared by adjusting PECVD process parameters. Certain SiN$_x$ dielectric film properties are more desirable for good a-Si TFT performance than others. Some SiN$_x$ film properties affect a-Si TFT characteristics significantly. Figure 12A.19 explains how to control and improve mobility of a-Si TFTs by adjusting gate insulator SiN$_x$ properties.

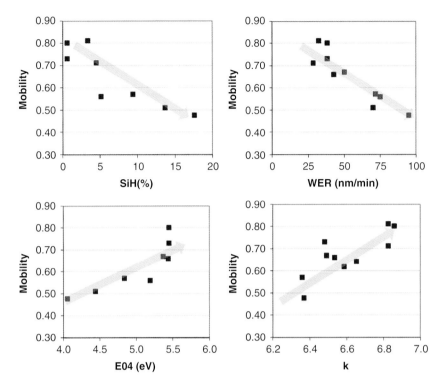

Figure 12A.19 Relationship between SiN$_x$ film properties and TFT mobility.

Generally, the Si-H content, wet etch rate (WER), optical band gap (E04), and dielectric constant (k) values are the most critical SiN$_x$ film properties for a-Si TFT performance. A lower Si-H content, lower WER, higher E04, and higher k produce higher mobility TFTs. The following SiN$_x$ film properties are required to achieve high mobility a-Si TFTs (e.g., >0.7 cm^2/Vs): <4% Si-H content, <50 nm/min WER, >5.3 eV optical bad gap (E04) and >6.8 dielectric constant (k).

In a-Si TFTs, SiN$_x$ film is used as a passivation layer on the back-channel area. In general, for good a-Si TFT performance, passivation SiN$_x$ properties should follow the same trends as the gate insulator SiN properties (lower Si-H content, lower WER, higher E04, and higher k).

12A.3.1.2 Amorphous Silicon (a-Si)

PECVD a-Si is used as an active layer of a-Si TFT devices. The dual layer structure (separate interfacial and bulk active layers) is often used in the large area TFT array for improving productivity and transistor characteristics. Table 12A.3 shows typical PECVD a-Si film properties for a-Si TFT application.

The most common source gases for PECVD a-Si film are SiH$_4$ and H$_2$, with Ar sometimes used instead of H2 for special purposes. The key PECVD process parameters for a-Si film deposition include gas flows for SiH$_4$, H$_2$, RF power, pressure, electrode spacing, and substrate temperature. Typical single parameter process trends for a 600 × 720 mm size substrate are shown in Figures 12A.20 through 22, with the following observed film property trends. As the SiH$_4$ flow increases, mainly the deposition rate increases and the film stress becomes closer to zero. Increasing H$_2$ flow reduces DR and increases compressive stress and [H] content. While more RF power significantly increases DR, stress and [H] content, higher pressure reduces stress, [H] content and FTIR Si-H peak position. Narrow electrode spacing shows the same effect as higher RF power. Therefore, a-Si film properties can be adjusted by controlling each process parameter to meet the target film requirements.

Table 12A.3 Typical PECVD a-Si properties used for a-Si TFT devices.

Film property	Typical value	Measurement method
Deposition rate	20–300 nm/min	Thickness
Refractive index	3.5–4.2	Ellipsometer
Film stress	−100 to 1000 MPa	Stress tool
Si-H peak position	1990–2000/cm	FT-IR
Si-H peak area (WH/T)*	0.4–0.5	FT-IR
H content	5%–10%	FT-IR
Optical band gap (E04)	1.8–1.9 eV	UV-Vis spectrometer
Activation energy (Ea)	0.75–0.85 eV	Conductivity
Dark conductivity	~5E−10 S/cm	Conductivity
Photo conductivity	~1E−4 S/cm	Conductivity
Photosensitivity	~1E6	Conductivity

* WH/T denotes W: Si-H peak width, H: Si-H peak height, T: film thickness

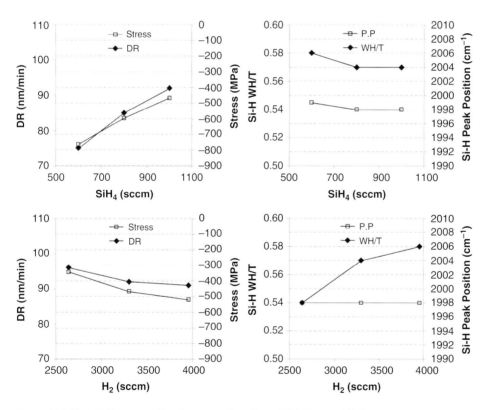

Figure 12A.20 a-Si film properties change as a function of SiH_4 flow and H_2 flow.

Generally, a high-quality a-Si film is characterized by low DR and adequate [H] content (5%–8%) to minimize the dangling bonds in the film. There is a linear-relationship between DR and [H] content (or WH/T) as shown in Figure 12A.23, and it can be created by reducing RF power and increasing pressure.

The relationship between a-Si TFT characteristics and a-Si film properties is very complicated. Physical and chemical properties of the film are all critical to the TFT performance. A wide range of a-Si film quality can

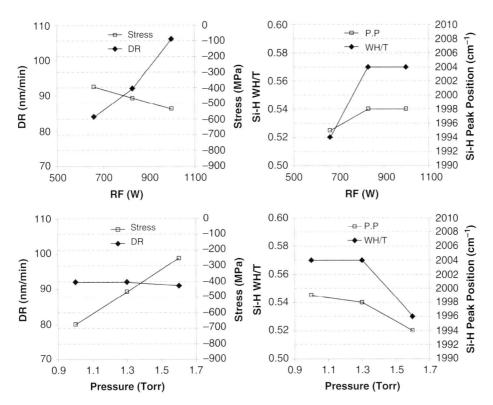

Figure 12A.21 a-Si film properties change as a function of RF power and pressure.

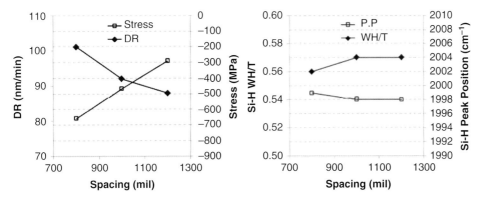

Figure 12A.22 a-Si film properties change as a function of spacing.

be prepared with PECVD by adjusting process parameters. Some of the a-Si film properties affect significantly the a-Si TFT characteristics, and Figure 12A.24 explains how to control and improve mobility of a-Si TFTs by adjusting a-Si properties.

Generally, deposition rate (DR), optical band gap (E04), and hydrogen content (WH/T or [H]) values are the most critical film properties for a-Si TFT performance. Lower DR, higher E04, lower [H], or WH/T produce higher mobility. The following film properties are required to get high mobility a-Si TFT (>0.7 cm^2/Vs): <500 nm/min DR, <1.84 eV optical band gap (E04), and <0.43 WH/T (6~8% [H]) values.

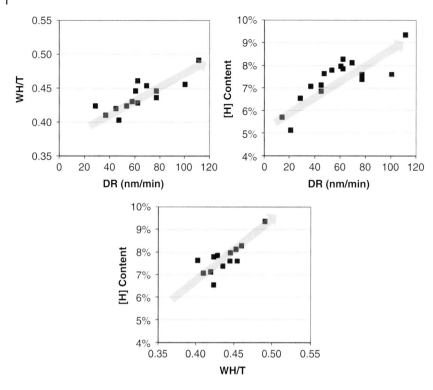

Figure 12A.23 Relationship between [H], WH/T, and DR.

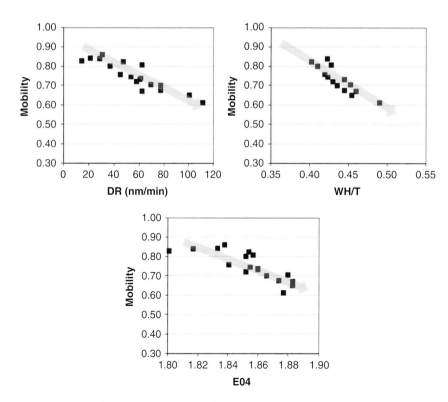

Figure 12A.24 Relationship between a-Si film properties and TFT mobility.

Table 12A.4 Typical PECVD n+ a-Si properties used for a-Si TFT devices.

Film property	Typical value	Measurement method
Deposition rate	50–200 nm/min	Thickness
Film stress	−100 to −300 MPa	Stress tool
Dark conductivity	~1E-2 S/cm	Conductivity
Dark resistivity	30–150 Ω cm	Conductivity
Activation energy (Ea)	~0.2 eV	Conductivity

12A.3.1.3 Phosphorus-Doped Amorphous Silicon (n$^+$ a-Si)

The n+ a-Si film can be deposited from PECVD by introducing PH_3 gas in the source gases for a-Si. A highly conductive n-type n+ a-Si film can be prepared over a large range of $PH_3/SiH_4/H_2$ ratios. The conductivity of the lightly doped film varies with the change of the $PH_3/SiH_4/H_2$ ratio. The film's conductivity can be influenced by factors such as the dopant concentration, the efficiency of dopant activation and defect states. These characteristics are dependent on deposition process parameters like source gas composition ($PH_3/SiH_4/H_2$), RF power, pressure, electrode spacing, and substrate temperature. Table 12A.4 shows typical PECVD n+ a-Si film property ranges for a-Si-TFT application.

The most common source gases for PECVD n+ a-Si film are PH_3, SiH_4 and H_2, and sometimes Ar is used instead of H_2 for special purposes. PECVD process parameters of n+ a-Si film include PH_3 gas flow, SiH_4 gas flow, H_2 gas flow, RF power, pressure, electrode spacing, and substrate temperature. Typical single parameter process trends of 600 × 720 mm size substrate and effects on the film properties are shown in Figures 12A.25

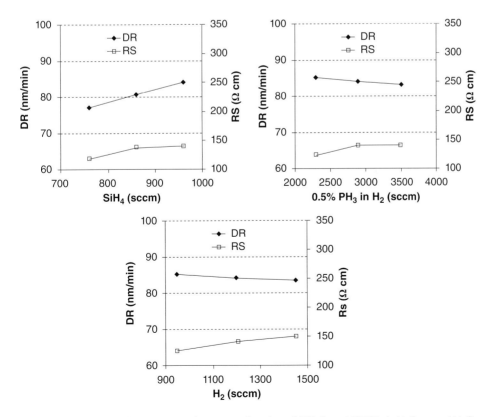

Figure 12A.25 n+ a-Si film property change as a function of SiH_4 flow, 0.5%PH_3 in H_2 flow, and H_2 flow.

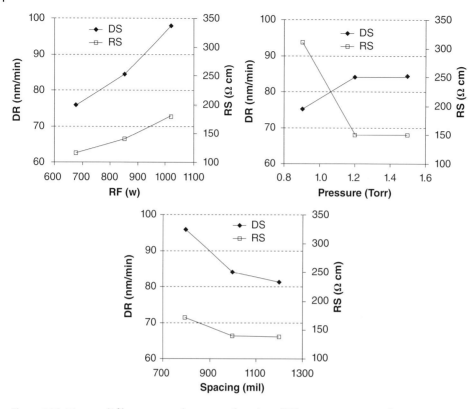

Figure 12A.26 n+ a-Si film property change as a function of RF power, pressure and spacing.

and 12A.26. The resistivity (or conductivity) of n+ a-Si film is mainly influenced by parameters such as RF power, pressure, and spacing. Lower RF power and higher pressure reduce resistivity significantly. Narrow spacing shows the same effect as higher RF power.

In order to use n+ a-Si film for a-Si TFT applications, the following requirements should be met: (1) a highly conductive n+ a-Si film needs to be deposited to obtain ohmic contacts at the source and drain regions, (2) the contact resistance between n+ a-Si and intrinsic a-Si has to be very low, (3) the n+ a-Si film deposition process should be compatible with other TFT processes. Figure 12A.27 shows the effect of n+ a-Si thickness on resistivity and how a certain minimum n+ a-Si film thickness is required to reduce contact resistance to the a-Si.

Generally, <100 Ωcm resistivity of n+ a-Si film is recommended to produce a high-performance a-Si TFT device. Therefore, about a 50 nm n+ a-Si film thickness produced with lower RF power, higher pressure and wider spacing process gives the best result.

12A.3.2 Low-Temperature Poly Silicon (LTPS) TFT

A significant technical challenge in making both high-resolution LCD and OLED is that the transistor material amorphous-Si must be replaced with a material having higher electron mobility, and, in the case of OLED, better V_{th} stability. Until recently, low-temperature poly-silicon (LTPS) was the only high-performance alternative to a-Si. LTPS, has excellent mobility and stability, but is considerably more expensive than a-Si, due to the need for additional process and masking steps, including ion implantation and expensive excimer laser annealing. LTPS is also difficult to scale to substrate sizes suitable for TVs. Figure 12A.28 shows a typical top gate LTPS TFT layer structure, which has been widely used in the industry.

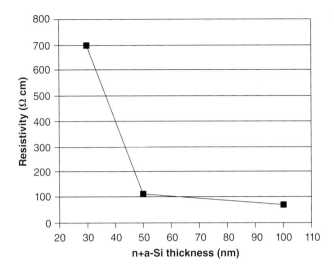

Figure 12A.27 n+ a-Si thickness versus resistivity.

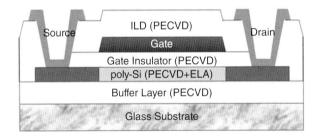

Figure 12A.28 LTPS TFT structure (top gate structure).

12A.3.2.1 Silicon Oxide (SiO)

Silicon oxide film is widely used as a gate insulator, buffer layer, and ILD layer in LTPS TFT devices. SiH_4/N_2O gas mixtures, and TEOS (Tetraethylorthosilane) / O_2 mixtures are typically used to deposit SiO films. Both SiH_4-SiO and TEOS-SiO have been well qualified in the LTPS TFT display industry.

Table 12A.5 presents typical PECVD SiO film properties used for LTPS TFT devices and Figures 12A.29 and 12A.30 show examples of C-V, I-V, and step coverage data for PECVD SiO films.

The most common source gases for PECVD SiO film are SiH_4 and N_2O or TEOS and O_2. PECVD process parameters for a TEOS-SiO film typically include TEOS gas flow, O_2 gas flow, RF power, pressure, electrode spacing, and substrate temperature. Typical single process parameter trends for a 730 × 920 mm size substrate (Gen 4.5 size) for a TEOS-SiO film are shown in Figures 12A.31 through 36, and the effects on film properties are as follows. As TEOS flow increases, deposition rate, WER, and V_{fb} increase while the RI and stress decrease, indicating the SiO film becomes less dense. Increasing O_2 flow produces higher RI and lower WER, indicating a denser film. While higher RF power and higher temperature significantly improve SiO quality by inducing higher RI/lower WER/lower V_{fb}, higher pressure degrades SiO quality, evidenced by lower RI/higher WER/higher V_{fb}. Generally, thermal oxide is recognized as the highest quality silicon oxide and the quality of a PECVD SiO is usually established by comparing its properties to those of thermal oxide, that is, 1.46RI/100 nm WER/1080~1095 cm^{-1} Si-O peak position. Therefore, lower WER/higher Si-O peak position/ lower V_{fb} are used as indications of a good-quality PECVD SiO film. According to the trend chart (Figures 12A.31 to 12A.36), low TEOS flow, high O2 flow, high power, low pressure, and high temperature can improve SiO quality. It is worth noting that TEOS-SiO films generally have much greater sensitivity to temperature than SiH4-SiO films, as indicated in Figure 12A.36.

Table 12A.5 PECVD SiO film property ranges (process dependent).

Film property	Typical value	Measurement method
Deposition rate	50~200 nm/min	Thickness
Refractive index	1.46~1.47	Ellipsometer
Film stress	−100~−300 MPa	Stress tool
Si-O peak position @1000 nm	1070~1080/cm	FT-IR
Wet etch rate (WER)	130~200 nm/min	6:1 BHF solution
H content	2~5%	HFS/RBS
Dielectric constant (κ)	4~5	Hg probe
Breakdown voltage (V_{bd})	>8 MV/cm	Hg probe
Leakage current @2MV/cm	<5E−10 A/cm^2	Hg probe
Flat band voltage (V_{fb})	<−1 V after annealed	Hg probe
Step coverage @500 nm step	>85%	SEM

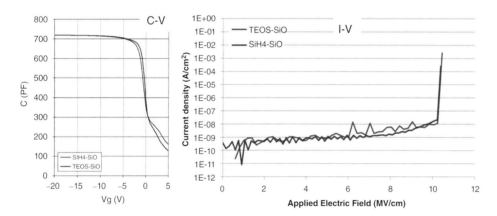

Figure 12A.29 C-V and I-V characteristics curves of SiO films.

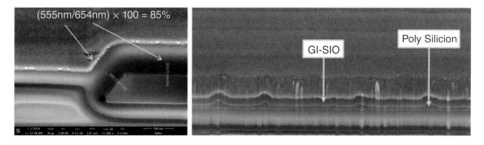

Figure 12A.30 Step coverage picture on 500 nm height pattern and laser-crystallized poly silicon.

SiO film properties can be adjusted by controlling each process parameter depending upon the film requirement for different applications like gate insulator, buffer layer and ILD layer.

12A.3.2.2 a-Si Precursor Film (Dehydrogenation)

A pulsed laser has been widely used to crystallize amorphous silicon precursor film into poly-silicon. The hydrogen content in PECVD a-Si precursor films is the major factor affecting the quality of crystallized

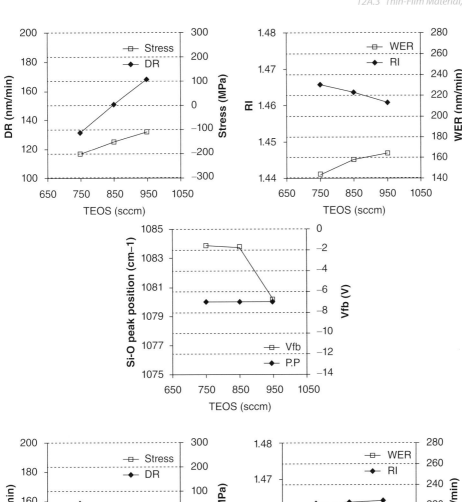

Figure 12A.31 SiO film properties change as a function of TEOS flow.

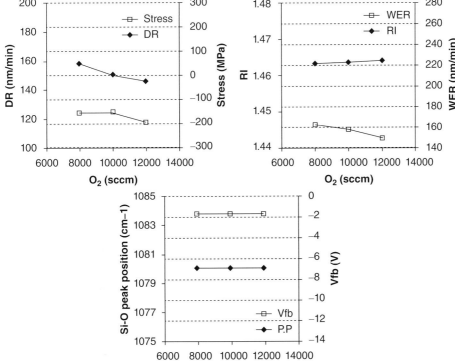

Figure 12A.32 SiO film properties change as a function of O_2 flow.

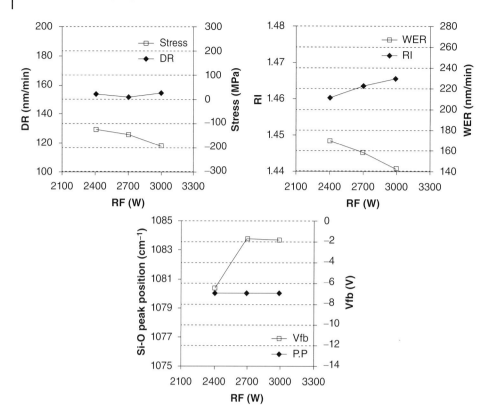

Figure 12A.33 SiO film properties change as a function of RF power.

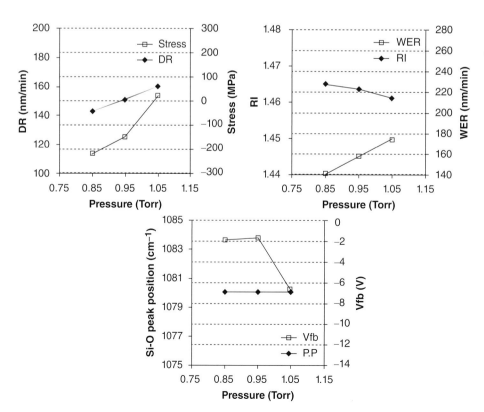

Figure 12A.34 SiO film properties change as a function of pressure.

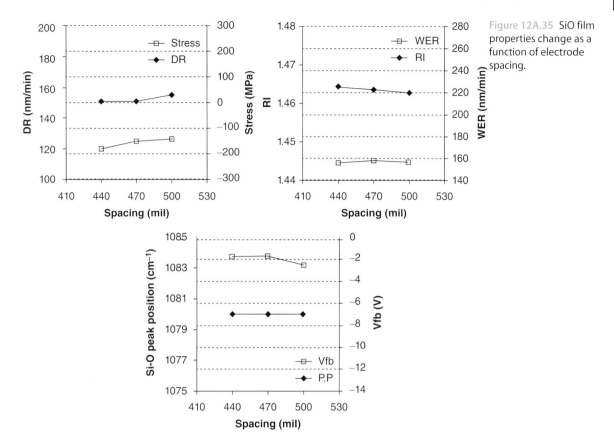

Figure 12A.35 SiO film properties change as a function of electrode spacing.

poly-Si [11]. In LTPS TFT preparation, an important factor is how to effectively reduce the hydrogen content in the film to <1% before the laser crystallization step. Otherwise, high hydrogen content creates bubbling or explosive evolution in the poly-Si surface during laser crystallization [12, 13]. Hydrogen content in the a-Si film can be lowered by using a high PECVD a-Si process temperature (>400 C), but hydrogen cannot be made lower than 4% to 5% as deposited. Therefore, additional heat treatment (annealing) in a heat chamber or furnace is required to reach below the target 1% [H] level.

The most common source gases for PECVD a-Si film are SiH_4 and H_2. Sometimes Ar is used instead of H_2 for special purposes. PECVD process parameters of a-Si film include SiH_4 gas flow, H_2 gas flow, RF power, pressure, electrode spacing, and substrate temperature. Typical single parameter process trends for source gases, RF power, pressure, spacing, are similar to the trends of a-Si film in a-Si TFT (refer to the a-Si TFT section). As stated previously, higher substrate temperature produces lower [H] content in the film. The relationship between deposition temperature, annealing time and [H] content is shown in Figure 12A.37. Higher deposition temperature and longer annealing time always reduce [H] content in the a-Si film and [H] content becomes less than 1% after more than 10 min annealing at 500 °C regardless of deposition temperature.

12A.3.3 Metal-Oxide (MO) TFT

Recently, metal oxide materials, particularly indium gallium zinc oxide (IGZO), have gained momentum as transistor materials due to the fact that they have over 10× higher mobility than a-Si while having lower cost of manufacturing and better scalability than LTPS [14]. Amorphous IGZO has typically >10 cm^2/Vs electron mobility, good uniformity, stability and a low-temperature fabrication process. Therefore, a-IGZO TFTs are used in large area, high-resolution, high frame rate liquid crystal displays (LCDs) and organic light-emitting diode displays (OLEDs). a-IGZO has the additional benefit of being highly compatible with existing a-Si fabs,

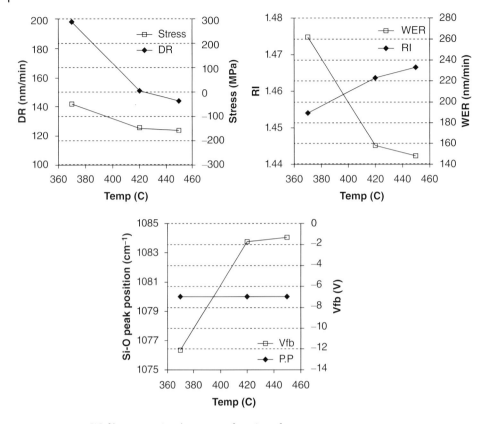

Figure 12A.36 SiO film properties change as a function of temperature.

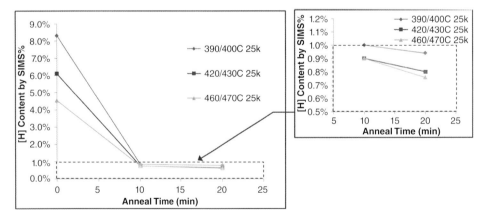

Figure 12A.37 Hydrogen content in a-Si precursor film change as a function of deposition temperature and time (annealing temp was done at 500 °C and [H] content was measured by SIMS at EAG in California, USA).

with only minor modifications. Further, the need for additional masking steps and laser annealing tools associated with LTPS can be avoided. It is important to note that the a-IGZO layer itself is deposited by PVD, but PECVD provides the critical insulating and barrier layers that protect the sensitive IGZO material, and help ensure the stability and performance of the device.

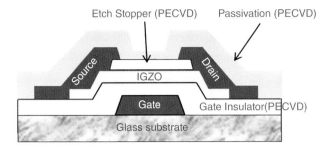

Figure 12A.38 Metal oxide TFT structure (etch stop type).

a-IGZO instabilities can lead to mura defects in large LCD TVs and all OLED devices. Stability performance degrades from impurities in the a-IGZO film such as hydrogen and moisture. To help guard against stability degradation, dielectric material of the highest quality must be used wherever contact is made with the a-IGZO, with good integrity and low hydrogen content [4]. Since SiN_x films have over 20% [H] atomic concentration, SiO with <5% [H] atomic concentration is typically chosen for the dielectric/metal oxide interface layers. To protect the a-IGZO from damage during subsequent process steps and improve the device stability, an etch stop layer is often required. For a bottom gate etch stop TFT structure (ES-TFT), the critical interface layers are at the top of the gate insulator, and the bottom of the etch stop. These layers can be entirely SiO, or they can be a SiO/SiN_x bilayer, as long as the low [H] content SiO is contacting the a-IGZO. Figure 12A.38 illustrates an example of a metal oxide Etch Stop-TFT structure and layers.

12A.3.3.1 Silicon Oxide (SiO)

Specific SiO film properties requirements vary by panel maker. Generally, the best achievable non-uniformity and lowest hydrogen content and defect level are desired. Table 12A.6 shows typical SiH_4-based PECVD SiO film properties ranges for MO-TFT application.

The most common source gases for PECVD SiO film are SiH_4 and N_2O and sometimes Ar is used. PECVD process parameters of SiO film include SiH_4 gas flow, N_2O gas flow, RF power, pressure, electrode spacing, and substrate temperature. Typical single process parameter trends of 680 × 880 mm size substrate are shown in Figures 12A.39 and 12A.40 and effects on the film properties are as follows. As SiH_4 flow increases, mainly deposition rate and WER increase, while RI and Si-O FTIR peak position decreases significantly; this indicates the SiO film becomes less dense. Increasing N_2O flow contributes to higher DR, higher Si-O peak position, and lower RI. While higher pressure produces lower DR, lower WER, lower RI, and higher Si-O FTIR peak position; higher temperature reduces WER and increases RI and Si-O peak position significantly.

Table 12A.6 SiO film properties ranges (process dependent).

Film property	Typical value	Measurement method
Deposition rate	50–200 nm/min	Thickness
Refractive index	1.46–1.47	Ellipsometer
Film stress	−100 to −500 MPa	Stress tool
Si-O peak position	1060–1080/cm @1000 nm	FT-IR
Wet etch rate (WER)	130–200 nm/min	6:1 BHF solution
H content	4%–5%	HFS/RBS
Dielectric constant (κ)	4–5	Hg probe
Breakdown voltage (V_{bd})	>7 MV/cm	Hg probe
Flat band voltage (V_{fb})	<−5 V after annealed	Hg probe

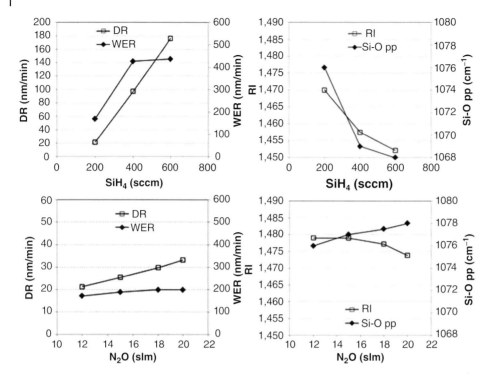

Figure 12A.39 SiO film properties change as a function of SiH$_4$ flow and N$_2$O flow.

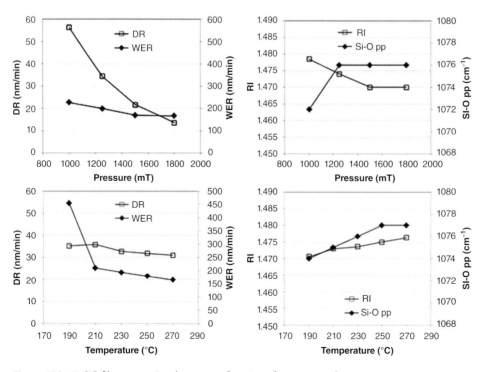

Figure 12A.40 SiO film properties change as a function of pressure and temperature.

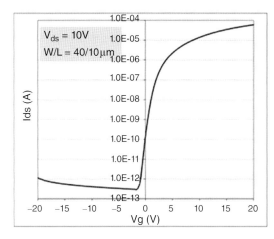

Figure 12A.41 Typical I-V characteristics of a-IGZO TFT and its structure.

Generally, thermal oxide is recognized as the best quality of silicon oxide and the quality of PECVD SiO films are usually determined by comparing properties with those of thermal oxide, 1.46RI/100 nm/min WER/1080-1095 cm^{-1} Si-O peak position. Lower WER and higher Si-O peak position are used as indications of good PECVD SiO film quality. As can be seen from the process trend charts (Figures 12A.39 and 12A.40), low SiH4 flow, high N2O flow, high pressure, and high temperature can improve SiO quality. However, as a-IGZO film is very sensitive to process temperature, temperature control during a-IGZO TFT fabrication is very critical to making stable a-IGZO TFTs. SiO film process temperature after a-IGZO deposition is usually limited to 250 °C due to instability of a-IGZO TFTs when exposed to higher SiO deposition temperature.

Amorphous metal oxide thin-film transistors in Figure 12A.38 have considerable benefits such as low-temperature processing (~230 °C), high mobility (~10 cm^2/Vsec), and low threshold voltage (V_{th}) compared to conventional a-Si TFT. However, a-IGZO-TFTs show poor V_{th} stability under electrical and light stresses. It is well known that V_{th} shift under positive bias temperature stress (PBTS) is caused by weak or unstable oxygen bonds in the IGZO film and V_{th} shift under negative bias illumination stress (NBIS) originated from excess oxygen vacancy sites and hydrogen in the IGZO film which creates charge trapping sites for electrons. Therefore, control of unstable oxygen bonds, interstitial oxygen, oxygen vacancies, and hydrogen content in the IGZO film and the IGZO/SiO interface is very critical to making stable a-IGZO-TFTs. The relationship between a-IGZO TFT characteristics and PECVD SiO film properties is very complicated because a-IGZO's electrical properties such as electron mobility, carrier concentration, and resistivity are changed and even after PECVD SiO deposition. Post-deposition plasma treatment steps and annealing steps are also critical to the final TFT performance. Figure 12A.41 shows structure and basic performance of the a-IGZO TFT used in the following stability study.

The trend charts in Figures 12A.42 and 12A.43 explain how to control the instability behavior of a-IGZO-TFTs by adjusting etch-stop SiO process conditions such as SiH$_4$ flow, N$_2$O flow, pressure, temperature, N$_2$O plasma treatment before SiO deposition, and N$_2$O plasma treatment after SiO deposition.

(1) General trend: V_{on}/PBTS/NBIS shift to the same direction together; more positive V_{on}/more positive PBTS/more positive NBIS or more negative V_{on}/more negative PBTS/more negative NBIS. Therefore, there is a clear trade-off between PBTS and NBIS: better PBTS/worse NBIS or worse PBTS/better NBIS (Figures 12A.42 and 12A.43).

(2) Strong total N$_2$O plasma time dependency: longer N$_2$O plasma time (lower DR process or longer pre-/post-N$_2$O plasma) induces higher PBTS (more unstable oxygen) and lower NBIS (less oxygen vacancy) and vice-versa for shorter N$_2$O plasma exposure time (Figures 12A.42 and 12A.43).

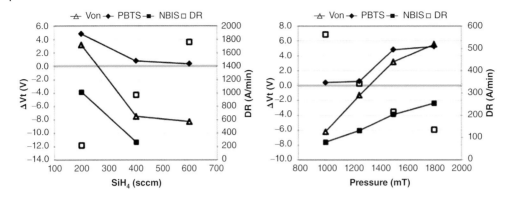

Figure 12A.42 IGZO-TFT characteristics and DR change as a function of SiH₄ and pressure.

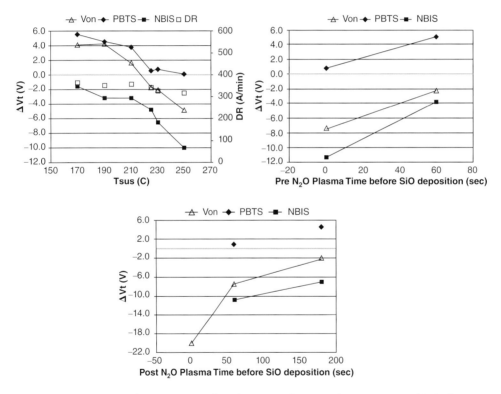

Figure 12A.43 IGZO-TFT characteristics and DR change as a function of temperature and N₂O plasma time (pre- and post-).

(3) Strong influence of hydrogen content on NBIS: more hydrogen, higher NBIS (Figure 12A.44). This trend is evident above 4% [H].
(4) Strong substrate temperature influence: higher SiO deposition temperature induces more negative V_{on}/lower PBTS/higher NBIS due to more oxygen vacancies (Figure 12A.43).

It is clear that control of oxygen vacancies, unstable oxygen, and [H] content in the IGZO film and IGZO-SiO interface is very important to minimize instability of IGZO-TFT [15]. In the PECVD etch-stop SiO process, deposition temperature and total N₂O plasma time are the most critical factors to control instability of IGZO-TFT and optimize PBTS and NBIS based on the TFT's application.

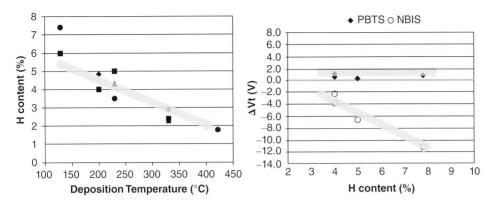

Figure 12A.44 [H] content change in SiO film as a function of deposition temperature and its effect on IGZO-TFT (PBTS under 60 °C/+30 V_g, NBIS under 60 °C/-30 V_g/1000 lux).

12A.3.4 Thin-Film Encapsulation (TFE)

OLED devices can be fabricated on either a rigid-substrate or on a flexible-substrate. However, OLED materials are easily degraded by water vapor and oxygen. The thin cathodes of OLED devices consist of reactive metals that can degrade even more rapidly than the organic EL material. Therefore, OLED require substrates and sealants that have extremely low permeability to water and oxygen. Simple metal or ceramic oxide barrier films for thin polyester or polypropylene substrate material have been developed and successfully utilized in flexible food packaging applications.

However, the substrate roughness and defects intrinsic to these materials limit the barrier performance of these systems. Most vapor deposition techniques are conformal, reproducing the substrate texture. Although vapor permeation rates fall as a function of deposited barrier film thickness due to the covering of substrate roughness and the filling of pinholes, breaks in continuity as the film traverses high aspect ratio features on rough substrates limit the effectiveness of vapor-deposited coatings on plastic. To be long-lived, OLEDs require a moisture barrier, which transmits $<1 \times 10^{-6}$ g/m^2/day of water. Vacuum-based multi-layer thin-film encapsulation is a PECVD technology that has been developed to create such superior vapor barrier architectures. The physical properties of the individual layers can be tailored as desired/required and used as building blocks to achieve the specific barrier performance requirements of various display applications on almost any display substrate. The process also has excellent economics because the individual PECVD precursor chemicals can be applied in a single processing sequence. In addition, because PECVD TFE is a vacuum-based process, display products can be made in mass production at the highest level of cleanliness.

12A.3.4.1 Barrier Layer (Silicon Nitride)

PECVD SiN_x, SiO, SiON dielectric films are commonly used in microelectronics fabrication, where SiN_x is used extensively as a final protective passivation layer for integrated circuits because it is an excellent diffusion barrier against moisture and alkali ions. By varying the flow rate of precursors, for example, SiH_4 and NH_3 in SiN deposition, along with RF power, pressure, and temperature, the resulting film properties can be optimized for specific requirements. Dielectric films such as SiN_x, SiO, SiON, and SiCN, prepared by PECVD at low temperature for OLED encapsulation, exhibit different degrees of barrier performance. Among them, SiN_x has been intensively studied and been applied in actual OLED display manufacturing. PECVD SiN_x having a water vapor transmission rate (WVTR) of below 10^{-5} g/m^2/day was demonstrated by Van Assche [16]. Other inorganic PECVD films such as a composite film of SiN_x and SiO [17, 18], stacked multi-layers of SiN_x of varying hardness, SiN_x/SiCN layer stacks, and SiN_x/SiON layer stacks [19] have also been studied and shown to have WVTRs meeting OLED encapsulation requirements.

Table 12A.7 Typical PECVD barrier-SiN properties used for OLED devices.

Film property	Typical value	Measurement method
Deposition rate	>300 nm/min	Thickness
Refractive index	1.8–1.9	Ellipsometer
Film stress	<100 MPa	Stress tool
Transmittance @400 nm	>90%	UV-Vis spectrometer
WVTR @85C/85%RH	~1.0E-3 g/m^2/day	Mocon tool

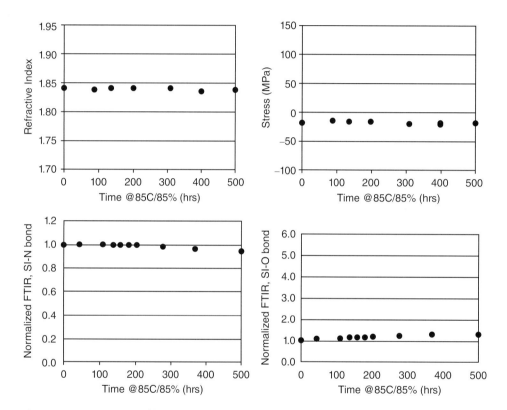

Figure 12A.45 Barrier SiN$_x$ film property change as a function of time at 85 °C/85% RH.

OLED barrier film must have low WVTR (<10^{-6} g/m^2/day under ambient environmental conditions), high optical transmission in visible range (>90%), low film stress (<100 MPa), high deposition rate, good film thickness uniformity, and low substrate temperature (<100 °C). Given these conditions, silicon nitride has barrier properties superior to any other PECVD film as indicated in Table 12A.7.

PECVD process parameters for barrier SiN film include SiH$_4$ gas flow, NH$_3$ gas flow, N$_2$ gas flow, RF power, pressure, electrode spacing, and substrate temperature. Typical single-parameter process trends are similar to the SiN trends in a-Si TFT devices (refer to SiN in the preceding a-Si TFT section), however, substrate temperature needs to be controlled to < 100 °C to prevent any degradation of the OLED and electrode materials.

Figure 12A.45 presents how barrier-SiN film properties change over time in an 85 °C/85% RH environment. As indicated, refractive index and stress do not change at all, and although the concentrations of Si-N and Si-O bonds in the film slightly change over time, this does not affect barrier performance.

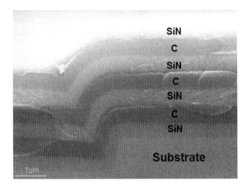 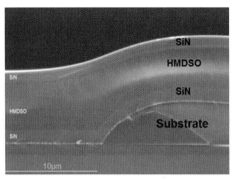

Figure 12A.46 (left) SiN$_x$/C multi-layer structure, (right) SiN$_x$/HMDSO multi-layer structure.

Table 12A.8 Typical PECVD buffer layer film properties used for OLED devices.

Film property	Amorphous Carbon	HMDSO
Deposition rate	~300 nm/min	500–1000 nm/min
Refractive index	1.5–1.6	1.4–1.5
Film stress	<100 MPa	<100 MPa
Transmittance @400 nm	>90%	>90%

12A.3.4.2 Buffer Layer

A buffer layer serves as a stress relaxation layer between the vapor barrier layers in the multi-layer encapsulation structure shown in Figure 12A.46 (left) and/or as a planarization layer to smooth the topology and cover particles on the surface, so that the inorganic barrier layer applied on top of the buffer layer will be a continuous, pinhole-free film as shown in Figure 12A.46 (right).

A PECVD amorphous carbon (a-C) film has been developed for the stress relaxation only, however, a-C has poor planarization performance. Therefore, a PECVD HMDSO (hexamethyldisiloxane) film has been developed to provide both planarization and stress relaxation. Key requirements for the buffer layer are high optical transmission in visible range (>90%), low film stress (<100 MPa), excellent particle coverage and planarization, good film-thickness uniformity and a high deposition rate, and a low-substrate-temperature deposition process (<100 °C). A PECVD amorphous carbon film or a HMDSO film can be the buffer layer of a multi-layer TFE for OLED display products, as shown in Figure 12A.46. The number of barrier and buffer layers and thickness of each in a multi-layer stack are the additional key factors to determine TFE performance, and those factors need to be optimized depending upon the particular type and structure of the OLED device. The SiN/a-C multi-layer TFE and the SiN/HMDSO multi-layer TFE, both have demonstrated >500 hrs OLED lifetime under 85 °C/85% RH conditions. Typical PECVD buffer layer's film properties are indicated in Table 12A.8.

PECVD process parameters of the buffer film include flow rates of source gases, RF power, pressure, electrode spacing, and substrate temperature. Film properties can be adjusted by controlling these process parameters according to film requirements for a specific device.

References

1 S. Y. Choi, et al., Large area PECVD technology, ECS Trans. 25(8) pp. 701–710 (2009).
2 Y. T. Yang, et al., The latest plasma-enhanced chemical-vapor deposition technology for large-size processing, IEEE J. of Display Tech., 3, pp. 386–390 (2007).

3 Q. Shang, et al., PECVD tool productivity enhancement with remote plasma source, Proceedings of Display Manufacture Technology conference, pp. 65–66 (1998).
4 B. S. Park, et al., Novel integration process for IGZO MO-TFT fabrication on Gen 8.5 PECVD and PVD systems a quest to improve TFT stability and mobility, ECS Transactions, 54(1), pp. 97–102 (2013).
5 L. Bardos, et al., Thin film processing by radio frequency hollow cathodes, Surface and Coatings Technology, 97, pp. 723–728 (1997).
6 T. Takehara, The latest PECVD technology for large-size processing, Proceedings of AM-FPD, pp. 95–98 (2006).
7 T. K. Won, et al., Qualification of LCD Grade NF3 for PECVD chamber cleaning, SID International Symposium, pp. 608–611 (2000).
8 C. C. Tsai, et al., Advanced PECVD technology for manufacturing AM LCDs, Proc. ASID'99, pp. 268–271 (1999).
9 A. Kuo, et al., Advanced multilayer amorphous silicon thin-film transistor structure: film thickness effect on its electrical performance and contact resistance, Japanese Journal of Applied Physics, 47(5), pp. 3362–3367 (2008).
10 A. Kuo, et al., Advanced amorphous silicon thin-film transistors for AM-OLEDs: electrical performance and stability, IEEE Transactions on Electron Devices, 55(7), pp. 1621–1629 (2008).
11 D. Toet, et al., Uniform high performance poly-Si TFTs fabricated by laser-xrystallization of PECVD grown a-Si:H, Materials Research Society Symp. Proc. 621 (2000).
12 M. Takahashi, et al., Influences of hydrogen in precursor Si film to excimer laser crystallization, Proc. SPIE, 3933, pp. 69–76 (2000).
13 K. Suzuki, et al., Influences of hydrogen contents in precursor Si film to excimer laser crystallization, Appl. Phys. A 69, pp. S263–S266 (1999).
14 H. Hosono, et al., Recent progress in transparent oxide semiconductors: Materials and device application, Thin Solid Films, 515, pp. 6000–6014 (2007).
15 D. K. Yim, et al., Process effects on instability in a-IGZO ES-TFT and oxygen role, The 14th International Meeting on Information Display (2014).
16 F. J. H. Van Assche, et al., A thin film encapsulation stack for PLED and OLED displays, Proc. SID International Symposium, pp. 695–697 (2004).
17 H. Lifka, et al., Thin film encapsulation of OLED Displays with a NONON stack, Proc. SID International Symposium, pp. 1384–1387 (2004).
18 J. J. W. M. Rosink, et al., Ultra-thin encapsulation for large-area OLED displays, Proc. SID International Symposium, pp. 1272–1275 (2005).
19 A. Yoshida, et al., 3-inch full-color OLED display using a plastic substrate, Proc. SID International Symposium, pp. 856–859 (2003).

12B

Thin-Film PECVD (Ulvac)

Masashi Kikuchi

Masashi Kikuchi, Ulvac Corp.

12B.1 INTRODUCTION

PECVD (plasma enhanced CVD) process has very important role for TFT-LCD fabrication, enabling the deposition of the semiconductor channel layer and of the gate insulator in TFT array. The quality of those layers critically determines the TFT performance and reliability. At the same time, despite the use of similar materials as in traditional silicon semiconductor processing, there are significant process and equipment differences. Si semiconductor processing uses high temperature, for instance up to 1000°C, while flat panel display processing requires much lower temperatures and glass substrates with far larger sizes. In this chapter, the plasma behavior in PECVD is discussed, followed by PECVD equipment architecture and thin-film deposition mechanism.

12B.2 PLASMA OF PECVD

Figure 12B.1 shows the plasma structure in CCP (capacitively coupled plasma) PECVD. Low density plasma (electron density = $10^8 \sim 10^{11}$ pcs/cm^3) is typically considered. If the plasma density is too high, overdissociation of precursor molecules happens resulting typically in a bad film quality. In plasma processing, deposition and etching (e.g., hydrogen etching) take place simultaneously. This is a very important aspect for understanding the PECVD process. For instance, SiH_4 dissociates to Si and hydrogen, where the latter can etch Si generating volatile SiH_2, SiH_3, and SiH_4 species. As shown in Figure 12B.1, ions collide with the substrate and the transferred energy affects the film stress and decreases the breakdown field of the film. High energy causes low break down field, while low energy causes low density film. Therefore, the process window in terms of energy setting should be properly considered during process recipe optimization.

12B.3 PLASMA MODES AND REACTOR CONFIGURATION

Different types of plasma modes are presently available for PECVD, however, the CCP type is dominantly used because it offers a simpler structure and has good scalability to larger substrate sizes.

Flat Panel Display Manufacturing, First Edition. Edited by Jun Souk, Shinji Morozumi, Fang-Chen Luo, and Ion Bita.
© 2018 John Wiley & Sons Ltd. Published 2018 by John Wiley & Sons Ltd.

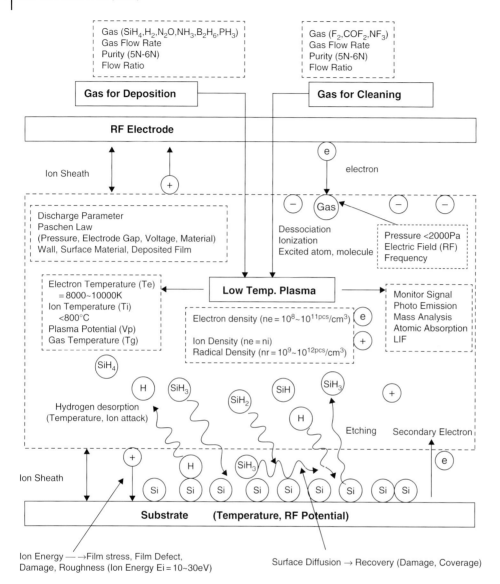

Figure 12B.1 Plasma structure in CCP PECVD.

12B.3.1 CCP-Type Reactor

Figure 12B.2 shows a CCP reactor schematic. The glass substrate is placed on the heater (also serving as an anode), while the cathode is located in the upper side. Reactive gas is introduced through this cathode, which has a showerhead design with openings allowing gas flow. RF (radio frequency at 13.56 or 27.12MHz) power is applied to cathode (RF electrode) and plasma can then be generated between the cathode and anode due to the RF power. The substrate temperature is maintained under 200 to 450°C.

12B.3.2 Microwave-Type Reactor

Microwaves can be used for plasma excitation since higher frequency waves contribute to higher plasma density. An example configuration for a reactor using microwave source (frequency 2.45GHz) [1] is shown in

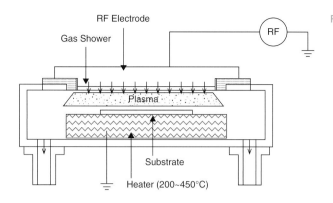

Figure 12B.2 CCP-type plasma reactor.

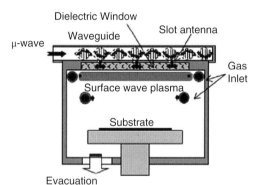

Figure 12B.3 Diagram for PECVD reactor with microwave excitation.

Figure 12B.3. Microwave power goes through the slot antenna of dielectric window made of aluminum oxide, and then high-density plasma is generated under that window. Thus, in this case the plasma does not use electrode discharge, that is, no cathode and anode.

This reactor is mainly used for passivation film deposition. This plasma has the characteristic of very low ion damage. These aspects will be discussed later in this chapter.

12B.3.3 ICP-Type Reactor

ICP (inductive coupled plasma), shown in Figure 12B.4, can also generate high-density plasma. ICP works under low pressure regime (<1 Pa) [2]. The gas dissociation level is high, while, in contrast, the plasma potential is low. For large substrate processing, ICP antennas are arranged in multiple numbers and connected to

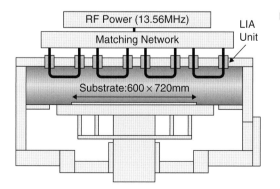

Figure 12B.4 ICP plasma reactor.

an impedance matching network. Using multiple antennas allows ICP to easily scale to larger substrates, including Gen 10 size. ICP demonstrated good effectiveness for insulator layer applications in IGZO TFT, however, the plasma uniformity can get worse in large sizes due to standing wave formation.

12B.4 PECVD PROCESS FOR DISPLAY

The films deposited by PECVD process are summarized in Table 12B.1 for both types of silicon TFTs: a-Si TFT and LTPS. In particular, a-Si films are used for both a-Si TFT and LTPS, however, the requirement film properties are totally different each other (Figure 12B.5).

12B.4.1 a-Si Film for a-Si TFT

a-Si TFT is very sensitive to the process condition parameters. Mono-silane (SiH_4) gas precursors dissociate as follows [3]:

$$SiH_4 \rightarrow SiH_3 + H$$
$$SiH_3 \rightarrow SiH_2 + H$$
$$SiH_2 \rightarrow SiH + H$$
$$SiH \rightarrow Si + H$$

When a-Si is deposited, the initial film has many defects. a-Si film has lone electron pairs called dangling bonds, shown in Figure 12B.5. Such defects can capture electrons when hydrogen forms a bond at the defect site (terminate the dangling bond). The temperature dependency of the dangling bond density is shown in Figure 12B.6 [3]. Below 250 °C, hydrogen atoms passivate the defect site, therefore, dangling bond is decreased with increasing temperature. Over 250°C, hydrogen dissociates and defect (dangling bond) increases again.

Table 12B.1 Overview of TFT layers and material choices.

	a-Si TFT	LTPS
Channel	a-Si / n+Si	a-Si + laser anneal
Gate	SiN_x	$TEOS-SiO_2$, SiH_4-SiO_2
Passivation	SiN_x	SiN_x/SiO_2

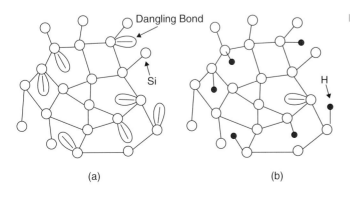

Figure 12B.5 a-Si film and dangling bond.

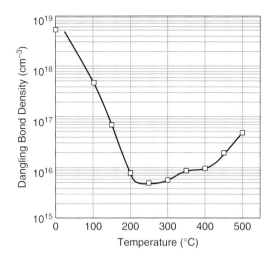

Figure 12B.6 Temperature dependence of dangling bond density.

12B.4.2 a-Si Film for LTPS

In the case of LTPS process, the a-Si deposition conditions are much different from those used for a-Si TFT, primarily since it is followed by laser annealing process. Laser anneal requires a-Si film with low hydrogen content. If the film contains too much hydrogen, Si film is ablated when the laser beam is radiated onto the film. In order to avoid this ablation, the content of hydrogen should be under 1%, whereas usual a-Si film has about 10% hydrogen after deposition by PECVD. Therefore, the process to eliminate excessive hydrogen from the film (called de-hydrogenation) is employed after a-Si film deposition at high temperatures 450–550°C (Figure 12B.7). Generally, there are two options for the dehydrogenation process.

(1) De-hydrogen is done in same chamber of CVD system.
(2) De-hydrogen is performed in another heating system.

The dehydrogenation heating chamber can be operated in vacuum or atmosphere. The chamber type is mainly batch type, because the annealing time is long (about 10 min). Furthermore, the risk of contamination should be considered in high temperature operation. Thickness uniformity is also important. Figure 12B.8 shows a-Si uniformity in a 1,500mm × 1,800mm substrate where the film uniformity is below 2.5%, above which the TFT uniformity will get worse.

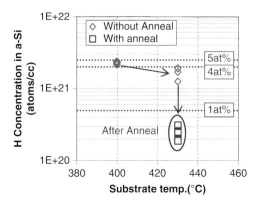

Figure 12B.7 Anneal temperature and hydrogen content.

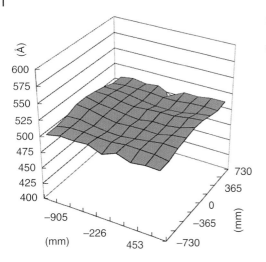

Figure 12B.8 Thickness uniformity for a-Si film deposited on 1,500 × 1,800 mm substrate (deposition rate 61 nm/min, uniformity < ±2.5%).

12B.4.3 SiN$_x$ Film

SiN$_x$ films provide high dielectric constant, good insulation property, and stability. It also works as barrier film against contamination by sodium or other contaminants for the TFT device. For SiN$_x$ films deposited at higher temperature, the wet-etching rate decreases due to higher film density. The typical deposition temperature is over 300°C. The NH$_3$/SiH$_4$ ratio in the deposition gases is important and is approximately 30 to 40. Typical properties of SiN$_x$ films are shown in Figure 12B.9 [4].

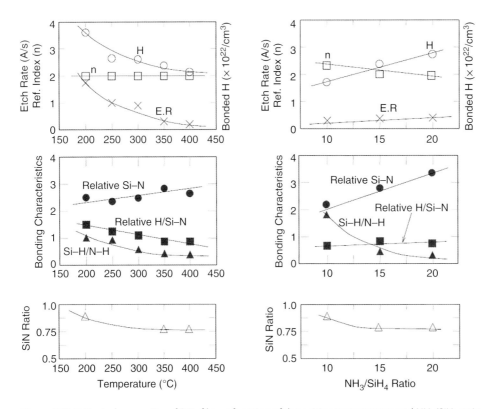

Figure 12B.9 Typical properties of SiN$_x$ film as function of deposition temperature and NH$_3$/SiH$_4$ ratio.

12B.4.4 TEOS SiO$_2$ Film

TEOS (tetraethylortho silicate) is a precursor for fabricating SiO$_2$ films used as planarization oxide and gate oxide for LTPS, due to limitations of SiN$_x$ for use for LTPS gate insulator. In the plasma state, TEOS reacts with oxygen as follows at 300 °C.

$$\mathrm{H_5C_2O-\underset{\underset{OC_2H_5}{|}}{\overset{\overset{OC_2H_5}{|}}{Si}}-OC_2H_5 + O \rightarrow H_5C_2O-\underset{\underset{OC_2H_5}{|}}{\overset{\overset{OC_2H_5}{|}}{Si}}-OH + CH_3CHO} \tag{1}$$

$$CH_3CHO + 3O \rightarrow HCHO + CO_2 + H_2O \tag{2}$$

$$HCHO + 2O \rightarrow CO_2 + H_2O \tag{3}$$

$$\equiv Si-OH + HO-Si \equiv \rightarrow \equiv Si-O-Si \equiv + H_2O \tag{4}$$

TEOS precursors generate dimmer, trimer, polymer of TEOS. These species have low gas pressures and are easily adsorbed around the substrate. This reaction contributes to good enough step coverage, since the adsorption assists the reaction on the surface. In contrast, those absorbed material has a lot of carbon, which may affect film quality, and in order to obtain excellent property of TFT, carbon should be excluded by the oxygen radicals.

12B.5 PECVD SYSTEM OVERVIEW

A typical cluster type PECVD is shown in Figure 12B.10. The design concept includes the following components:

(1) Reactive gas supply system;
   Two gases; flammability gas and combustion-supporting gas are isolated.
(2) Exhaust system;
   Two gases are isolated same as above system.
(3) In situ cleaning system;
   In situ cleaning system is adopted. It can be done at every 10-20pcs/substrate. In situ cleaning is very important for getting the high yield, which related to the particle and the stability of deposition rate.
(4) The precise control can be done in each chamber by different process recipe. This is important to get high repeatability of deposition.

In Figure 12B.11, the requirements for the main chambers of a Gen 6 size LTPS PECVD production system are shown. This system is operated at high temperature. Therefore, materials, handling robot and heater are totally different from conventional a-Si PECVD system.

12B.6 REMOTE PLASMA CLEANING

For PECVD, chamber cleaning is very important because particles can be produced from deposits on the chamber walls. While chamber cleaning could use conventional plasma etching, this is not typically preferred due to accumulated damage to the chamber walls. To avoid such damage, in situ chamber cleaning is preferred to use remote plasma (plasma generated in an external chamber and delivered to the main process chamber). Figure 12B.12 shows a plasma cleaning system.

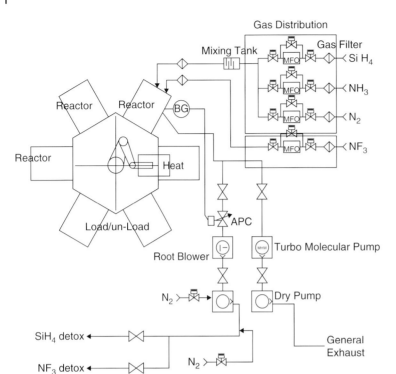

Figure 12B.10 Total system of CVD.

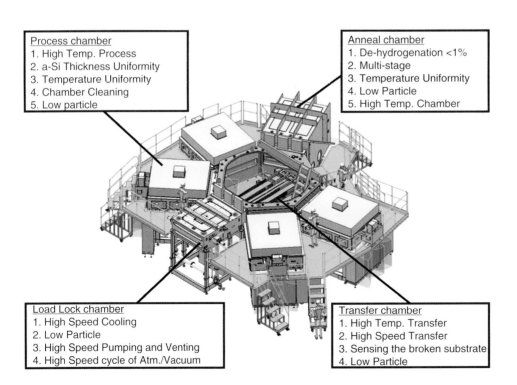

Figure 12B.11 Overview of Gen 6 size LTPS PECVD system and characteristics of the main chambers (*Source:* Ulvac Corporation).

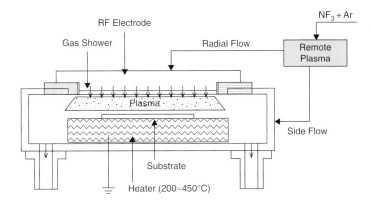

Figure 12B.12 Total system of remote plasma cleaning.

Table 12B.2 Comparison of side and radial gas flow of remote plasma cleaning.

	Side Flow	Radial Flow
Advantage	Uniform cleaning	High radical utilization
Disadvantage	Low radical utilization	Non-uniform cleaning

12B.6.1 Gas Flow Style of Remote Plasma Cleaning

Two cleaning gas flow styles are available in PECVD as shown in Figure 12B.12. NF_3 gas is used for cleaning. Here fluorine radicals are generated in remote plasma source and they flow in the chamber from the side or the radial from the gas shower center. Side flow type is effective for the small loss of radical while radial flow type can have loss of radicals. However, radial flow type is easy to get symmetrical cleaning, though it requires high input energy. Comparison between radial and side flow types are summarized in Table 12B.2.

12B.6.2 Cleaning and Corrosion

The main material used to fabricate the process chamber is typically an Al alloy because it offers good electrical and thermal conductivity. However, Al alloys are susceptible to chemical attack by fluorine radicals. The radicals react with the Al material to generate AlF_3, which can become a source of particle generation. Therefore, the chamber surface should be covered with anodized Al (alumite). However, after long-time use, even anodized Al can convert to AlF_3 as shown in Figure 12B.13. The AlF_3 residues are granular particles

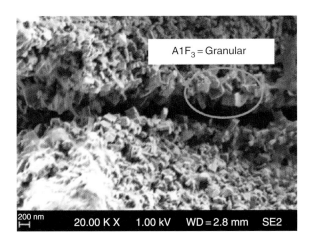

Figure 12B.13 AlF_3 particles formed on the chamber wall.

(resembling sugar crystals) and because they bond to the alumite surface only weakly, they can detach and generate particles.

Thus, the anodized Al parts are regularly changed during preventive maintenance. It should be noted that there are multiple candidates for ceramic materials for passivating the process chamber surfaces, with different levels of durability: yttrium oxide (Y_2O_3) > aluminum oxide (Al_2O_3) > aluminum nitride (AlN) > silicon carbide (SiC) > quartz (SiO_2). From the point of view of cost savings, aluminum oxide Al_2O_3 is commonly used but it should be taken into consideration it is susceptible to form AlF_3 on the chamber surface.

12B.7 PASSIVATION LAYER FOR OLED

As introduced in earlier chapters, thin-film passivation is very important for OLED device manufacturing. The WVTR (Water Vapor Transmittance) is adopted as indicator of passivation capability.

12B.7.1 Passivation by Single/Double/Multi-Layer

Single-layer SiN_x film deposited by microwave plasma system has been used for OLED passivation layer [5]. On the other hand, we proposed a double passivated film (SiN_x/Polyimide/SiNx) [6]. This film demonstrated good WVTR capabilities reaching 4.0×10^{-7} g/m^2/day. The dependency of the WVTR on the gap between electrode and substrate is shown in Figure 12B.14. This curve shows an interesting result. The double layer under discussion is easy to apply the system design.

Usually, the step coverage of the SiN_x film is not good enough. And sometime, the SiN_x film can be peeled off or develops cracks since thicker SiNx films build larger stresses. We explored a SiN_x/acryl multi-layer passivation. Figure 12B.15 shows the step coverage profile of SiNx/acryl double layer. Acryl layer is filled in the step corner. Due to the excellent step coverage, WVTR of 5.2×10^{-5} g/m^2/day was achieved. Figure 12B.16 shows the proposed SiN_x/acryl multi-layer deposition system combined with PECVD and acryl deposition chambers.

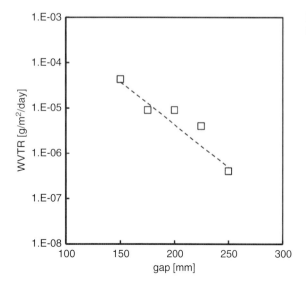

Figure 12B.14 WVTR versus gap between electrode and substrate.

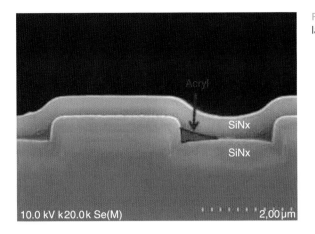

Figure 12B.15 The step coverage profile of SiN$_x$/acryl double layer.

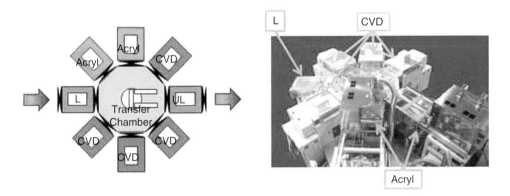

Figure 12B.16 SiNx/acryl multi-layer deposition cluster system.

12B.8 PECVD DEPOSITION FOR IGZO TFT

The active layer in IGZO TFT is very sensitive to O_2 or H_2O. PECVD SiN$_x$ can be suitable gate insulator and passivation of IGZO TFT [2].

12B.8.1 Gate Insulator for IGZO TFT

The SiN$_x$ film deposited by conventional PECVD contains a lot of hydrogen and it may provide negative effect on the TFT stability. Here, we proposed no hydrogen plasma deposition using only SiF_4 and N_2 as source gases.

Since the decomposition of nitrogen gas requires a high plasma power, we employed inductively-coupled CVD (ICP-CVD) method for the deposition of SiN$_x$:F, shown previously in Figure 12B.4. The maximum deposition rate was as high as 100nm/min. This system is also able to introduce hydrogen gas during the deposition to control the film quality.

The TFT with SiN$_x$:F with no hydrogen shows very stable Vth shift (ΔVth) in transfer characteristics against positive bias stress, as shown in Figure 12B.17(b). High concentration of hydrogen deteriorated ΔVth

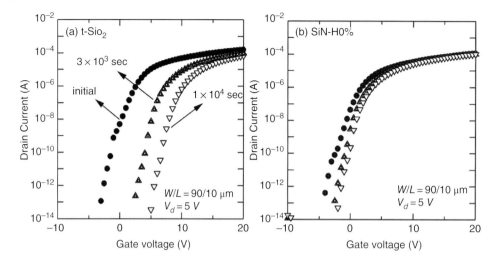

Figure 12B.17 Transfer curves after positive bias stress for IGZO TFT with (a) thermal oxide SiO_2 gate insulator layer, and (b) hydrogen free SiN_x:F gate insulator.

increasing the density of state, however, that was not the dominant factor considering the high ΔVth in the TFT with thermal oxide (Figure 12B.17(a)). It can be explained that fluorine in the SiN_x:F film plays an important role to improve the stability against the electrical stresses.

12B.8.2 Passivation Film for IGZO TFT

Oxygen and hydrogen should be controlled for highly reliable IGZO TFT. Furuta et al. [7] used SiNx:F as passivation film, as shown in Figure 12B.18. The fluorine passivated IGZO TFT has extended operation temperature, and is also advantageous in achieving high performance and high reliability oxide TFTs.

12B.9 PARTICLE GENERATION

To achieve a high yield process, particle management is quite important. In other systems using vacuum chambers, such as sputtering or dry etcher, the particles are generated at hardware, such as the chamber wall, electrodes, and the substrate. However, in the case of PECVD, the particles come from the plasma space, which is a different phenomenon from other vacuum systems. As an example, the particle map in SiH_4/H_2 plasma is shown in Figure 12B.19 [8]. It was observed that Si particles including polysilazane gather near the cathode sheath.

In order to reduce those particles, the pulse plasma, instead of RF plasma was applied. PECVD, which used a high voltage pulse applied between the cathode and anode showed an effective particle reduction. Especially, particles could be reduced by the pulse plasma, especially when using 1 kHz pulse as shown in Figure 12B.19. This is due to the flowing out of particles during the plasma-off period.

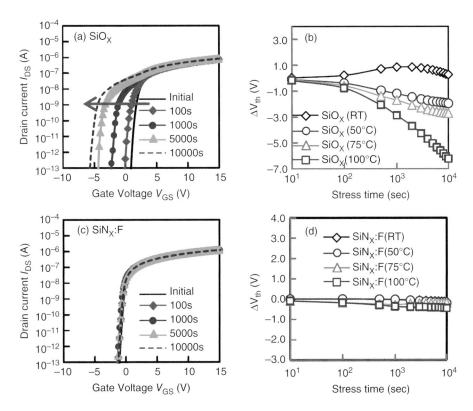

Figure 12B.18 Change in transfer characteristics and Vth shift during PBTS for the IGZO TFT with (a)(b) SiO$_x$ and (c)(d) SiN$_x$:F passivation after N$_2$ annealing at 350°C for 3 h.

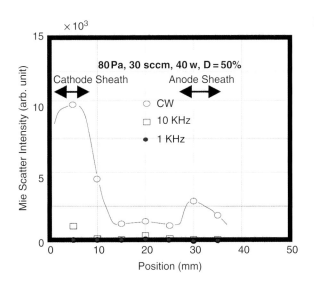

Figure 12B.19 Particle generation location.

References

1. S. Ueno, Y. Konishi, K. Azuma, Thin Solid Film, pp. 116–119 (2015).
2. E. Takahashi, et al., Jpn. J. Appl. Phys., 46(3B) (2007).
3. A. Matsuda, Proc. International Seminar on Reactive Plasmas (ed. T. Goto, Nagoya) p. 405 (1991).
4. R. E. Rocheleau and Z. Zhang, Thin Solid Film, 220, p. 73 (1992).
5. S. Ueno, Y. Konishi, K. Azuma, Thin Solid Film, pp. 116–119 (2015).
6. Y. Kato, T. Yajima et. al., Ulvac Journal, 79, pp. 9–13 (2015).
7. M. Furuta, J. Jiang, T. Toda, D. Wang, 24th Annual Meeting MRS Japan, XA-010-007 (2014).
8. Y. Watanabe, M. Shiratani, Japan Academy, 153 committee meeting pp. 31–38 (1990).

13

Photolithography

Yasunori Nishimura[1], Kozo Yano[2], Masataka Itoh[3], and Masahiro Ito[4]

[1] *FPD Consultancy, 8-10-4 Nishitomigaoka, Nara, Japan, 631-0006*
[2] *Foxconn Japan RD Co., Ltd., 3-5-24 Nishi-Miyahara, Yodogawa-Ku, Osaka, Japan, 532-0003*
[3] *Crystage Inc., 2-7-38 Nishi-Miyahara, Yodogawa-Ku, Osaka, Japan, 532-0004*
[4] *Electronics Division, Toppan Printing Co. Ltd., 5-1, Taito 1-chome, Taito-Ku, Tokyo, Japan, 110-8560*

13.1 INTRODUCTION

Photolithography is a technique that can be applied to form precise patterns in metal, oxide, nitride, and semiconductor films during the fabrication of the TFT arrays and color filters. The precise and fine patterns on a photomask are accurately transferred by UV irradiation through the photomask plate into to a photoresist material coated as a layer over the film to be patterned. The patterns recorded into the irradiated (exposed) photoresist appear after the development process, and then wet or dry etching is employed to etch out the underlying film through the window areas created in the locations where the photoresist was removed during the development process. The photoresist acts as a mask during the etch step protecting the underlying portions of the film. The final patterned film with the desired structural design is obtained after the remaining photoresist mask is completely removed by a photoresist stripping process.

The main processing steps in photolithography include (1) pre-cleaning, (2) photoresist (PR) coating, (3) pre-baking for PR hardening, (4) exposure, (5) development, (6) post-baking, and it is completed with (7) etching, (8) PR stripping and (9) post-cleaning. For high volume manufacturing operation, the photolithography equipment usually adopts a design with an in-line architecture where all the stations for pre-cleaning, PR coating, pre-baking, exposure, PR development through post-baking are connected together in a so-called "photoresist track system" to allow high throughput inline motion of glass substrates. One end of the photoresist-track system is connected to the exposure system, and the loader/unloader station is installed at the other end. The equipment for wet/dry etching and stripping equipment is installed in separated zones, to which the glass substrates are automatically handled and conveyed.

In this chapter, the main in-line processes from photoresist coating to development for performing photolithography for the TFT arrays and color filters are discussed. The key materials used in photolithography are photoresists and development solutions. First of all, an overview of photolithography for TFT arrays is given, followed by photoresist coating methods and equipment, focusing on the slit coating that is suitable for over Gen 6 glass substrate sizes. Then, the exposure process is discussed with detailed explanations on the four main types of exposure equipment (stepper (Nikon), multi-lens scan (Nikon), mirror projection (Canon), and proximity), as well as photoresist materials and UV light sources. For the large size exposure in TFT arrays over Gen 6 sizes, both lens scan and mirror projection are available. Puddle development is the major approach for over Gen 6 size TFT arrays. Finally, photolithography for color filter manufacturing is discussed.

13.2 PHOTOLITHOGRAPHY PROCESS OVERVIEW

Photolithography is the process which forms the masking layer for etching and ion doping using photosensitive materials called photoresist (PR), in the so-called "patterning" process. Using photolithography, the pattern of the photomask can be transferred into a desired film that is coated on the substrate. Photolithography can be applied for the patterning of TFTs, bus-lines, via-holes, pixel electrodes, and so on, for TFT arrays, for the patterning black matrix (BM), R G B color filters, photo-spacer for the color filters arrays, as well as patterning for touch panels. The key process steps in photolithography are shown in Figure 13.1. In this section we discuss in details the process flow and each of the unit processes in photolithography.

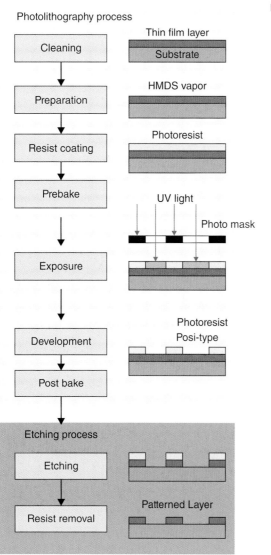

Figure 13.1 Overview of photolithography process flow.

13.2.1 Cleaning

The cleaning process is employed to remove particles and contamination from the surfaces of the substrates. A suitable cleaning process can be selected from various washing methods including mechanical cleaning (e.g., brushes, high-pressure sprays), chemical cleaning, UV or plasma cleaning, deionized water (DIW) rinsing and others, depending on the conditions of the glass substrate surfaces.

13.2.2 Preparation

The substrate is initially heated up to a sufficiently high temperature to drive off any moisture that may be present on the substrate surface (dehydration baking). Further, a liquid or gaseous "adhesion promoter," such as HMDS (hexamethyl-di-silazane) is typically applied to promote adhesion of the photoresist to the substrate.

13.2.3 Photoresist Coating

The photoresist layer is uniformly coated on the substrate by spin or slit coating a photoresist solution. Positive-tone photoresist, the most common type, is used for TFT array patterning to get fine patterns. For color filter fabrication, a colored resist (usually negative tone) is commonly employed.

The photoresist-coated substrate is dried in vacuum for accelerated processing, and then prebaked to complete removing the excess solvent, typically at 90–100 °C for 30–60 seconds on a hotplate.

13.2.4 Exposure

After pre-baking, the photoresist layer is irradiated with intensive UV light through a photomask having an opaque layer patterned with the same pattern as the image intended to be transferred onto the substrate using the exposure system (photoresist exposure step). For positive-tone photoresist materials, the UV irradiated area becomes soluble in the basic developer solution, whereas for negative tone photoresists the exposed region becomes insoluble in the organic developer used for the negative photoresist.

13.2.5 Development

After exposure, the photoresist layer is developed using a developing solution (or developer). For positive photoresists, TMAH (tetra-methyl-ammonium hydroxide) solutions are used as the developer, and an appropriate organic solvent is used for the color resist. Upon development, the pattern designed on the photomask is formed as a physical PR pattern, after regions of the PR material are dissolved away. Then, the substrate is "post-baked," typically at 120 180 °C. The post-baking solidifies the remaining photoresist, to increase its robustness as a protecting layer during subsequent wet or dry chemical etching steps.

13.2.6 Etching

The wet and dry etching processes are discussed in Chapters 14A and 14B, respectively.

13.2.7 Resist Removal

After etching the film covered by photoresist mask, the hardened PR material is stripped away (removed) using a resist stripping chemical solution. Subsequently, the substrate is cleaned again with a rinsing agent and DIW.

13.3 PHOTORESIST COATING

13.3.1 Evolution of Photoresist Coating

The methods used for photoresist coating on glass substrates have evolved from (1) spin-coating, (2) slit and spin-coating, to (3) slit-coating, as shown in Figure 13.2, as the dimensions of the display glass substrates increased through manufacturing generations.

Spin-coating has been used in up to Gen 4 size plants. It employs dropping an appropriate volume of PR solution onto the center of a spinning substrate, relying on the centrifugal force to spread the solution and leave on the substrate a uniform film coating with thickness dependent on the spinning speed. However, since only about 10% of the dispensed photoresist solution remains on the substrate after spin-coating, the wasted 90% of PR becomes a serious issue as the size of the glass substrate increase.

Slit and spin-coating has been developed for Gen 4/Gen 5 size plants. In this method a slit coater is used to first coat the photoresist over the substrate with approximate thickness, and then spinning is applied to produce a photoresist layer with a highly uniform thickness. Photoresist consumption by using slit and spin-coating can be reduced down to one third of that by using spin-coating.

As glass sizes processes in Gen 6 size plants and beyond have become even larger, the methods involving spinning are is no longer used. Later on, improved slit-coaters using newly developed slit nozzles with high-precision mechanisms have been developed. Such slit-coaters are capable of not only generating highly uniform photoresist layers, but also reducing photoresist consumption down to as low as less than one-tenth of that by spin-coating. The details of such slit-coating are further discussed in the next section.

13.3.2 Slit Coating

13.3.2.1 Principles of Slit Coating

Figure 13.3 shows the schematic diagram of a slit-coating system. In this system, the photoresist solution is dispensed onto a glass substrate through a moving slit nozzle (with a linear opening as wide as the target-coating width on the substrate). First, the nozzle is positioned at the start position on the substrate such that a photoresist solution bead forms in the precisely controlled gap between the nozzle and substrate. The photoresist, as it is pumped through the nozzle, stays around the bead on the substrate for a short period of time and then is carried away with the motion of the substrate at a predetermined constant speed that instantly matches the volume of the photoresist that has just been carried away with the moving substrate. This bead structure and behavior, which is dependent on the up- and downstream menisci and on the gap thickness, are critical to achieving high stability and uniformity of the resultant photoresist films. To achieve best optimized results, delicate balance among all the factors related to the width of the slit nozzle, coating gap, moving

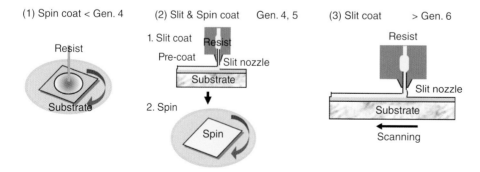

Figure 13.2 Evolution of photoresist coating processes for glass panel substrates.

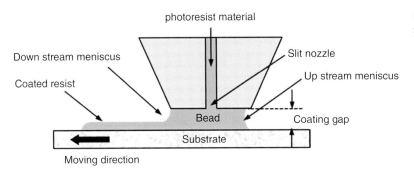

Figure 13.3 Conceptual diagram of slit-coating process.

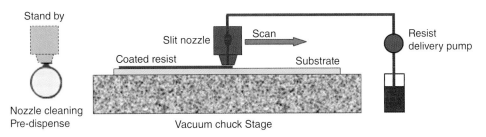

Figure 13.4 Diagram of a slit-coating system (*Source:* Courtesy of Screen Finetech Solutions).

speed, viscosity of the photoresist, and coating atmosphere is required. In slit-coating it is necessary to use PR solutions with a lower viscosity than those used in spin-coating (Figure 13.4).

13.3.2.2 Slit-Coating System

A slit-coating system can uniformly coat the dispensed PR solution onto a glass substrate (typically vacuum-chucked) by dispensing the solution with a predetermined flow rate, keeping the gap between the nozzle and substrate constant when the slit nozzle is scanned with constant speed. The detailed structure of the slit coating system is shown in Figure 13.4. The system consists of the vacuum-chuck stage, X-Z gantry, slit nozzle, photoresist delivery system, and standby stage. The vacuum-chuck stage has a precisely leveled flat surface for fixing the substrate on the stage. The X-Z gantry can move along the horizontal direction (X-axis) with an accurate velocity and maintain the height positioning accuracy with respect to up/down direction (Z-axis) of the nozzle and a fixed gap between the substrate and nozzle head during the coating step. The slit nozzle is designed to have a constant dispensing volume with respect to the longitude direction (across the width of the substrate), and a high gap accuracy. The PR delivery system provides the photoresist fluid to the slit nozzle with a constant flow rate and fast rising response. The stand-by stage is used for maintaining the high reproducibility process, by cleaning the nozzle head after coating and pre-dispense before coating.

To perform slit-coating, the system starts with cleaning the slit nozzle and predispensing PR solution for the forthcoming coating in the standby stage. Then the slit nozzle moves to the starting point on the substrate, and the nozzle head moves again to approach the substrate and reach a narrow gap to form the PR solution bead. The photoresist is dispensed through the nozzle as the head is scanned maintain a constant gap by the system's automatic adjustment mechanisms. When the nozzle is scanned to the end point, the whole slit-coating process is completed. The nozzle head moves up, and then moves back to the stand-by stage.

With the development of slit-coating systems, it was possible to eliminate the back-rinse process needed after spin-coating for removing unnecessary PR deposited around the edges and the backside of the glass plate. Thus, by employing slit-coating, the coating cycle time can be reduced and PR consumption can be cut down to one third of that used in slit- and spin-coating.

13.4 EXPOSURE

13.4.1 Photoresist and Exposure

13.4.1.1 Photoresist

Depending on the working photochemistry, photoresist materials can be categorized into positive-tone photoresists and negative-tone photoresists. In the photochemistry engineered for positive PR, the positive-tone material exposed to the UV light undergoes chemical changes and becomes soluble in the developer liquid. Thus, the positive PR material exposed to the UV light can be easily removed by rinsing with the developer solution. Since the positive PR is suitable for obtaining finer patterns than the negative PR, it is usually employed for fine-patterning in TFT arrays. The photochemistry for negative PR materials works in the opposite way: when the negative PR is exposed to UV light, the material polymerizes and becomes insoluble in the target developer liquid. Therefore, the negative PR material exposed to UV light pattern remains on the substrate surface, while the unexposed negative PR material is removed after the rinsing process with the developer liquid.

Positive PR is a material consisting typically of a photoactive compound (PAC), base polymer, solvents and photosensitizer (PAC: NQ, amino ketone, benzo phenone etc.; base polymer: Novolac resin; solvents: PGMEA, NMP, MBA, etc.). The developer liquid is a solution based on TMAH (tetramethyl-ammonium hydroxide).

In general, the photochemistry of the negative PR involves radiation induced crosslinking. The photoactive materials employed in negative PR include vinyl, epoxy, halogen-containing polymers. Such parent polymers are usually soluble in organic solvents. Therefore, if not exposed to the UV light, the parent polymers can be removed by rinsing with the developer liquid.

13.4.1.2 Color Resist

The color resist is used to form (1) the R, G, and B pixel and (2) BM on the color filter plate. It is a negative tone PR with colored pigments dispersed in the color mill base. BM resin is used to form the black matrix pattern, and is also a negative tone PR but with dispersed light absorbers such as carbon black. Ketone, ether, and ester are the main solvents for the color resist.

The details in photolithography for the color filter fabrication will be discussed in section 13.8.

13.4.1.3 UV Light Source for Exposure

Both positive and negative PR are designed to be sensitive to UV light radiation. Mercury arc lamps are the most popular light source for exposure in the photolithography process. The emission wavelengths from the mercury arc lamp, 436 nm (*g*-line), 405 nm (*h*-line), and 365 nm (*i*-line), can work effectively for achieving the pattern resolution required for TFT-LCDs/OLEDs using the conventional photoresists.

13.4.2 General Aspects of Exposure Systems

There are three primary exposure methods: contact exposure, proximity exposure, and projection exposure. The technical factors that can affect the performance of exposure systems for TFT displays include:

(1) Image field and exposing area: area that can be exposed per shot or scan, and size of the substrate that can be exposed by a single step or multiple steps
(2) Resolution of projection optics: resolved line width and line space
(3) Alignment or overlay accuracy: the repositioning accuracy of successive exposures.
(4) Productivity (throughput): capability of processing substrates per unit time

The contact exposure method is not available for TFT display lithography due to the limitation in the size of the photomask and the damage of the photomask that can be caused by its contact with the substrate.

Table 13.1 Overview of exposure systems and key process requirements.

Process	Mask layer	Substrate size	Resolution overlay	Exposure system
a-TFT array	TFT, pixel, etc.	Any G.	3 µm, < 1.0 µm (3σ)	Mirror projection Multi-lens projection
LTPS	TFT, pixel, etc.	< G.5	1.5 µm, 0.3 µm (3σ)	Stepper
		> G.6	2 µm, ≦ 0.5 µm)	Multi-lens projection Mirror projection
Color Filter	High resolution BM	Any G.	3 µm, < 1.0 µm (3σ)	Multi-lens projection Mirror projection
	BM, R, G, B, PS, etc.	Any G.	~8 µm, ≦ 1 µm)	Proximity exposure

The proximity exposure method offers high throughput processing and is less expensive due to the system's simple architecture, but its resolution is over 8 µm. Thus, it is used for exposing layers with coarse features.

The projection exposure method, which is the dominant approach in TFT display manufacturing, has three types of systems: stepper system, mirror projection system, and multi-lens projection system. The stepper system gives the highest resolution and overlay accuracy among the three exposure systems. However, as the image field of the stepper system is smaller than other systems, the throughput can be severely affected for processing large-area substrates. On the other hand, the mirror projection system and multi-lens projection system offer < 3 µm resolution, < 1.0 µm overlay accuracy, and larger image fields. Thus, the mirror projection system and multi-lens projection system are more suitable for high-throughput processing of large-area substrates.

One of the most important requirements in the exposure process is to ensure the alignment accuracy (overlay accuracy) of the photomask to allow precise fabrication of multi-layer structures such as TFTs, and CF pixel. The photomask must be precisely aligned with the substrate after the first patterning step is completed.

The exposure systems and their applications are summarized in Table 13.1.

13.4.3 Stepper

Figure 13.5 shows the schematic diagram of a stepper.

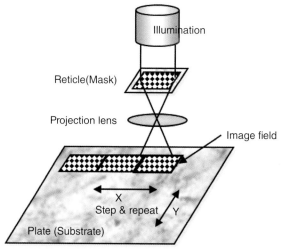

Figure 13.5 Schematic diagram of the stepper.

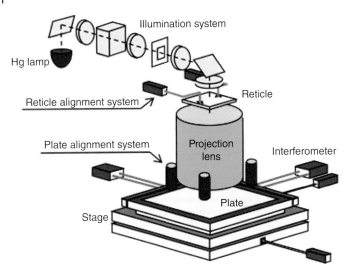

Figure 13.6 Detailed configuration of a stepper (*Source:* Courtesy of Nikon).

The pattern image on the photomask (or reticle) is projected onto the photoresist-coated substrate. In order to achieve high resolution, the projected image field per shot is limited to a small area, and this small image field is stepped and repeated over the substrate surface.

Figure 13.6 shows the fundamental architecture of a stepper provided by Nikon. It consists of multiple optical systems, such as illumination optics, projection optics, optical system to measure the focal and stage positions, and alignment optics that detects the alignment mark on the substrate, as well as the mechanical stage being capable of high accuracy motion. In this system, the UV light from the light source (high-pressure mercury lamp) passes through the illumination optics and is incident on a square reticle with the size of 6 inches. The image of the reticle is projected onto the substrate by the large aperture projection lens. The size of the projected image field is 132 × 132 mm, and that image exposes the photoresist on the substrate. This exposure is repeated by stepping in (x,y) directions to cover the entire region on the plate. At each step, the alignment between mask and substrate, and focusing on the substrate surface are controlled by the alignment and focusing systems. Thus, < 3 μm resolution and < 0.3 μm overlay accuracy can be achieved.

Resolution (R), corresponding to the accuracy of the resolving dimension, is expressed as:

$$R = k_1 \times \lambda / NA \tag{13.1}$$

k_1: process factor
NA: numerical aperture of the projection lens

In the case that NA is 0.08, k_1 is 0.6, and λ is 400 nm as a mixture light of g-h lines, the expected resolution R is around 3 μm. As shown in Equation 13.1, the resolution (R) can be improved by the increasing the aperture of the projection lens and reducing the wavelength of the illumination source. As example, 1.5 μm resolution is achieved in the Nikon FX-903N advanced stepper, based on a large aperture lens with i-line. These capabilities are especially required for small- and middle-sized display panels used for mobile applications. Table 13.2 shows the corresponding performance metrics for that stepper.

13.4.4 Projection Scanning Exposure System

For large-sized glass plates, the exposure process is achieved by multiple cycles of exposure, stepping and repeating to cover the entire glass plate. The processing throughput, especially for stepper tools, can be very

Table 13.2 Performance of Nikon FX-903N stepper system (*Source:* Courtesy of Nikon).

FX-903N Performance	
Resolution (L/S)	1.5 µm
Image field	132 × 132 mm
Wavelength foe exposure	I-line
Projection magnification	1 : 1.25
Alignment accuracy	0.3 µm (3 σ)
Max. plate size	730 mm 920 mm
Reticle size	6 inches
Take time	61 sec (30 shots, 50 mJ/cm^2)

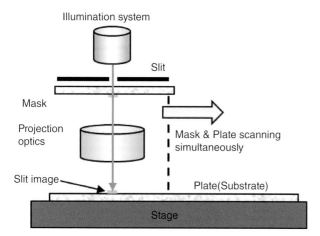

Figure 13.7 Concept of scanning exposure system.

low. With the demand for larger screen sizes as well as increase in sizes of the glass substrates, it is necessary to develop exposure systems with larger projection optics and lithography system for higher throughput. Thus, the projection scanning exposure system was devised to replace the stepper exposure system.

The concept of the projection scanning exposure is shown Figure 13.7, and the exposure procedure for a Gen 7 size substrate (1,850 × 2,200 mm) with six-up panelization of 47-inch diagonal panels is also shown in Figure 13.8. The UV light from the light source is irradiated onto a mask (850 × 1,200 mm) through the circular arc or linear slit with a narrow aperture as long the width the mask. The mask image with the slit shape is projected onto the substrate and 1:1 aperture image is exposed. During exposure, the mask and the substrate are synchronously scanned toward the end of the panel region. This scanning system can expose the whole area of a single display panel (area = exposure width by scan distance) without any visible seam. Subsequently, the system is stepped to the next display panel. By such step and repeat motion, the entire substrate can be exposed, as shown in Figure 13.8. In this example, the Gen 7 size plate is exposed by six scans. The feature of this system is to realize large-area exposure with a resolution as high as 3 µm. With high-precision measurement systems and adjusting mechanism of projection optics, it is possible to achieve 0.5 µm overlay accuracy and satisfy the requirement for TFT array patterns.

The details of scanning exposure systems using the mirror projection or the multi-lens projection are discussed in the following section.

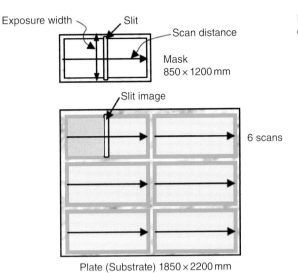

Figure 13.8 Example of Gen 7 47-inch panel by six-shot exposure.

13.4.5 Mirror Projection Scan System (Canon)

The Canon mirror projection aligner is a projection scan exposure system with mirror projection optics. The reflective optical systems using mirrors have simpler components than transparent optical systems using lenses, providing such benefits as no chromatic aberration and degradation in image performance. The optics consisting of the concave and convex mirror has an advantageous feature that high-quality image areas with arc shapes, which is cut out from the circular belt zone along the inner circumference on the concave mirror in order to minimize the distortion error, can be easily obtained. The slit mask image made from the arc-shape slit is projected onto the plate. That system can achieve the large area exposure with high resolution.

Figure 13.9 shows the conceptual diagram of the mirror projection aligner. Its optics consists of the trapezoidal, concave, and convex mirrors. The large concave mirror, having a large diameter to allow exposing a large panel seamlessly in a single pass with a wide enough exposure area, enables a significant increase in throughput and overall productivity.

As an example, the specifications of Gen 6 size and Gen 8 size mirror projection aligners by Canon are summarized in Table 13.3, and a photo of the Gen 6 size Canon mirror projection aligner MPAsp-E813H is shown in Figure 13.10.

13.4.6 Multi-Lens Projection System (Nikon)

13.4.6.1 Multi-Lens Optics

Figure 13.11 shows the principle of the multi-lens optics. The high resolution property of each projection lens is maintained across the entire image field, with this two-row lens array arrangement allowing for exposure with 2–3 μm high resolution. This lens array precisely works as the giant lens with the same NA that each small lens has, and it offers such an advantage that the increase in the number of the projection lens on the array can easily scale up the capability for glass substrates with further increased sizes. In fact, the Gen 10 size system (Nikon FX-101S) utilizes a 14-lens array for exposing in one scan displays panels over 60 inches in size.

13.4.6.2 Multi-Lens Projection System

Figure 13.12 shows the configuration of the exposure system with a multi-lens optical system.

The bending of the large-sized photomask is compensated by the focusing mechanism. Less than 0.5 μm overlay accuracy is achieved by the alignment sensors and simultaneous measurement systems.

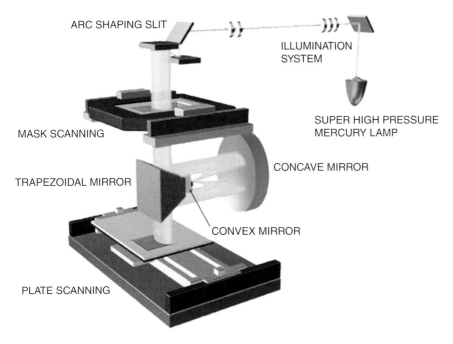

Figure 13.9 Conceptual diagram of mirror projection optical system (*Source:* Courtesy of Canon).

Table 13.3 Specifications of MPAsp-E813H (Gen 6 size) and MPAsp-H803T (Gen 8 size) mirror projection scan systems.

	3MPAsp-E813H	MPAsp-H803T
Generation	G6	G8
Exposure System	Mirror: 1:1	Mirror: 1:1
Mask Size	850 mm × 1,200 mm	850 mm × 1,400 mm
Plate Size	1,500 mm × 1,850 mm	2,200 mm × 2,500 mm
Resolution	1.5 μmL/S	2.0 μmL/S
Application	High-resolution small-medium LCD&OLED panels	High-resolution large LCD&OLED TV panels (e.g. UHD)

(*Source:* Courtesy of Canon)

The specification of Nikon FX-86SH2 for G8 is summarized in Table 13.4, and its exterior view is shown in Figure 13.13.

13.4.7 Proximity Exposure

In the proximity exposure system, the photomask plate and photoresist-coated substrate are placed in parallel with a very narrow gap (proximity positioning), and then a collimated UV beam is illuminated through the photomask so that the mask image is transferred to the photoresist layer. The image size of the photomask and that formed on the glass plate are exactly the same. Using a close contact between substrate and photomask in a contact exposure system can facilitate high image resolving capability and avoid diffractive scattering light. However, the photoresist and photomask are vulnerable to damages caused by small particles or

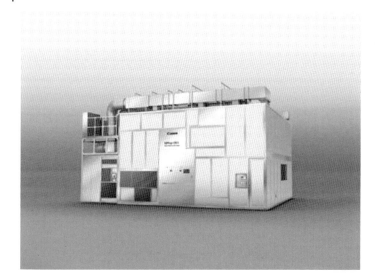

Figure 13.10 Exterior view of a Gen 6 size mirror projection scan system, Canon MPAsp-E813H (*Source:* Courtesy of Canon).

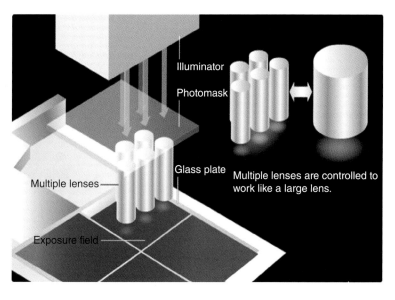

Figure 13.11 Principle of multi-lens optics (*Source:* Courtesy of Nikon).

dust on the substrate. Thus, only proximity systems were introduced for large-sized exposure, despite the slight degradation of the resolving capability.

Figure 13.14 shows the conceptual architecture of a proximity exposure system. The UV light beam from a Hg lamp source passes through Mirror 1, fly-eye lens, and collimation mirror to form a parallel beam. That beam irradiates the photoresist coating by passing through a photomask placed in proximity of the glass substrate surface. For large-sized substrates with multiple display panels, the stage is stepped in (x,y) directions for multiple shots, as shown in Figure 13.15. The resolution in such systems is defined by the source collimation quality and the proximity gap. Typical performance-results for proximity exposure systems using ~100 μm proximity gap are: (1) proximity gap accuracy: ±10 μm, (2) resolution: ~8 μm, (3) alignment accuracy (overlay): ±1.0 μm, and (4) total pitch accuracy: ±3.0 μm.

Proximity exposure systems are typically used for BM (black matrix), RGB color layers, photo spacers, and VA rib fabrication, which are relatively low-resolution patterns.

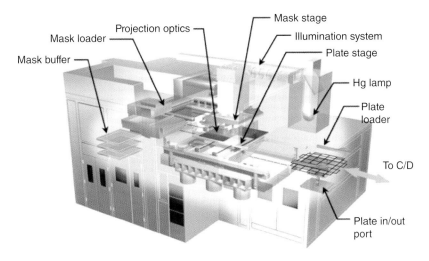

Figure 13.12 Exposure configuration with multi-lens optical system (*Source:* Courtesy of Nikon).

Table 13.4 Specifications of multi-lens projection system, Nikon FX-86SH2.

	FX-86SH2 Performance
Resolution (L/S)	2.2 μm (g+h+I–line)
Projection magnification	1:1
Alignment accuracy	$\leqq 0.5$ μm)
Max. plate size	2,200 mm × 2,500 mm
Takt time	49 sec./plate Conditions: 2,200 mm × 2,500 mm, 4 scans, g+h+l-line, 30 mJ/cm^2

(*Source:* Courtesy of Nikon)

Figure 13.13 Exterior view of multi-lens projection system, Nikon FX-86SH2 (*Source:* Courtesy of Nikon).

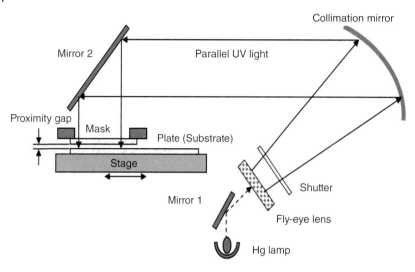

Figure 13.14 Diagram of a typical proximity exposure system.

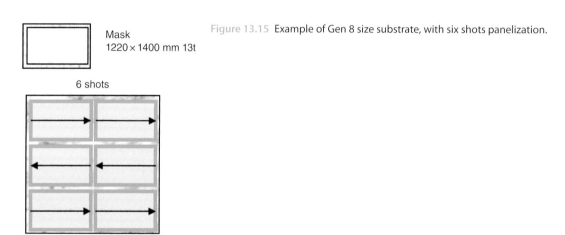

Figure 13.15 Example of Gen 8 size substrate, with six shots panelization.

13.5 PHOTORESIST DEVELOPMENT

In the development process, the latent image formed by UV exposure in the photoresist layer is converted into a physical pattern as portions of the photoresist material are dissolved in the development chemical solution (developer). The development process consists of (1) development, (2) deionized water (DIW) rinse, (3) dry, and (4) post-bake, as shown in Figure 13.16.

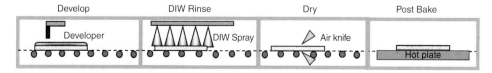

Figure 13.16 Schematic of developing process flow.

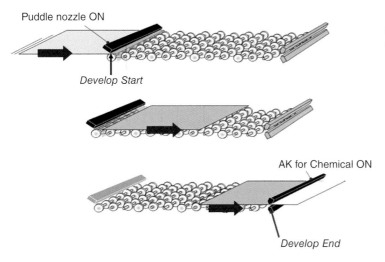

Figure 13.17 Principle of puddle development equipment operation: the substrate moves inline on rollers and a linear nozzle dispenses developer solution. An air knife (AK) is used to remove the developed resist solution (*Source:* Courtesy of Screen Finetech Solutions).

There are three main approaches for applying developer solution onto a PR coated substrate during the development process: liquid shower (spray), dipping, and puddle. The puddle method can achieve high development uniformity even for large-sized substrates, and it is mostly employed for substrate sizes over 5G. The puddle development method utilizes a thin liquid layer of developer solution that is dispensed on the PR layer through a slit nozzle. In general, the puddle development method is considered the best approach in terms of uniformity.

The principle of the puddle development is shown in Figure 13.17. The development process begins when the substrate moves just beneath the puddle nozzle where the developer solution is dispensed onto the glass substrate. The development process continues while the substrate travels in-line through the equipment until the substrate reaches the AK (air knife) position and the solution is blown away completing the process. As a puddle nozzle with similar width as the substrate is used, a highly uniformity can be achieved even for the large-sized substrates. Compared to other methods such as shower and dipping, puddle process can significantly minimize the consumption of development chemicals. Furthermore, the running cost can be reduced even more by recirculating the developer solution and controlling the concentration of the chemical during the recycling procedure. For example, aqueous TMAH (tetra-methyl ammonium hydroxide) solution of typically 2.38% concentration is used as the development chemical for positive photoresist.

The developed substrate is then post-baked at temperatures typically between 120 to 180 °C on a hot plate or in an oven chamber. The post-bake procedure solidifies the remaining photoresist to increase its durability for the subsequent process, such as wet etching, dry etching, or ion implantation. In some cases, UV curing is additionally applied to obtain harder layers.

13.6 INLINE PHOTOLITHOGRAPHY PROCESSING EQUIPMENT

In a typical manufacturing process, all steps for substrate pre-cleaning, dehydration bake, photoresist coating, pre-bake, exposure, and post-bake are performed in an integrated system in which the unit process stations are connected together so that the substrate can be continuously and sequentially processed in a single flowing line. The typical structure of an inline system is shown in Figure 13.18. Substrates are loaded into the loader unit, and then is move sequentially through the cleaner, dehydration bake, photoresist coater, and pre-bake stations stations. After exposure, the substrate is moved back through developer and post-bake stations, and finally unloaded. In some cases, an AOI (automatic optical inspection) station is included after post-bake. The substrate transportation is achieved by robotic arms. Since each station is designed to have the

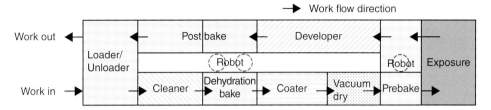

Figure 13.18 Schematic diagram of inline photolithography system (*Source:* Courtesy of Screen Finetech Solutions).

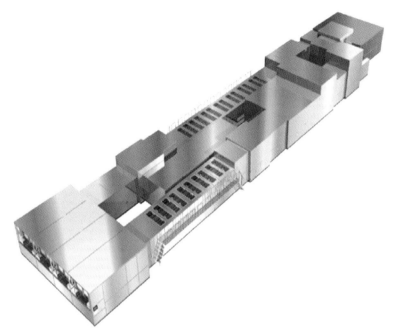

Figure 13.19 Exterior view of a Gen 8 size inline photolithography system (*Source:* Courtesy of Screen Finetech Solutions).

same cycle time, the inline system is quite efficient in terms of footprint and throughput. Figure 13.19 shows the exterior view of a Gen 8 size inline photolithography system provided by Screen Finetech Solutions.

13.7 PHOTORESIST STRIPPING

After etching or ion doping through the patterned photoresist mask, the photoresist material is removed with the help of a chemical solution (wet removing) or with a combination of dry (plasma/ashing) and wet processes.

As the surface of a photoresist coating can become carbonized or hardened during use as mask for etching or ion doping, it can be difficult to strip the photoresist by wet processes alone. Therefore, O_2 ashing is employed to remove the carbonized and hardened photoresist. Once this surface layer is removed, the remaining photoresist material can be removed and cleaned by wet processes. Table 13.5 shows etching processes and stripping methods for a-Si TFT and LTPS.

Usually O_2 ashing is achieved using the same equipment as that used for dry etching, where O_2 plasma is excited in the dry etching chamber for PR removal. The equipment for wet removing is similar to that for a

Table 13.5 Photoresist removal processes for a-Si and LTPS TFT.

		a-Si TFT		LTPS	
layer	material	process	PR removal	etching	PR removal
Channel S/D	a-Si n+Si Poly-Si	Dry	O$_2$ ashing + Wet	Dry	O$_2$ ashing + Wet
Gate electrode	Mo/Al Cu/Mo	Wet	Wet removal	Wet Dry	Wet removal O$_2$ ashing +Wet
Contact hole	SiN SiO$_2$	Dry	O$_2$ ashing + Wet	Dry Wet	O$_2$ ashing +Wet Wet removal
S/D	P+ B+ Doping	-	-	Ion Dope	O$_2$ ashing + Wet
Pixel	ITO	Wet	Wet removal	Wet	Wet removal

wet etching process. It consists of shower (spray) removing, rinse, post cleaning, and drying zones, typically with a tilted substrate to help removing chemical and water.

13.8 PHOTOLITHOGRAPHY FOR COLOR FILTERS

Manufacturing of CF plates has evolved with that of TFT arrays by concomitantly adapting to the increase of glass substrate sizes and the scale of production equipment so that the two substrates can be successfully coupled for filling the thin liquid crystal layers. As the patterns of BM (black matrix) and RGB (red, green, and blue) layers are prepared by photolithography, photo-sensitive materials based on photoresists with pigment particles dispersed within are used for patterning these layers. CF manufacturing has remarkably advanced with the increase in the size of glass substrate for TFT arrays, starting with G1 size (320 × 400 mm) in the mid-1980s and gradually going up to Gen 10 size (2,880 × 3,130 mm) as result of progress in production equipment and manufacturing technology.

13.8.1 Color Filter Structures

A photo of two overlapping color filter plates is shown in Figure 13.20(a). In this photo, the Moiré interference color pattern can be observed in the overlapping area. Figure 13.20(b) shows a magnified image in which the

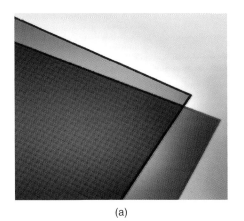
(a)

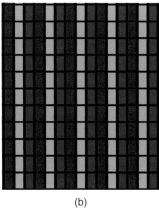
(b)

Figure 13.20 (a) Photograph of two color filter plates showing Moiré pattern in the overlap area, (b) RGB pixel arrangement on the color filter array.

RGB color layers are patterned in the pixel area including the black matrix (BM) pattern that blocks the undesired light between color pixels and shields the TFT being placed between the color layers.

The fundamental function of a color filter is to provide color images on the display screen. The essential structure of a color filter consists of RGB primary color cells and the black matrix (BM), which separates each RGB cell to prevent mixing of RGB colors and prevents undesired light leaking from the backlight. While the basic structure and function of BM and RGB patterns are similar in different liquid crystal display modes (i.e., TN (twisted-nematic) [1], VA (vertical alignment) [2], and IPS (in-plane switching) [3]), the detailed structure of the color filters array varies in each liquid crystal mode.

13.8.1.1 TN

TN is the original mode created for LCDs, and still offers the most modular basics for color TFT-LCDs. Figure 13.21 shows the CF structure for TN mode, which also represents the initial and conventional architecture for CF array. The BM layer blocks the light passing through areas between the pixels where the LC orientation is not controllable, in order to enhance the contrast ratio. To avoid undesired reflection at the panel surface, BM must be a low-reflection component. To electrically control the liquid crystal molecules, transparent conductive films based on ITO (indium-tin-oxide) are deposited across the color filter area.

The pixel sizes for recent high-end display models are 230 μm for 60-inch 4k (3,840 × 2,160 mm) TVs, and 50 μm for 5.5-inch mobile phones with WQHD (2,560 × 1,440 mm) resolution. Typically, the color layers are 1.5–2.5 μm thick and black matrix layers are 1.0–1.5 μm thick. In general, RGB patterns are overlapped with the black matrix layers by a few μm to avoid color degradation by light leakage.

13.8.1.2 VA

The vertical alignment (VA) mode was created to improve the viewing angle properties. In TN mode, liquid crystal molecules are horizontally aligned with respect to the glass substrate, but in VA mode the nematic molecules are vertically aligned by the unique protrusion structure. The color filter structure for VA cell is shown in Figure 13.22.

Usually the protrusion (VA pattern or "VA Rib") is formed on the ITO layer, where the LC molecules are aligned along the vertical direction and can be driven to the oblique directions by the application of an electrical filed. The VA pixel is divided into four domains resulting in negligibly small viewing angle dependency. The "VA Rib" is also prepared by photolithography.

13.8.1.3 IPS

Another LC mode that offers wider viewing angle is IPS (IPS is a registered trademark of Japan Display Inc). In IPS mode, two counter electrodes are formed on the TFT substrate and switching is achieved by the

Figure 13.21 Color filter structure for TN mode.

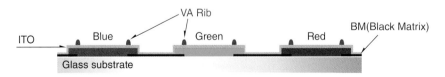

Figure 13.22 Color filter structure for VA.

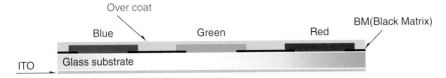

Figure 13.23 Color filter structure for IPS.

rotation of the liquid crystal molecules. Thus, an ITO electrode is not required on the CF array since the molecules are controlled by the laterally structured electrodes on TFT plate (Figure 13.23). However, an overcoat layer passivating the CF array becomes indispensable for meeting stricter requirements for surface flatness to eliminate orientation defects of LC molecules and for preventing impurity diffusion from color layers into the LC layer.

13.8.2 Materials for Color Filters

The main materials used for color filters are: (1) black matrix, (2) RGB color materials, and (3) photo spacers (PS) materials. The properties of those materials are described below in this section.

13.8.2.1 Black Matrix Materials

The main function of BM is to block unnecessary light from the backlight. One of the critical properties of BM is black density, which is defined as OD (optical density) value. So far, two types of BM materials are available, based on metal and resin.

The metal type BM consists of double or triple layers formed by Cr and its oxide [4]. The metal type BM features higher OD value, which can reach as high as over 3.5 for a < 0.2-μm-thick BM layer. However the metal type BM is not in use anymore due to environmental protection regulations.

For the latest color filters, the resin-type BM is dominantly used. Recently, the improved resin-type BM materials offer high OD value, high volume resistivity, better adhesion, and low reflection. Especially in IPS mode, BM must have high enough electrical resistance to avoid any impact on the molecular alignment. Adhesion to the glass is also important to prevent any failures caused by detached resin during or after the cell process. Low reflectance is another key requirement for the LCD panel to increase visibility, as reflection by BM may create image degradation under constant exposure to ambient conditions.

Typical OD value and volume resistivity are shown in Table 13.6. It should be remarked that, in general, materials with higher OD tend to show lower resistivity. The BM material is the resin (photoresist) containing carbon particles, and its properties can be controlled by the surface structure and size of the carbon particles. For the small/middle size mobile displays using IPS, BM with high resistance (over 10^{10} cm-Ω) and low reflectance (less than 2%) is required. Those mobile applications also require high-adhesion BM to keep good enough shock-proof property.

13.8.2.2 RGB Color Materials

The property of the color filter mainly depends on the RGB color materials and their optical properties. The RGB color layers are formed by the photoresist (or color resist) within which color pigment particles are

Table 13.6 Properties of resin-type BM.

	High Resistive Material	Low Resistive Material	Remark
OD Value (/mm)	2.4 ~ 3.7	3.0 ~ 4.5	
Resistivity (cm-W)	$10^{10} \sim 10^{15}$	$10^6 \sim 10^9$	DC1V

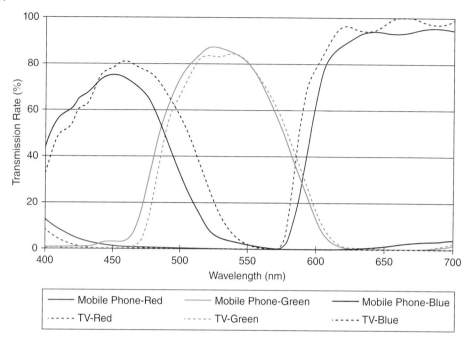

Figure 13.24 Transmission properties of RGB color layers.

dispersed. The key functions of those color resists are to realize the high-quality image with high brightness and contrast ratio on the LCD panel. The color resist is required to achieve high transmission and color purity, which are opposing properties.

A typical color resist consists of RGB materials dispersed into a negative type photoresist. The pigment type RGB materials used to be the major color materials. Due to the requirements for higher chromaticity and transmission, the dye or dye/pigment mixture type RGB materials have become the most advanced materials with improved temperature durability of the dyes.

Figure 13.24 shows the typical color transmission properties for RGB color layers. Both TV and mobile applications basically use the same color photoresist.

Recently "high contrast" and "high-color purity" versions with respect to the "standard" version have been developed. High-color purity version offers larger NTSC area ratio in chromaticity diagram shown in Figure 13.25. A "super high color purity" version, which satisfies EBU requirements, has recently become available. Besides those TV requirements, recent mobile phone applications strongly demand higher transmission and color purity for higher image quality and brightness performance.

13.8.2.3 PS (Photo Spacer) Materials

PS is a material patterned to form spacers ~3 μm in height to maintain a uniform gap between TFT and color filter glass substrates where the liquid crystal is filled. The gap is very important to achieve sufficient uniform optical properties of the liquid crystal layer. Therefore, PS height variation must be controlled to be less than 0.1 μm over the entire substrate. Otherwise, LC cell thickness cannot be precisely controlled and may also result in leakage of the LC material or air intrusion into the cell. Furthermore, PS material is required to have proper elasticity to compensate the expansion and shrinkage of the LC layer caused by temperature changes and the stress forces applied from the outside of the display panel.

To keep the uniform cell space over the entire cell area, typical PS materials used are negative type photoresist based on acrylic resins with suitable elasticity and mechanical strength.

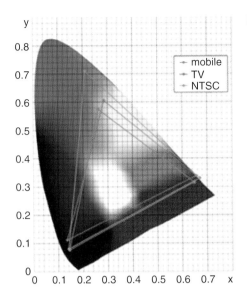

Figure 13.25 Color properties in chromaticity diagram.

13.8.3 Photolithography Process for Color Filters

Various methods, such as inkjet, printing, and electro-deposition [5], have been used for fabricating color filters in the past. Today, the dominant color filter fabrication method employed in manufacturing plants is the photolithography process by utilizing photoresist containing dispersed color pigments [6].

The color filter plates used to be provided by color filter vendors from their sites to TFT manufacturers. However, with the increase in the glass substrate sizes, color filters are now being made on-site in TFT manufactures' plants.

The color filter manufacturing process is shown in Figure 13.26. As can be seen in the figure, after the BM layer is formed, the R, G and B color layers are subsequently fabricated. Then, the ITO layer or overcoat is applied to the top surface. As an additional process, the VA Rib or PS is formed for specific applications. For the patterning processes of BM, R, G, B, PS and VA, the microfabrication technology based on "photolithography method" is applied, which consists of color resist coating, exposure, development and color resist post-bake, as shown in Figure 13.26(b).

13.8.3.1 Color Resist Coating

After cleaning of the glass surface, the color resist is coated (typical thickness: 1.0–1.5 μm). In order to keep sufficient uniformity in thickness, "spin coating," the conventionally developed method, used to be frequently applied. With the increase of the glass substrate size up to 6G or over, slit die coating [7] has become the commonly applied method.

After coating, the color resist is dried under reduced pressure and pre-baked to harden the layer to endure the processes that follow.

13.8.3.2 Exposure

After color resist coating, photo-exposure is applied to form the precise and accurate patterns. Usually, the negative-type color resist is used. Exposure is performed using the proximity exposure system by UV light (see section 13.4.7). Upon exposure to UV light, the exposed area of the negative-type color resist can undergo photochemical reactions to form the patterns to be remained on the glass substrate. For Gen 8 to Gen 10 glass sizes, "step and repeat" exposure, in which the photomask is moved by step motion with a precise alignment between the photomask and substrate at each step, is used.

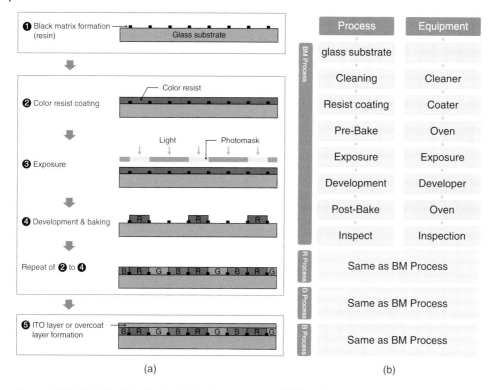

Figure 13.26 (a) Color filter device fabrication process and (b) flow chart.

13.8.3.3 Development

The color resist in the unexposed area is removed by spray shower of alkali-type development chemical solution. After the residual solution on the substrate is washed away with deionized water, the color resist on the substrate is post-baked in the oven.

Fabrication of BM and RGB patterns are repetitively performed by the same processing steps. The BM layer is prepared first and then the RGB layers are subsequently formed by precise error-minimizing alignment with the BM patterns.

The equipment for completing those coating, exposure, and development processes is similar to that for TFT-array production. Each process station is designated as a specific inline setting, which can minimize substrate handling and the footprint of equipment. Figure 13.27 shows the conceptual layout for color filter manufacturing in the cleanroom. Each dedicated line handles the BM, R, G, or B layer fabrication process. Such equipment is similar to that for TFT array photolithography process. BM and RGB lines are connected by the automatic cassette transportation system. After the cassette is loaded into the entrance of the BM station, a single plate is moved to the cleaner, the resist coater, and the pre-bake station by the robotic handler, and then exposure is carried out. After the exposure process, development and post-baking are applied, and then optical inspection is done. Those processes are automatically and precisely controlled to promote defect-free manufacturing quality. As an alternative layout, the complete inline structure, in which the BM and RGB lines are connected in-line and the glass substrates and cassettes are transported to one direction, is available.

The overcoat or PS layers are formed using the same photolithography technique as that for the color resist. Materials for the overcoat have two curing types, that is, light curing and thermal curing. In the fabrication process by using the light curing materials, the photoresist for the overcoat is coated over the color layer, followed by low-pressure drying, pre-baking, exposure, development, and post-baking procedures. In the case

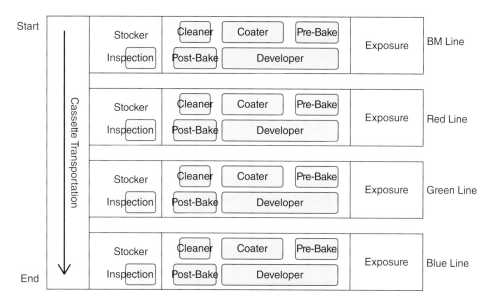

Figure 13.27 Conceptual layout of color filter manufacturing line.

of using thermal curing type overcoat materials, exposure and development procedures are not available. The PS is formed by the same procedures using the light curing type overcoat materials.

13.8.4 Higher-Performance Color Filters

13.8.4.1 Mobile Applications

Mobile applications, typically represented by smartphones, require high pixel density to display picture-like images on the panels. Presently, 300 ppi (pixel-per-inch) is the most common pixel density for mobile applications, and 400-450 ppi will dominate in the near future. In the further future, up to 600 ppi may be required.

As higher pixel density is expected for mobile displays, color filters with narrower and thinner BM patterns are required. For high density pixels, the aperture ratio, which is the ratio of the opening area to the total pixel area, may become smaller and smaller. The aperture ratio is directly related to the transmission efficiency of the backlight, which modulates the brightness of displays. When the pixel density is enhanced up to 600 ppi, the aperture ratio becomes so small that the brightness significantly goes down. Therefore in the high-density pixel arrangement, it is necessary to decrease the stripe width of the BM to utilize the most of the backlight. Presently, the width of BM is over 5 μm. For 450 ppi and 600 ppi density, the stripe width of the BM should be reduced to around 4 μm and 3 μm, respectively. To meet the technical requirements, the development of new BM materials with the smaller thickness keeping the same OD value is desired. Besides, PS materials will become so narrow that similar issues must be resolved for PS as well.

13.8.4.2 TV Applications

With the infrastructure development for higher definition image broadcasting, LCD panels for TV applications are also moving along that development toward higher quality images with higher color reproducibility. Recently, the backlight generated by narrow spectrum LED light sources and quantum-dot LEDs [8] for each RGB color has been developed.

To better fulfil such requirements, fundamental properties of color materials must be improved. The ideal color filter is required to exhibit both higher brightness and high color purity, but brightness (light transmission) and color purity are properties of conflicting nature. In order to resolve that issue, 100% dye materials, instead of the conventionally used hybrid materials based on the mixture of pigments and dyes, are recently evaluated for improving the color performance. Though, in general, dyes are more vulnerable than pigments to degradation induced by light and temperature, the development of highly durable 100% dye materials would be the key to open up future TV applications.

References

1 M. Schadt, W. Helfrich, Voltage-dependent optical activity of a twisted nematic liquid crystal, Appl. Phys. Lett. 18, 127 (1971).
2 A. Takeda, et al., A super-high-image-quality multi-domain vertical alignment LCD by new rubbing-less technology, Digest SID Symposium, 98, p. 1077 (1998).
3 M. Oh-e, et al., Principles and characteristics of electro-optical behaviour with in-plane switching mode, Proc. of the 15th International Display Research Conference (Asia Display 95), p. 577 (1997).
4 M. Bender, et al., Economically sputtered black matrix systems for FPD applications, Digest 1996 Display Manufacturing Technology Conference, p. 71 (1996).
5 A. Matsumura, et al., The new process for preparing multicolor filter for LCDs, Proc. Eurodisplayb'90, p. 240 (1990).
6 T. Sugiura, Development of pigment-dispersed-type color filters for LCDs, Journal of SID, 1/3, p. 341 (1993).
7 D. Schurig, et al., High performance capillary coating system, Digest 1994 Display Manufacturing Technology Conference, p. 99 (1994).
8 H. Ishino, et al., Novel wide-color-gamut led backlight for 4k LCD embedded with mixing cup structure for isotropic light source, Digest SID Symposium 2014, p. 241 (2014).

14A

Wet Etching Processes and Equipment
Kazuo Jodai

SCREEN Finetech Solutios Co.,LTD., Tenjinkita-machi 1-1 Teranouchi-agaru 4-chome Horikawa-dori, Kamigyo-ku, Kyoto, Japan

14A.1 INTRODUCTION

Wet etching is characterized by lower running costs than dry etching. While dry etching is a vacuum process, wet etching is conducted under normal pressure. This makes its equipment configuration relatively simple compared to dry etching. In addition, the process gases used in dry etching are discharged. However, the chemicals used for wet etching are reused, without discharge at a fixed interval. This significantly reduces the cost of chemicals used in etching as well as its environmental load.

The sizes of the mother glass used in LCD manufacturing are increasing. Currently the predominant size for TV displays is Gen 8 size (2,200 × 2,800 mm), and Gen 10.5/11 size factory (3,000 mm × 3,300~3,400 mm) is also started with the goal of operation during 2019. Even low-temperature polycrystalline silicon (LTPS), which is used for mobile displays, is currently being produced at Gen 6 size (1,500 × 1,800 mm). The increasing size of the mother glass is one factor that has brought about an increase in the running costs of wet etching.

As the mother glass becomes larger, the capacities of the processing chambers for wet etching also increase. This increase in chamber size, in turn, leads to a rise in the volume of etchant used. In addition, the sizes of the glass substrates themselves also change the volume of chemicals consumed. These two factors are contributing to higher running costs and also compounding the problems currently being experienced with wet etching. Thus, that increase in running costs is becoming a major problem for wet etching. In order to reduce the running cost, chemical concentration control system is used to extending the life cycle of etching chemical. This technology is one of the solutions that adopted to cost control.

The principal characteristic of wet etching is that it is isotropic etching, which means that it is unsuitable for microfabrication. The number of pixels in mobile LCDs has been increasing each year, and at the same time the mobile LCDs are moving to higher resolution. In order to follow that increase in resolution, etching is being performed using the shower method by improving the etch factor.

This chapter will cover the current applications for wet etching, as well as the issues currently faced by this technology and the ways of resolving them.

Flat Panel Display Manufacturing, First Edition. Edited by Jun Souk, Shinji Morozumi, Fang-Chen Luo, and Ion Bita.
© 2018 John Wiley & Sons Ltd. Published 2018 by John Wiley & Sons Ltd.

14A.2 OVERVIEW OF TFT PROCESS

Figure 14A.1 shows a pattern diagram of the photolithography. Etching is one part of the photolithographic process. An etching mask is formed by photoresist patterning process. Etching is the process of removing the parts of the metal film that are not necessary in forming the electrode patterns. The principle of wet etching is very simple. An etchant composed primarily of acid is used to dissolve the unmasked parts of the substrate and form the objective pattern.

The series of processes shown in Figure 14A.1 constitutes the basic routine of photolithography. This routine is repeated to form thin-film transistors (TFT) on the mother glass. These are then combined with backlights and color filters to create the liquid crystal displays that we see everywhere around us.

In recent years the touch panel is built in the display panel, and in the most cases photolithography is used to form the sensor electrodes for touch panels. Also in the organic electroluminescent (OLED) displays, which

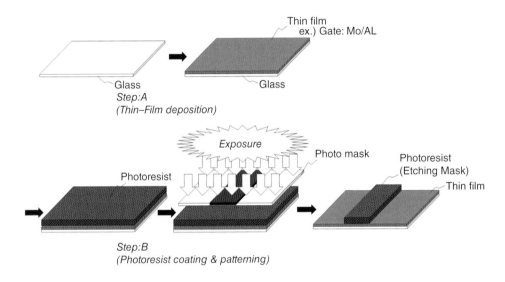

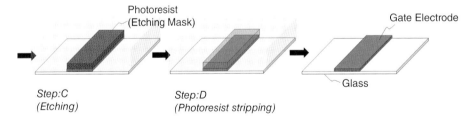

- Film formation: Deposition of the thin film => See Step A
- Etching mask creation:
  (1) Coating with photoresist (2) Pattern printing (exposure) (3) Developing (4) Baking => See Step B
- Etching
  Etchant is used to dissolve and remove the unnecessary parts of the thin metal film (the parts other than the sections covered with the photomask formed in Step B) => See Step C
- Removal of photoresist pattern (photoresist stripping) => See Step D

Figure 14A.1 Process from film formation through pattern formation.

recently have come into use, the backplane that drives the luminescent layer in an OLED display utilizes the LTPS-TFT. The basic fabrication process routine for it is same as that shown in Figure 14A.1.

14A.3 APPLICATIONS AND EQUIPMENT OF WET ETCHING

14A.3.1 Applications

In the LCD manufacturing process, wet etching is still used in the process for gate electrodes (in the case of LTPS, dry etching is also used), via holes, S/D wiring, and pixel electrodes. Figure 14A.2 shows an example of the cross-sectional structure of an LTPS-LCD. And Table 14A.1 shows the types and temperatures of etchants and the etching methods used for each etching process performed for TFT formation.

As is noted at the outset, wet etching is an isotropic process, and therefore it is not suitable for use in microfabrication with the precise dimensions.

In isotropic etching, as shown in Figure 14A.3, an undercut is produced, where the line dimensions become thinner than the photomask and spatial dimensions larger than the photomask. For mobile and 4K TV, microfabrication is required in gate electrode and SD layer. In microfabrication, the etch factor is required to increase. The etch factor is ratio of "depth" to "undercut," as shown in Figure 14A.3. In general, the etch factor of shower etching is superior to dipping, but the linewidth uniformity of dipping is better. So use of shower or dipping depends on requirement. In contrast, pixel electrodes have been a large-area pattern without requiring the precise dimensions from the beginning, and wet etching is expected to continue to be the primary etching method used for the time being.

14A.3.2 Equipment (Outline)

As shown in Table 14A.1, the etchant for wet etching is commonly heated to approximately 40 °C and sprayed from a nozzle. As a result, large quantities of fumes and mist are produced inside the etching chamber. The chamber must be airtight enough to prevent these chemical fumes from escaping.

Unlike dry etching techniques, however, wet etching is performed at atmospheric pressure, and there is no need to ensure airtightness and a pressure-resistant configuration on the order of a vacuum chamber. Accordingly, the etching chamber is able to make of polyvinyl chloride (PVC). This makes it possible to keep equipment costs lower than in the case of dry etching.

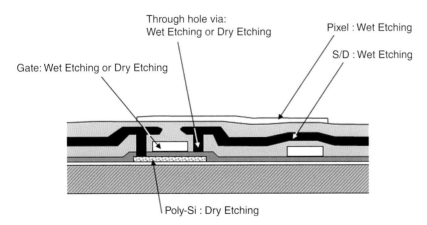

Figure 14A.2 Cross-sectional structure of LTPS-LCD (example).

Table 14A.1 Chemicals of wet etching in TFT process.

Layer	Etching material	Chemical/Chemical temperature	Method of immersion
Gate	Mo/AL	$H_3PO_4/CH_3COOH/HNO_3$ $(\sim 40°C)$	Dip or shower
	Cu/Mo	H_2O_2 based chemical $(\sim 30°C)$	shower
Through hole via	SiN/SiO	BHF: HF/NH_4F (room temp.)	shower
Source/Drain	Mo/AL/Mo	$H_3PO_4/CH_3COOH/HNO_3$ $(\sim 40°C)$	Dip or shower
	Cu/Mo	H_2O_2 based chemical $(\sim 30°C)$	shower
Pixel	ITO	Oxallic Acid: $(COOH)_2$ $(\sim 40°C)$	shower

Note: The table above shows the usual etching materials and examples of chemicals for etching those materials. In the case of crystallized indium tin oxide (ITO) for pixel, hydrochloric acid/salt iron or aqua regia are used.

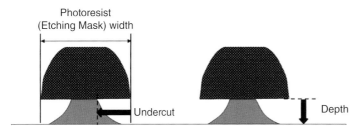

Figure 14A.3 Example of isotropic etching.

Other advantages of wet etching include the long equipment maintenance cycle and simple maintenance tasks. The major maintenance tasks are replacement of deteriorated etchant and replacement of the filters used for etchant and rinse fluid.

An example of the configuration of this system is shown in Figure 14A.4. As shown this system has a series of dedicated chambers in which the individual etching, rinsing, and drying processes are performed. A glass substrate enters the etching chamber from the loader via the excimer UV. After this, the substrate is oscillated for a fixed period inside the etching chamber and then discharged to the rinse chamber.

The processing method inside the etching chamber is shown in Table 14A.1. In the rinse chamber, spray rinsing is conducted using a DI water (DIW) shower before drying is finally performed using an air knife. The length of the etching chamber is determined by the processing capacity of the system (tact time) and the etchant processing time. When the tact time is short and the etching time is long, the length of the etching chamber must be increased. In the system shown in Figure 14A.4, it is not possible to increase the length of the etching chamber alone, so the size (footprint) of the entire system must be made larger.

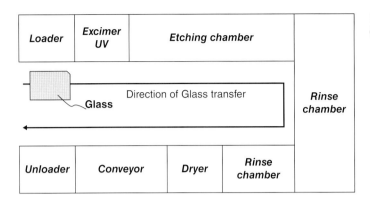

Figure 14A.4 Configuration of wet etching system (example).

14A.3.3 Substrate Transferring System

In a wet etching system, the mother glass is transferred with supported by a roller type conveyor unit. The various types of etchant are sprayed in shower onto the glass substrate. In some systems, the roller conveyor itself is tilted. The reason is that, due to the increasing size of the mother glass, the chemical tends to accumulate in the center of the substrate [1]. This accumulation causes two problems:

(1) At the etching stage, it is difficult to eliminate etchant whose chemical reaction is complete and etchant that includes residue. In the locations where such etchants have accumulated, the etch factor will be reduced and the etching rate will become uneven.
(2) At the rinse stage, fluid replacement is adversely affected and rinse time is prolonged. This will adversely affect the rinse efficiency and increase DI water consumption.

These are quite particular in the case of larger mother glass sizes.

The tilted system helps to eliminate the fluid as shown in the Figure 14A.5, and it is used as a resolving means the aforementioned problems.

Figure 14A.6 shows the efficiency measurement in the rinse section following etching. (The changes in the pH value over time on the glass substrate were measured; replacement is complete at pH = 7.) The pH value is restored to pH = 7 more quickly in the case of the tilted system, indicating that the rinse efficiency is better. The tilted transferring system plays a major role in reducing the running costs that result from larger mother glass sizes.

However, in the case of processes that require highly precise dimensions in the metallization width and wiring cross-sectional profile in the gate and source/drain electrodes, the tilted transfer method may actually be disadvantageous.

In such cases, the dip method (covered in section 14A.3.4) is used, even though it has the disadvantage of reducing the etch factor.

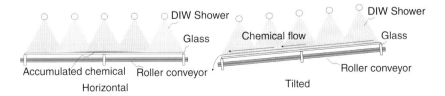

Figure 14A.5 Overview of horizontal and tilted system.

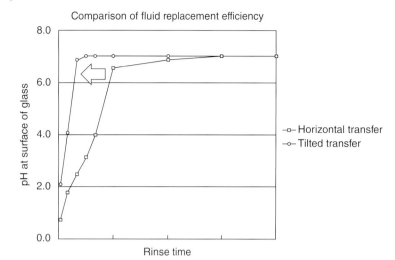

Figure 14A.6 Comparison of horizontal system and tilted system (pH in replacement rinse section).

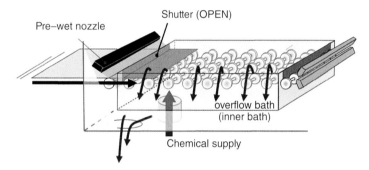

Figure 14A.7 Dip treatment system.

14A.3.4 Dip Etching System

The dip processing system has an overflow bath (inner bath) with a roller conveyor transfer system built into the overflow bath. The overflow bath has glass entrance and exit ports that are equipped with shutters. A sufficient quantity of etchant is supplied to ensure that the overflow status is maintained even when the shutters are opened to insert and remove glass substrates (see Figure 14A.7).

During the etching process, the glass substrate is oscillated inside the overflow bath to agitate the chemical on the surface of the substrate. It helps agitation of chemicals by moving the substrate back and forth to the transporting direction inside the dip bath. The greatest advantage of the dip method is that it ensures that the chemical reaction conditions are uniform across the surface of the glass substrate.

As described in the preceding section, line width is required to be uniform over the entire area of the substrate in the gate and S/D wiring process. With dip etching, the glass substrate is horizontal and the etching reaction continues under the same conditions for every position on the substrate surface. For this reason, dip etching does not cause the differences in line widths or inconsistencies in the shape of the taper on the substrate surface. This is one of the acknowledged advantages of dip processing. As a result, dip processing is frequently used for gate and S/D wiring patterning process with respect to the shower method.

14A.3.5 Cascade Rinse System

In general, multiple rinse chambers are used in common. The final rinse is a direct rinse using DI water provided directly from the facility. For the other rinses, the DI water is provided using a cascade system, using rinse tanks on the upstream side as shown in Figure 14A.8.

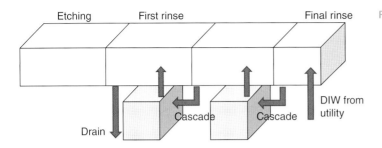

Figure 14A.8 Example of DI water use for rinse.

For this reason, drainage is needed only for the rinse conducted immediately after the etching process. Despite this fact, however, the large DI water shower flow rate that is needed due to the increasing size of the mother glass means that the quantity of DI water replenished by the cascade from the final rinse tank is insufficient.

14A.4 PROBLEMS DUE TO INCREASED MOTHER GLASS SIZE AND SOLUTIONS

14A.4.1 Etchant Concentration Management

With the increase in mother glass size, the volume of the process chambers has also been increased. This means that the discharge volume of the spray and the quantity of fluid in the dip tank become larger. The quantities of etching chemical have also been increased, resulted in the increase of the chemical tank.

Etchant deteriorates after a certain number of mother glass substrates have been processed, and thus decreases the etching rate since the chemical reactions or evaporation of only specific components may deteriorate the etchant quality. For example, oxalate indium is created in ITO etching, thereby decreasing the oxalic acid component of the etchant. With AL etchant, the vapor pressure of each component is different, resulting in variable vaporization speeds. Each of these cases unfortunately leads to a situation in which the ratio of the components that constitute the etchant gradually change over time. That's why, that the etchant is generally replaced before deterioration can affect panel quality.

The larger volume of chemical tanks increases the quantity of etchant needed for replacement and the quantity for waste. This pushes up running costs and is also a negative factor in terms of the environment. For these reasons, efforts are made to control the etchant concentration and extend the period of time before replacement is needed.

The ratio of components in the etchant inside the fluid tank is measured by means of titration, absorbance and so on. And then components that are insufficient are replenished. This method ensures a stable ratio of, for example, the phosphoric acid, acetic acid and nitric acid components that are used in aluminium etching. This also increases chemical life, and the etching process is stabilized.

14A.4.2 Quick Rinse

In addition to the increased volume of etchant, the increase of DI water that is used for rinsing after etching has also become a problem due to the mother glass becomes larger. This is a common problem to all etching methods. In the case of aluminium etching, in which the horizontal transfer is used, it is more serious since it needs larger quantity of DI water for rinsing than in the case of etching with the tilted transfer. The quick rinse system is one way of resolving this problem.

The rinse process is same as infinite dilution of the remaining chemicals on the glass substrate by DI water. Even though rinsing reduces the etching rate, the chemical reaction will proceed until rinse is completed. This leads to the following process problems.

(1) Non-uniformity of electrode line width.
(2) Non-uniformity of electrode cross-sectional structure (e.g., taper angle).

When such problems happen, the unevenness is visible to our eyes as a result. To solve these problems, spraying a large quantity of DI water is conventionally employed.

The DI water discharge quantity is estimated. For example, Gen 8 size display will need to 400 litters per minute [1]. The drained water from first rinse chamber includes etchant and will require the same type of liquid waste treatment as etchant. So in addition to the cost of the DI water in use, the cost of liquid waste treatment will also be additionally required.

A rapid substitution rinse system is one method that can resolve both panel quality and cost problems.

It consists of a high-pressure DI water spray and Air Knife.

The substrates quickly pass through a quick rinse chamber. And rinse is almost completed during substrates pass through. Conversely, the high-pressure rinse is turned off when the substrates are not being passed through the chamber. This reduces the total time that DI water is being discharged. The etchant on the glass substrate to be replaced with approximately one third of the quantity of DI water needed for a conventional system [1].

Although the quick rinse system discharges the entire quantity of DI water used in this process as wastewater, it can reduce the total quantity of DI water used by the system and thereby further reduce its footprint.

14A.4.3 Other Issues

Etching is a chemical reaction that dissolves metal with acid or other types of etchant. Various by-products are produced in this process, and these by-products cause a number of problems. In the case of Mo/Al films that are commonly used for gate and S/D electrodes, hydrogen gas is produced as a by-product of the etching reaction. As shown in Table 14A.1, the etchant used for aluminium etching is generally a mixture of acids based on phosphoric acid. This etchant is characterized by its high viscosity, and for this reason it is difficult to eliminate the gaseous hydrogen that is a by-product of the reaction. Failure to eliminate the hydrogen gas will cause the reaction to stop at that point, creating a short in the writing pattern. To solve this problem, prior to the etching process, the surface must be modified with an excimer UV light and atmospheric-pressure plasma to reduce the angle of contact (to create a surface from which the hydrogen can be removed easily).

14A.5 CONCLUSION

The increase in the size of the mother glass for TV displays as to Gen 8 size and even larger is a hot topic of discussion. The issues discussed in this paper coming from such size increase will become more and more conspicuous. Efforts to extend the life of the etchants and minimize the consumed quantity of DI water used in the rinse process are likely to continue for the time being.

In addition, the higher definition panels for mobile display require an improvement in the wet etching process to meet fine patterning. It is the quite tough challenge for wet etching.

However, the increasing size of mother glasses used for mobile devices, the advantage of wet etching in terms of controlling running costs is still significant. Wet etching equipment manufacturers will continue to play a major role in the development of methods that resolve the issues involved in the processing of substrates for high-definition displays, while keeping running costs low.

Reference

1 E. Yamashita, Prospects for FPD manufacturing technology in 2007: The wet process, Monthly Display, 13(1), p. 21 (2007).

14B

Dry Etching Processes and Equipment

Ippei Horikoshi

Tokyo Electron Limited, Akasaka Biz Tower, 5-3-1 Akasaka Minato-ku, Tokyo, Japan, 107-6325

14B.1 INTRODUCTION

There are two types of etching system, dry and wet etchings, are available for the TFT fabrication process. Dry etching that is processed in the vacuum chamber using the plasma gas, whereas the wet etcher uses the liquid chemicals, such as acid. Dry etching is featured by good enough uniformity, high taper controllability, less CD (critical dimension) loss, and one-time-etch capability of multi-stuck layer, and therefore it is recently dominantly used in TFT process.

In terms of cost comparison, the initial cost of wet etcher (i.e., equipment and its installation cost) is lower than dry etch, on the contrary the running cost of dry etcher is lower than wet etcher since dry system does not requires the costly supporting system unlike the wet etching system which needs the huge volume liquid chemical supply and treatment facility.

Table 14B.1 shows the comparison between wet and dry etchers. The wet etcher is usually dominant for materials having a low vapor pressure such as metals, and dry etcher mainly used for processing semiconductor, oxide, and nitride films with precise dimensions. In general the dry etcher is advantageous over wet in E/R (etching rate) uniformity, edge taper controllability, CD loss, one-time-etch of multi-layer, and running cost. On the other hand, wet etcher offers better performance in selectivity, particle contamination, throughput and initial cost.

Nowadays, the demand for high functionality and high-definition displays, especially for the smart phones and tablets, is booming, as shown in Figure 14B.1, "Trends of Display Panels." IGZO (metal oxide), LTPS and OLED are expected to replace the current a-Si TFT-based displays, at first for the mobile and tablet applications, followed by TV displays in the near future.

In this chapter, the principle and detailed information in the plasma etching on the major thin-film materials used in an a-Si, IGZO and LTPS TFTs will be discussed.

14B.2 PRINCIPLE OF DRY ETCHING

Dry etcher system is characterized by utilization of the low temperature plasma by glow discharge in the etching gas, and that plasma etches the target layer by ions and radicals from the plasma [1, 2].

Flat Panel Display Manufacturing, First Edition. Edited by Jun Souk, Shinji Morozumi, Fang-Chen Luo, and Ion Bita.
© 2018 John Wiley & Sons Ltd. Published 2018 by John Wiley & Sons Ltd.

Table 14B.1 Comparison between wet and dry etchings.

	Comparison item	Wet Etcher	Dry Etcher
1	E/R Uniformity	Fair	Good
2	Taper Controllability	Poor	Good
3	Selectivity	Good	Fair
4	CD Loss	Poor	Good
5	Particle	Good	Fair
6	Multi-layer Etch	Bad	Good
7	Throughput	Good	Fair
8	Initial Cost	Good	Poor
9	Running Cost	Poor	Good

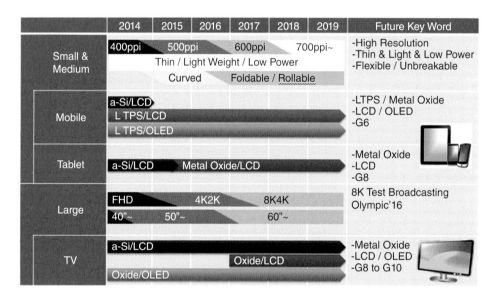

Figure 14B.1 Trends in display panels and technologies.

14B.2.1 Plasma

Plasma condition may be called as the fourth condition other than three; solid, liquid, and gas. Molecules and atoms in the gas are electrically neutral in the steady state, since plus charges in the atomic nucleus and minus charges of electrons are balanced. If an electron collides with a molecule or an atom, the outside electron may be removed from it and it may be separated into a positively charged ion (refer later part) and an electron. (this phenomenon is called "ionization"). In that state the neutral molecules, atoms, positively charged ions, and negatively charged electrons co-exist, keeping them electrically neutral as a whole state. That is called as "plasma" state (see Figure 14B.2).

When all molecules and atoms are ionized, we call it "fully ionized collision-less plasma," whereas in the state in which only part of them are ionized yet retaining the same quantity of the neutral molecules and atoms, it is defined as "weakly ionized plasma." Dry etching system uses weakly ionized plasma. The ionization rates, which stand for the rate of ions to the total numbers of molecules or atoms for the PE/RIE and ICP modes, are usually less than 0.01% (less than 1 per 10,000 is ionized) and less than 1% (less than 1 per 100 is

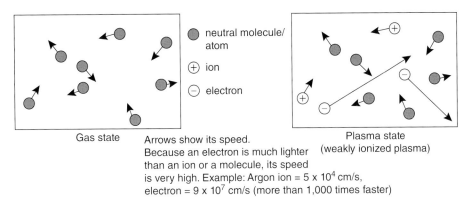

Figure 14B.2 State of the gas and plasma.

ionized), respectively. Here PE/RIE is "plasma etching/reactive ion etching," and ICP is "inductively coupled plasma."

14B.2.2 Ions

This is a condition that an electron removed from a neutral molecule or atom. It loses negatively charged electron, resulted in plus-charged, and thus it is attracted to the cathode electrode across the applied electric field. At the same time an electron, which is reflected or removed from molecule, is accelerated by the anode electrode, causing other molecule or atom's ionization to maintain the glow discharge (see Figure 14B.3).

14B.2.3 Radicals

The radical is the state that the molecule is broken apart by electron collisions. It is electrically neutral and is unstable and very reactive due to the broken electron pairs. The regular stable molecules have electron pairs, while radicals have non-pair electrons since the collision hit by electrons breaks the pair and searches the counter part of it. Since it is electrically neutral, it receives no effect from the electric field. (They always behaves with random movement—thermal movement) (see Figure 14B.4).

In dry etching system, electrons that are accelerated by the electrical field applied by the voltage potential across the electrodes collide with the gas molecules and ionize them or change the stable molecule into radicals, which are unstable and reactive. Our system uses weakly ionized and low-temperature plasma, which includes ions, electrons, electrically neutral gas molecules, and radicals.

In dry etching system, the reactive radicals and ions are attracted vertically to the panel by the cathode electrode with self-bias. Radicals are electrically neutral and diffuse to the film surface at random angles,

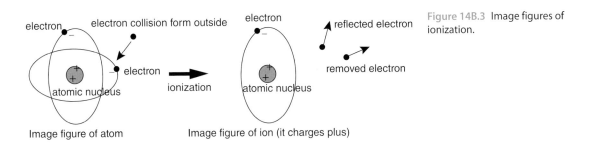

Figure 14B.3 Image figures of ionization.

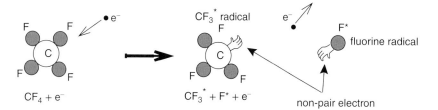

Figure 14B.4 Conceptual drawing of radical generation.

reacting with molecules on the surface because they are very reactive. Volatile reaction by-products are removed by the vacuum pump. The positively charged ions are attracted to the panel by the negatively charged self-bias and collide with the film surface. This collision hit by ions promotes the chemical reaction by radicals and also removes the generated materials on the surface.

14B.3 ARCHITECTURE FOR DRY ETCHING EQUIPMENT

The dry etching system "Impressio^{TM}" series originated from Tokyo Electron Limited (TEL^{TM}) (Figure 14B.5) is multi-chamber plasma etch/ash system, available for sixth- through tenth-generation substrates. The system offers a high uniformity in plasma density within the process chamber and high etching efficiency for smartphone, tablet, and large-sized TV displays.

- Load Lock Chamber(L/L):
  In order to maximize throughput, "Impressio^{TM}" is equipped with two load lock modules, which execute vent and pumping continuously. Also to minimize load lock chamber capacity, mechanism inside the chamber is of extremely simple design.
- Transfer Module (T/M):
  The transfer module is equipped with the ultra-fast motion vacuum robot to transfer substrate to process chamber.
- Process Chamber (P/C):
  They are the vacuum chambers using for etching process. We have ECCP (enhanced capacitive coupled plasma) mode that is the bestseller for the a-Si LCD manufacturing of laptop PC, monitor, and TV display

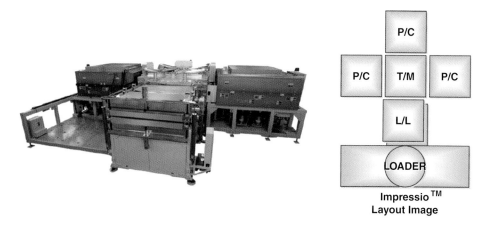

Figure 14B.5 Outline of Tokyo electron limited/dry etching system "Impressio^{TM}" series

Table 14B.2 Specification list of "Impressio™" series.

	Impressio™ 1800	Impressio™ 2400	Impressio™ 3300
Substrate size	<1,500 × 1,850 mm	<2,200 × 2,500 mm	<2,940 × 3,370 mm
Application	a-Si / LTPS	a-Si / Metal Oxide	a-Si
Transfer system		L/L+T/M	
P/C Unit Q'ty		MAX 3P/C	
TEL P/C mode	ECCP,ICP	ECCP,ICP	ECCP

panels. The maximum size of equipment is now available up to tenth generation. Also, TEL dry etcher with ICP (inductive coupled plasma) mode, which is featured by high-density plasma, has the top share record for the LTPS mass-production line using 4.5th generation substrate. ICP etcher using for LTPS and metal oxide TFTs for sixth to eighth generation sizes has also been developed. It has an excellent uniformity and etching rate to meet the needs for the high-function and high-definition displays.

The specification of "Impressio™" is summarized in Table 14B.2. It is equipped with the EPD (end point detector), which detects the best ending point of etching by monitoring the emission at the specific wavelength from the plasma. End point is detected by the change of emission when etching ends. Recent improvement for that detector is the addition of the differential algorithm, where the point that shows the maximum slope in the emission waveform corresponds to the end point.

14B.4 DRY ETCHING MODES

In the dry etching mode, both chemical and ion-assisted modes are available.

(d) Chemical etch (etch mostly by radicals)

It etches the substrate surface by a reaction between the radicals and material on the surface, resulting in gas by-product. Since radicals move at random by thermal motion, the etching property is isotropic, as shown in the Figures 14B.6 and 14B.7. Also the etching is selective due to chemical reaction etch, showing the low damage to the substrate.

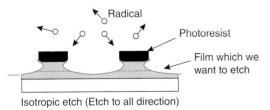

Figure 14B.6 Concept of isotropic etching.

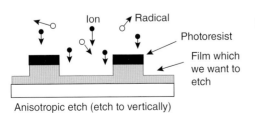

Figure 14B.7 Concept of anisotropic etching.

(e) Ion assist etch (etch by ions and radicals)

Figure 14B.7 shows a concept of an anisotropic etches,

Etching is achieved by activating the chemical reaction by the ion kinetic energy with acceleration by self-bias (explained in later). Since the ions are accelerated to one direction, anisotropic etching can be achieved. However the damage for substrate by ion bombardment is considerable.

14B.4.1 Conventional Etching Mode and Each Characteristic

In the conventional dry etching system, both PE and ICP modes have been available.

PE (plasma etching) mode:

Application: a-Si TFT process (island, channel, and passivation)

In this mode (see Table 14B.3), RF power is connected to the upper electrode and the substrate is placed on the lower electrode. Since the substrate is placed on the anode side (ground), it is defined as "anode coupling." In this mode, radicals mainly etch the film, so the etching type is isotropic. The damage to the substrate is slight.

RIE (reactive ion etching) mode:

Application: a-Si TFT process (island, channel, and passivation)

In this mode (see Table 14B.3), RF power is connect to the lower electrode and the substrate is placed on it. Since the substrate is paced on the cathode side, it is defined as "cathode coupling." In the RIE mode, the accelerated ions by self-bias to incident the substrate almost vertically. Here self-bias is created by RF (high frequency) power applied to the lower electrode and its ion attack intensity is proportionate to RF power. The chemical reaction with radicals is activated by the ion collision energy, resulted in anisotropic etching. The damage to the substrate is considerably higher.

Table 14B.3 Etching modes and applications.

	PE	RIE	ECCP	ICP
Etching Mode Concept	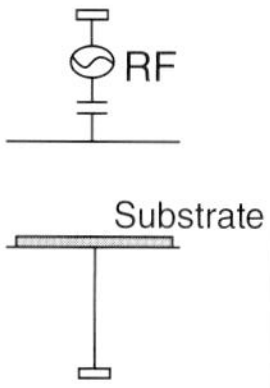 RF / Substrate	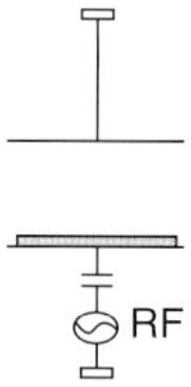 RF	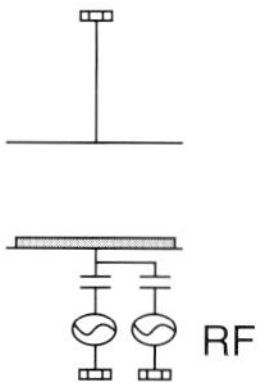 RF	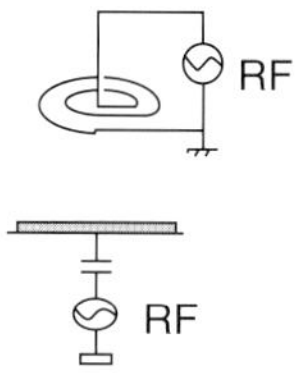 RF
Electron Density	$10^{9\sim10}/cm^3$	$10^{9\sim10}/cm^3$	$10^{9\sim10}/cm^3$	$10^{11\sim12}/cm^3$
Chemical Etch Radicals	Good	Good	Good	Excellent High DensityPlasma
Ion Assist Etch Bombardment Control	Bad No Bias	Good Control only by Self Bias	Excellent Controllable by Independent Bias	Excellent Controllable by Independent Bias
Uniformity Control for Large Size	Fair	Fair	Good	Excellent Plasma Distribution Control
Etching Application	a-Si TFT a-Si, SiN, Ashing	a-Si TFT a-Si, SiN, Ashing	a-Si TFT a-Si, SiN, Ashing Metals	LTPS/IGZO TFT p-Si, SiO, SiN Metals, Full Ashing

14B.4.2 Current Etching Mode and Each Characteristic

Currently ECCP and ICP modes instead of the conventional etching modes are under use mostly for over 6G TFT manufacturing. In those modes, enormous ion attack energy is utilized by RF (low frequency) power, which is called "independent bias." "Bias" is defined by RF power application to the lower electrode that sustains the glass

<u>ECCP (enhanced capacitive coupled plasma) mode</u>:
Application: a-Si TFT process (island, channel, contact, metals), and LTPS TFT process (metals)

In this mode, as shown in Table 14B.3, RF power is connected to the lower electrode where the substrate is placed. The plasma density of HF (high frequency) is higher than LF (low frequency) plasma, and therefore, chemical reaction rate in HF plasma is much higher. On the other hand, LF can have variable bias so that anisotropic etching is possible. Since the electron mass is lighter than ion and electron is attracted to the cathode electrode, the self-bias voltage at the cathode electrode is increased, resulted in enhancement of ion bombardment. Moreover, because the plasma uniformity of LF is better than HF, the combination of high frequency and low frequency, which allows both chemical etching and anisotropic etching with good enough uniformity is quite reasonable for the larger substrate like FPD applications.

Now the flat display panel size for TVs is getting larger and larger, in which mother glass size for manufacturing is also expanding up to now over 3 meters, so called generation 10. Tokyo Electron Limited (TEL) have developed ECCP etching mode for a-Si TFT and been supplying to TV panel manufacturers to produce a large panel efficiently.

<u>ICP mode—inductively coupled plasma mode</u>:
Application: IGZO TFT process, and LTPS TFT process (p-Si, SiOx, SiNx, metals, and full ashing)

In this mode, as shown in Table 14B.3, the upper electrode has a coil shaped and the lower electrode connects biased RF power. The substrate is placed on the lower electrode. The coil-shaped upper electrode generates the inductive electric field to transfer the electrons horizontally, in which few electrons are caught by the electrode, making its ionization rate about hundred times higher than the other modes with generating high density plasma. The lower biased electrode attracts ions and causes collision on the substrate, so the etching type is ion-assisted mode.

The screen resolutions in FPD is moving up to the levels of 500, 600, and even 800 pixels per inch (ppi) for smart-phones, with driving innovations in power saving and user interface technologies. LTPS and IGZO TFT backplanes are suitable for such high-resolution displays. Accordingly the dry etch systems for FPDs are now required to offer even more refined processing, higher productivity, and better yield than ever. TEL has developed the new advanced PICP™ with the innovative concept corresponding to those high-spec demands.

14B.5 TFT PROCESS

14B.5.1 a-Si Process

The typical a-Si TFT structure is shown in Figure 14B.8. An etching system/mode for each etching steps and materials are summarized in Table 14B.4. In recent years, the dry etching process mode has switched from the conventional PE/RIE to ECCP/ICP as the substrate has been upsized. In addition, the "half-tone photo-resist ashing" process [3] is employed for saving the total mask layers. Since its process requires higher A/R (aspect ratio) with improved uniformity, ICP is suitable for that application.

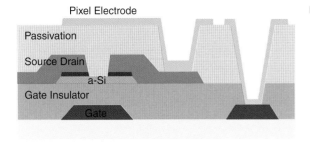

Figure 14B.8 a-Si TFT structure.

Table 14B.4 Appropriate etching modes for each etching step and material in a-Si TFT.

Etching Step	Material	Plasma Mode		Main Etch Gas
		Conventional	Current	
Gate line	Al	Wet	Wet	
	Cu	Wet	Wet	
Island/active	a-Si	PE, RIE	ECCP	$SF_6 + Cl_2$
Half-tone ashing	Photo Resist	RIE	ECCP, ICP	O_2
Source drain	Al	Wet	Wet	
		RIE	ECCP, ICP	Cl_2
	Cu	Wet	Wet	
Channel	N+	PE, RIE	ECCP, ICP	Cl_2
Contact hole	SiN	PE, RIE	ECCP, ICP	$SF_6 + O_2, CF_4 + O_2$
Pixel electrode	ITO	Wet	Wet	
			ICP	Cl_2

14B.5.2 LTPS Process

The typical LTPS structure is shown in Fig. 14B.9, and the etching systems and modes at each layer is also shown in Table 14B.5. Usually p-Si, contact hole, metallization and S/D, and passivation layers are dry etched. In recent years LTPS have been applied to the high resolution display panels that require high performance ICP etching system with improved productivity. Recently TEL has proposed the advanced ICP system.

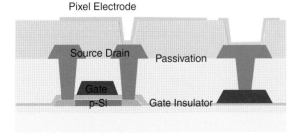

Figure 14B.9 LTPS structure.

Table 14B.5 Appropriate etching modes for each etching step and material in LTPS.

Part of TFT	Material	Plasma Mode		Main Etch Gas
		Conventional	Current	
Channel etch	P-Si	ECCP	ICP	Cl_2
Gate etch	Mo, MoW	ECCP,ICP	ICP	SF_6, Cl_2
Contact etch	SiO/ SiN	ICP	ICP	$SF_6, CF_4, CHF_3, C_4F_8$,
Source drain etch	Al	ECCP,ICP	ICP	Cl_2
Passivation	SiN	ECCP,ICP	ICP	SF_6, CF_4
Pixel electrode etch	ITO	Wet	Wet	
			ICP	
Doped ash	Photo resist	ICP	ICP	O_2

14B.5.3 Oxide Process

The typical oxide TFT structures are summarized in Table 14B.6. And the etching modes at each layer are shown in Table 14B.7. Oxide process with the several different MO × TFT structure is introduced in recent years. The features in each structure are shown in the Table. MO × TFT for LCD backplane can use ES (etch-stop) and BCE (back-channel etch), while OLED may use ES and top gate structures. TEL continues technical development to meet these new TFT fabrications.

The critical issues in manufacturing MO × TFT [4] are:

(1) The deep hole etching in all structure is necessary. Higher E/R of SiO_2 is newly required.
(2) The BCE structure allows small TFT with the short channel length. However, BCE structure requires the low damage in the channel layer during metal dry etching. On the other hand, the parasitic capacitance in BCE is comparatively large due to mask misalignment, which makes driving OLED difficult.

Table 14B.6 Typical oxide structures and features.

	MOx ES	MOx BCE	MOx Top Gate
Structure			
Reliability	Good	Fair	Fair ~ Good
Channel L	Fair	Good : Short Channel	Good
Aperture	Fair	Good : Small TFT	Fair
Parasitic C	Poor	Poor	Excellent : Self Align
Mask #	Fair	Good	Fair
Others		Wet Channel Damage	Back Side Light Shield
Application	LCD/OLED	LCD	OLED
Dry Etch	High E/R CNT Hole	Low Damage S/D Etch High E/R CNT Hole	Uniform Etch Gate / GI High E/R CNT Hole

Table 14B.7 Appropriate etching modes.

Part of TFT	Material	Plasma Mode		Etch Gas
		Conventional	Current	
Gate etch	Al	Wet	Wet	
		RIE	ECCP,ICP	Cl_2
	Cu	Wet	Wet	
Dielectric – Etch stopper – Gate insulator – Passivation	SiNx	ECCP	ICP	$SF_6, CF_4, CHF_3, C_4F_8$
	SiO	ECCP	ICP	
Ash	Half-Tone	ECCP	ECCP,ICP	O_2
Source drain etch	Al	Wet	Wet	
			ECCP,ICP	Cl_2
	Cu	Wet	Wet	
Pixel electrode etch	ITO	Wet	Wet	
			ICP	

(3) The TOP gate structure can minimize parasitic capacitance by self-alignment and is most suitable for driving OLED. In gate and gate insulator layer in this structure, etching uniformity is particularly important.
   (a) Recently TEL has released Gen 8.5 size PICP™ system for application to those new etching processes in metal-oxide TFT.

References

1 B. Chapman, Glow discharge processes sputtering and plasma etching, New York: John-Wily & Sons (1980).
2 Y. Kuo, Plasma technologies in the fabrication of thin film transistors for liquid crystal displays, American Vacuum Society National Short Course Book (1999).
3 C. W. Kim, et al., A novel four-mask-count process architecture for TFT-LCDs, SID Digest Tech. Papers, p. 1006 (2000).
4 Z. Yanbin, et al., Dry etching characteristics of amorphous indium-gallium-zinc-oxide thin films, Plasma Science Technology, 14(10), p. 915 (2012).

15

TFT Array: Inspection, Testing, and Repair

Shulik Leshem[1], Noam Cohen[1], Savier Pham[2], Mike Lim[3], and Amir Peled[1]

[1] *Orbotech Ltd., Yavne, Israel*
[2] *Photon Dynamics, Inc., San Jose, California, USA*
[3] *Orbotech Pacific Ltd., Bundang-Gu, Sungnam City, Kyoungki Do, Korea*

An FPD manufacturing fab is efficient and successful when its yields are high, that is, when the fab produces a high percentage of good ("GO") panels without defects, and minimum scrap. Implementation of various inspection, metrology, testing, repair, and software analysis tools along the entire production line allows FPD manufacturers to manage and increase their yields (Figure 15.1).

Production defects are the main cause of low yields. Production defects can have a wide variety of causes and sources, such as incorrect calibration or failure of the production equipment, inaccurate setup of the production process conditions and parameters, defective materials, contamination with particles, and so on. In the following section we will discuss the main types of production defects that are common to all technologies, FPD device types and processes, and also the detection and repair of these defects.

15.1 DEFECT THEORY

15.1.1 Typical Production Defects

The defects that appear during the production process can be classified into two main groups: pattern defects and foreign particles, as shown in Figure 15.2.

15.1.1.1 Pattern Defects

Pattern defects involve a distinct change of the layer pattern and as such, are easily recognizable as defects.
Such defects can appear on any production layer.
There are two types of pattern defects:

- Excess material defects (islands, shorts, protrusions, material residues)—these defects usually have the same color as the deposition material
- Missing material defects (pinholes, opens, nicks)

Some pattern defects cause a complete violation of the pattern; other pattern defects are partial (they cause a potential risk of a violation of the pattern) (Figure 15.3).
Often the pattern defects appear during the photolithography process (Figure 15.4).
Sometimes the pattern defects can appear during the etching or deposition processes (Figure 15.5).

Flat Panel Display Manufacturing, First Edition. Edited by Jun Souk, Shinji Morozumi, Fang-Chen Luo, and Ion Bita.
© 2018 John Wiley & Sons Ltd. Published 2018 by John Wiley & Sons Ltd.

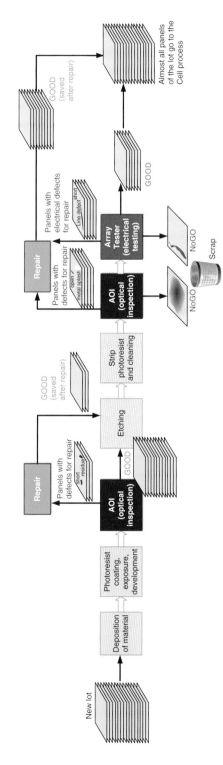

Figure 15.1 Inspection, testing and repair—for maximum yields of the production line.

Pattern defects: **Foreign particles:**

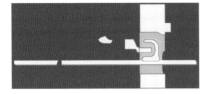 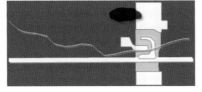

Figure 15.2 Main types of production defects.

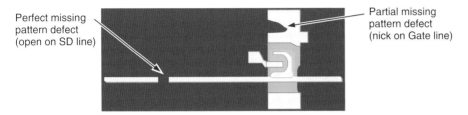

Figure 15.3 Complete and partial pattern defects.

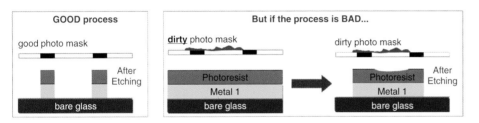

Figure 15.4 Excess pattern defect of the photolithography process (short between gate and common lines on gate layer after photoresist development).

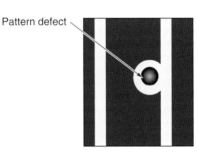

"Splash" defects can cause pattern violations and z-axis shorts between layers.

Figure 15.5 Excess pattern defect of the deposition process.

As modern display technologies introduce tighter design rules and stricter tolerances, finer features and denser, more complex patterns, the issue of pattern defects becomes even more critical.

15.1.1.2 Foreign Particles

Foreign particles vary in shape, size, color, and functional location. These defects are not made of the materials used in the production process, but of foreign matter, so their color and texture are typically different from the color and texture of the deposited material. Approximately 80% of all production defects are caused by particles. The main sources of foreign particles are:

- The fab environment (e.g., dust, debris, particles of glass or human skin)—usually such defects are large
- Process machinery (e.g., contamination with carbon, oil, grease, glue)—such defects look similar to other foreign particles, but can be very small

There are two types of foreign particles:

- Surface particles—Surface particles detected on the surface of any layer at any production stage
- Buried (or covered) particles—particles that adhered onto the previous layer due to electrostatic attraction and then got buried in the material deposited during the next process step

Both surface and buried particles can cause critical production defects.

The example in Figure 15.6 shows a perfect short within the ITO layer that was probably caused by a foreign particle during the ITO deposition step, before photoresist coating. Because of the photoresist thickness variation at the location of the particle, photoresist was not fully removed during the development. As a result, the ITO pattern was not etched at this location.

On OLED displays where the organic light-emissive material layer is very thin and is very sensitive to the environmental damage, the issue of foreign particles becomes even more significant: here a tiny submicron particle can cause a critical defect, for example, a vertical short between layers.

15.1.2 Understanding the Nature of Defects

15.1.2.1 Critical and Non-Critical Defects

The defects detected by an AOI (Automated Optical Inspection) system can usually be classified as critical or non-critical.

In order to control the production process and manage yields efficiently, the nature of the defect should be analyzed and understood correctly. A particle might be considered as non-critical if situated on the top surface of a layer, since it will be removed during panel cleaning before the next deposition stage. However, if the same particle is buried, it can become a killer defect (if it causes a vertical short between the two conductive layers). For example, a short (e.g., caused by excess material in the gate area) on the transistor is a killer defect, while a short on the common line (capacitor) might be considered as less critical.

Thus, there is a need to understand how any given defect affects the functionality of the final product. This need raises the importance of accurate defect classification and correct Go/NoGo/Repair judgment of defects.

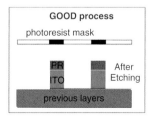 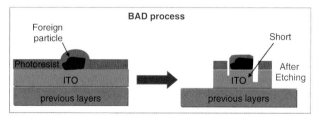

Figure 15.6 Surface particle causes a short between ITO fingers.

15.1.2.2 Electrical and Non-Electrical Defects

The defects detected by the array tester (electrical testing) system can usually be classified into electrical and non-electrical types, with most of the critical defects being classified as electrical defects.

Electrical defects are usually caused by the pattern defects such as opens, shorts, residues, and so on. Electrical defects will directly affect the light-on performance of the pixel, so these are critical defects that must be repaired, if possible (Figure 15.7).

Non-electrical defects are usually caused by foreign particles. Such defects may not be critical and may not be found during the cell light-on test. However, some optical defects such as partial ITO defects may be critical, as they might partially diminish the intensity of the "on" pixels.

15.1.3 Effect of Defects on Final FPD Devices and Yields

A small local defect such as a short or open on a transistor can lead to defective pixels on the end product display of the following types:

- Dead pixel (always off)—will appear as a black spot on a white background of a display device
- Hot pixel (always on)—will appear as a white shining spot on a black background of a display device
- Stuck pixel—will appear as a constantly lit colored spot (this is the case when one of the RGB sub-pixels is faulty)

A single bad pixel may not be noticeable on most backgrounds or in most lighting conditions. However, if there are a lot of defective pixels on a display, the entire display panel has to be scrapped.

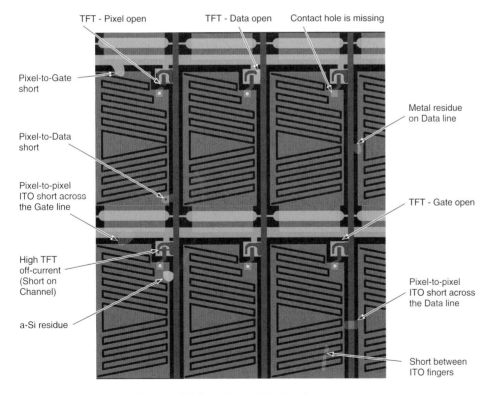

Figure 15.7 Main types of electrical defects detected during the array test.

Some defects affect an entire row or column of display pixels—these are called line defects. Some fatal defects can affect the pattern, functionality, optical, and electronic performance of a whole display.

Leading FPD manufacturers tend to adopt a "zero defect" policy, driven by high demand for displays with "zero" bad pixels, as well as high-quality and performance requirements.

However, this "zero defect" policy might be too costly for mass production. It is impossible to entirely eliminate defect generation, regardless of the effort, time, and cost invested in the process quality control and yield management. The production process conditions are constantly changing over time, production equipment sometimes fails, and particles are present even in a clean room environment.

To turn the potentially defective displays into saleable products and to increase yields, FPD manufacturers try to detect the defects early in the production process, identify the functional effect of the defect and repair the defects or rework the panel. Thus, inspection for defects and testing of the display panels at various production stages becomes critical for yield management.

15.2 AOI (AUTOMATED OPTICAL INSPECTION)

15.2.1 The Need

Use of AOI (Automated Optical Inspection) is a key part of the fab's yield management activity. The AOI system is used in order to either reduce the statistical occurrence of defects (e.g., by identifying their sources for subsequent corrective action); or for supporting repair activity on a specific display panel, which might otherwise be rendered "scrap."

AOI machines use a non-contact optical system to acquire an image of the panel, and image processing algorithms to detect abnormalities on the panel image. These abnormalities are further analysed, sorted and classified by the AOI software, and parameters for the final defect reports are calculated (defect coordinates, size, functional location, defect type, etc.). The inspection data that includes the final defect reports, defect images, defect maps, and so on, is transferred to the fab host computer for further processing and action.

Benefits of AOI in comparison with other testing tools:

- AOI systems detect defects early in the production process, before the defects are covered with further materials and become inaccessible in the next process steps. This allows the manufacturers to take corrective action when repair is still possible and save the display from scrap (electrical testing is possible only when the display panel is fully, or in some cases partially, electrically functional)
- Many defects affect optical performance of the display but are non-electrical in nature, so they cannot be found by electrical testing equipment
- AOI is the optimal solution for inspection of high-resolution small displays that cannot be inspected efficiently by contact testing equipment (e.g., due to very high contacting pitch)

Figure 15.8 Example of an AOI system: Orbotech Quantum^{TM}.

- Even defects that have no *functional* (i.e., electrical nor optical) impact are detectable, and by doing so contribute to the formation of a *statistical* picture of manufacturing process health, enabling the engineers to, for example, stop and service a machine or production module which has begun to deteriorate, before it starts to damage the manufactured displays

15.2.2 AOI Tasks, Functions, and Sequences

Figure 15.9 shows the basic tasks of a typical AOI system:

15.2.2.1 Image Acquisition

Image acquisition is the first step and the core of an automated optical inspection (Figure 15.10).

The basic image acquisition system includes:

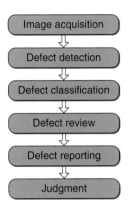

Figure 15.9 Main AOI Tasks.

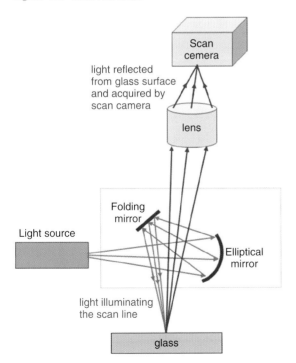

Figure 15.10 Example of the basic image acquisition system.

- A light source that illuminates the panel
- A camera sensor that acquires the light reflected from the illuminated panel surface
- Additional optical elements (lenses, mirrors, beam splitters, etc.) that align the light, focus and magnify the image, and improve the illumination and imaging properties
- The image acquired by the camera is transferred to the image processors (electronic boards and computers)

To increase the glass coverage and inspection throughput, most AOI systems use multiple optical channels and multiple image processors. Each of the cameras acquires a section of the panel image. Each computer processes the image data of one or several optical channels at the same time (Figure 15.11).

15.2.2.2 Defect Detection

Image processing hardware and software algorithms are used to process the panel image and find defects according to the detection sensitivity parameters specified by the user.

Advanced AOI systems can detect defects of various types in different panel areas. For example, during inspection of the TFT array panels such AOI systems can detect defects in both cell areas and connector (peripheral) areas, as well as both local micro defects that affect one or several display pixels and global macro defects that affect the whole display or the whole glass substrate.

15.2.2.3 Defect Classification

Most AOI systems include a defect classification mechanism for automatic identification of defect *types* according to the user-defined examples or rules.

Defect classification can be performed online (during the scan) or offline.

Basic defect classification mechanisms are based on defect parameters such as size, gray level, and functional location. More advanced classification mechanisms involve processing and analysis of the scan images or video images of defects. If defect classification is based on the review images, it is performed after the image grabbing.

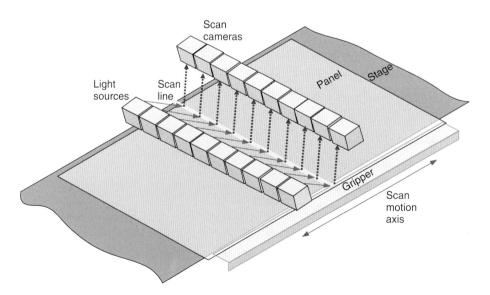

Figure 15.11 Image acquisition with multiple optical channels.

15.2.2.4 Review Image Grabbing

Most AOI systems include high-resolution color video cameras that grab images of detected defects, for subsequent analysis, review, and verification.

In some AOI systems, review image grabbing is performed after the defect detection scan, as an additional process, on a dedicated review/metrology platform.

Advanced AOI systems can perform review image grabbing on the fly, in parallel with the defect detection scan. While the system scans across the glass, acquires a panel image and detects defects on one portion of the glass, the video system grabs images of the defects that were detected on parts of the glass that have already been scanned.

In addition to video grabbing of defect images, an advanced AOI systems can also grab video images from predefined locations on the panel, for sampling, testing, or measurement purposes. For example, such a system can grab video images of predefined CD/O (Critical Dimension and Overlay) test locations, to measure the pattern features on the images and detect production problems such as line width non-uniformity or misalignment of different layers.

15.2.2.5 Defect Reporting and Judgment

For each defect, the system calculates the defect properties and generates a defect report. Once the panel has been inspected, the AOI system outputs the following data to the fab host computer (Figure 15.12):

- A defect file that contains defect reports (defect coordinates, defect size, defect gray level parameters, functional location, etc.)
- Defect images
- If the AOI system includes an automatic defect classification mechanism, defect classification results are also included in the output data
- Some AOI systems also provide digital macro images of the panel and metrology (measurement) results
- In addition, the AOI output can include the results of the automatic auto-stack comparison and judgment. (These mechanisms will be discussed in detail later in this chapter.)
- Some AOI systems also provide various statistical data for process control, for example, defect distribution maps and defect trend charts

The inspection data output by the AOI system is used for further analysis, process monitoring, and control.

15.2.3 AOI Optical Concept

The optical system of an AOI machine is designed based on the following FPD production process parameters:

- Size of the glass substrates to be inspected
- Panel design: display pixel size, dimensions (critical dimensions) and density of pattern features, minimum defect that must be detected

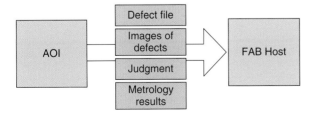

Figure 15.12 AOI outputs.

- Types of production defects that must be detected
- Materials used in the process and contrast between layers and materials

New display technologies introduce tighter design rules, finer features and denser, more complex patterns where very high scan resolutions are required. In addition, display panels have a complex multi-layer structure, with a mix of transparent and reflective materials: a modern AOI must be able to detect defects on a wide range of photoresists, on highly reflective metals, on matte plasticmaterials and on transparent materials.

The optical system of an AOI machine must:

- Use different illumination types to suit the diversity of applications
- Produce intense, concentrated light
- Use fast, stable, low-noise camera sensors
- Provide good image quality
- Provide high-resolution images

15.2.3.1 Image Quality Criteria

Before designing an AOI optical system to meet requirements, we must first define various image quality criteria:

- Focus
- Image resolution—strongly correlated to the physical dimensions of the inspected devices (e.g., the "design rules") but also to other parameters such as the sensitivity to different types and sizes of defects, engineering trade-offs (e.g., level of "noise" inherent in the camera sensor) and more. Modern AOIs used for FPD array processes will often employ resolutions ranging from around one micron to a few microns
- Image brightness (e.g., correct exposure—not over/under exposed)
- Contrast between layers: An optimal image contrast will produce the maximum gray level difference between layers/materials in which we want to inspect for defects (Figure 15.13)

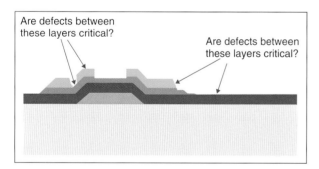

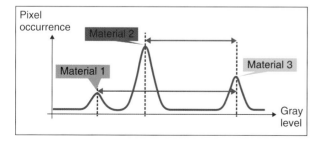

Figure 15.13 Image contrast concept: optimizing contrast between materials 1 and 3.

The image brightness and contrast are affected by the exposure and integration time of the camera sensor, illumination type and intensity, and other factors related to optical design.

15.2.3.2 Scan Cameras
15.2.3.2.1 Camera Type
Different types of scan cameras can be employed in an AOI system, the choice of which is part of a complex interplay of system requirements, illumination design approach, and engineering trade-offs:

- CCD (Charge-coupled Device) line cameras, which have one row of pixels and allow a single line exposure. To reduce noise the scan pixels are large, limiting the ability to achieve very high scan resolutions
- TDI (Time Delay and Integration) cameras, that have multiple rows of pixels and accumulate multiple exposures of the moving (scanned) panel. The initial charge is passed from one pixel row to the next pixel row and is accumulated synchronously with the continuous movement of the panel along the scan axis. For example, if the TDI camera has 256 pixel rows, the same line on the panel is exposed 256 times and the 256 pixel rows of the same panel line are merged into one image row
- An area camera, which has a two-dimensional matrix of pixels that are exposed simultaneously so that an image frame is acquired instantaneously. This enables a short exposure time and significant reduction of image smear caused by mechanical vibrations and velocity variations (but at the expense of stricter requirements on the illumination system)

15.2.3.2.2 Resolution Changer
Some AOIs support several optical resolutions on the same machine. Such systems include a resolution changer (i.e., a lens changer or zoom lens) that moves a selected lens into the optical path, providing the detection resolution specified for the panel being inspected.

15.2.3.2.3 Backside Inspection
In standard AOI systems the cameras scan the panels from above.

However, there are AOI systems that are used for backside inspection: the illumination source and camera are located below the panel so as to acquire an image of the bottom side of the panel. Backside inspection systems are used to detect scratches and large defects on the bottom side of panels, at relatively low resolutions.

Backside inspection optical heads and standard scan optics can be installed on the same AOI machine, on separate inspection decks.

15.2.3.3 Scan Illumination
The current trend in illumination design is a shift from conventional metal halide or halogen lamps to LED illumination.

The advantages of the LED illumination are:

- Long life time
- LED illumination does not become weaker over time
- Illumination uniformity and spatial symmetry in all directions (e.g., in the center and on the sides of the scan line)
- Strong light intensity
- Enhanced image contrast (e.g., better emphasis of small defects, better contrast on photoresist layers, etc.)
- Short-pulsed, high-powered LED illumination (flashes of light) is less sensitive to mechanical vibrations and significantly reduces the smear of pixels

15.2.3.3.1 Types of Illumination
Different materials and production layers have different reflectivity. "Tunable" illumination enables the use of various wavelengths, and selecting the settings that are optimal for a panel type being inspected.

- Some AOI systems use white LEDs only or RGB LEDs as a bright-field light source. There are AOI systems that implement multi-spectral RGB illumination technology. The AOI users can select a combination of colors ("color mix") that provides the best image contrast, according to the dominant colors of the panel materials. Bright-field illumination is optimal for detection on bright reflective materials and detection of surface defects
- In addition to the bright-field illumination, some AOI systems use dark-field illumination. Dark-field illumination emphasizes the edges of the pattern and defects and is optimal for detection of "missing pattern" defects (cuts) and "extra pattern" defects (shorts). This light type is also optimal for detection of 3D particles. The dark-field illumination helps to differentiate between killer open defects and dust defects that are less critical
- Backlight illumination (light source is below the panel) improves detection of pinholes and metal peeling on metal-coated layers after the deposition process, and enhances detection on color filter panels. This light type is optimal for detection of "backside" defects and improves the image contrast on transparent layers

The need to detect a wide variety of defects on multi-layer panels with a mix of transparent and reflective materials and to differentiate between killer defects and non-critical defects has led leading AOI vendors to develop the multi-imaging or "multi-modality imaging" concept. Using this technology, the system can acquire several different images of the same panel location, instantaneously and in parallel, as part of the same scan. Each image is acquired very rapidly, with a different light type and light parameters. AOI users can select several light types that are optimal for the specific detection needs (i.e., to maximize contrast), and set different parameters for each light type. Multi-modality imaging technology allows seeing more properties of the inspected panel and understanding the true nature of the defects. The added dimensions of image data that are produced significantly increases the detection and defect classification capabilities of AOI. The Orbotech Quantum system is an example of an AOI system using multi-modality imaging technology (Figure 15.14).

15.2.3.4 Video Grabbing for Defect Review and Metrology

15.2.3.4.1 Review/Metrology Cameras

AOI systems are usually equipped with a microscope or a high-resolution color video camera for grabbing of defect images at predefined test and metrology locations on the panel. The images are used for manual review and verification, and for manual or automatic defect classification and judgment.

To increase the inspection throughput, multiple review cameras can be installed on the same machine. During the video grabbing, the video cameras must move to the defect location, focus and capture an image at very high speeds.

15.2.3.4.2 On-the-Fly Video Grabbing

To allow very fast review image grabbing (e.g., many dozens of images within a short cycle time), some AOI systems introduce an on-the-fly video grabbing concept. Review images of defects are grabbed during the defect detection scan as the panel moves.

The on-the-fly video grabbing concept includes:

- Very short pulses
- Very large amount of light, per pulse
- Very fast exposure to prevent image smear

15.2.3.4.3 Alternative to Video Images

Some AOI systems support more limited defect classification based on the scan panel image acquired by the line CCD, TDI, or area scan cameras. These lower resolution "scan data" images can be used for defect review and classification in addition to or instead of the video images, which allows the AOI users to increase inspection throughput by eliminating the video grabbing scan or by capturing less video images.

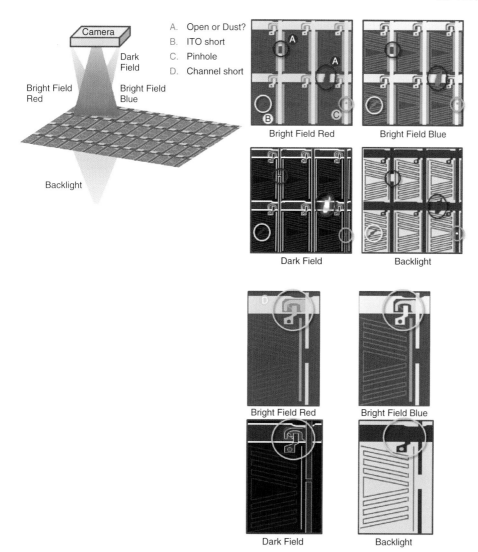

Figure 15.14 Defect detection and classification with multi-modality imaging.

15.2.4 AOI Defect Detection Principles

After the scanning optical system acquires an image of the panel, the image is processed to detect irregularities or abnormal pixels that are considered as potential defects.

Defect detection on AOI is less "obvious" than on electrical or functional test equipment. An electrical tester tests the actual circuitry of the panel pattern and detects defects that are obvious violations of the circuitry. A standard AOI detects abnormal pixels on the image that may or may not be killer defects. That's why the balance between true defect detection and false defect detection (false alarms) is so important to AOI performance. If the sensitivity is insufficient, the system can miss critical defects. On the other hand, if the sensitivity is too high and differentiation between the defect types is insufficient, non-critical harmless abnormalities can be reported as defects, introducing large amounts of "nuisance" data.

AOI detection performance depends on the image quality (which was discussed in section 15.2.3), the performance of the image processing algorithms and optimal setup of detection thresholds. In the following

15.2.4.1 Gray Level Concept

A scan pixel is a pixel on the scan image of the panel. Each scan pixel has a gray level value that indicates the amount of light in the pixel. The darker the scan pixel, the lower its gray value is (Figure 15.15).

15.2.4.2 Comparison of Gray Level Values Between Neighboring Cells

The conventional detection method is based on gray level comparison of pixels in neighboring cells.

A typical pattern of the cell area on the TFT array panel contains a spatially periodic, repetitive pattern of cells. One may assume that the gray level of a non-defective pixel must be the same as the corresponding pixel in the neighboring cells with the same cell pattern. If the gray level of a pixel is significantly different from the gray level of the reference pixels in the neighboring cells, this pixel is considered as abnormal (potential defect) (Figure 15.16).

15.2.4.3 Detection Sensitivity

As mentioned before, the detection sensitivity set by the AOI users is an important factor in AOI detection performance. The detection algorithm runs on all pixels of the image, and each time compares the current pixel with the reference pixels in the neighboring cells. If the gray level of the current pixel differs from the gray level of the corresponding pixel by more than the detection threshold, the current pixel is considered as a defect (Figure 15.17).

For example, if an AOI user sets the gray level detection threshold to "12" (the value is in gray levels):

If a user sets a higher detection threshold in the setup, the detection algorithm works in lower sensitivity and detects less defects, but at the same time is less sensitive to false "nuisance defects" (e.g., the 210→220 GL pixels, at the left side of Figure 15.17).

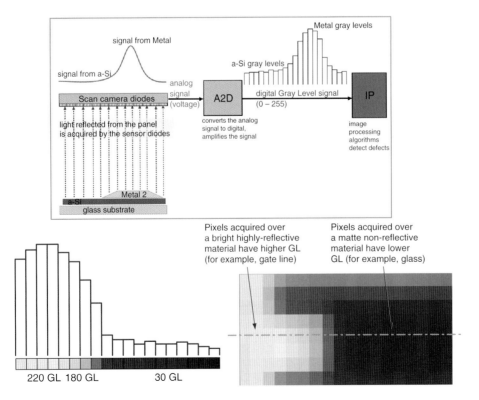

Figure 15.15 Gray level concept.

15.2 AOI (Automated Optical Inspection) | 343

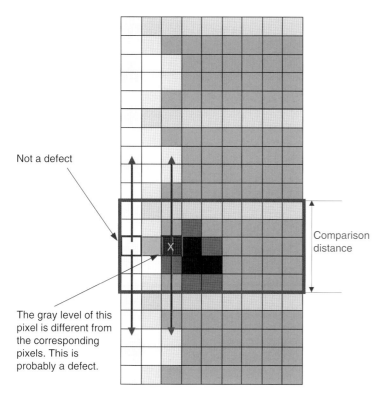

Figure 15.16 Basic detection concept: gray level comparison of pixels in neighboring cells.

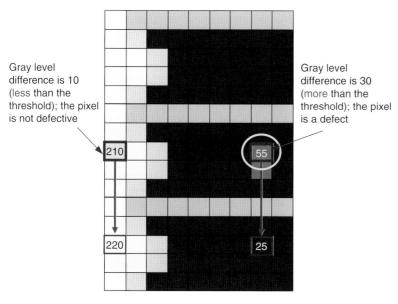

Figure 15.17 Detection threshold concept.

15.2.4.4 Detection Selectivity

Some AOI systems allow the users to set different detection sensitivity for different areas in the cell, depending on the functionality of the cell pattern and on detection needs (Figure 15.18).

The selective detection mechanism enables the system to focus on cell areas that are very sensitive to killer defects (e.g., TFT channel and data lines) and to ignore unimportant defects on cell areas that are less critical.

15.2.5 AOI Special Features

Various additional special features may allow extended AOI functionality.

15.2.5.1 Detection of Special Defect Types

In addition to conventional defect types such as shorts, cuts, pinholes, scratches and particles, some AOI systems are able to detect special defects, for example:

- Glass breakage defects
  The AOI system can detect broken glass edges that allows the FPD manufacturers to identify critical panel damage at an early stage of production and prevent serious damage to the clean room environment (Figure 15.19)
- Exposure mask defects
  The AOI system can detect an incorrect exposure mask (e.g., a shifted exposure mask or an exposure mask of an incorrect product, which creates a global change in pattern across the entire glass panel, in one of the layers) (Figure 15.20)
- Poor coating defects
  During inspection of panels after photoresist development (ADI), the AOI system can detect locations where the photoresist coating is insufficient (Figure 15.21)

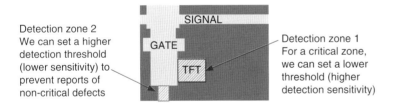

Figure 15.18 Selective detection control concept.

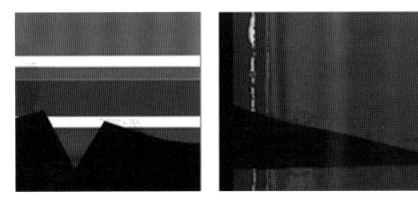

Figure 15.19 Examples of the glass breakage defects.

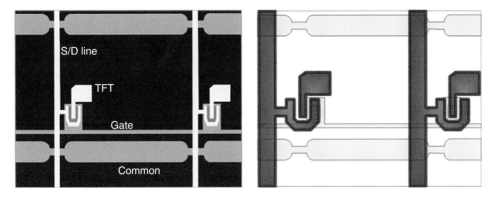

Figure 15.20 Example of the wrong exposure mask (oversized and shifted brown pattern, on right image).

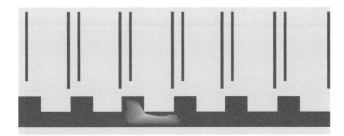

Figure 15.21 Example of the poor coating defect.

15.2.5.2 Inspection of In-Cell Touch Panels

In the standard process, touch panels and display panels are manufactured separately and later combined into one unit.

The touch panel is located above the display panel, with a protective cover layer on the top.

In contrast, when using in-cell touch technology, the touch sensors are integrated directly in the cell area on the TFT array panel of the display. As a result, the active area includes very big display pixel structures and very long comparison periods (up to 10 mm pitch). Each display pixel structure contains irregular and asymmetric patterns of both touch electrodes and display pixels. For example, one display pixel structure can have thousands of TFT pixels of several different types (with different patterns) that are arranged in multiple locations and with a different pitch (Figure 15.22).

An attempt to inspect such active areas using a conventional method of gray level comparison of neighbor cells would result in a huge number of false defect reports.

Some AOI systems provide advanced inspection solutions addressing this challenge that include:

- Automatic or semi-automatic learning of the display pixel structure and pixel types
- Image registration with sub-pixel accuracy based on automatic recognition of unique, irregular pattern features
- Advance algorithms that improved registration on irregular patterns at long periodicity
- Combining short-pitch comparisons for common periodic patterns and long-pitch comparisons for irregular patterns
- Detection of defects in pixels of all types, even on irregular patterns, and applying a different detection sensitivity in different pixel types and areas inside each pixel

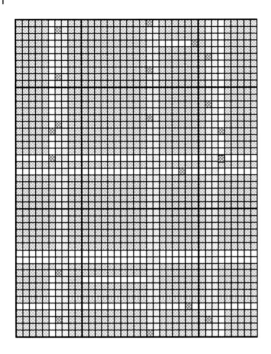

Figure 15.22 Example of the structure of one display pixel in in-cell touch technology (each small square indicates a TFT pixel, each color indicates a different pixel type).

15.2.5.3 Peripheral Area Inspection

Most discussion on the use of AOI in display manufacturing focuses on the "active area," or pixel part of the display panels. Indeed, this portion makes up for the vast majority of the glass area, and, moreover, due to its repeating nature lends itself well to the comparison-based detection approach previously described.

Yet, equally critical to yield management is the tiny narrow area that surrounds the array—perhaps 2 or 3 millimetres wide—referred to as "peripheral area" (or also "drivers area" or "fan-out area"). This is the portion of the display that contains various elements of circuitry that control the flow of data to, and the electrical activation of, the rows and columns of pixels in the display. As such, a circuit defect in this region might affect the function of a whole row or column of pixels, easily rendering an entire display "scrap," and well-worth the investment of time to detect and if possible repair it.

Optical detection of defects in the peripheral area might be more difficult compared to the pixels/array area. For one thing, in the pixels area, neighboring comparisons can be made in multiple directions, whereas in each peripheral area (rows/columns) there is only one direction for comparisons, which potentially limits the effectiveness of the detection algorithms. Some structures in the peripheral area appear less often and at greater intervals, making comparisons more difficult, and some patterns have no repetition at all (appear once per display). Obviously defects in such structures cannot be detected using neighboring comparisons, so the AOI systems use other inspection methods, such as comparison of structures across neighboring displays or comparison with a "master" reference (Figure 15.23).

15.2.5.4 Mura Defects

So far we have discussed mostly "micro defects"—which are small (from a few microns to several cells) and cause local malfunction of one or some display pixels.

Macro ("Mura," or "cloud") defects are large, very low-contrast defects, that cause non-uniformity and irregularity of a display or several displays on the panel, and in the end can cause faulty visual effects on a final display product. Mura defects are hard to define and detect because they occur on different layers and during

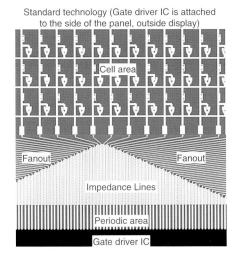 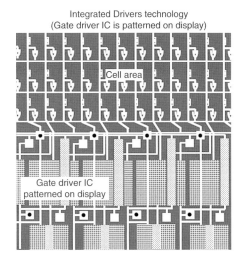

Figure 15.23 Schematic drawing of the peripheral area with fanout and peripheral area with integrated drivers.

different process steps, they can vary in color, gray level, shape, size, height, and type, appear at random locations on the panel, and can be specific to a production process or manufacturing equipment.

The most common cause of Mura defects is a physical change of the material during the production process (e.g., a minute and gradual change in the thickness of a layer, or in uniformity of the etch process).

Macro defect inspection methodologies and systems help the FPD manufacturers to identify defective panels that can be repaired or reworked, and control the problematic production equipment or process conditions that led to the Mura or macro defect.

Conventional macro defect inspection is manual and is performed on dedicated off-line review stations. Operators review each glass substrate at different angles and in different illumination conditions and look for large low-contrast non-uniform areas on the panel. Manual review is usually based on panel sampling.

Some FPD manufacturers use automatic inspection systems that are dedicated to macro inspection.

As an alternative to the dedicated macro review stations, some AOI systems provide a "digital macro" inspection functionality. The digital macro inspection is performed in parallel with the micro defect detection and is based on the same scan image of the panel that is acquired for micro defect detection. Despite the very low contrast of Mura defects, special image processing approaches are applied to selectively enhance contrast and enable the Mura to be clearly seen or detected in the digital image (Figure 15.24).

15.2.5.5 Cell Process Inspection

The AOI systems are successfully implemented in the cell process, and are used to inspect the polyimide-coated layers for such defects as (Figure 15.25):

- Polyimide edge shift (that can result in functionally important connectors being covered with polyimide)
- Pinholes in polyimide (that can cause local misalignment of LCD molecules in the final product)

AOI systems can also be used to provide the metrology data for calibration and control of the production equipment used in the cell process, for example, for calibration of the NIP rollers.

15.2.5.6 Defect Classification

Defect classification is essential to differentiate between critical and non-critical defect types, to identify defects that are repairable; and also to build meaningful defect statistics, required for process monitoring and control.

Different AOI systems offer different methods of automatic defect classification. Defect classification can be performed offline or online (during inspection), can be based on gray scale or color images, on scan images

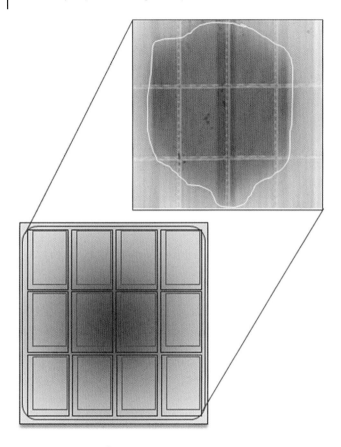

Figure 15.24 Digital macro inspection concept.

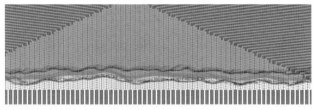 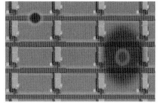

Shifted polyimide edge: polyimide covers part of the connector area   Pinholes in polyimide

Figure 15.25 Examples of the cell process defects detected by AOI.

or high-resolution review images, and can be done according to "pre-learned" examples of each defect type or according to predefined rules. The defects are automatically classified according to various defect properties extracted from the defect images: size, functional location, gray level, connectivity, topology, color variations, and so on.

Classification performance can be defined by how often the system correctly assigns a defect to its proper category, and is measured by two parameters (for each of the defect categories):

- Capture rate—the absence of false negatives (100% minus the rate of false-negative indications)
- Purity—the absence of false positives (100% minus the rate of false-positive indications)

Often, the choice of illumination is critical to classification performance. As illustrated by the example in Figure 15.26, it is easier to determine if the defect is an open circuit in the metal line or a dust particle resting on the line when dark field images are used.

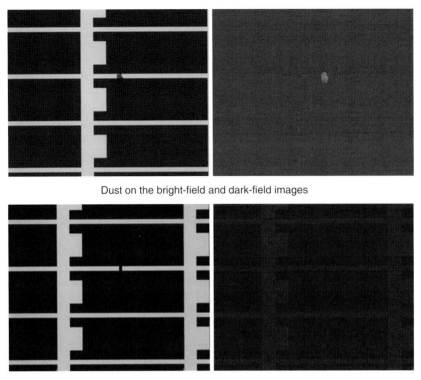

Dust on the bright-field and dark-field images

Open on the bright-field and dark-field images

Figure 15.26 Using different illumination types for defect classification.

15.2.5.7 Metrology: CD/O Measurement

"Critical dimension and overlay" measurements are a vital tool in effective monitoring of the photo-etch process. Often done using dedicated stand-alone measuring tools, very high accuracy measurements are taken at specific points across the panel, compared to specifications, and the result tracked over time. An alternative approach to dedicated systems is to use the data gathered by an AOI from the many review images that it captures anyway on every glass. Using the AOI in this way minimizes the expense, footprint, and maintenance and avoids the extra glass handling that would be needed in a dedicated system. Although each individual AOI measurement is of lower accuracy compared to a dedicated machine, statistical theory shows that the _larger quantity_ of CD/O measurements from an AOI leads to equally effective process monitoring (quantity can compensate for the lower accuracy). Moreover, performing the measurements "in situ" (e.g., at the "after develop" stage AOI located within the coater/developer) will help generate a faster alarm if the process goes wrong, shortening time-to-correction and saving a lot of valuable panels (raising yield).

Advantages of this approach include:

- Very short time needed for each measurement
- If three to four points are measured per glass, and location is changed every substrate, then after 30 or 40 panels, all key points have been measured
- CD/overlay sampling rates are greatly increased compared to the conventional approach
- This easily allows relaxing the measurement repeatability from, say, 30 nm to 100 nm (for 3 sigma), without sacrificing statistical validity of the process monitoring
- Much faster time to alarm compared to conventional technique, because the CD/O is monitored non-stop
- History-based selection of measurement sites enables fast and effective determination of out-of-spec CD/O

- Transport and handling of glass is reduced because there is no need to send to metrology department
- Fewer types of equipment are installed, maintained and operated
- With CD/O metrology on AOI, there is no need to transport glass to test department; every substrate can be measured

15.2.5.8 Automatic Judgment

Go/NoGo testing is a part of yield management that continuously tests in-process substrates or panels to determine if they should be passed on to the next process step, or scrapped/repaired. Automatic pass (Go) or fail (NoGo) judgment of defects, panels and production lots, is based on defect classification combined with additional user criteria (such as size + location of a certain defect type).

If the panel does not meet specifications, a decision to repair, rework or scrap is made. In the case of repair or rework, the panel or substrate is re-routed, a repair is attempted, and then retest is performed again. Go/NoGo testing is performed on 100% of all samples.

Some AOI systems will support user definitions of criteria for automatic judgment, allowing the machine to flag glasses or individual panels within a glass for Go/NoGo.

15.2.6 Offline Versus Inline AOI

Originally AOIs were deployed as "offline" or "stand-alone" tools, equipped with a dedicated load/unload robot to which cassettes with substrates to be inspected were routed by the fabs' host computer (Figure 15.27). The offline configuration is still used today, mostly for engineering analysis and process monitoring activities, and is characterized by:

- Single deck (Scan cameras and Microscope Review cameras)
- Multiple detection sensitivities
- Glass loading/unloading by robot, from the same side

As yield management practices evolved, the need for tighter process control and Go/NoGo inspection brought about inline configuration AOIs: inline AOI systems are more complex and more tightly integrated with production equipment, often consisting of more than one deck and having distinct and separate load/unload sides (Figure 15.28). Several panels can be located on the machine at the same time: the system scans one panel, loads the next panel, and unloads the previous panel. Most inline AOI systems support a "bypass" mode too. They are characterized by:

- Scan deck: Scan cameras for micro/macro inspection
- Review deck: Video cameras for microscopic review images and metrology
- Single detection sensitivity
- Glass loading/unloading via conveyor or robot from either side

An additional configuration is a "fast stand-alone" AOI where glass loading/unloading is performed by the same robot or conveyor, but the inspection is performed with the throughput of an inline machine. In these cases, the glasses are loaded on one side of the AOI and unloaded on the other side. A set of elevators or conveyors will return the glass to the upstream side for unloading.

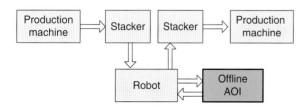

Figure 15.27 Example of offline AOI configuration.

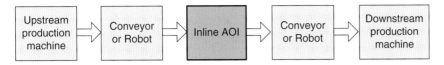

Figure 15.28 Example of inline AOI configuration.

Main considerations of the FPD manufacturer when choosing an inline or offline AOI:

- Capacity (TACT)
- Detection performance (sensitivity)
- Inspection sampling policy—criticality of 100% inspection and early alarm in a given process module
- Glass handling

TACT for the AOI will consist of scan time and overhead time (load/unload, gripping, align, etc.), and depends on:

- Glass size
- Sensitivity (detection resolution)
- Number of scan cameras (which significantly affects the cost of the AOI) and their performance

15.2.7 AOI Usage, Application and Trends

AOI usage varies significantly depending on the type of AOI employed, the type of device being inspected, the specific technology/design, and the particular manufacturer involved. The number and type of AOI systems, sampling rates, sampling locations, and so on, are all part of a particular manufacturer's yield management strategy, integrated with their specific test, repair and review practices. For offline inspection, sampling rates and inspection strategies vary. Manufacturers may choose to inspect several substrates per lot, several complete lots per run, or some combination thereof.

When ramping up a new fab or new product design, AOI equipment is used to inspect the first substrates, quickly detect problems, and feed information back to adjust the process. In engineering analysis mode, operators carefully review selected substrates at fine resolutions and capture many video images; throughput is usually low priority. For large glass sizes, inline systems have increased inspection rates, in some cases to 100% of substrates. In these cases, AOI throughput is a high priority as systems cannot operate at anything less than line speed, but resolution requirements tend not to be as tight.

As the process is debugged and a fab ramps toward steady-state mass production, AOI is used less for engineering analysis and more for process monitoring and Go/NoGo sorting.

Several AOI systems are installed in the production line, and the same layer is inspected for defects after different production processes. The AOI systems are usually installed after photoresist development (**ADI AOI**) and after panel cleaning (**ACI AOI**). As an alternative to the ACI AOI (after cleaning) installation, some FPD manufacturers install the AOI systems after etching (**AEI AOI**) or after photoresist stripping (**ASI AOI**). The AOI systems can be also used after deposition of material (**AOI AFI**—after film inspection) (Figure 15.29).

The number of machines actually used will be based on the decisions taken above as to when, and how much to inspect—multiplied by the number of overall photomask cycles present in the manufacturing process. For example, a critical layer such as gate might employ more AFI/ADI/ACI inspections, some of which are performed at 100% sampling; whereas a less critical layer will require less inspection points, and at lower sampling rates. Such choices—as well as choice of inspection resolution—vary greatly by FPD maker, manufacturing process technology, device design and maturity of the process technology.

AOI systems are also used for color filter inspection, polyimide coating inspection, and so on.

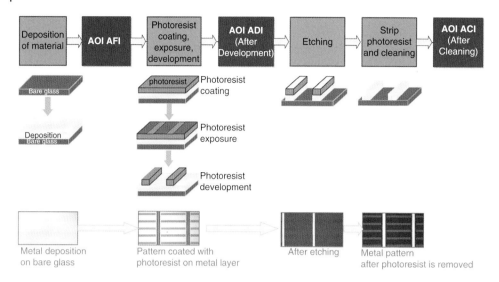

Figure 15.29 AOI systems in the production line: example of the gate layer inspection at different production stages.

As new FPD technologies such as OLED and flexible displays enter the mass production realm, they trigger the following AOI usage trends:

- Use of higher inspection resolutions (sometimes in the sub-micron range); this trend is also driven by the increasing use of very high display resolutions (e.g., 500 ppi and above) in small yet very high-resolution displays, for smartphones, smart watches, and others
- Inspection of the polyimide layer often used to implement flexible substrates
- Inspection of the "barrier," or "encapsulation" layers used to seal sensitive OLED layers from the environment
- Various OLED "Cell"/frontplane inspection points such as after OLED RGB deposition, "fine metal mask" inspection
- Increased inspection of peripheral area/integrated driver circuitry

15.3 ELECTRICAL TESTING

15.3.1 The Need

The array tester system detects electrical defects at the S/D and final array production stages, when the defects can be repaired or when the defect data can be used for process monitoring purposes. The array tester defect detection results closely match the results of the cell light-on testing, which verifies the final electrical performance of the display.

Some optical defects, including buried particles, can also be detected by the array tester system, as long as such defects affect the light-on performance of the pixel.

Array tester systems use either contact or non-contact sensors to measure the electrical characteristics of the pixel to determine whether the pixel or display is abnormal or not.

The data on the electrical characteristics of the pixel can be processed directly by the software algorithms for defect analysis, or can be output as optical images by the image processor. These images of abnormalities are further analysed and classified by the software algorithms that calculate the data reported in the final defect reports (defect coordinates, size, type, etc.).

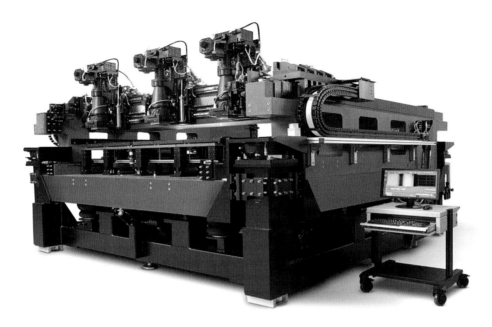

Figure 15.30 Example of the array tester system: Orbotech Array Checker™.

The inspection data (final defect reports, defect images, flaw maps, etc.) is transferred to the fab host computer for further analysis and corrective action.

Benefits of the array tester in comparison with other testing tools include:

- An array tester system detects defects at the S/D and final array process stage. This is the last gating stage (mostly electrical) that allows for repair and correction of previous process errors, before the panel is transferred to the cell process. The array test allows the manufacturers to take corrective action when repair is possible and save the display from scrap, or to receive an alarm about process problems
- An array tester can detect many critical electrical defects, which will greatly affect yield but are hard to detect using optical inspection methods. This applies also to high-resolution and ultra-high-resolution small displays
- An array tester can also serve as a pre-cell light-on tester, which can evaluate the overall panel light-on performance at the array stage. The array testing results (after repair) should closely match the final cell testing result. All additional defects detected during the final cell test can be easily classified as caused by the cell process
- An array tester can also provide an overall electrical characteristic map of the whole display or glass substrate. This can be very useful as an alarm about an abnormal process, for root cause and engineering analysis. The overall electrical characteristic map can also help to identify large-scale visual defects (Mura) that are not detectable by an optical or electrical tester (Figure 15.31)

15.3.2 Array Tester Tasks, Functions and Sequences

Figure 15.32 shows the basic tasks of a typical array tester system:

15.3.2.1 Panel Signal Driving

In order for the array tester system to perform defect detection, the first step is to drive a signal into the display to charge the pixels and to check their electrical properties.

Figure 15.31 Example of an electrical characteristic map.

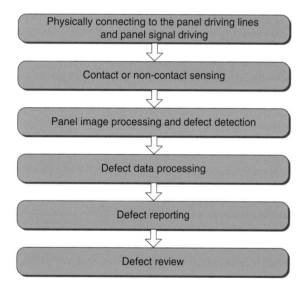

Figure 15.32 Main tasks of the array tester system.

By applying different sets of signals, array tester is able to detect different types of defects.

There are different methods of signal driving, but all of them require direct probing on the contact pads to input the signal into the display. The most common methods of signal probing are shorting bar probing and full contact probing. For small and medium displays, cell contact pads are used.

15.3.2.1.1 Shorting Bar Probing Method

Shorting bar is a method of shorting certain gate or data lines together and driving the signal through the combined contact pad. This method ensures the minimum number of pads required to input the signal into the panel. The shorting bar pads are typically large, in order to provide easy and fast probing for signal input. The FPD designer must design the panel layout according to the shorting bar design rules (Figure 15.33).

15.3.2.1.2 Full Contact Probing Method

Instead of shorting the gate or data lines together, the full contact method probes individual gate and data lines. This can be more challenging on high-resolution and ultra-high-resolution displays, and may require combining some of the lines together to reduce the number of pads.

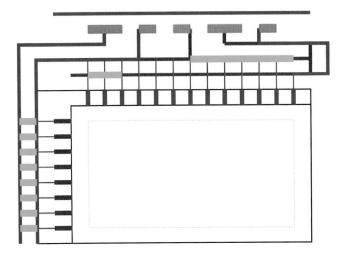

Figure 15.33 Example for shorting bar design.

15.3.2.2 Contact or Non-Contact Sensing

After a signal is input into the display, the array tester must sense and collect characteristics of the pixels that allow the tester to fully distinguish between a good pixel and a bad pixel.

This is a data acquisition step.

There are contact and non-contact sensing methods.

15.3.2.2.1 Contact Sensing

When the contact sensing method is used, direct contact is made with the sensing pads, for direct electrical signal measurement.

15.3.2.2.2 Non-Contact Sensing Methods

There are two main non-contact sensing methods:

- The electron beam sensing method measures the electrical characteristics of the pixel using secondary electrons emitted from the pixel electrode after scanning it with a primary electron beam inside a vacuum chamber
- The VIOS (Voltage image optical system) sensing method operates in normal ambient atmospheric conditions. A sensor floats closely above the panel, separated by a small air gap. The sensor and the pixel create an electrical field that can be measured and later analyzed by the image processing algorithms (Figure 15.34)

15.3.2.3 Panel Image Processing and Defect Detection

After collecting the electrical characteristics data, the array tester processes the data, including converting it into an "optical" gray-scale image. Defect detection takes place using dedicated detection algorithms. The user can set an optimal recipe to detect the desired defects by increasing the contrast between the good or bad pixels. The recipe includes such setup parameters as signal pattern type, number of signal patterns, signal pattern magnitude, pattern hold time, and so on (Figure 15.35).

15.3.2.4 Post-Defect Detection Processes

After defect detection is completed, the array tester system can further process the defect data for classification, correlation and analysis, and then output the final defect report file with defect parameters (defect type, location, size, etc.). The operator can also use the built-in review camera to review and verify the defects of interest.

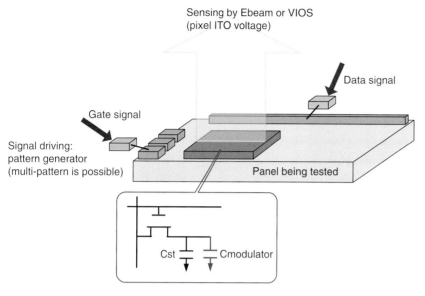

Figure 15.34 Non-contact sensing concept.

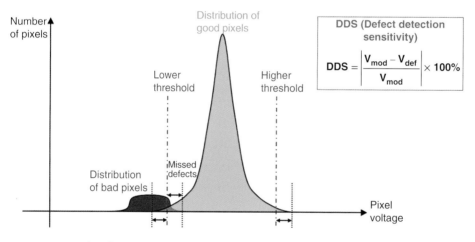

Figure 15.35 Defect detection sensitivity concept.

15.3.3 Array Tester System Design Concept

The design of an array tester system is based on the following FPD production process parameters:

- Size of the glass substrates to be inspected
- Panel probing design: panel probing methods (shorting bar, full contact, etc.)
- Panel resolution: pixel size parameters (FHD – 4k2k – 8k4k, sub 20 um pixel testing)
- Panel channel driving requirements: the number of channels that are required for panel signal driving
- Panel processing technology: a-Si, IGZO, LTPS, AMOLED, Flex

New display technologies introduce tighter design rules, finer features, and denser patterns, with requirements for very high testing resolutions and very accurate probing. More complex driving patterns also require a higher number of signal driving channels.

An array tester system must:

- Adapt to the new design rules and be able to probe small contact pads for signal driving
- Perform defect detection at high resolution (>550 ppi) on small and medium displays (watches, mobile devices, tablets) and also on large IT and TV displays
- Inspect in short TACT regardless of panel and inspection resolutions for high-volume production
- Provide a sufficient number of pattern generation channels
- Test all types of displays regardless of panel processing technology (a-Si, Oxide, LTPS, AMOLED, Flex)

15.3.3.1 Signal Driving Probing

The main considerations for signal driving probing are a fast panel model switching time, low running cost for probing, and the accuracy of the probing. The latest features of shorting bar probing include fully automated probing changers, probing for high speed and low running cost, and high probing precision (Figure 15.36). Full-contact probing systems require a special probing device for each different panel model, and thus a longer probing set up time and more time to switch between panel models.

15.3.3.2 Ultra-High-Resolution Testing

As new display technologies call for tighter design rules and finer features, mobile display has already stepped into the sub 20 um production phase (mainly for LTPS panels).

Capability of testing sub 20 um pixel testing has become a requirement for array testers. There are also requirements for upgrading the previous models of array tester for testing display pixels in the 20–32 um range (mainly for a-Si panels).

Ultra-high-resolution testing requirements include:

- High sensing sensitivity and resolution
- Software algorithms that increase the signal-to-noise ratio and reduce the rate of false defects
- Signal pattern driving flexibility
- Application signal driving setup capability
- Ability to design the ultra-high-resolution probing layout without glass utilization reduction
- High testing speed regardless of the high number of display pixels

Although it is clear that the resolution requirements will reach >700 ppi, the highest resolution being tested at the time of writing is 592 ppi for LTPS products (Figure 15.37).

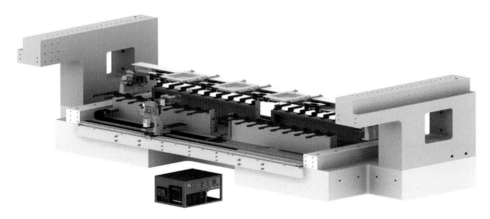

Figure 15.36 Automated probing changers in Orbotech Array Checker™.

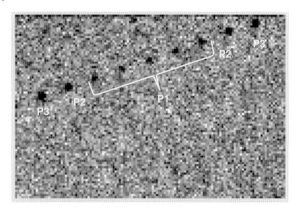

Figure 15.37 Example of the ultra-high-resolution LTPS testing.

15.3.3.3 System TACT

System capacity is one of the major considerations for array tester systems. FPD manufacturers will require 100% panel testing capacity for full yield improvement (repair/rework). Slower array tester systems can only do sampling testing and have limited usage for process monitoring.

FPD manufacturers will require an array tester with constant TACT/capacity regardless of the display resolution and panel design.

Depending on the panel design, testing conditions and glass size, a typical average TACT for a shorting bar array tester system is in the range of 100–350 seconds, while a typical average TACT for a full contact array tester system is usually in the range of 7–15 minutes.

15.3.3.4 "High-Channel" Testing

Due to the complexity of the pixel circuit design (OLED, GOA/ASG, etc.), an array tester system must be able to test a large number of different channels in order to fully implement the pixel light-on testing and detection. Following recent panel design developments, the number of channels to be tested has increased from 6 CH to 24 CH with a growing need for more than 24 CH (Figure 15.38). For small and medium displays, the typical number of channels to be tested is 48 CH.

15.3.3.5 Advanced Process Technology Testing (AMOLED, FLEX OLED)

As new display processing technologies are developed for better display performance, the array tester system is also required to adapt to new technology concepts. Modern FPD manufacturing lines will require their array tester to test such displays as AMOLED backplane and Flex OLED on rigid glass.

The main considerations for testing of AMOLED backplane are:

- OLED pixel circuitry is a current driver, therefore, array testing technologies based on current testing, and not voltage testing, must be used
- Due to multiple transistors and capacitors in AMOLED design, direct contact and sensing methods may be challenging
- In some AMOLED pixel design, some of the non-contact sensing testers will not have sufficient electron data for sensing
- Multiple patterns and runs for AMOLED backplane testing
- Electro Mura testing capability

Flex OLED on rigid glass testing is typically regarded as similar to AMOLED testing. However, other special conditions need to be considered for Flex OLED testing:

- PI layer probing issue
- Potential bump or dimple under the PI layer
- ESD issue for special PI layer characteristics

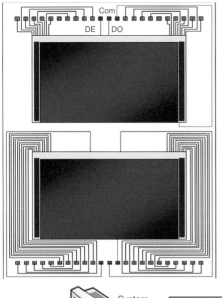

Pin No.	Symbol	Description
1	Rest	Gate driver reset pulse input
2	CLK1	Gate driver clock input
3	CLK2	Gate driver clock input
4	CLK3	Gate driver clock input
5	CLK4	Gate driver clock input
6	Vdd	Gate driver power supply voltage
7	Vdd1	Gate driver power supply voltage
8	Vdd2	Gate driver power supply voltage
9	Vgl	Gate driver low voltage
10	Vst	Gate driver start pulse input
11	DE	Data even signal input
12	Com	Pixel common signal input
13	DO	Data odd signal input

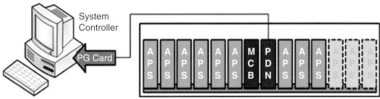

Figure 15.38 High-channel tester and pattern generator of GOA panels.

15.3.4 Array Tester Special Features

Various additional special capabilities allow extended array tester functionality.

15.3.4.1 GOA, ASG, and IGD Testing

GOA (Gate on Array), ASG (Amorphous Silicon TFT Gate Drive Circuit) and IGD (Integrated Gate Driver) have become a common panel design practice in recent years. On such panels, the array tester system is required to test these peripheral areas for open and short circuits (Figure 15.39).

There are two methods of peripheral area testing: indirect active area testing and direct GOA/ASG area testing.

- The direct GOA/ASG Area testing method is based on direct testing of the GOA/area circuits with e-beam, voltage image, or full contact sensing (Figure 15.40)
- The indirect active area testing method is based on testing of the active area instead of the GOA peripheral area. If there are defective GOA/ASG circuits in the peripheral area, these defects will prevent the gate driver from turning on the transistors of the pixel row. As a result, by detection of a gate line defect along the entire pixel row in the active area, the array tester system can alarm on a defect in the GOA/ASG circuit in the peripheral area (Figure 15.41)

15.3.4.2 Electro Mura Monitoring

As described in the previous AOI section, Mura defects are large, very low-contrast defects that cause non-uniformity and irregularity of a display or several displays on the glass. With the capability of the electrical characteristic (voltage) measurement of the whole display or glass substrate, the array tester system is able to create a voltage distribution map of the whole display or glass. The voltage distribution map can be then visually inspected to detect electro Mura defects shown as the electrical characteristic variation on the map. Mura

360 *15 TFT Array: Inspection, Testing, and Repair*

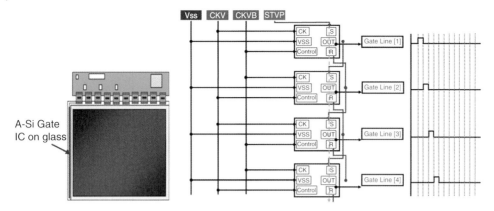

Figure 15.39 Example of the GOA design and pattern signal data.

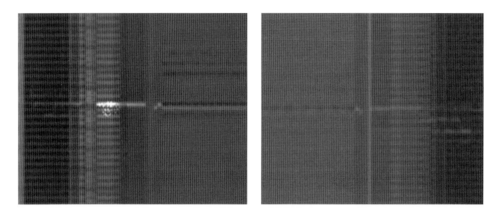

Figure 15.40 Direct testing of the GOA area.

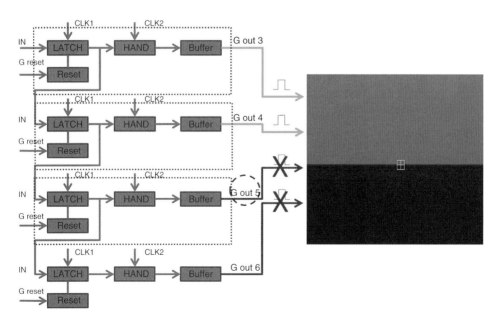

Figure 15.41 GOA testing by sensing on the active area (the pixel of the active will not turn on because of the defect in the peripheral area).

defects have become more critical, especially on the OLED display panels where the uniformity requirements are higher (Figure 15.42).

15.3.4.3 Free-Form Panel Testing

As modern display applications are expanded, the array tester system is required to test free-form displays such as round displays for watch and automotive applications. Free-form testing typically requires new software algorithm developments rather than hardware changes. However, the panel resolution and probing pad design must be considered for defect detection performance (Figure 15.43).

15.3.5 Array Tester Usage, Application and Trends

Array tester systems have been changing and developing to match the new types of the devices, new technologies, and panel designs. In a traditional production line of LCD IT and TV panels, the array tester is typically used in the final array process, before the final repair step. The array tester system must detect the critical defects for repair, and thus improve the yield.

For small and medium panels, especially on ultra-high-resolution displays, the array tester system is typically used for process monitoring and sampling as a GO/NoGo system, mainly due to TACT considerations and non-repairable characteristics of the defects on such displays. However, some advanced array tester systems provide high capacity, 100% testing capabilities for modern small and medium ultra-high-resolution displays. Combined with modern repair system, array testers can play a critical role in 100% testing for rework of advanced small and medium panels.

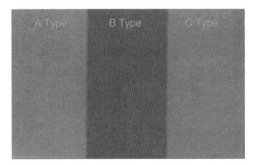

Figure 15.42 Example of an electro Mura defect due to process variations.

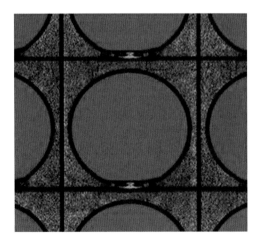

Figure 15.43 Example of free-form display testing.

There are a few major trends in array testing requirements:

- Testing the panels at the early production stages, before the final array process, to further improve the rework/repair capabilities and yields
- Providing array testing probing design that can maximize the glass utilization, especially on ultra-high-resolution small displays
- Testing panels manufactured using newly developed touch technologies
- Detecting Mura defects

The new display trends and applications require expanding of array tester usage and testing capabilities:

- Source/drain layer testing for LTPS LCD/OLED display
- Advanced FPC probing method that maximizes the glass utilization rate and uniform driving
- In-cell-touch panel testing

15.3.5.1 Source/Drain Layer Testing for LTPS LCD/OLED

Using an array tester to test panels before the final array production steps allow more repair options and thus improved yields. Certain array tester systems can now test the source/drain M2 layer for point, line, and GOA defects. This capability has been developed for LTPS LCD and OLED devices that have the highest material cost and complex processes. On modern ultra-high-resolution displays where utilization of a traditional open/short tester is difficult, array testing of source/drain layers will provide the best testing solution for FPD manufacturers (Figure 15.44).

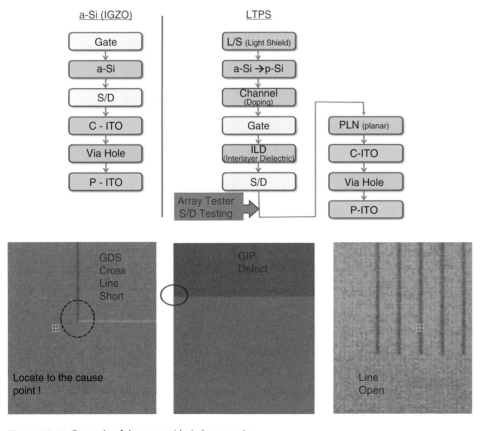

Figure 15.44 Example of the source/drain layer testing.

15.3.5.2 New Probing Concept

As the display design becomes more complicated, previous signal driving concepts (shorting bar and full contact) show their limitations. There is a need for a new probing method that will allow a panel designer to design the probing pads on an ultra-high-resolution panel, so as to obtain the highest glass utilization rate while providing sufficient signal driving performance. In addition, any new probing concept should also optimize the signal driving uniformity.

15.3.5.3 In-Cell Touch Panel Testing

In-cell touch technology is a newly developed touch panel technology implemented in the latest mobile and tablet displays. There is a need to develop a testing method for in-cell touch displays, where both display pattern and touch pattern testing is achieved simultaneously.

15.4 DEFECT REPAIR

15.4.1 The Need

In FPD manufacturing, defects caused by the production process, equipment, or particles can lead to malfunctioning of the TFT circuits and yield loss.

The main purpose of the repair system is to repair the production defects in order to recover the electrical functionality of the TFT array. The repair can be done by:

- Removal of excess, unwanted material (laser ablation)
- Deposition of the missing material or corrective patterning, based on such technologies as LCVD (Laser Chemical Vapor Deposition), metal inkjet printing, LIFT (Laser-Induced Forward Transfer), and so on

Basically, a repair system can be used at all production process stages if the defect data is provided. The defect data input to the repair process is usually based on AOI and electrical testing results.

Figure 15.45 Example of a repair system: Orbotech Process Saver™.

In addition to its main purpose of defect repair and recovery of the electrical functionality of the TFT array, the repair system also provides enhanced defect review and verification options.

The repair process is the final step following inspection and testing, and the final panel judgment stage of the whole TFT array process. Therefore, advanced, high-resolution review capabilities are a critical element for a repair system as the key to analysis of the root cause of the defects and for process monitoring. In addition, this is a critical step in creating detailed, high quality inspection data output to the customer fab (defect images before and after repair, defect files with defect type and judgment, etc.).

15.4.2 Repair System in the Production Process

As mentioned previously, the repair system can be used at any stage of the production process (Figure 15.46).

Depending on its place in the production line, the repair system can be used for in-process repair or final repair.

15.4.2.1 In-Process Repair

In-process repair is the repair of defects at each production stage (repair of gate layers, repair of source/drain layers, etc.), such as:

- Photoresist in-process repair—repair of photolithography defects on panels after photoresist development
- Metal in-process repair—repair of defects on metal layers on panels after etching, that includes metal cutting repair of shorts and residues, or metal connecting repair of opens

In-process repair systems are usually used for repair of line defects and residue defects.

15.4.2.2 Final Repair

Final repair is the repair of defects at the final step or at the end of the TFT array production process. Final repair systems are used for repair of all types of electrical defects that can cause malfunction of the thin-film transistors and pixels.

15.4.3 Repair Sequence

At the beginning of the repair sequence, a glass substrate is loaded onto the machine. At the same time, the relevant files are loaded: map data with predefined glass parameters (glass size, thickness, etc.), recipe with predefined repair parameters (laser intensity, wavelength, etc.), defect data from AOI and/or electrical tester (defect code, coordinates, type, etc.).

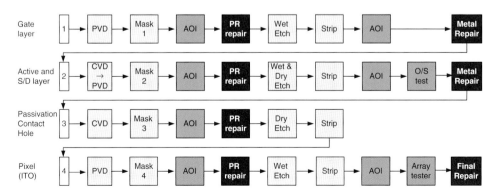

Figure 15.46 Repair system in the TFT array production line (example of a four-mask process).

Following the loading steps, there is an iterative process of moving to the coordinates of the next defect, grabbing the defect image, reviewing and judging the defect, marking the defect area, performing the repair, judging the results and then continuing to the next defect, until the review and repair of all defects on the currently loaded glass is completed.

The current glass substrate is then unloaded, the data of the current glass is sent to the fab host, and the next glass substrate is loaded (Figure 15.47).

15.4.4 Short-Circuit Repair Method

The short-circuit method is used to repair excess material defects (shorts, residues of material). Laser ablation is used to remove the excess, unwanted material.

15.4.4.1 Laser Ablation Concept

Laser ablation is the process of removing material (excess material defect) from the panel surface by irradiating the defect location with a laser beam.

Several methods of laser ablation can be used for defect repair on flat panel displays:

- Thermal ablation
- Cold ablation
- Photochemical ablation

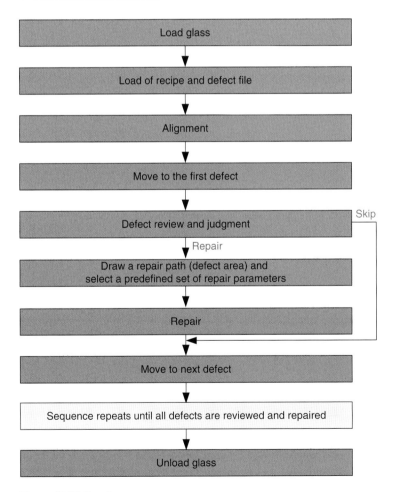

Figure 15.47 Repair sequence.

15.4.4.1.1 Thermal Ablation

During thermal ablation, the material is removed as a result of thermal impact of the laser energy on the ablated material. A pulsed laser beam is applied to a very small area of the panel surface. The laser energy heats the exposed area, the material in the heated area melts and evaporates.

Most manufacturers use lasers with a pulse width of at least 1 nanosecond.

Although the thermal ablation method is widely used, it has some disadvantages:

- The thermal impact of the laser energy can cause local damage to the areas surrounding the defect (panel pattern and previous layers under the defect)
- Debris (particles of melted material) can be generated during ablation

15.4.4.1.2 Cold Ablation

Cold ablation uses ultra-short laser pulses.

A very short laser pulse cause electron emission and ionization. This leads to removal of material so quickly that the surrounding areas absorb very little heat, and the panel pattern and previous layers under the defect are not damaged. Cold ablation is also discussed later in this chapter.

15.4.4.1.3 Photochemical Ablation

During photochemical ablation, laser energy breaks the chemical bonds in the molecules of the exposed material and causes decomposition of the material.

While lasers operating at visible or infrared wavelengths are used for thermal ablation, photochemical ablation method is tailored for a specific wavelength and specific materials: deep UV lasers (wavelengths of approximately 200 nm) are used for ablation of organic materials.

Although during photochemical ablation the laser energy also heats the sample, the thermal impact is not dominant. A significantly less amount of debris can be generated during photochemical ablation versus thermal ablation, and the risk of damage to the defect surroundings is minimal.

15.4.4.2 Laser Light Wavelengths and their Typical Applications

15.4.4.2.1 Laser Matter Interaction

To determine which laser light wavelength is optimal for repair of a given defect on a given material and production layer, we must understand how each material "behaves" when exposed to the laser light (Figure 15.48).

The ratio of laser light absorption, reflectance, and transmission depends on the optical properties of a specific material, the layer thickness, and on light wavelength.

- Transparent materials such as ITO and polyimides have a very high transmission level and can absorb only a small amount of laser energy. Such materials have a very low reflectance level; for example, ITO reflectance is much lower than reflectance of most metals, but higher than reflectance of most transparent materials used as passivation layers

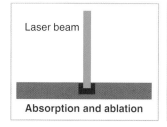
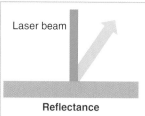
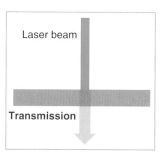

Figure 15.48 Laser matter interaction: material optical properties.

- Most metal layers used in the FPD manufacturing are relatively thick layers. Thick metal layers have a high reflectance level and a sufficient absorption level. Such materials do not transmit laser light

Laser ablation results are greatly affected by the ability of an **ablated material** to absorb laser energy and, at the same time, the ability of the **surrounding materials and previous layers** to reflect or transmit laser energy.

15.4.4.2.2 Using DUV Laser Light (266 nm) for Short-Circuit Defect Repair

As mentioned earlier in this chapter, the DUV wavelength is optimal for ablation of **organic materials**, such as polyimide, photoresist, and OLED. The advantages of the DUV photochemical ablation on organic materials are:

- Good absorption of DUV laser energy by organic materials (the DUV laser energy easily breaks the chemical bonds in the organic material molecules)
- Minimum amount of ablation debris

15.4.4.2.3 Using Infrared Laser Light (1,064 nm) for Short-Circuit Defect Repair

The infrared wavelength is considered as optimal for ablation of **ITO with passivation layer underneath**: ITO absorbs the infrared laser light well, on the other hand, the organic materials used as passivation layers are transparent to Infrared light (transmit the infrared light and do not absorb it).

15.4.4.3.4 Using Green Laser Light (532 nm) for Short-Circuit Defect Repair

- The green wavelength is considered as optimal for ablation of **metal with passivation layer underneath**: metal absorbs the green laser light well, on the other hand, the passivation material is not damaged
- **Polysilicon (p-Si) materials** absorb DUV light better than green light. However, if the DUV light is used, the passivation layers under the p-Si material may get damaged. Thus usage of green light is preferable for ablation of p-Si layers (Figure 15.49)

To summarize, when deciding which laser light wavelength is optimal for repair, we must take into account the following factors and parameters:

- Optical properties of the material that must be ablated
- Structure of the production layer (materials of previous layers that are under the defect)
- Pattern geometry (defect location on the panel pattern) (Figure 15.50)

15.4.4.3 Typical Applications of the Short-Circuit Repair Method

15.4.4.3.1 Cutting

Laser ablation is used to cut a line in the photoresist or metal to disconnect the excess material from the pattern, in order to repair a short circuit (Figure 15.51).

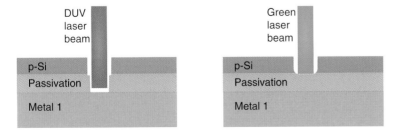

Figure 15.49 Example of the p-Si Ablation by DUV Laser and Green Laser (Cross-Section of the Production Layer).

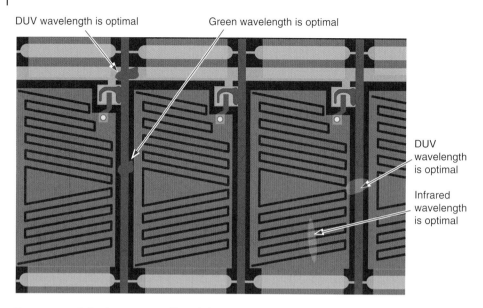

Figure 15.50 Selecting an optimal laser light wavelength for repair (application example).

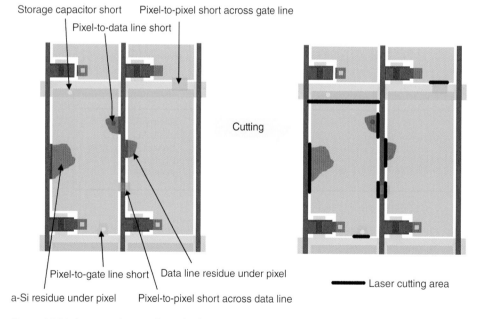

Figure 15.51 Laser cutting repair method.

15.4.4.3.2 Welding

In case of a missing contact hole or certain types of open defects, the laser can be used to create small welding (inter-layer contact) repairs (Figure 15.52).

An additional application can be to simply remove large particles that might be embedded in potentially sensitive areas of the design.

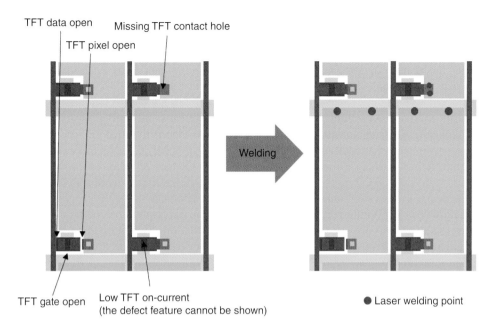

Figure 15.52 Laser welding repair method.

15.4.5 Open-Circuit Repair Method

15.4.5.1 LCVD (Laser Chemical Vapor Deposition)

LCVD is a traditional method for producing thin deposition layers. LCVD technology is widely used for repair of the pattern of flat panel display panels and for addition of missing metal wiring.

Organometallic gases that usually contain tungsten (W) or molybdenum (Mo) are used for deposition.

During the laser chemical vapor deposition process, a focused laser beam is used to heat the substrate at the location of a defect. This causes a pyrolytic chemical reaction (chemical decomposition of organometallic gas), which leads to deposition of the metal onto the substrate (Figure 15.53).

The main advantages of the LCVD method are:

- High resolution
- Uniformity—ability to produce a thin uniform layer of the deposited material

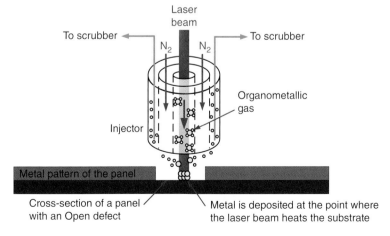

Figure 15.53 LCVD deposition—basic principle.

However, the LCVD method has some critical disadvantages:

- High resistivity of metals used for deposition (this is a major reason why the LCVD method is nor optimal for repair of OLED display panels since high current cannot be used)
- Very high local temperatures are applied during the process—this thermal effect might cause local damage to the panel pattern and previous layers in the vicinity of the defect (this is a major reason why the LCVD method is not optimal for repair of Flex display panels)

The high resistivity of metals that are traditionally used for chemical vapor deposition has led some manufacturers to try to use metals with better resistivity properties, for example, to use cobalt precursors (instead of tungsten and molybdenum) or to use organic liquids that contain metals (instead of organometallic gases).

The disadvantages of the traditional LCVD method and the growing needs for higher conductivity, faster process, and minimum effect on the areas surrounding the defect have led advanced manufacturers of repair systems to develop new metal ink deposition technologies.

15.4.5.2 Metal Ink Deposition Repair

The following deposition methods are used for repair of the Open Circuit defects:

- Dispensing
- Inkjet and super inkjet (used for high-resolution flat panel displays)
- Laser transfer

Metal nanoparticle inks or inks of metal complexes are used as deposition materials.

15.4.5.2.1 Dispensing

Dispensing is a contact transfer method somewhat similar to using a micro-pipette; the material to be deposited is carried by a probe that is brought into contact with the desired location, leading to the transfer of a small amount of material. The method is simple and robust, but suffers from limited accuracy and flexibility.

15.4.5.2.2 Metal Inkjet Deposition

Metal inkjet deposition is performed at very high speeds. However, this may render the inkjet deposition method as less effective for repair of very small defects on fine patterns, where only a few drops of ink are required (Figure 15.54).

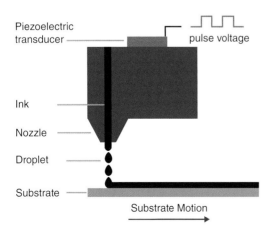

Figure 15.54 Metal inkjet deposition—basic principle.

15.4.5.2.3 LIFT (Laser-Induced Forward Transfer) Deposition

The process has two main steps (Figures 15.55 and 15.57):

(1) Deposition of metal nanoparticle inks
(2) Sintering in order for the deposited material to attain optimal conduction.

Silver nanoparticle ink is considered as optimal for deposition on flat panel display layers, because of its high viscosity, low resistivity, high conductivity after sintering and good adhesion.

The main advantages of the laser transfer deposition method are:

- High resolution
- High viscosity and good wetting properties of the metal inks used as a deposition material (this allows for good step coverage and high printing accuracy, as shown in Figure 15.56)
- Although high local temperature is applied to the defect location during sintering, the thermal impact on the defect surrounding is significantly smaller in comparison with the LCVD process where very high local temperatures are applied
- Simpler handling and lower costs

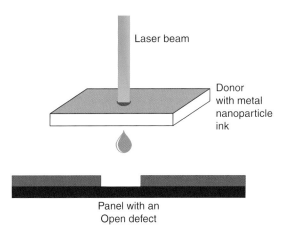

Figure 15.55 LIFT deposition—basic principle.

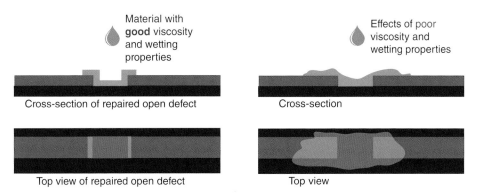

Figure 15.56 Illustration of the high viscosity properties of metal nanoparticle inks used for LIFT deposition.

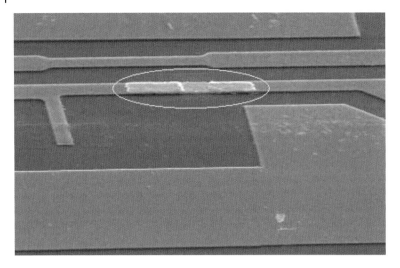

Figure 15.57 Example of the LIFT deposition results.

15.4.5.3 Main Applications of the Deposition Repair (Open-Circuit Repair)

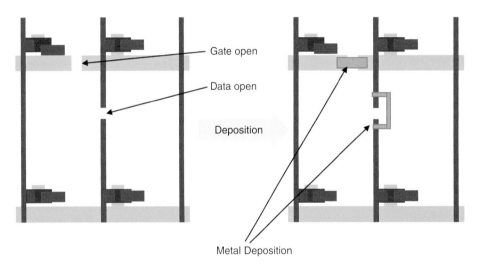

Figure 15.58 Deposition repair applications.

15.4.6 Photoresist (PR) Repair

Several FPD manufacturers have tried to implement repair systems at the photoresist stages of the TFT LCD production, because of significant advantages of photoresist repair in comparison with metal repair. These attempts have not always been successful due to the complex requirements of the application and equipment challenges.

However, as LTPS, OLED, and Flex display technologies become dominant on the FPD market, the efficiency of photoresist repair has regained attention of FPD manufacturers.

Benefits of the photoresist repair:

- Recovery of the initial mask (photoresist) pattern
- Damage-free repair

- 100% selective repair
- No debris issue
- Full automatic repair
- Cost saving for mask rework

15.4.6.1 Main Applications of the Photoresist Repair

Recovery of the missing pattern on the layers after photoresist development is one of the typical applications of the photoresist repair (Figure 15.59).

15.4.6.2 Photoresist Repair Technology

As mentioned earlier, photochemical ablation with the **deep UV laser light** is considered optimal for defect repair and re-patterning on photoresist layers.

In photoresist repair, the purpose of the application is to generate a pattern of laser illumination that matches the desired pattern of the photoresist, in order to remove excess photoresist and re-create the desired photomasking pattern. Re-patterning is usually done using a programmable **DMD** or **FSM** device.

15.4.6.2.1 Using DMD for Patterning

DMD (Digital Micro-mirror Device) is an opto-electro-mechanical device that is used for projection and patterning. A DMD has millions of very small mirrors, arranged in a matrix of rows and columns. Each DMD micro-mirror can move to two different positions (mirror states). DMD is actually a light modulator: all DMD micro-mirrors receive laser light, but because each micro-mirror moves to the ON or OFF position, this creates a pattern of "light-ON" (exposed) and "light-OFF" (unexposed) pixels:

- If a DMD micro-mirror moves to the OFF position (tilted away from the panel): the laser light is deflected to the "beam dump" (result: unexposed pixel on the sample)
- If a DMD micro-mirror moves to the ON position (tilted toward the panel): the laser light is directed to the sample (result: exposed pixel on the sample) (Figures 15.60 and 15.61)

15.4.6.2.2 Using FSM for Patterning

FSM (Fast Steering Mirror) is an advanced device used for laser beam steering and positioning: a dedicated motor moves the two-axis Galvo mirror at different angles, to set the laser beam spot position on the sample, at very high speed and with very high accuracy (Figure 15.62).

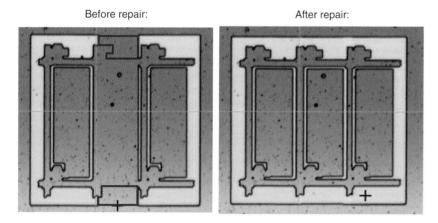

Figure 15.59 Repair of the photoresist layer.

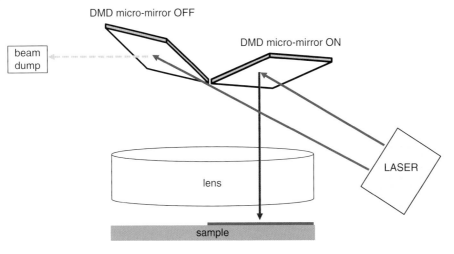

Figure 15.60 DMD principle of operation on the example of two micro-mirrors.

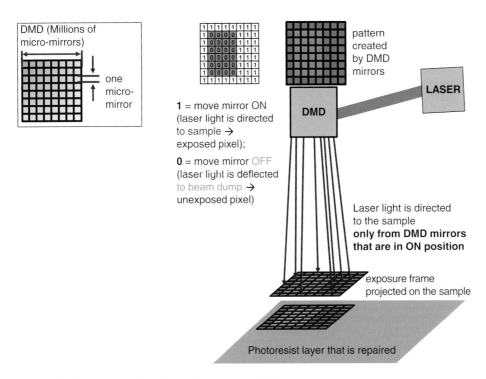

Figure 15.61 Patterning of photoresist layers using DMD.

Advantages of FSM:

- Wide working range and wide angular coverage
- High ablation accuracy, even in high velocity
- Multiple repetition of small laser shots, at high speed

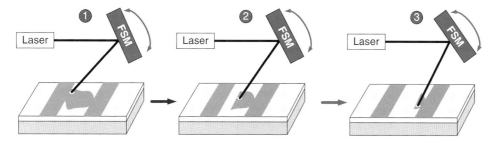

Figure 15.62 FSM patterning principle.

- High selectivity and minimum risk of damage to the panel pattern and to the previous layers
- Uniformity of laser energy over the defect area
- Repair and re-patterning of any shape is possible
- Stability
- FSM is a digital device that enables accurate synchronization with the laser

An FSM provides much higher resolution than a DMD (where the pattern is "pixelized" based on the binary grid of DMD micro-mirrors and the resolution is limited to the size of one DMD micro-mirror). In addition, with DMD, the whole optical head or the stage with the glass substrate must move to the next patterning location on the same defect. On the other hand, with FSM, the optical head and the stage positions are fixed during the repair process, and only the fast steering mirror moves to set the next beam position over the defect. This enables very high ablation speed.

In comparison with FSM devices that have two steering mirrors (a mirror per motion axis), the two-axis FSM mirror (one mirror with two motion axes) is compact, and provides very high velocity and accuracy (because synchronization between the two mirrors and two axes is unnecessary).

FSM is successfully used in defect repair on various materials and production layers.

15.4.7 Special Features of the Repair System

15.4.7.1 Line Defect Locator (LDL)

Reports of defects that require repair are often received from the electrical testers (such as open/short testers). Sometimes the open/short tester can electrically detect that a short or an open circuit exists on a line, but cannot determine—based on the electrical test—where precisely a small defect that causes a short or open circuit is located. The LDL feature is intended to optically scan the panel along such lines where the electrical tester reported a line defect, and to search for the precise location of the short or open (Figure 15.63).

The LDL mechanism is applicable to both active area and integrated gate driver (pad) area.

15.4.7.2 Parallel Repair Mode for Maximum System Throughput

In traditional repair systems, the repair flow is sequential: the system grabs a defect image, an operator reviews and judges the defect and marks the repair path, and then the system repairs the defect.

Advanced repair systems implement various methods of parallel repair that maximize system utilization and allow optimal throughput. While the system automatically repairs the previous defect that has already been reviewed by an operator, the operator can already review and judge the next defect. At the same time, the system grabs and saves the images of next defects, using the fast image grabbing mode with real-time tracking autofocus and advanced motion control.

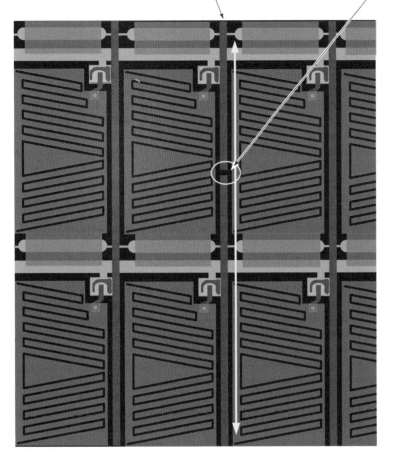

Figure 15.63 LDL principle.

15.4.8 Repair Technology Trends

As modern display technologies introduce tighter design rules and stricter tolerances, finer features and denser, more complex patterns, the issues of high laser performance, advanced ablation and deposition repair technologies become even more critical.

- Fine cutting capability (sub-micron resolution)
- Prevention of under-layer damage
- Selective metal cutting

15.4.8.1 Cold Ablation

All laser ablation techniques involve the precise application of laser energy in a small location, resulting in evaporation/ablation of the material that needs to be removed.

However, the precision of this technique is limited by the "surrounding damage," which often occurs as a result of the spreading of the heat and/or mechanical shock waves that are formed. With the advance of

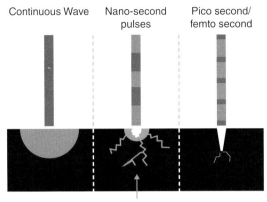

Figure 15.64 Cold ablation.

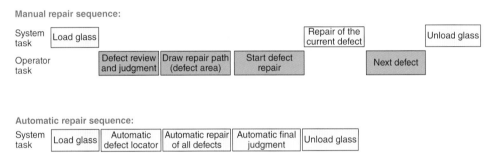

Figure 15.65 Automatic repair sequence versus manual repair sequence.

ultra-short pulse width lasers, the use of nanosecond pulses—and more recently Pico or Femto second pulses, allows extreme precision and avoidance of surrounding damage (Figure 15.64).

15.4.8.2 Full Automatic Repair Solution

Most of the FPD production and testing machines operate automatically, however, most repair systems must be still operated manually. Some advanced manufacturers of repair systems have introduced various automatic mechanisms such as an automatic defect locator, and have developed solutions for fully automatic repair sequence (Figure 15.65).

15.4.9 Summary

In the mid-2000s, as the flat panel display industry grew and undertook the task of producing larger and larger screens—initially monitors, and later TVs—the role of inspection, testing and repair became critical.

As was established in the semiconductor industry 40 years ago, the ratio of "good" versus "bad" (defective) dies—also known as the line yield—is strongly dependent on the size of the die. In semiconductors, it was a question of how many millimeters large the chip was, whereas in display the difference is from a few inches diagonal to tens of inches. However, the mathematics are the same: from a purely statistical perspective, the larger the die or display, the higher the chance it will contain one or more "killer defects." These "killer defects" must be repaired, otherwise the large and expensive display in which they are located will be scrapped. Thus, as LCD TV fabs were built, the repair tool became the single most common tool on the production floor, backed up by numerous AOI and test tools that provided it with the data on defects to be repaired. This investment in inspection, testing and repair allowed the LCD TV manufacturers to increase yields of large panels from 60%–70% to well over 95% and has been a decisive factor in the success of the flat panel display industry.

16

LCM Inspection and Repair

Chun Chang Hung

AU Optronics, Taoyuan City, Taiwan, ROC, 32543

16.1 INTRODUCTION

After finishing the module assembly (MA) process, a detailed test is necessary before packing and shipping. The test can effectively screen out nonqualified panels by simulating customer's requirements and testing conditions in order to avoid defective modules being shipped out.

Panel defects can be divided into two categories: cosmetic defects and functional defects. Cosmetic defects include for example trapped particles or air bubbles in the panel stack (Figure 16.1), scratches, and so on, and they generally can be found directly by visual inspection. Although cosmetic defects won't affect display functions, they may induce uncomfortable feelings to the client and have to be screened out.

Functional defects occur during display operation typically manifesting as flaws of the image, including for example bright spots, defective lines, and so on, as shown in Figure 16.2. It is more complicated to find out functional defects. Panels have to be powered on and, in order to simulate all the situations when display is in use, specialized equipment with capabilities of generating different display patterns are utilized. Furthermore, since some of the defects may not be easily seen by the user, overkill should be avoided from cost point of view. Achieving a balance between overkill and leakage of defect containing product is one of the arts in manufacturing.

In this chapter, some commonly used test methods for detecting panel defects in mobile display and in TV LCMs, and the matters needing attention will be introduced. A brief introduction of new trends in inspection such as AOI (automatic optical inspection) is also included. At the end of the chapter, some commonly used repair methods for defective panels are introduced.

16.2 FUNCTIONAL DEFECTS INSPECTION

As mentioned in the introduction, specialized testing fixtures are necessary for connecting to the display panel and enabling functional defects detection. Figure 16.3 shows a general structure and composition of the test system. It includes three major submodules: host, signal module, and power supply.

Flat Panel Display Manufacturing, First Edition. Edited by Jun Souk, Shinji Morozumi, Fang-Chen Luo, and Ion Bita.
© 2018 John Wiley & Sons Ltd. Published 2018 by John Wiley & Sons Ltd.

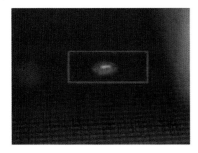

Figure 16.1 Cosmetic defect: bubble on polarizer.

Figure 16.2 Functional defect: bright point in dark image.

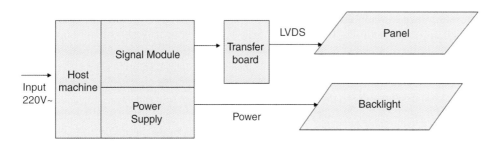

Figure 16.3 Test system configuration.

Key functions for the HOST are:

(a) To edit display patterns for test.
(b) Control power supply for the panel under test.
(c) To input panel parameters like timing, resolution, frequency, and so on.
(d) Output control signals to panel drivers.
(e) Edit and control the sequence of output of data, power, and so on.

The signal module is controlled by the host machine and generates LVDS type signals to panel. Typically, since the physical connector of different LCM products may have different pin assignments, a transfer board is included between the signal module and the LCM for proper signal transfer.

The power supply provides power to both panel and backlight unit, with the host controlling the time sequence during testing.

Table 16.1 shows examples of typical test image patterns and the defects to be detected for mobile phone displays, while Table 16.2 we show representative test images and defects of interest for TV displays. As shown in these tables, each testing pattern has its special purpose for defect detection. Such functional tests are time consuming and require significant labor during operation.

Table 16.1 Display patterns for defect inspection in mobile phone LCM [1].

Pattern No	1	2	3	4	5
Pattern name	White	Black	64 Gray	Waku	B
Defect Detection	B/L Defect, Polarizer Defect	particles, Cell scratches, line open	Mura, Cell bright point, dark point, Polarizer Defect	Waku defect	bright point, dark point, Cell particles, Line Defect, abnormal color, Polarizer Defect
Screen					

Pattern No	6	7	8	9	10
Pattern name	R	G	RGB Strip	128 Gray	192 Gray
Defect Detection	bright point, dark point, Cell Particle, Line Defect, abnormal color, polarizer Defect		abnormal color, Line Defect	Mura, bright point, dark point, Polarizer Defect	
Screen					

Pattern No	11	12	13	14	15
Pattern name	Window	Checkerboard	Gray Level H	Gray Level V	Sleep in
Defect Detection	Crosstalk	Image Sticking	Line Defect	Abnormal Display (Broken)	
Screen					

16.3 COSMETIC DEFECTS INSPECTION

Detecting cosmetic defects relies mainly on direct visual inspection in controlled test environments and associated visual specification criteria. For some defects, additional testing may be required according to specification, in order to avoid product over kill. For example, upon initial detection of particle defects gauges or rules are used to compare with specifications by defect size and length.

Cosmetic defects can be divided into several categories according to different components of LCM. The categories include cell, polarizer, display IC, glue, flexible printed circuit (FPC), and backlight (B/L). By their appearance, cosmetic defects can also be classified as panel broken, scratch, dirt, bubble, abnormal color, and size NG (not good, or out of spec), and so on. Table 16.3 shows some typical cosmetic defects test items for mobile phone LCM, while Table 16.4 shows typical cosmetic defects test items for TV LCM.

Table 16.2 Display patterns for defect inspection in TV LCM [2].

Pattern No	1	2	3	4	5
Pattern name	White	Black	Check Sub-pixel	Gamma Pattern	Blue
defect Detection	Polarizer Defect, Dark Point	Particle, Cell Scratch, Line Open, Edge Mura, Gap Mura	Line Defect	Pool Gray Level	bright point, dark point, Polarizer Defect, Cell Particle, Line Defect, Color Mura
Screen					
Pattern No	6	7	8	9	10
Pattern name	Red	Green	3V Color Bar	128 Gray Level	192 Gray Level
Defect Detection	bright point, dark point, Polarizer Defect, Cell Particle, Line Defect, Color Mura		Color Mora, bright point, dark point, Line Defect	Mura, Cell White, Black point, bright point, dark point, Polarizer Defect, Line Defect, Function Defect	
Screen				128 Gray	192 Gray
Pattern No	11	12	13	14	15
Pattern name	Window	Checkerboard	Gray Level H	Gray Level V	Image Tail
Defect Detection	Crosstalk	Image Sticking	PoolGray Level, Line Defect, Function Defect		Horizontal Image Tail
Screen					

Table 16.3 Examples of cosmetic defect inspection for mobile phone LCM [1].

Defect	types	images
Cell	Cell scratch, broken, lead scratch, Dimples	
Polarizer	Bubbles, Dimples	
IC	IC broken	
FPC	Components miss, FPC Fold	
Glue	Missing of Tuffy, UV, Ag Glues; Tuffy Bubble	
B/L	Barrel Deformed, B/L Broken, Folded Reflector	

16.4 KEY FACTORS FOR PROPER INSPECTION

16.4.1 Variation Between Inspectors

The results of visual defect detection are strongly dependent on the identification capability of human eyes, which is why it is important to understand the associated sources of variation and to develop appropriate test methodologies.

Table 16.4 Examples of cosmetic defect inspection for TV LCM [2].

Defect	Types	images
Cell	Cell scratch, Bright Point, Mura, Cell Leakage	
Polarizer	Particles, Bubbles, Dimples, Scald	
IC	crushed, damage, Scratch	
FPC	Bend, Scratch	
Glue	Tuffy Missing, Tuffy Bubble, Tuffy on Polarizer, Tuffy on COF	
PCB	Elements Missing, Crushed, Scratch, Poor Soldering	

Human eyes are very intelligent sensors. However, the eyes' sensibility to color and brightness for different people are not the same. The same defects judged by different inspectors may get different results. For example, when checking cell particles, optical filters are usually used. The filters have different transmittance and allow defect judement for various degrees of visibility. As shown in Figure 16.4, the appraiser holds the filter

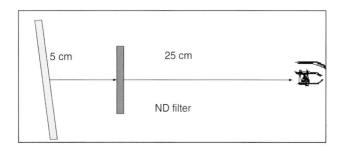

Figure 16.4 Visual inspection.

between screen and eyes for two to three seconds. If the defects can be seen, the panel will be judged as NG. Since different people have different eye sensitivity, the results may be different for those not-so-obvious defects. Some people are more sensitive to scratch defects, while others are less sensitive to these defects while at the same time being more sensitive to other defect patterns such as cell particles, dimples, or muras. Furthermore, skill, mood, and fatigue of the inspector are also factors that may influence the inspection results.

Good news is that the sensitivities for different defect modes can be both improved and also aligned by training to minimize inspectors' variations. Proper training program for inspectors is important and necessary.

16.4.2 Testing Environments

Three major environmental factors may impact the testing results. They are humidity, environmental cleanliness, and ambient light illumination.

Environmental humidity may influence the dissipation of electrostatic charge in panels. Low humidity may cause electrostatic muras, or even much more severe damage the panel or electronic parts. The humidity for testing environment is usually required to be higher than 60%.

Environmental cleanliness is another factor that may impact the testing results. Particles or dirt may fall and stick on the testing surfaces for inspection and subsequently affect judgments. The panel may thus be overkilled due to misjudgment.

Ambient light illumination is the most obvious factor that can affect the results of inspection, due to its direct impact on the contrast and visibility of defects. Consider vertical Mura band inspection as an example. Table 16.5 shows how ambient light can influence the testing results of a mobile phone LCM. Vertical Mura band will become insignificant when increasing the ambient light level.

16.4.3 Inspection Distance, Viewing Angle, and Sequence of Test Patterns

The inspection distance, viewing angle, and test image sequence are additional key factors that can affect the results of visual inspection results. For example, the longer the inspection distance, the lower possibility to find some defects like dark points, bright points, and dimples, and so on. For mobile phone LCM inspection, the distance is usually defined within 25 to 35 cm as shown in Figure 16.4.

In general, viewing angles for inspection are defined as 45 degrees with four viewing directions: left, right, upper, and lower. Figure 16.5 shows examples of skew inspection. The purpose of skew inspection is to find

Table 16.5 Impact of ambient illumination (lux) level on defect inspection. JND = just noticeable difference, a quantity used to judge the visibility of mura to human eyes.

Defect	Lux	JND Value	Photo
Vertical Band Mura	3	2.3	
	15	2.3	
	25	2.3	
	36	2.3	
	45	2.2	
	56	2.2	
	65	2.2	

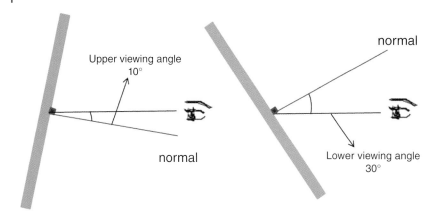

Figure 16.5 Skew inspection geometry.

some special defects like Waku (backlight light leakage), mura, and so on. Table 16.6 shows an example of Waku inspection. In this case, Waku can only be found when viewing angles are larger than 35 degrees.

To find out defects in displayed images, several predefined test patterns have to be shown on the screen. The sequence arrangement for these patterns is very important because it may also affect the visibility of certain defects. Generally, the higher the defect ratio a particular pattern can detect, the earlier this pattern is included in the test sequence to be displayed . Such arrangements results in higher inspection accuracy and efficiency.

The consideration of interference between two sequential patterns is also important. Figure 16.6 shows two different arrangements of test pattern sequencing. The arrangement of Black => 128 Gray => White is preferred compared to Black => White =>128 Gray. When the screen changes from Black pattern to White directly, the stimulation is too strong to human eyes and some of the defects become easier to be undetected. On the contrary, the stimulation for Black to Gray transition is much softer to human eyes and will be better for defect detection.

Table 16.6 Impact of viewing angle on defect inspection.

Defect	Specification	View Angle	Condition	Photo
Waku Defects (White Line Checking)	No Waku Defects	Front View	No Waku	
		Left 5°~30°	No Waku	
		Left 35°	Bad Waku (Color Change)	
		Left 40°	Bad Waku (Color Change)	

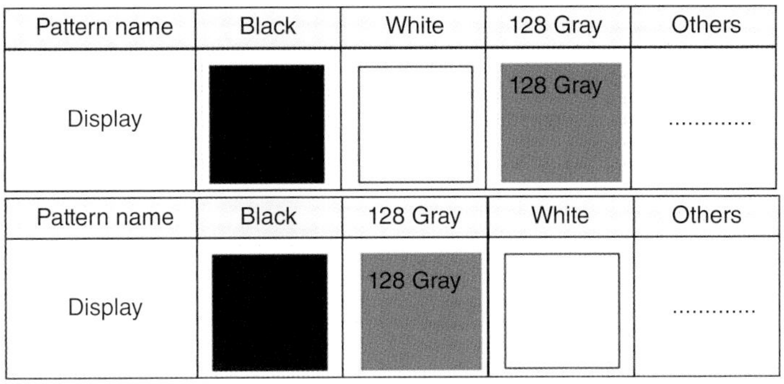

Pattern name	Black	White	128 Gray	Others
Display	■	□	128 Gray	....

Pattern name	Black	128 Gray	White	Others
Display	■	128 Gray	□	....

Figure 16.6 Different sequences of test patterns.

16.4.4 Characteristics of Product and Components

The characteristics of the LCM product itself and of its specific components can also influence the inspection. A few examples include backlight luminance, polarizer film transmisivity, presence of protection films, and the resolution of display (Figures 16.7 and 16.8).

Figure 16.7 shows how backlight luminance level affects the visual detection of cell particles, defect points, and mura. Normally, higer luminence of backlight will obviously enhance those defects. In some situation, we can raise the luminance of backlight for increased detection capability.

Figure 16.8 shows the effects of panel resolution on defect detection. For a given panel size, the higher the resolution is, the smaller the size of the corresponding pixels. Pixel defects such as dark point and bright point

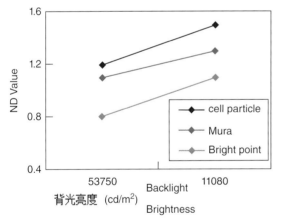

Figure 16.7 Influence of backlight luminance.

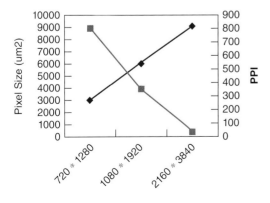

Figure 16.8 Influence of resolution.

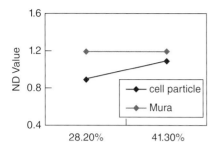

Figure 16.9 Influence of polarizer transmissivity.

will be unobvious when pixel size become small. Generally, when the resolution of mobile phone is higher than 300 PPI, or the area of a single pixel is smaller than 3,000 um^2, pixel defects will be unobvious to human eyes and consequently the capability of the defect inspection step may become worse.

Figure 16.9 shows the influence of polarizer transmittance on the detection of point type defects. A higher transmittance will enhance visibility of point defects, including pixel defects, cell particles, and backlight particles. On the other hand, the variation of polarizer transmittance has no significant effect on mura judgment.

16.5 AUTOMATIC OPTICAL INSPECTION (AOI)

Visual inspection by human eyes is labor intensive. In traditional visual inspection lines, lots of employees are necessary for balancing the inspection with the production line. As described in the earlier sections, the results of defect detection strongly rely on human eyes and inspector's skill, mood, and status of fatigue. These factors are sometimes difficult to control and may induce unexpected leakage of defective product and customer's complains. Additionally, continuous and extended visual inspection on the screen may be harmful to inspector's eyes.

In recent years, the progress in development of testing software, imaging systems, and light sources has been immense. The application of production level automatic optical inspection (AOI) equipment in the inspection line has become mature. Using AOIs to replace human eyes and solve the problems mentioned above becomes more and more popular with LCM manufacturers.

Figure 16.10 shows a typical AOI system and its configuration. It includes CCD camera, lens, inspection light source, and image processing software. Light sources may be projected from back, front, or sides to enhance defects on the LCM. The selection of CCD camera and lens should carefully match the defect measurement objectives. For example, the resolution of the CCD camera usually is selected to be higher than the minimum defect size. Additionally, the focal length of the lens should be uniform to whole area of the specimen being tested.

At present, most of the defects like point, line, scratch, and particles can be easily found by AOI. The capability of AOI for LCM inspection is generally good, but not yet completely meeting all the requirements in production lines. Nevertheless, successful AOI adoption and labor reduction has been undergoing in many LCM factories.

16.6 LCM DEFECT REPAIR

Some of the LCM defects can be repaired before packing and shipping. For convenience, defects are divided into three categories. They are open cell, polarizer, and backlight.

Open cell defects like circuit line short, line open, abnormal COF, abnormal PCBA (PC board assembly), and polarizer defects can generally be repaired. Circuit line short and open will cause abnormal display

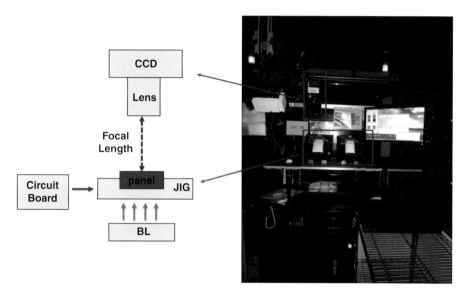

Figure 16.10 AOI equipment [1].

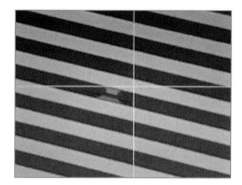

Figure 16.11 Laser-cut short circuit.

Figure 16.12 COF replacement.

operation. They can be fixed by using specialized laser cutting or laser welding equipment. Figure 16.11 shows an example of laser cutting application. For abnormal COFs and PCBs, the replacement is possible by using off-line equipments or fixtures. Figure 16.12 shows an example of COF replacement by off-line bonding machine. Figure 16.13 shows an example of PCBA repair.

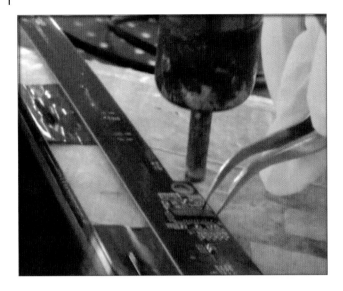

Figure 16.13 PCBA repair.

Figure 16.14 Polarizer detach [3].

For defective polarizer films, the only way to fix them is to replace with new parts. The process flow consists of detach, clean, and attach. The corresponding flow is similar for both small and large panels. Figure 16.14 shows an example of detach equipment and Figure 16.15 shows an example of lamination equipment for attach step.

To fix backlight defects, the LCD module has to be disassembled first followed by cleaning of particles or replacement of the defective optical films. For some LCMs such repair processes may not be available. For example, some of the mobile phone LCMs are very thin and require using adhesive tape or glue for attaching

Figure 16.15 Polarizer attach [3].

Figure 16.16 LCM disassembly.

Figure 16.17 Fixing optical film.

backlight unit to the panel. This makes reworking difficult, with possible breakage of the display panel or damage to the backlight upon separation. Even if the panel is not broken after detachment, the possibility to fix the backlight itself is very low. Usually the whole backlight has to be replaced. Figure 16.16 shows an example for disassembling large panel LCM. Figure 16.17 shows an example of fixing optical film in backlight module.

Finally, upon repair all the defective panels have to be tested again before shipping in order to ensure they are qualified.

References

1 Mobile phone product inspection SOP of AU Optronics (Xiamen).
2 TV product inspection SOP of AU Optronics (Xiamen).
3 Internal repair procedures of AU Optronics (Xiamen).

17

Productivity and Quality Control Overview

Kozo Yano[1], Yasunori Nishimura[2], and Masataka Itoh[3]

[1] Foxconn Japan RD Co., Ltd., 3-5-24 Nishi-Miyahara, Yodogawa-Ku, Osaka, Japan, 532-0003
[2] FPD Consultancy, 8-10-4 Nishitomigaoka, Nara, Japan, 631-0006
[3] Crystage Inc., 2-7-38 Nishi-Miyahara, Yodogawa-Ku, Osaka, Japan, 532-0004

17.1 INTRODUCTION

In this chapter, we introduce the topic of production quality control in terms of yield and productivity issues, which are indispensable cost factors for LCD. At first, the historical evolution of productivity is discussed, followed by yield management and improvement. Finally, quality control systems are reviewed tracing back to the world's first a-Si TFT factory operated by Sharp in the 1980s.

The authors' first work on LCDs was in 1972 at Sharp. At that time, Sharp successfully started the world's first LCD production in April 1973 for hand calculators and watches, and later for hand translators using 4- to 10-digit segment type screens based on simple passive-matrix LCD. That plant consists of simple clean booths that were far from the real cleanrooms available today. Then, the new "active-matrix LCDs" technology burst onto the scene. Since the presentation by Le Comber, Dundee University, on a-Si film in 1979 [1], many activities started aimed at utilizing a-Si TFT arrays for LCD backplane. Sharp also participated in such development of active-matrix using a-Si TFT in a quite unqualified cleanroom. In 1983, Epson announced color TV using TFT-LCDs [2], and many Japanese companies joined the industrialization of LCD color TVs using a-Si TFT. Sharp completed a prototype 3.2-inch a-Si TFT LCD [3], followed by construction of a-Si TFT LCD plant with the capacity of 50k/month for 3-inch diagonal screens and commercialization in May 1987.

The initial mother glass size for those small-size TV screens was settled at 320 mm × 300 mm owing to the substrate size capability of the photolithography exposure machine by Nikon. However, for the next target of a full page size display, it was obvious that the mother glass size should be much larger and ever since that time the remarkable plate glass size evolution commenced. The mother glass size expansion enabled reduction of the production cost, a critical barrier for the industry, with improvement in both productivity and production yield a first priority.

Productivity had increased through glass size expansion as well as tact time reduction. In terms of the production yield, during the primitive stages in late 1980s most of equipment was transferred from IC industry and not dedicated to LCD production, lacking the full spectrum of instruments for inspection and quality check. Therefore, new knowledge and skills related to equipment, cleanroom structure, anti-particle countermeasures, UPW (ultra-pure water) preparation system, non-stop power supply, and bulk/specialty gas supply system, were necessary throughout the LCD industry and high-tech community. Since then, as more and more LCD plants were built, the quality control and yield improvement have been continuous challenges.

Flat Panel Display Manufacturing, First Edition. Edited by Jun Souk, Shinji Morozumi, Fang-Chen Luo, and Ion Bita.
© 2018 John Wiley & Sons Ltd. Published 2018 by John Wiley & Sons Ltd.

17.2 PRODUCTIVITY IMPROVEMENT

In LCD production, the improvements of productivity and yield necessary to deliver display panels with reasonable cost have been impressive. Productivity is the index that shows how efficiently display panels are manufactured, and is defined as the number of panels produced per unit time and per capital investment. In other words, productivity improvement action is represented by maximization of the count of producible panels (production capacity) in the LCD plant made with significant amount of capital investment, for example, over $5BUS for an 8G plant with 90k mother glass input.

17.2.1 Challenges for Productivity Improvement

Productivity is determined by several different key factors, as summarized below. Productivity improvement is the challenge to optimize such key factors.

Enlargement of the mother glass size
Multi-panel assignment (panelization) on one mother glass substrate had contributed to significant increase of production volume, improvement of investment efficiency, and reduction of depreciation cost. That means the size of the mother glass should reasonably be as large as possible, and therefore the mother glass size enlargement have been the first priority in achieving productivity improvement, as represented by evolution from Gen 1 to Gen 10 sizes. They will be discussed again in section 17.2.2.

Reduction of tact by high-speed process and high throughput equipment
Tact (turn around cycle time) is defined by the time to finish one process step, and thus a smaller tact allows for a larger processing capacity. High processing speed is quite useful for tact time reduction and had been implemented for large size substrates in TFT array production, as well as in the cell process by series of innovation, such as one-drop filling (ODF) and photo-alignment in cell process.

Commoditization of TFT device and process, and reduction of process steps
TFT structure and deposition, patterning, etching and cleaning processes are standardized and shared in TFT array manufacturing. Also a-Si TFT array process was simplified by reduction of photomask layers from seven to five, and further down to four by introduction of half-tone exposure [4].

Stabilization of process (Figure 17.1)
Batch procedures had been replaced by the single plate processing, in which each glass plate is processed under precise control. It had achieved the highly uniformity and reproducibility, minimizing the fluctuation in process and system. Presently, single-plate processing has become standard.

Lead-time minimization
Inline connection of each process equipment and stations, and process integration of etching and ashing in dry processing have been useful to avoid unnecessary time loss from glass transportation between the process steps.

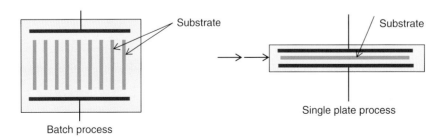

Figure 17.1 Batch to single plate processing.

Optimization of WIP (work-in-process) is also important in preparing for scheduled and unscheduled downtime of equipment.

Process automation

Automatic handling and transportation of the glass plate instructed by CIM (computer-integrated-manufacturing) is quite efficient to optimize the total glass flow through the process stations.

Uptime maximization (downtime minimization)

Downtime minimization by CVD tool in situ cleaning [5], improvement of PVD target utilization, reduction of photomask interchange in exposure, as well minimization of the equipment failure probability (maximization of MTBF) and reduction of recovery time (MTTR) are additional key factors to improve productivity.

17.2.2 Enlargement of Glass Substrate

Productivity improvement by the mother glass enlargement

In display panel production, the typical approach is to design multiple panels per each large glass substrate as shown in Figure 17.2 and then cutting into individual display panels after the process is completed. Since the size of display panel products has grown over time, the mother glass sizes had to become larger and larger by necessity.

The size of the glass substrate is defined as "Generation," for standardization of production equipment. The evolution of the mother glass size (G) is shown in Figure 17.3: it begun with the first generation (1G) of

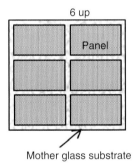

Figure 17.2 Six-up panel layout.

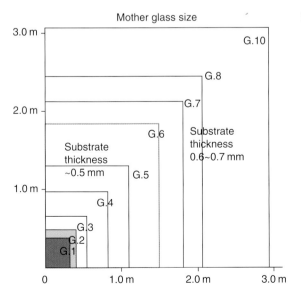

Figure 17.3 Mother glass substrate sizes.

Table 17.1 Relation between panel sizes and panel counts for each substrate generation.

		Panel size (Diagonal inches)															
Gen.	Mother glass size mm	10.4	12.1	14.1	15	17	20	23	28	32	37	42	47	52	55	60	65	70
1	300 × 350, 320 × 400	2	8.4″ × 4															
2	360 × 465	4																
3	550 × 650	6	6	4	4	2												
3.5	600 × 720	9	6	6	4	4	2											
4	680 × 880		9	6	6	4	4	2										
	730 × 920		9	9	6	6	4	2	2									
5	1000 × 1200		16	15	12	9	6	6	3	2	2	2						
5.5	1300 × 1500					20	12	8	8	6	3	2	2	2				
6	1500 × 1800							12	8	8	6	3	3	2				
7	1870 × 2200							20	15	12	8	6	6	3				
8	2160 × 2460									18	10	8	8	6	3			
	2200 × 2500									18	10	8	8	6	6	3		
10	2880 × 3130											15	10	8	8	8	6	6

300 × 350 mm, and it advanced to 10G (2,880 × 3,130 mm) used presently in production. As the glass size is enlarged, the number of panels from one glass substrate is also increased, leading to significant productivity gains. In Table 17.1, the panel count for representative display sizes is shown at each generation. For each generation there are several best panel sizes, that is, not just one size. Panel makers have been producing those sizes of the display panels as their strategic products.

Mother glass size evolution (generation advancement)

The evolution of mother glass sizes and the year for first operation is shown below in Figure 17.4.

Square markers indicate the production start in each generation, with almost a linear trend being observed for the increase of substrate area versus time at the rate of 1.9× increase every three years.

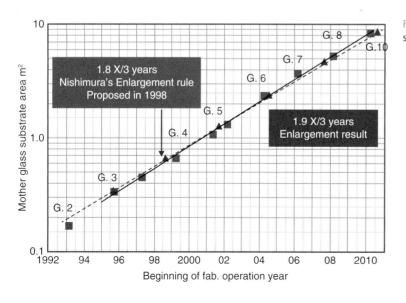

Figure 17.4 Transition of mother glass substrate enlargement.

In 1998, the authors proposed the enlargement rule of "the glass substrate size (area) will grow by 1.8× in each 3 years" as a roadmap of mother glass substrate area enlargement and to predict the upsizing of future LCD panels and the manufacturing panel cost [6]. The background for the "1.8x in 3 years" rule are:

(1) The development of next-generation production requires collaboration among panel makers, equipment vendors, and material suppliers, and typically needs 3 years to prepare the necessary total infrastructure in technologies and designs.

(2) The panel counts in one mother glass are typically 4, 6, 8, 9, or 12, 15, 18, and so on. When the glass size is enlarged, panel count is also increased. There were two cases for increasing panel count according to the panel size. In two possible cases in panel count increase, 2→4, 3→6 (the multiple factor of 2) and 4→6, 6 →9 (the multiple factor of 1.5), the average multiple factor of 1.8 is obtained.

The dotted line (triangle marks) corresponds to the proposed rule of "1.8× growth in each 3 years", and it is very close to the real evolution shown by the "1.9× in 3 years" line which is slightly larger than the proposal.

17.2.2.1 Productivity Improvement and Cost Reduction by Glass Size Enlargement

In order to improve the productivity through glass size enlargement and resulting panel count increase, the equipment in the new generation should have similar or better performance in terms of throughput, process capabilities and cost of ownership compared to the original generation. Equipment vendors had encountered plenty of problems during such development (glass bending, glass break, deterioration of patterning accuracy and of process uniformity across the entire substrate, processing speed, and throughput, as shown in Table 17.2). Those problems were solved systematically with various countermeasures. The equipment footprint also grew during the generation advancements. Figure 17.5 shows the example such evolution in CVD equipment.

Table 17.2 Technology challenges for scaling up the mother glass and panel sizes.

Process Equipment	Mother glass substrate enlargement	Panel size enlarge-ment	Throughput Compare to Previous Gen.	Accuracy Uniformity improvement
CVD	Scale up		same	Uniform plasma density
PVD	Electrode enlargement			
Dry Etching	Target enlargement			
	Improvement of RF power supply			
Photo-lithography	Image field enlargement	Image field-enlargement	~same	Pattern accuracy
	Scan exposure	Photomask-enlargement		Flatness
	Spin-less processing			
Wet	Batch → Single plate process		same	Uniform wet process
	Treatment area enlargement			
	Dry cleaning, Free of acid wet process			
	Chemical & Water saving process			
LC cell	Scale up	One drop filling	same	Pattern accuracy
	Inkjet coating	Photo alignment		Flatness

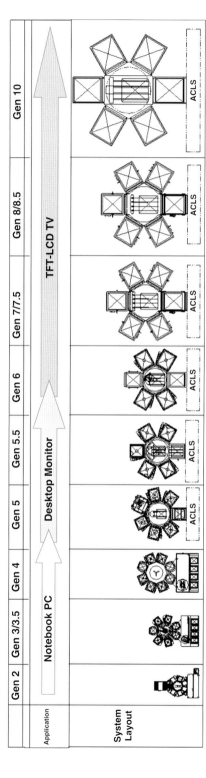

Figure 17.5 Example of equipment (CVD) advancement corresponding to mother glass enlargement (*Source*: Provided by AKT).

Thus the ratio of the cost increase of newly developed equipment was kept less than the square root of the glass size increase without degradation of throughput, as a result of the efforts of equipment vendors.

$$\textbf{\textit{Equipment cost}} \; C1 \leq C0\sqrt{A1/A0} \tag{17.1}$$

Here, C0 is the cost of original generation, A1 is the area of new generation and A0 is substrate area of the original generation.

Productivity to investment (PI) is defined as:

$$\textbf{\textit{PI}} = \textbf{\textit{Glass Area}}\,(A) \times \textbf{\textit{Throughput}}(Th) \times \textbf{\textit{Yield}}\,(Y)/\textbf{\textit{Capital Investment}} \tag{17.2}$$

And PI ratio between the new and original generation is:

$$\textbf{\textit{PI Ratio}} = \left(A1 \times Th1 \times Y1/C1\right)/\left(A0 \times Th0 \times Y0/C0\right) = \left(A1 \times C0\right)/\left(A0 \times C1\right) = \left(A1 \times C0\right)/\left(A0 \times C1\right) \tag{17.3}$$

under the condition that Th0 and Th1, Y0 and Y1, are the same.

$$\textbf{\textit{PI Ratio}} \geq \sqrt{A1/A0} \tag{17.4}$$

This means that the productivity-to-investment factor is improved in accordance with the square root of the glass area ratio. For instance, the area of the glass plate becomes twice, that factor is increased by 1.4, and it contributed to the remarkable cost reduction by reduction of depreciation and initial capital investment.

17.3 YIELD MANAGEMENT

Yield is defined as the percentage of the output product counts with respect to the input numbers. In the case of 90% yield, 10% is rejected and lost at the final inspection or during in-process inspection, though some of the failures can be repaired and recovered.

At the starting phase of new plants the production yield is quite low, typically less than 50%, because of multiple out-of-control issues such as cleanroom instability, facility down, equipment malfunction, human errors, and so on. With an energetic effort yield is gradually improved within a year up to levels as high as 90% to 95%, which is adequate to have a cost-effective operation. Yield improvement activities are required all over the plant including human resources and it starts with failure analysis, collecting information on failure rate and failure modes from every quality checking step in the process flows for TFT array, color filter, cell making, and module assembly processes. Once those failures are fixed and the plant operation has been stabilized to a good enough yield, all the necessary procedures to maintain yield are consistently and strictly applied to the daily operation with rigid quality control system and organization.

17.3.1 Yield Analysis

17.3.1.1 Inspection and Yield

The TFT-LCD production process consists of three major sections: array, LC cell, and module assembly, as shown in Figure 17.6. In the TFT array process, after unit processes such as film deposition, photolithography, etching, and so on, inspection steps are performed and then after TFT process completion an array test is applied. In cell processing, inspections are included after the unit process of PI coating and cell making, and a final panel test is done at the end of cell process. Module assembly includes similar inspection steps.

During the inspections after each unit process, the corresponding process yields (Y_{ai}, Y_{cj}, etc.) are determined. The array process yield (Y_{ap}) is defined by the product of each unit process yields: $Y_{ap} = Y_{a1} \cdot Y_{ai}$, and

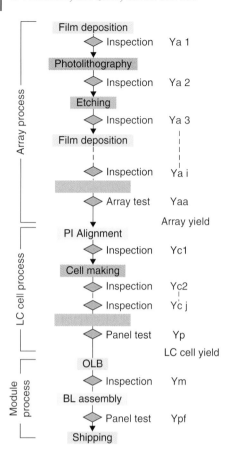

Figure 17.6 Process flow and yield definition.

array total yield (Ya) is defined as (Yap·Yaa) by also including the array test yield (Yaa). The LC cell process yield (Ycp) is the product of (Yc1) thru (Ycj), and by including the panel test yield (Yp) the LC cell total yield (Yc) is obtained as well, Yc = (Ycp·Yp). The module yield follows a same approach. Thus, the total yield (Y) is defined as the overall product: Y = Ya·Yc·Ym = Yap·Yaa·Ycp·Yp·Ymp·Ypf.

Table 17.3 shows the typical yield summary in a mature a-Si TFT LCD production line.

Table 17.3 Typical yield structure.

Process		Yield	
Array	Process yield	Yap	99%
	Array test yield	Yaa	96%
	Array total yield	Ya	95%
LC cell	Process yield	Ycp	99%
	Panel test yield	Yp	97%
	LC cell total yield	Yc	96%
Module	Process yield	Ymp	99.8%
	Panel final test yield	Ypf	99.2%
	Module total yield	Ym	99%
	Yield (Total)		90%

17.3.1.2 Failure Mode Analysis

In order to improve the yield, analytical and total approaches to the process procedures, equipment, materials, and design factors are required in order to find the root-causes for yield losses. The typical failure modes and root-causes in a-Si TFT-LCD production line are summarized in Table 17.4, including the predetermined inspection criteria for each inspection step. The failed panels with, for example, glass crack/break, out of specification, cell gap failure, etc. are rejected after inspection, and only the good units continue to the next step. Thus the panels output after the end of process are the final good ones, and represent the process yield (line yield).

The major reasons for failures found during TFT array testing come from TFT pattern, pixel pattern, line open/short, and TFT characteristics. For panel testing, the most common failures are point(pixel) or line defects and display mura (non-uniformity). Those failures are categorized into systematic/ parametric modes related to the process, equipment and designs, and random mode whose defects dominantly come from particles and contaminations as shown in Figure 17.7.

The array test yield (Yaa) and panel test yield (Yp) are expressed as:

$$Yaa, Yp = Ysys \times Yrandom \quad (17.5)$$

where Ysys is yield by systematic failure and Yrandom by random failure.

Systematic failures mostly come from non-uniformity error of TFT characteristics and panel display mura caused by fluctuations in film thickness, line width, cell gap, and process parameters. On the other hand, Yrandom is described by Poisson probability distribution formula;

$$Yrandom = e^{-DA} \quad (17.6)$$

where D is the defect density and A is the panel area. Thus Yrandom greatly depends on the panel size. The particles and dusts attached onto the glass plate during processing can introduce pixel or line defects resulting

Table 17.4 Failure modes in a-Si TFT-LCDs.

Process	Inspection	Checking Items	Failure Modes	Root-Causes
Array	Thin-Film Deposition (Sputtering, CVD)	Post-Process Inspection	Film Thickness, Resistance and Uniformity Particles Glass Breaking and Clipping	Out of Spec. in Thickness, Resistance and Uniformity, Film Peeling-off Particles Glass Breaking and Clipping	Equipment Malfunction, Abnormal Discharge, Non-uniform Plasma Density, Splash, Foreign Materials Bi-Product Generation
	Photo-lithography		Pattern Check, Line Width, Overlay, Edge profile and Their Uniformities, Film Peeling-off Particles Glass Breaking and Clipping	Pattern Defects, Particles, Line Width Error, Overlay Error, Unusual Edge profile, Uneven Uniformities, Glass Breaking and Clipping	Resist Coating Failure, Resist Mura & Scratch, Exposure and Developing Failure Particles and Foreign Material Immixture, Contamination Film Peeling-off
	Etching Resist Removal				Etching Failure, Removal Failure
		Array Test	TFT Ion/Ioff, Metallization, Pixel, Open/Short TFT Characteristics	TFT Ion/Ioff, Metallization, Pixel, Failure Open/Short TFT Characteristics Failure	Pattern Defects, Particles Film Peeling-off, Abnormal Discharge, Electrostatics, Damage, Out of Specification in Film Thickness, Line Width, etc., Uniformity Error

(Continued)

Table 17.4 (Continued)

Process		Inspection	Checking Items	Failure Modes	Root-Causes
LC Cell	PI Coating	Inspection	PI Thickness and Uniformity Scratch, Pinhole	Coating Failure, PI Mura, Scratch, Pinhole	Coater Malfunction and Failure, Disqualified PI Material
	Alignment	Inspection	Rubbing Condition	Mura, Scratch, Streak	Rubbing Equipment and Rubbing Cloth Error
	Seal Print	Inspection	Seal Pattern and Shape	Dealing failure, Seal Width Error	Air Bubble, Dispenser Failure
	LC Filling	Inspection	LC Filling/Dropping Volume	Over/Under Volume	Equipment Condition Failure
	Cell Making	Cell Gap Measurement	Cell Gap and its Uniformity	Cell Mura	Lamination Equipment Failure, Vacuum Chuck, LC Drops, Spacer Height Distribution Error
		TFT Alignment Inspection	TFT/CF Alignment	TFT Array/CF Alignment Error	Total Pitch Error in TFT or CF Plate, or Mismatch Lamination Equipment Failure
		Alignment Check	Alignment Appearance	Photo Leakage, Disclination	PI and Rubbing Failure
	Scribe & Break	Inspection	Cut Appearance	Cut Failure	PI and Rubbing Failure
	Polarizer Sticking	Inspection	Outview Check	Cut Failure	Equipment Failure
		Panel Test (Visual Check)	Failure in Pixel and Metallization	Line Defect, Block Line Defect Pixel Defect(Bright, Half-Bright	TFT/CF Plate Defects, Particles
			Failure Display Image	Brightness Mura, Color Mura, Stain, Filling Hole Mura, LC Drop Stain, Light Leakage, Discrimination, Contrast Failure, etc.	PI Non-Uniformity, Rubbing Mura, Scratch, Steak, Cell Gap Non-Uniformity, Contamination on TFT/CF, Glass Water Mark, LC/Seal Material and Polarizer Failure, etc.
Module	AB/NBL Assy Inspection		TAB Connection	TAB Connection Error, BL Failure	TAB Equipment Failure Dusts in BL
	Final Panel Test Burn-in Test		Same Check as for Panel Test Test Results, Image Sticking	Same as Panel Test Pixel Defect, Flickering, Image Sticking	TFT Array/LC Cell Process Error

in the observed random failures. Those particles/contaminations can remain on the film or in some cases they introduce pattern failures during photolithography process.

Figure 17.8 shows examples of various defects in TFT array. Those defects such as line, pixel, TFT defects are caused by pattern defects mostly from particles, dusts, and contaminations in the array process. Those defects are categorized random failures.

In the study of yield analysis, failures must be categorized into the systematic and random modes. Systematic failures takes place in the events that process optimization is insufficient or process parameters for film thickness and pattern dimension are not satisfactorily controlled, while random failures mostly caused by pattern defects by the particles. Such modes are varying depending on the phase of operation; the

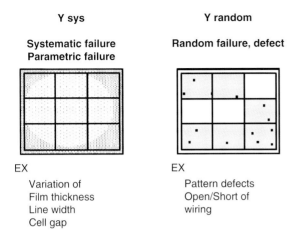

Figure 17.7 Systematic and random failures.

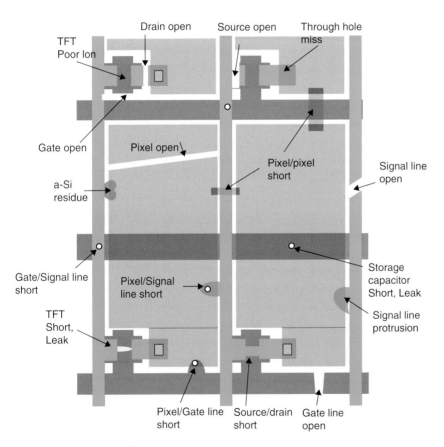

Figure 17.8 TFT array defect examples.

beginning, learning, and matured stages. At the beginning phase, systematic failures tend to dominate due to insufficient process optimization. However, after production learning when the process is stabilized with a minimization of systematic failures, it is random failures coming from pattern defects that become dominant.

17.3.2 Yield Improvement Activity

In this section we discuss yield improvement actions for systematic failures related to the process, equipment, and design optimization at the beginning stage, yield improvement for random mode failures associated with particles coming from the process, equipment, materials, and clean room environment, as well as the unique yield improvement approach in TFT-LCD production: repair process.

17.3.2.1 Process Yield Improvement

The failure modes that degrade the process (or line-) yield at the beginning stage come primarily from glass chipping/breaking and out-of-specification processes such as film thickness, line width, and cell gap, and so on. In the adjustment phase, optimization of process and equipment must be continuously improved to approach the matured stage step by step. After the learnings of those failure modes are incorporated, process yield gradually improved up to satisfactory extent.

17.3.2.2 Systematic Failure Minimization

The major cause of systematic failure is the fluctuation of the panel characteristic within or between panels. Such fluctuations are brought by variation or non-uniformity in process, equipment or panel design. The following issues are the major actions in improvement of systematic failures.

- Photolithography process:
  Minimization of variation and improvement of uniformity of critical dimensions (line width and overlay), and cross-sectional profile
- CVD:
  Control and stabilization of film thickness and quality (refractive index, H concentration, etching rate, etc.) and improvement of uniformity
- PVD (sputtering):
  Control of film thickness and resistivity, minimization of variation, and improvement of uniformity
- PI coat:
  Thickness control, minimization of fluctuation, and improvement of uniformity
- LC drop volume:
  Control of the photo spacer height and minimization of fluctuation
- Molecular alignment:
  Control of alignment strength and condition
- Cell making:
  Control of the LC drop volume and the photo spacer height with minimization of fluctuation
  Cell gap control to keep the uniform gap

These process controls are strongly related to the process window, which is defined as the allowable range of variation in process or equipment. Therefore, minimization of systematic failures can also be achieved by widening of the process window. Once the process window is firmed up, small change of process or equipment will no longer create failures and subsequently yield will be highly stabilized at a high and satisfactory level.

In some cases, systematic failures can come from cross-interactions between different processes or equipment. It is also important that the adjustment and optimization between processes or equipment are explored.

Design verification in terms of yield is another factor that must be checked. All designs must consider that the chosen processes are capable of high yield.

17.3.2.3 Random Failure Minimization by Clean Process

There can be several root causes for random failures: particles, contaminations, and electrostatic damages.

Particle Control and Minimization

Particles (including dust and foreign contamination) are the major root cause of random defects. Sources of particles are located everywhere, such as equipment (by-product generated during CVD, peeling-off of deposited films, nodule during ITO deposition, splash, and abnormal discharge, etc.), frictional wear of equipment, human body during transportation and handling, glass edge chipping, materials used in the process, environment in stocker, and loader/unloader.

The countermeasures to prevent the particle, contaminations are listed below.

– Substrate environment control:
Clean room, clean air flow, clean design in equipment, cleaning, clean transportation, and stocker
– Clean deposition process:
CVD (down-deposition): in situ chamber cleaning, remote plasma cleaning, by-product free process
Sputtering (side-deposition): manual cleaning, nodule suppression in ITO deposition, recipe to prevent splash and abnormal discharge
– Clean etching process:
By-product free dry-etching, clean ashing
– Cleaning:
Pre-cleaning before film deposition and photolithography, pre-cleaning before PI coating and LC cell making
– Filtration of materials such as gas, chemicals, resist, and so on.
– Electrostatic: exclusion and removal of electrostatic charge build-up

Contamination Minimization

There are many sources of contamination from the clean room environment, equipment, and wet processing that can lead to organic and inorganic contaminants accumulation on the substrate. Additionally, re-deposition of contaminants during cleaning and rinsing processes can also happen. Contamination from ambient during storage in stocker and loader/unloader is another potential issue that needs attention.

The organic contaminants can be removed by UV and ozone treatment, while inorganic contaminants can be removed by liquid such as acid, alkali, ionized water, and so on. Clean drying is very important after chemical/water cleaning. To prevent re-deposition of contaminants during cleaning /rinsing, the zeta potential of cleaning liquid and water should be controlled effectively.

Materials and Environment Cleanliness Control

There are a lot of chances of contamination from materials (chemical solutions, gases, purified water, liquid-crystal, PI, seal resin, etc.) and environment (process chambers, vessels, and pipes).

Many countermeasures, such as use of >5 N purity sputter targets, semiconductor-grade gases, highly pure photoresist/developer/organic solvent, highly clean process chamber/vessel/piping, can be very effective to minimize particle generation. As another example in cell process, the LC material quality should always be checked to maintain high resistivity and eliminate the residual DC voltage and smear in the LC cell.

Electrostatic Control

Electrostatic charging and discharging are phenomena that can cause fatal failures during production. There are various modes of charging, such as friction, exfoliation and induction charging and discharging. Electrostatic charging and discharging causes many kinds of defects such as cross short, TFT failure, line defects, bright/dark spot defects, display defects, and so on.

In order to exclude such electrostatic damages, many countermeasures can be implemented:

– Improvement of substrate handling to reduce friction during transfer, using slow transfer speed and substrate proximity holding
– Prevention of static charge by installation of charge neutralizers, such as ionizer blower, soft x-ray static eliminator, and anti-static devices

– Electrostatic discharge prevention through panel design with discharge circuit, shortening bars, and dummy wiring patterns

17.3.2.4 Yield Improvement by Repairing

Since the design pattern dimensions in TFT-LCDs are not so precise as in the case of semiconductor ICs, the failures coming from pattern defects in specific layers can be economically repaired. After the inspection step and decision whether the defect is repairable, the defective panels are moved to repairing process. If they are non-repairable, the defective panels are rejected as a fatal failure.

The main repairing procedures and methods are summarized in Table 17.5. Due to employment of the repairing process, process yield is improved by 5–10% in large sized panel at the premature phase.

In the case that the defects are detected by ADI macro-inspection after the development step in photolithography, the substrate or the entire lot can be returned for photolithography after removal of the photoresist and cleaning. This procedure is so called rework. If the defect is found after PI coating process, PI film is removed and cleaned by O_2 plasma and PI is re-coated.

To maintain a high yield over long term and continuously improve it, yield management under QC system is required, which is the subject of the next section.

17.4 QUALITY CONTROL SYSTEM

The production yield and the quality of LCDs are decided by the quality control capability in the manufacturing factory, as well as other electronic device and component manufacturing. In other words, excellent quality is created by the qualified plant. However, it cannot be achieved solely through use of qualified equipment provided by fully experienced vendors, but it is also due to incidental facility/supporting systems, auxiliary system including the substrate stockers, and the people who precisely follow the recipe determined by the process engineers. This total plant capability is closely connected to the quality of the products. Qualified products and high yields can be achieved by combination of knowledgeable development engineers, highly experienced process engineers, and well-trained operators in highly qualified plant. Thus, the definition of "yield and quality control" is to monitor all the plant operations and functions and ensure they are working according to predetermined flows and that the production equipment are yielding the specified products

Table 17.5 Defect repairing procedures and methods.

Process		Failure Mode	Repairing Procedures	Repairing Method
Array	Gate	Short	Disconnection/Removal	Laser cutting
		Open	Reconnection	Laser CVD, Metal thermal transfer
	Source/drain	Short	Disconnection/Removal	Laser cutting
		Open	Reconnection	Laser CVD, Metal thermal transfer
	Pixel	Short	Disconnection/Removal	Laser cutting
		Particle Bump	Removal	Tape polishing
LC cell	PI coating	PI Pinhole	PI Filling	PI Dispense
Panel		Bright Shot	Convert to Dark Spot	Connection/Disconnection of Redundant Circuit, Laser Cutting, Laser welding, PI destruction, Opaque material dispense

based on the assigned recipe. If any process step falls out of its controllable ranges they must be immediately corrected.

17.4.1 Materials (IQC)

The raw materials consumed in the plant are one of the keys to maintain high yield and quality. A system to consistently procure materials with guaranteed quality is necessary, and that is IQC (input quality control). Material vendors are selected through the inspection and verification procedures performed by IQC team members. In the case that the vendor is changed, a "material change committee" that includes the third party experts is assembled to avoid the changes causing a quality problem. Thus, IQC system secures the material quality.

Table 17.6 shows a representative list of raw materials consumed for LCD production. It includes over 50 various materials for chemicals, optical component, electric devices, and so on. Those materials are delivered

Table 17.6 Materials used for TFT-LCD manufacturing.

Process	Material	Specification
Photolithography	Photo resist	PEGMES: 45%~95 wt%, Res in 4~40 wt%
	Dilution Liquid	PGME: 70%, PGNSAE: 30%
	Development	20% TMAH
	Surfactant	Hexamethyldisilazane (HMDS)
Specialty Gases	SiH_4	over 99.9996 vol%
	$PH_3(1\%)/SiH_4$	over 99.9996 vol%, 1.0 ± 0.05
	NH3	over 99.99 vol%
Etching Gas	Cl_2, SF_6, BCl_3, CF_4	over 99.999 vol%
	NF3	over 99.999 vol%
PVD Target	Cu, Al, TiMo,	over 99.9% purity
	ITO	over 99.9% purity, In_2O_3 : 90.9 ± 0.5 wt%
Etching Chemical	Metal Etching	H_2O_2 under 6%, HNO_3 under 3.5%
	Oxalic Acid	Organic Acid:under 5%, Additives:under 1%
	Stripper	$HOCH_2CH_2NH_2$: $70 \pm 5\%(CH_3)_2SO$
	H_2O_2	50%
	HNO_3	50% G
Substrate	Glass Plate	0.7 mm thickness
Other 23 materials		
Cell	Alignment	
	Surfactant	
	Alkali Aqua-Remover	
	Epoxy Adhesive	
		Au Ball
	LM Material	
	Other Chemicals	NMP, Acetone, IPA, TMAH
Panel Inspection	Cut/Break	
	Packing	

(Continued)

Table 17.6 (Continued)

Process	Material	Specification
Module	Polarizer	
	TAB	Source/Gate Driver
	PCB/FPC	
	ACF	TAB/PCB connection
	C-PWB	
	Backlight Unit	
	Bezel	
	Packing	
Color Filter	Glass Plate	0.7 mm thickness
	Surfactant	
	Photo resist	PEGA 44–55%, Cyclohexanone 30–40%
	Dilution Liquid	Cyclohexanone 60%, Propylene Glycelite 20%, Aromatic Hydrocarbon 15–25%
	Development 1	KOH system
	Development 2	Sodium Carbonate System
	ITO Target	ITO (SnO_2 10 wt% 4 N)
	Polishing Media	Aluminium Oxide 15%
	ITO for rework	$FeCl_3$ + HCl
	RGB for rework	PGME/KOH

to the plant under predetermined IQC system decided and agreed between the vendor and IQC staff. Usually those materials are delivered with quality-check documentation from each material supplier factory, which was audited and authorized upon regular factory visits by IQC team. And furthermore, IQC step includes a sampling check of the delivered material batches with appropriate frequency.

17.4.2 Facility Control

The facility control organization monitors the status of all operations, as shown in Table 17.7. Those operating conditions are strictly determined, and the staff is always watching if they are set within the safe range. Monitoring gauges and inspection equipment are verified and adjusted. The production equipment and verifications results are monitored and recorded in the control center.

17.4.3 Process Quality Control

Examples of manufacturing process flows and inspection flows used for quality control and yield improvement were shown in Figures 17.9, 17.11, 17.12, and 17.14 for TFT array, color filter, LC cell, and modulization, respectively. Those process flows are always reviewed and modified based on production experience and can vary for each plant. Also, the flows at the starting phase with yields, for example, of about 85% are reasonably different from the flows in the matured plant with optimized yield, for example, about 95%. Also, the flows in factories of small-/medium- sized displays, where over 200 display panels can be assigned in one plate, will have differences from the flows for TV plants with only eight panels on one plate.

The detailed process flows discussed in this section are examples for TV production in Gen 8 size plant. In the figures below, the mark ▨ denotes a process that includes go/no-go test for all the panels, where once a

Table 17.7 Facility monitoring items.

Facility Segment	Object	Conditions/Items
Electricity Power	Operation Status	24 hours/365 days
	Specification	Voltage, Frequency, Leakage
	Momentary Voltage Drop	
Bulk Gases	N_2, O_2, H_2, Air Gas	Purity, Pressure
	Steam (Heat Source)	Temperature, Pressure
Air Conditioner	Air Supply	Temperature, Contamination (Organic and Inorganic Chemicals)
	Exhauster	Anti-Air Pollution Regulation, Smell
	Luminance, Noise	
Cleanroom	Cleanliness, Temperature, Humidity, Pressure Balance	
	Chemical Contamination	
Purified Water Supply	Resistance, Particles in Water	
Drain Water	Anti-Pollution Regulation	

failure is found then rework or repair is done for recovery. The mark ▦ corresponds to a process step that counts the numbers of processed glasses and regularly applies the periodic maintenance (PM: preventive maintenance) stopping the operation. This stop is promptly decided by the authorized process engineers assigned to that machine in order to avoid failure from continuing production without notice. Thus, the production yield and quality are monitored for 24 hours by those systems and organizations.

17.4.3.1 TFT Array Process

The TFT array process starts with input of the glass substrate, as shown in Figures 17.9 and 17.10. The glass is supplied from the vendor after fine cleaning. Quality of the glass is guaranteed by its specification and the IQC organization is always checking the quality by sampling.

Just before film deposition, the cleaning equipment directly connected to the deposition equipment is employed for removing particles attached during the plate transportation and storage in stocker. There are several different cleaning approaches including wet and dry, and recently brush cleaning with purified water has become the main stream.

The sputtering equipment (PVD) is employed for Metal Film 1, Metal Film 2, and ITO layers. Since in situ cleaning cannot be used for this equipment, particle generation from the vacuum chamber is the most critical issue for yield control. Those particles come from films deposited on the protection shield, holder and tray after considerable deposition cycles have been performed, even though the film removal and dropping-away countermeasure is applied. Consistent monitoring of particle count, pinhole count, and sheet resistance, as well as application of PM under Cpk (process capability index) management [7] are necessary.

The photolithography process takes place within inline integrated equipment performing the flow sequence of cleaning, pre-bake, temperature control, exposure, photoresist development, and post-bake. After the process is completed, all substrates are inspected by AOI (automated optical inspection) machine, and the defective plates are reworked. Recently it was replaced by the photoresist repair by advancement of machine technology. Also CD (critical dimensions), TP (total pitch), and photoresist thickness are measured and input into data base for monitoring the process capability by statistical management. Then Metal 1 or 2 are wet-etched and the photoresist is removed, followed by open/short test for all the plates. If a failed substrate is found, it is then moved to the repair process.

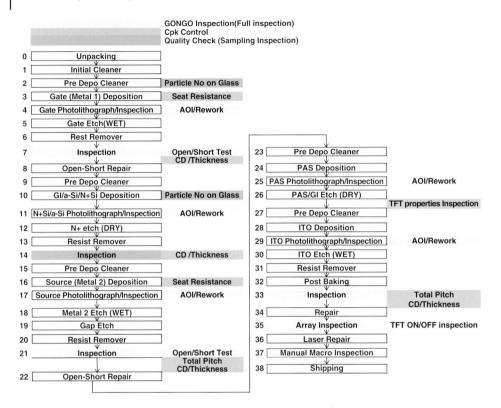

Figure 17.9 Process flow and quality check steps for TFT array manufacturing.

The tri-layer (N⁺ a-Si, i a-Si, and SiN) stack and the passivation layer (PAS) depositions are done using plasma CVD equipment. Since the quality of these films directly impacts the device reliability, process monitoring systems are used to accurately monitor the operation after each step. Examples of Cpk management for SiN_x film thickness is shown in Figure 17.10. At the same time, all the parameters such as time vacuum, plate temperature, and gas flows are monitored at the central system. Film quality is also sampled using FTIR (Fourier transfer infrared spectroscopy) and ellipsometer. Particle generation inside the CVD chamber is cleaned in situ using NF_3 gas.

Dry-etching the tri-layer stack has a similar particle problem as in the case of sputtering equipment, that is undesirable particles falling onto the plate as well as abnormal plasma discharge. The PM (preventive maintenance) timing is decided by the statistical estimation of the predicted particle counts. In terms of abnormal discharge PM timing is also set up by statistic calculation using AOI result.

After the PAS and ITO films are processed, array testing is performed and laser repair is employed for the failed TFT.

17.4.3.2 Color Filter Process

The flow of the manufacturing and inspection steps for the color filter substrate is shown in Figure 17.11. The manufacturing process is completed by the five-film coating/photolithography steps of BM (black matrix), R (red), G (green), B (blue), and PS (photo spacer) and ITO deposition. Each photolithography step is monitored by AOI visual inspection and Cpk management for CD and TP as for the TFT array.

After AOI, a repair procedure is used at failed sites for yield improvement. Optical density (OD) check for BM and color properties for RGB layers are obtained by sampling. One of the critical issues in this process is the reference standard for visual inspection by human eyes. Since appropriate inspection machines that can

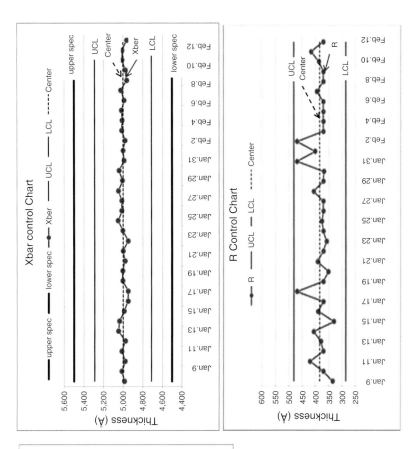
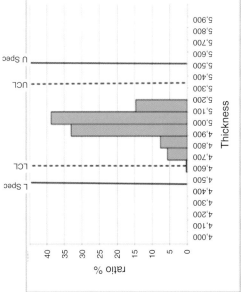

Figure 17.10 Example of Cpk control for SiNx thickness.

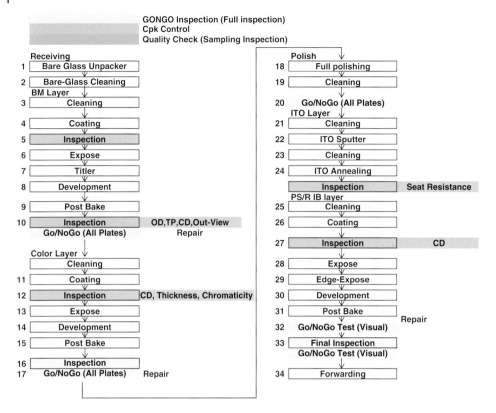

Figure 17.11 Process flow and quality check steps for color filter manufacturing.

completely replace human eyes are not available, development of inspection standards is quite difficult and subject to variability, remaining under continuous development.

17.4.3.3 LCD Cell Process

The process and inspection flow for LCD cell fabrication is shown in Figure 17.12. At first, the cell ID is marked on the substrate and then cleaning is applied for both TFT and color filter plates before printing of PI (polyimide) film. After the PI layer is coated, AOI is used to check the uniformity and pattern. Printing alignment accuracy requirement becomes quite high as the pixel density gets higher. The alignment layer for liquid crystal (LC) layer is changed from mechanical to an optical process based on new material and equipment that enable getting higher LCD contrast ratio.

The ODF (one-drop filling) process is performed after calculating the optimal dropping volume based on estimating the PS height and required cell gap. The examples of Cpk management for cell gap is shown in Figure 17.13. The panel edge seal is prepared by a printing method using dispenser, followed by AOI inspection of the seal width and its relative location with respect to PI-coated region across all the cells.

The lamination accuracy between the TFT array and color filter plates can lead to degradation of contrast ratio and photo leakage through the cell. Therefore, the accuracy of all the cells is necessary to be checked.

At the last step, all the cells are electrically contacted for display drive and image test.

17.4.3.4 Modulization Process

Figure 17.14 shows the process flows for module assembly and related inspections, including backlight unit, IC driver, PCB, and mechanical parts. At first, after reading the cell ID, a control number is written down onto

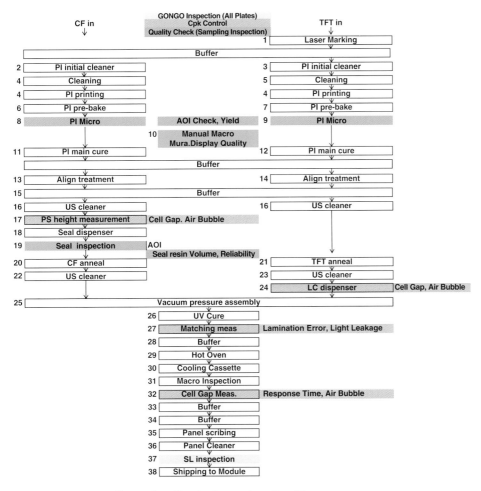

Figure 17.12 Process flow and quality check steps for cell-making process.

the cell. Then the cell is cleaned up to remove particles. After the cell is separated, the polarizer films are applied. Polarizer lamination needs a careful placement of the polarizing axis to meet the LC alignment. Furthermore, failures caused by particles under the polarizers are visually checked. If such failure is found, the polarizers are detached and a new one is applied again after removal of particles.

Driver ICs on TAB (tape-automated bonding) are connected to the cell using ACF (anisotropic conductive film). In order to achieve the regular connection, distortion (crushing) of the conductive particle in the ACF layer is observed by sampling. Additionally, the alignment accuracy between the TAB lead and electrodes on the cell is monitored by observing a dummy, alignment test pattern.

In the process step 12 of Figure 17.14 brighter backlight than the designed BL is applied for inspection purposes only. After the module is completed, aging test is applied. It was originally taken for 12 hours, and now reduced to 2 hours.

17.4.4 Organization and Key Issues for Quality Control

The quality control system is represented with an organization chart to clearly show how it works in the plant. One example is shown in Figure 17.15.

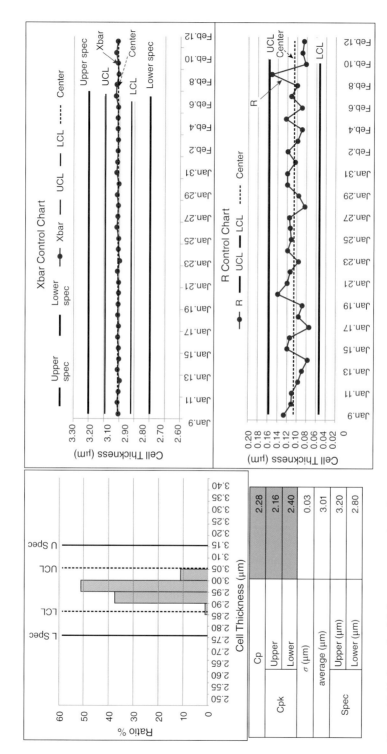

Figure 17.13 Example of Cpk management for cell thickness.

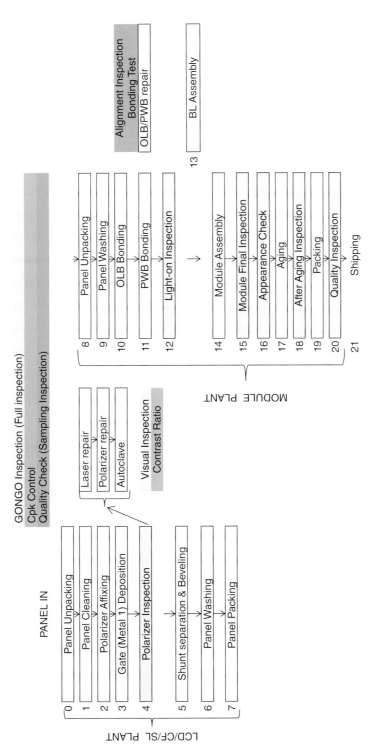

Figure 17.14 Process flow and quality check steps for modulization.

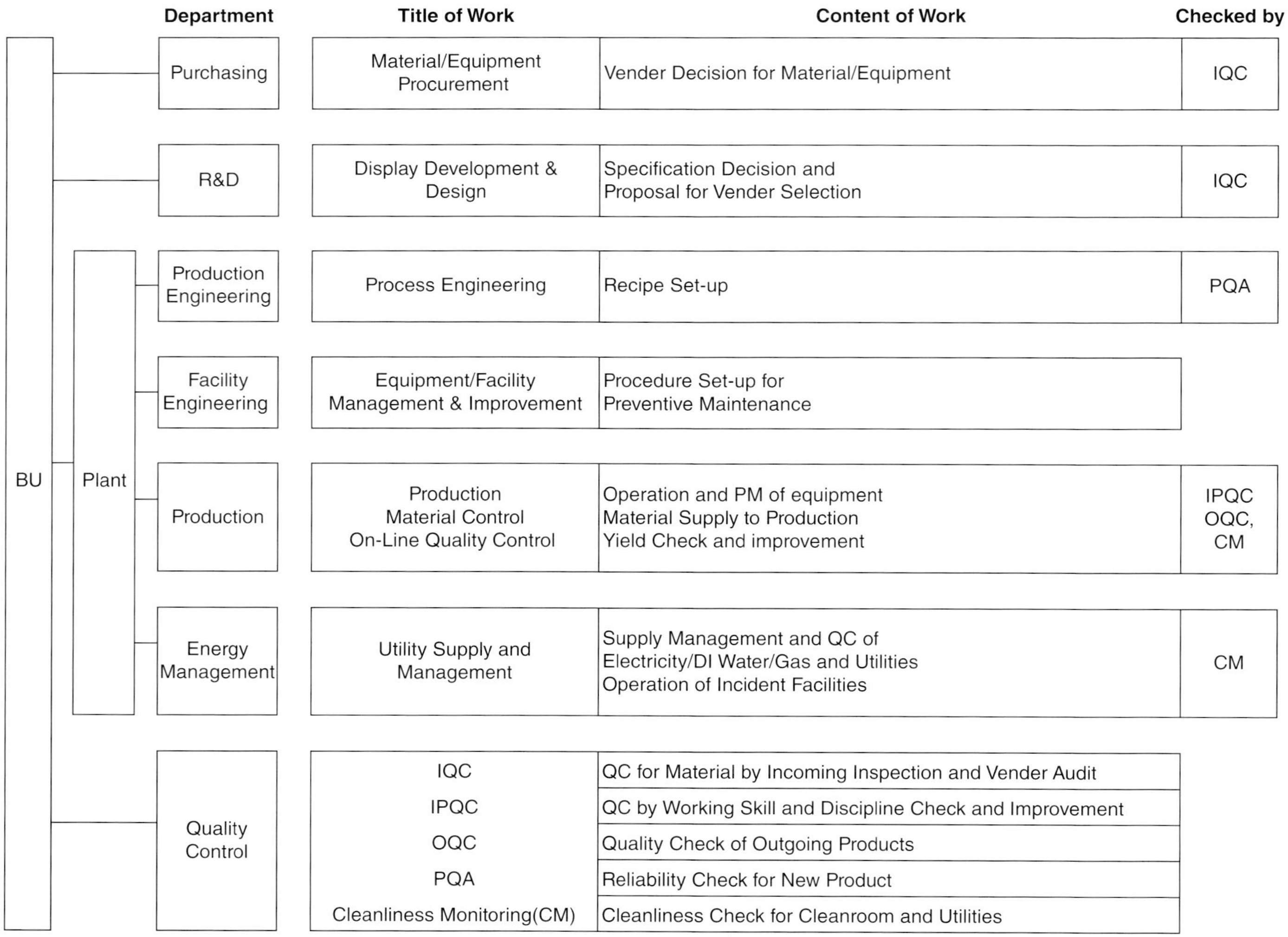

Department	Title of Work	Content of Work	Checked by
Purchasing	Material/Equipment Procurement	Vender Decision for Material/Equipment	IQC
R&D	Display Development & Design	Specification Decision and Proposal for Vender Selection	IQC
Production Engineering	Process Engineering	Recipe Set-up	PQA
Facility Engineering	Equipment/Facility Management & Improvement	Procedure Set-up for Preventive Maintenance	
Production	Production Material Control On-Line Quality Control	Operation and PM of equipment Material Supply to Production Yield Check and improvement	IPQC OQC, CM
Energy Management	Utility Supply and Management	Supply Management and QC of Electricity/DI Water/Gas and Utilities Operation of Incident Facilities	CM
Quality Control	IQC	QC for Material by Incoming Inspection and Vender Audit	
	IPQC	QC by Working Skill and Discipline Check and Improvement	
	OQC	Quality Check of Outgoing Products	
	PQA	Reliability Check for New Product	
	Cleanliness Monitoring(CM)	Cleanliness Check for Cleanroom and Utilities	

Figure 17.15 Organization in quality control.

Usually the QC department is independent from the production operation and always monitoring the production and product status in terms of quality. The QC department consists of five functions: IQC (incoming QC), IPQC (in-process QC), OQC (outgoing QC), PQA (product quality assurance), and CM (monitoring of cleanroom and clean operation). IQC is in charge of the materials used for production and working with purchasing and R&D departments. IQC inspects the incoming materials, and if necessary audits the vendor facility. IPQC is responsible for process operation quality. OQC is monitoring the quality level of the shipping products. PQA checks the quality and reliability of the new products. CM is also monitoring the cleanliness in cleanroom operation and environmental protection.

The critical issues in quality control and yield management are listed below. It is very important to follow the predetermined procedures in order to maintain high quality and production yield.

(1) Facility control organization must always monitor the particles, air chemical contamination, electro statics and its behavior, UPW (ultra-pure water), WWT (water and waste-water treatment), bulk and specialty gases in the incidental facility and equipment.
(2) Material and vendor changes must be approved under the control of "material change committee," which is organized across the different engineering departments.
(3) Process recipe changes must be done under the control of "process change committee," which is organized across the different engineering departments.
(4) PM timing is decided by authorized process engineers under Cpk management for 24-hour monitoring and quick action.
(5) All the monitoring and inspection tools must be checked periodically.
(6) Start of the new product production, production capability is reviewed by both development and production departments.
(7) Design rules and production capability are improved based on the roadmap prepared by development and marketing departments.

References

1 P. G. Le Comber, et al., Amorphous-silicon field-effect device and possible applications, Electron. Lett., 15, p. 179 (1979).
2 S. Morozumi, et al., B/W and color LC video display addressed by Poly-Si TFTs, 1983 SID Digest Tech. Papers, p. 156 (1983).
3 F. Funada, et al., An amorphous-Si TFT addressed 3.2-in. full-color LCD, SID Digest Tech. Papers, p. 293 (1986).
4 C. W. Kim, et al., A novel four-mask-count process architecture for TFT-LCDs, SID Digest Tech. Papers, p. 1006 (2000).
5 Q. Shang, et al., PECVD tool productivity enhancement with remote plasma source, Proceedings of Display Manufacture Technology Conference, p. 65 (1998).
6 Semi Japan (Semiconductor Equipment and Materials International) Production Cost Saving (PCS) Forum FPD-Phase-IV Roadmap Report (2002) (in Japanese).
7 R. Boyles, The Taguchi Capability Index, Journal of Quality Technology, 23(1) (1991).

18

Plant Architectures and Supporting Systems

Kozo Yano[1] *and Michihiro Yamakawa*[2]

[1] *Foxconn Japan RD Co., Ltd., 3-5-24 Nishi-Miyahara, Yodogawa-Ku, Osaka, Japan, 532-0003*
[2] *Industrial Infrastructure Business Group, Fuji Electric Co., Ltd., Gate City Osaki East Tower 11-2, Osaki-1, Shinagawa, Tokyo, Japan, 141-0032*

18.1 INTRODUCTION

This chapter introduces the typical TFT LCD plant architecture and supporting systems, pointing out the key issues when designing the new facility. First, the total structure of the cleanroom fab, where all the manufacturing processes are performed, is explained for a Gen 8 size plant example. The critical issues affecting the cleanroom; floor structure, air-flow control, particle generation countermeasures, and chemical contaminations are discussed in detail. Then the optimization and countermeasures for environmental issues in the supporting systems, including the incident facilities are introduced. Finally, the production control system CIM (computer-integrated manufacturing) is briefly summarized.

The first a-Si TFT LCD Gen 0 size plant (300 × 320 mm glass size) was built in April 1985 and used a cleanroom structure mimicking those using in the semiconductor industry at that time. Since then, the authors have studied cleanroom technologies aiming for optimizing the cleanroom architecture for LCD manufacturing for more than 30 years. One of the text books was "Ultra-Clean Room Technologies" [1], published in April 1988.

Until the Gen 4-Gen 5 era, most of cleanroom technologies were still transferred from semiconductor (integrated-circuit: IC) industry with slight modifications. However, after Gen 6, since the glass plate size was so much beyond IC industry standards, new cleanroom concepts were investigated in order to solve critical problems unique to LCD manufacturing. The key was to prepare cleanrooms which offer highly efficient production. Those developments and achievements were fulfilled with plenty of assistance and support by many vendors related to the cleanroom.

For progress of cleanroom structure for LCD manufacturing, close collaboration between the LCD company engineers who know well the manufacturing processes/equipment and vendors who can provide the necessary materials and know-hows for cleanroom is really important to create the best cleanroom technologies. That goal was always achieved by teamwork in interdisciplinary environment. All decisions made between those two parties were based on "fair and reasonable" discipline and "share and respect" spirit to maintain highly professional team-work.

In this chapter, the concept of the TFT plant design and construction is described, reviewing past teamwork with the vendors.

Flat Panel Display Manufacturing, First Edition. Edited by Jun Souk, Shinji Morozumi, Fang-Chen Luo, and Ion Bita.
© 2018 John Wiley & Sons Ltd. Published 2018 by John Wiley & Sons Ltd.

18.2 GENERAL ISSUES IN PLANT ARCHITECTURE

18.2.1 Plant Overview

Figure 18.1 shows an example of a cleanroom layout used for TFT array production. It consists of multiple process bays (photolithography, film deposition, wet-etching, dry-etching, wet-cleaning, and inspection), which are connected together by a cassette transportation system (roof-transportation). The loader/unloaders of the glass substrate are installed in buffer zones, which are also connected to the roof transportation system. Each processing bay has multiple production equipments, in order to share the process and maintain production operation when one system is down. The production capacity is estimated by the number of equipment units installed in parallel, factoring in the scheduled and unscheduled down time.

A typical cross-sectional view of the cleanroom fab is shown in Figure 18.2. The main structures of a clean room include the "clean area," the "sub-fab," and the "FFU" (fan filter unit). The process equipment is installed on the operation floor of the clean area, the location where the highest cleanliness is maintained. The under-the-floor area is called "sub-fab," where the systems delivering water, chemicals, and gases for process equipment operation and maintenance are separately located since they do not need the high cleanliness level required in the process operation space. Outside air (OA), after chemical contaminants and particles are removed, is ducted into the inside of the cleanroom through by FFU using the ULPA (ultra low penetration air) filter for particle removal. The cleaned air is circulated inside the cleanroom with temperature and humidity control. Clean air that might be chemically polluted is exhausted after removal of those contaminants by

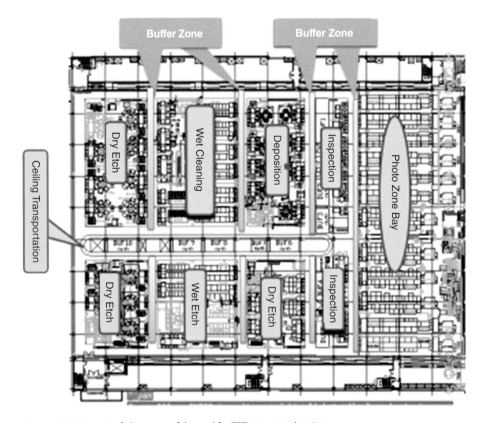

Figure 18.1 Layout of cleanroom fab used for TFT array production.

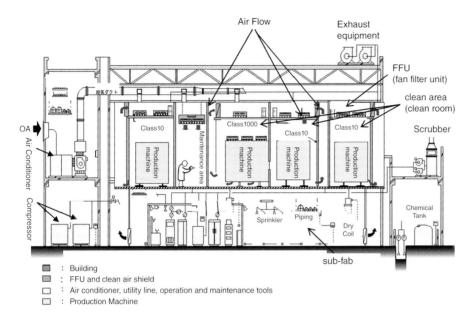

Figure 18.2 Cross-sectional view of a typical TFT LCD cleanroom fab. OA= outside air.

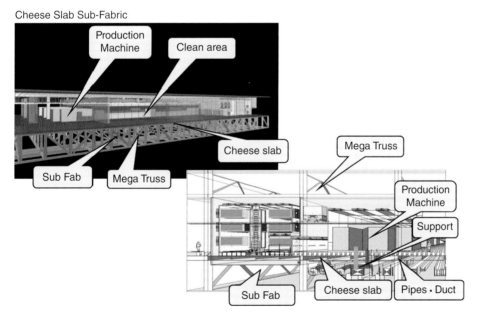

Figure 18.3 Cross-sectional structure of cleanroom.

scrubber. The typical incident facilities, including gas, chemical supply systems, as well as high-pressure air supply systems are shown in Figure 18.2.

The cross-sectional structure of a typical cleanroom fab is shown in Figure 18.3. The cleanroom floor that sustains the production equipment and transportation system is supported by a set of mega-trusses necessary for accepting heavy weight. The "sub-fab" spaces are used as the service chase for equipment,

Figure 18.4 Cassette and glass substrate transportation systems (*Source:* Courtesy of Muratec).

piping, and duct. The air in the sub-fab and the cleanroom floor is connected through perforated flooring called "cheese-slab."

The mechanisms and vehicles for the automated transportation system of cassettes with glass substrates are shown in Figure 18.4. In the early stages of Gen 2 to Gen 3 glass processing, the particle contamination introduced during glass transportation and cassette storage was a serious problem that degraded yield. In the phase of Gen 4.5 to Gen 5, AGV (automated ground vehicle, which moves along line markers on the floor) and RGV (railed ground vehicle) were the main approaches for cassette transportation in cleanrooms, however yield degradation due to the airborne particulates was still a major issue. From Gen 6 on, these systems were replaced with a ceiling transportation system connecting storage buffers and process equipment loader/unloaders, as the glass substrates become so large and heavy. Inside the buffers cranes are used for glass handling. In this system, since the glass is perfectly isolated from the open environment in the cleanroom, the chance of particle contamination is greatly reduced resulting in the stabilization of production yield. Such transportation systems have been commonly utilized up to Gen 10.

18.2.2 Plant Design Procedure and Baseline

When building up a new plant, the design work commences with organizing the project team, as shown in Figure 18.5. It is supported by multiple task force teams: plant building construction, cleanroom and inside arrangement, electricity, gas supply, water supply, wasted water treatment, and so on. The close collaboration between the teams involved is the key to make the project successful.

The starting point of the project is to clearly define the desire (goal) and demand (market). As one more important key, the plant for LCDs production should not provide any serious load to the environment since LCDs themselves are ecological devices that save space and energy. The plant should incorporate the latest environmental regulations as well as be cost-competitive in capabilities.

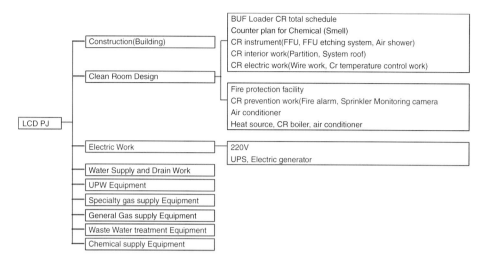

Figure 18.5 Project organization for new plant construction.

In the following discussions of plant architectures, the virtual plant of the most popular Gen 8.5 plants with 90k sheets/month input capacity will be referenced, comparing to a primitive Gen 0 plant.

Regarding the glass substrate transportation system, a full-automatic plate transportation system between equipment using AGV (automatic guided vehicle) controlled by CIM (computer-integrated manufacturing) have been introduced since Gen 2.5. Afterward, a combination of ceiling transportation system for bay-to-bay transfer and RGV for within bay transfer has been in use since G5. This system makes the operation rate (uptime) much higher due to the remarkable transportation efficiency.

The recently built Gen 8.5 plants with capacity >90k substrates/month have a high production efficiency due to reduction of TAT (turn-around time), avoiding process step bottlenecks by including three to four equipment units for parallel processing with the same process recipe. In fact, in the most advanced TFT-LCD plants, multiple equipment tools for each process are installed within the same bay using a centralized disposition of L/UL(loader/unloader), stocker with buffer cassette, and stacker crane for cassette transportation.

18.3 CLEAN ROOM DESIGN

18.3.1 Clean Room Evolution

Table 18.1 shows the evolution of LCD production from Gen 0 to Gen 8. The display size grew more than 10 times, from 3-4 inches to 50 inches diagonal, and at the same the time production capacity increased by more than 10 times. The plant scale and related floor area also greatly expanded in accordance with each substrate size enlargement.

Table 18.1 Evolution of TFT-LCD plant.

	G0-G1	G8-G10
Application	3″ TVs	50″-70″ TVs
Production Capacity	100k-500k/year	5.5 M/year
Time Frame	1988~	G8/2006, G10/2009
Plant Scale Ratio	1	240

Table 18.2 Scale example for Gen 8 plant.

	TFT	Cell	Color Filter	Inspection	Modulization	Total
Floor Size (m × m)	270 × 220 (2 floors)	300 × 160	300 × 160	140 × 140	140 × 100	
Area(mm^2)	108k	48k	48k	19.6k	14k	237.6k

Table 18.3 Comparison of fab requirements for exposure equipment by Nikon.

	G0 First Version(Stepper)	G8 (Lens Scan)	Ratio
Size (m)	3.3 × 2.1 2(height)	3.3 × 2.1 2(height)	23
Weight(ton)	3.6	130	36
Floor Strength	1 (ton/m^2)	3 (ton/m^2)	3
Plate Size (mm)	300 × 320	2200 × 2500	57

The first TFT plant reused an idle IC factory for producing 3-inch LCD screen TVs with a capacity of 50kpcs/month, and had an area only a little over 1,000 m^2. On the other hand, a typical G8 plant currently uses by comparison a huge clean room area as large as 240,000 m^2, as shown in Table 18.2.

The equipment sizes have become gigantic as well. Table 18.3 shows an example of the outer dimensions, weight, and the required floor strength vs. weight loading for a photolithography exposure machine provided by Nikon. The weight and size grew 37 and 20 times larger, respectively, with the glass substrate size more than 50 times larger, while maintaining the same highly precise operation conditions, such as anti-shock and anti-vibration capabilities. It is clear that clean room technology transferred from IC industry was no longer useful or reasonable beyond Gen 5 factory.

18.3.2 Floor Structure for Clean Room

In the past, the raised floor structure with a grating floor was used to maintain a high cleanliness with layered air-stream, as well as flexible equipment layout. Figure 18.6 shows a picture for a typical grating floor, and Figure 18.7 shows a schematic explanation of the air flow pathways inside grating system. Clean air flows from the ceiling to the grating floor, making for smooth down-stream returns from the under-floor through the returning shaft zone.

As the glass substrate size got larger, the production equipment also became heavier and bigger resulting in requirement of tougher floor strength and higher performance against shocks and vibrations. The cheese-slab floor shown in Figure 18.8 replaced the grating system. The cheese-slab structure employed round cheese-shaped openings over the concrete floor to achieve the layered stream. The floor itself is supported by a truss structure over the concrete floor as shown in Figure 18.9. The concept of the air flow is exactly the same as for grating floor system. One advantage of the cheese-slab structure, besides high strength, is the large space in the truss area, which can be used for sub-equipment, duct, and pipes in the sub-fab, making construction and maintenance easier.

18.3.3 Clean Room Ceiling Height

The ceiling height for clean rooms before Gen 6 was mainly determined by the height of the production equipment. In those days, glass plate transportations between equipment were achieved by AGV, so those

Figure 18.6 Photo of grating floor structure.

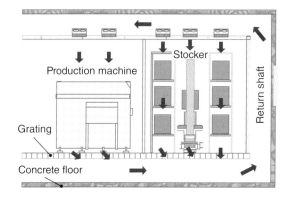

Figure 18.7 Airflows in grating floor structure.

transportation systems were not impacted by ceiling height. As example, the ceiling height was 2.4 m and 2.5 m for Gen 2 and Gen 3, respectively.

On the other hand, the ceiling height of clean rooms after Gen 6 is determined by the stocker arrangement rather than the equipment height. The stocker must store a predetermined number of substrates in the clean room as a buffer in order to maintain good enough operation rate against downtime for scheduled maintenance or unpredictable accidents. The location of the buffering stocker is designed to allow continuous total plant processing operation for total plant processing even if the process is stopped at a particular equipment or zone. The capacity of that buffering stocker is determined considering the parameters with a dynamic simulation:

- Processing time and cassette transportation time
- Scheduled maintenance time (estimated by MTBF) and recovery time (MTTR)
- Mean time during failure and recovery

The usual cassette contains 25 or 30 glass substrates, and the typical stocker can accept a stack of three cassettes height. The path line of the glass substrate is 1.5 m long in general (decided by the plant designer),

Figure 18.8 Photo of cheese-shaped slab floor structure.

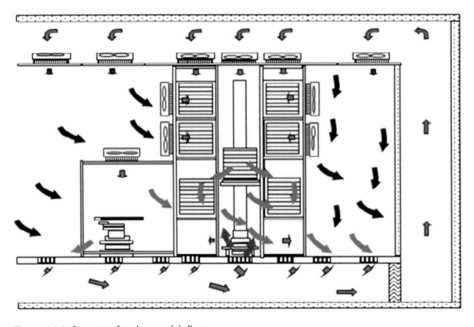

Figure 18.9 Structure for cheese-slab floor.

the cassette height is 1.7 m including handling space, thus the total height of the stocker becomes 7-8 m with three stacked cassettes. This height becomes the factor that determines the clean room overall height—in most cases, the height of equipment is lower.

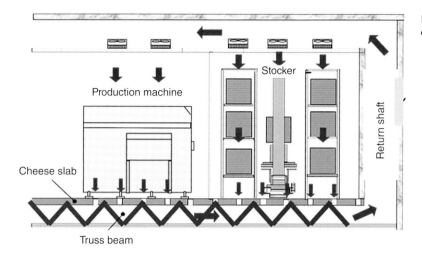

Figure 18.10 Total space air flow and circulation.

18.3.4 Air Flow and Circulation Design

In the design of air flow and circulation, two different concepts have to be investigated: total circulation and separated circulation

- **Total Circulation:** air flow circulates in the whole cleanroom as one space (Figure 18.10)
  (1) All the space in the building is clean room
  (2) Common air-return shaft is simple and flexible in location, and possibly contributes to space saving
  (3) Cross-contamination risk due to air pressure fluctuation and filter leakage is high
  (4) Chemical contamination from other equipment may happen
  (5) Energy for air-conditioning is reduced because outer static pressure is lower
  (6) Slant turbulence may occur due to relatively long distance from returning shaft
  (7) Cross-contamination during glass transport may happen due to air-flow diffusion

- **Separated Circulation:** Each processing area has separate and independent air-flow circulation (Figure 18.11)
  (1) Each area has independent returning shaft
  (2) Returning shaft has certain restriction and is not flexible in location
  (3) Contamination risk is low even with air pressure fluctuation and filter leakage
  (4) Chemical contamination risk is low, and countermeasure is easily done
  (5) High air-conditioning energy is required because the outer static pressure is relatively high
  (6) Smooth air-flow is easily achieved due to closely located returning shaft
  (7) Large enough air-flow volume within controllable range allows efficient particle-prevention

In smaller plants till Gen 6, total circulation was a useful choice. In Gen 8 to Gen 10 plants, since the cleanroom space has become much larger and therefore air-return on the four sides of one big cleanroom has become extremely difficult, each process zone is separated into small rooms employing partition walls and independent air-return in each smaller space. Thus, in the recent Gen 8 to Gen 10 plants a separated air-circulation (small-unit space circulation) became the major structure for the cleanroom. Besides a-Si TFT LCD plants, in LTPS and optical alignment processes, which have relatively weak molecular enforcement, chemical contamination has become more serious and perfect isolation of those processing rooms is required. Partitioning into small processing rooms to avoid chemical cross-contamination is advantageous, and thus the separated air-circulation has become the norm.

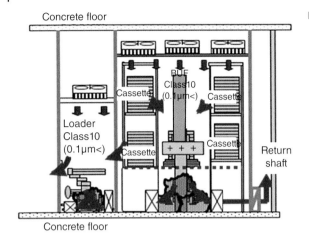

Figure 18.11 Separated space air-flow and circulation.

18.3.5 Cleanliness Control

Before Gen 5 plants, the entire processing spaces operated under high cleanliness control. However, beyond those generations, the plant building size has escalated enormously causing extremely increasing construction expense as well as running costs. Therefore, the specification of the cleanroom was reviewed and modified to save initial and running costs. As one useful approach, a high cleanliness specification was used for the areas where the glass substrate is exposed to air, with a lower cleanliness for the areas where the glass is not exposed. Thus, in Gen 8 plants, the area for maintenance and for the glass-unexposed zone is designed to be class 1,000 (over 0.3 μm size particles).

As an example, Table 18.4 shows a comparison of the cleanliness settings of the Gen 1 plant (320 × 400 mm) built up by Sharp as the world's first factory dedicated TFT-LCD and those for a Gen 8 plant (2,160 × 2,460 mm), including a breakdown into each process step. In the G1 plant, the entire processing area is designed to be Class 10 (0.3 μm), whereas in a Gen 8 plant each processing area is independently controlled, making the clean room operation cost-effective and energy-efficient.

18.3.6 Air Flow Control Against Particle

The high cleanliness of the clean room fab environment is maintained by clean air flows. Unfortunately, in practice unnecessary particles and dusts are always generated by movement of the transportation system, and thus controlling and minimizing this particle generation is the most important preventive action for quality control and associated production yield improvement. The major root causes of this particle generation are:

(1) Particles coming from the motion of the plate transportation vehicle
(2) Particles generated by the air turbulence when transportation vehicle is in motion
(3) Particles generated by up/down motion of the robot arms

Therefore, in order to keep high cleanliness, minimization of those particle generation has to be considered.

(1) Employment of high air flow FFU (fan filter unit): the air velocity of the conventional FFU was about 0.4 m/sec, whereas 0.6 m/sec FFU was used in the high ceiling area, preventing the anti-directional air flow from the floor when handling robot arm moving up or down.
(2) Balanced air return layout
(3) Duct installed underneath the stacker crane: avoiding air turbulence when crane moves

Table 18.4 Evolution of cleanliness requirement for Gen 1 and Gen 8.

G1

Process	Buffer	Loader	Equipment	Maintenance
Cleanliness (Class)	10(0.3 μm)	10(0.3 μm)	10(0.3 μm)	100(0.3 μm)
FFU Air Velocity (m/s)	0.4	0.4	0.4	0.4
FFU Size (mm × mm)	not specified	not specified	not specified	not specified
Sensible Heat (Δ t/deg)	4	4	4	4

G8

Process	Buffer	Loader	OHV	Exposure	Wet/Vacuum Maintenance	Inspection	CA: Inspection	Photo Cleaner Dev.	CA: I/F CT	Cell Equipment	Cell Maintenance
Cleanliness (Class)	10(0.1 μm)	10(0.1 μm)	10(0.1 μm)	1,000(0.3 μm)	5,000(0.3 μm)	10(0.1 μm)	100(0.1 μm)	1,000(0.3 μm)	10(0.1 μm)	10(0.3 μm)	5,000(0.3 μm)
FFU Air Velocity (m/s)	0.6	0.4	0.4	0.4	0.4	0.4	0.4	0.4	0.6	0.4	0.4
FFU Size (mm × mm)	750 × 1800	750 × 1800	750 × 1800	750 × 1800	750 × 1800	750 × 1800	750 × 1800	750 × 1800	750 × 1800	750 × 1800	750 × 1800
Sensible Heat (Δ t/deg)	1.5	2	1.5	2.4	2	4	2	2	1.5	2.4	2.4

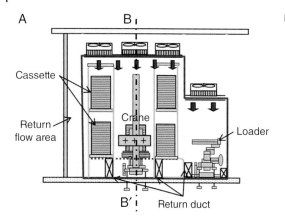

Figure 18.12 Model for simulation.

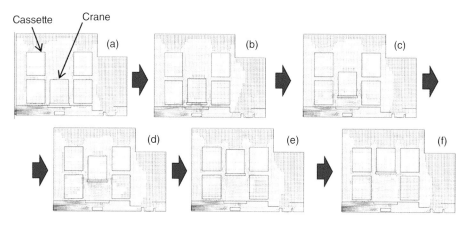

Figure 18.13 Model results for vertical motion of the crane (1).

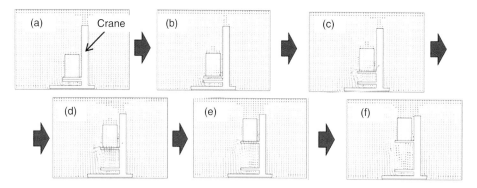

Figure 18.14 Vertical motion model of the crane (2).

In order to verify the impact of the above countermeasures, computer simulations of the air flows and turbulence were carried out. An example for such simulations is shown in Figures 18.12 to 18.14, including the cassette stocker model, computer simulation results of air flow in the direction of plane A and results in the vertical direction, respectively (B-B' cross section, see Figure 18.12 diagram).

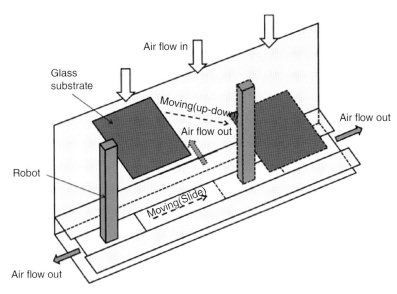

Figure 18.15 Model for robot motion.

Here, the computed air flow is shown when the stacker crane moves up under the predetermined air flow velocity from FFU on the ceiling (Figure 18.14(a) → (f) and Figure 18.15(a) → (f)). The floor is a grating system and supplemental duct is used for air-exhaust. The simulation results show:

(1) Under the crane, whirlwind turbulence is generated but anti-directional air flow from the floor does not occur
(2) The velocity of the whirlwind can be high in areas, but the impact on the cassette is negligible small
(3) Particles from the crane are also negligibly small

It was proven using computer simulations that the impact of the crane motion in the stocker can be minimized down to a negligible extent (Figure 18.13). Thus, computer simulations are quite useful in building up a highly qualified clean room operation.

Afterward, the simulation of robot motion during glass transportation has been done, as shown in Figure 18.15. The simulation of air turbulence with horizontal motion and slide moving of the robot is shown in Figure 18.16. Here the appropriate number of openings on the floor is optimised in the simulation model to offer a smooth and balanced air flow. The results show:

(1) A slight upward air flow was observed around the side of the glass plate when robot is moving (Figure 18.16A)
(2) The air flows into openings on the floor when robot stops, and no upward-air flow was observed (Figure 18.16B)

Thus, in order to achieve an optimized cleanroom design, a series of action programs starting with verification of customer request on the specification, and then settling on the particle generation mechanism for cleanroom structure investigation on how to minimize such particle generation, and execution of the computer simulation. Finally, verification of the cleanroom performance after construction is necessary.

18.3.7 Chemical Contamination Countermeasures

Once chemical contamination happens in the cleanroom, degradation of the glass substrate or formation of an oxidation layer can be observed, resulting in product disqualification. Accordingly, chemical contamination as well as particle generation must be avoided.

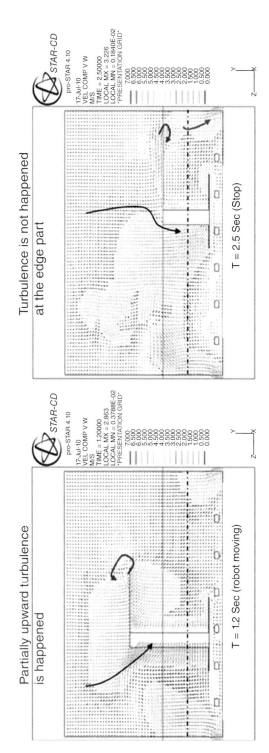

Figure 18.16 Air Turbulence when horizontal motion of the robot.

The principles for excluding such contamination is almost the same as for the particle control:

(1) Prevent contaminants from getting into the cleanroom; incorporate the washer and chemical filter into the air-conditioner
(2) Prevent generation in the cleanroom; use materials with minimized outgas
(3) Prevent diffusion of chemical contaminants; use separated space layout
(4) Remove the contaminants; promotion of the air circulation using chemical filter

In terms of classification of chemical contamination zones, there are two kinds: chemical contamination generation zone and contamination avoiding zones. In the chemical contamination generation area, such as photoresist coater and developer zone, the air pressure must be set negative and exhausters must be employed so as not to transfer the contaminants into other zones. On the other hand, in the chemical contamination avoiding area, such as the exposure zone, invasion of the contaminants must be absolutely avoided. Therefore, a large enough volume of chemically decontaminated air must be provided into that zone and that area must maintain a positive air pressure. Anyway, in order to avoid those chemical contaminations, air-shielding is the important countermeasure using airtight means in the doors or walls which separate the zones.

The buffering zones (stockers) in particular require the upmost care for chemical contamination since many glass substrates are stored in stockers for relatively long time compared to processing areas. Once such contamination affects the substrates, it may lead to a big financial loss. Therefore, in order to avoid even small chance of chemical contamination, double shielding method (double skinning) as shown in Figure 18.17 is a useful approach.

(1) Applying an outer partition over the buffering area, which makes double shielding
(2) Chemically decontaminated air is induced into the buffering zone with positive air pressure
(3) Air-return for the buffering area is located in independent area avoiding the air intrusion from neighboring the chemical zones
(4) As an internal countermeasure, chemical filter is applied into the return path

18.3.8 Energy Saving in FFU

In order to keep high enough cleanliness in the cleanroom, the air flow volume must be high. The ratio of FFU area with respect to total ceiling area used to be around 70% until the Gen 6 size plant. That means the high numbers of FFU is installed into the ceiling, resulting in high running cost as well as the high initial charge. After the Gen 8 size plant, due to computer simulation and test experiments before installation, the real installed number of FFU to keep the reasonable air flow was minimized. In fact, it was shown that even 40% FFU area ratio can maintain enough cleanliness. It has made the Gen 8 size plant cost effective and energy efficient.

18.4 SUPPORTING SYSTEMS WITH ENVIRONMENTAL CONSIDERATION

18.4.1 Incidental Facilities

The incidental facilities consist of water, chemical and gas supply systems, waste treatment, and electricity (part of them are previously shown in Figure 18.2)

When a new LCD factory is planned and designed, the plant engineers have to consider environmental issues as much as possible to minimize the load to the environment. Besides the operation efficiency discussed in the previous section, environmental consideration is another important key factor to achieve the consistent and economical operation. In this section, water usage and its saving are the first issue to discuss, followed by the topics on chemicals, waste treatment, gases, and electricity.

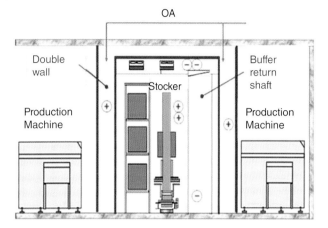

Figure 18.17 Double skinning structure.

18.4.2 Water and Its Recycle

The water supply for the plant usually depends on the local society and community, and therefore the location of the newly built-LCD plant is carefully selected after investigating water source feasibility and availability.

In the LCD factory, especially at the early stages, it was common sense to use a huge volume of water for cleaning, air conditioning, and so on. Furthermore, the wasted water drained out from cleaning processes was immediately released to environment after cleaned up without recycle, resulting in high-level of water consumption. The first Sharp-Tenri LCD plant was located inland and accordingly water for the plant depended on the municipal water supply because no dedicated water source was available within the factory territory, and it always suffered from water shortage. In fact, in the 1990s, water shortage season at mid-summer, water was conveyed from the neighboring factories using tank trucks. From those experiences and background it is clear that the "recycle of water" became the definite trend in the next LCD plant.

An example of the water capacity is summarized in Table 18.5 showing the most advanced water treatment technology for Gen 8 size fab with 90k-plate start meeting the ecological and economical requirements.

Table 18.5 Estimation of water usage for Gen 8 size 90k fab.

Input Volume (/day)	tons
Required Volume for Cleaning (purified water)	A 47,500
Source Supply Volume (Rate of purification 90%)	B 53,000 (47,500/0.9)
Recycled Water from Cleaning Process and Scrubber	C 37,500
Supplemental Volume from Water Supply (B-C)	D 15,500
Water Usage for Cooling and other Purposes	E 14,500
Total Water Consumption (D + E)	F 30,000
(in the case without Recycle System) (C + D + E)	(G) (67,500)
Water Recycle Rate (B/C)	70%
Released Water by Piping-out	H 19,950
Release Water to Air	I 10,000
Contaminated by Sludge	J 50

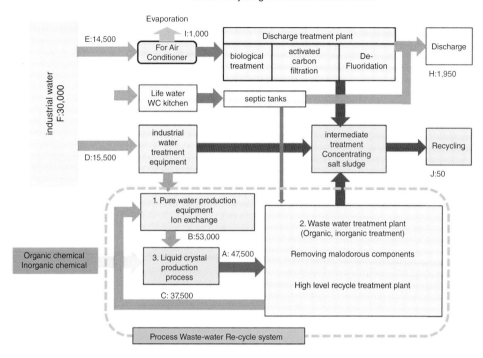

Figure 18.18 Water treatment and recycling system.

The water usage for cleaning was estimated to be 47,500 tons/day (A) based on the specification of equipment and process conditions. Water released to the air in cooling devices for process equipment and cooling towers in air-conditioners was estimated to be 14,500 tons/day (E). Here wasted water is cleaned and purified, and then returned to the purified water source within an economical range. Water used by the cleaning processes is recycled with an appropriate water treatment. Thus, the wasted water is released either by discharging or recycling depending on the situation. Of course, discharged water is also treated and purified up to extent that environment is required. The total structure of water treatment and recycling system is shown in Figure 18.18.

The wasted water recycle system for processing is shown in Figure 18.18. In-process cleaning water (A) is mostly recycled back to purifier equipment after applying the appropriate treatment (chemical treatment, biological treatment, reverse osmosis membrane treatment) depending on processes and equipment. Then purified water using ion exchange and UV radiation is re-used for liquid crystal production.

Source supply volume for process is estimated to be 53,000 tons/days (B) (A: 47,500/0.9) and recycled water is 37,500 tons/day water (C). Therefore 15,500 tons/day water (D) (B: 53,000-C: 37,500) should be supplemented from industrial water source. Industrial water needed for cooling and other purposes consumes 14,500 tons/day (E). Necessary total industrial water is 15,500 (D) + 14,500 (E) = 30,000 ton/days, whereas it needs 53,000 + 14,500 = 67,500 ton/day without using recycle system. Thus 55% of water can be saved by applying water recycle system.

In the future, some of discharged water could be reused with further progress in water treatment technologies such as:

(1) Arrangement of dedicated pipe line in anticipation of future improvement: wasted water should be sorted as much as possible with the dedicated pipe lines

(2) Concentration control: equipment design for avoiding mixing-up of cleaning water with the liquid chemical in equipment
(3) Retrieval of phosphoric acid from water for recycle: RO membrane now comes to feasible stage

Thus, it is necessary not only to employ all of the currently available water treatment technology, but also to leave spaces to accept future improvement.

18.4.3 Chemicals

Ten years ago the target price of LCDs was $100/inch, while presently the price of 32-inch LCD TVs is only around 300$. These price reductions were enabled by material cost saving as well as productivity improvements. In such a sense, the saving of chemicals is an urgent issue to solve, and the following recycling programs have been taken:

(1) Recycling photolithography developing chemicals: The lifetime of developing chemicals was extremely extended by removing the polymers coming from photoresist using reverse osmosis (RO) membrane and adjusting alkali concentration by adding alkali TMAH liquid
(2) Recycling of etching chemicals: The lifetime was also extended by monitoring and adjusting the liquid concentration. It also contributed to fine CD control
(3) Recycling of photoresist remover chemical (stripper): using evaporation refining either on-site or remote.
(4) Recycle of photoresist: Recycle of the photoresist is technically possible by viscosity adjustment removal of particles, optimization of photo-sensitivity approaches. However, photoresist consumption was extremely reduced by the slit coating technology, replacing spin-coating used in the primitive stages, and subsequently interest in the recycling of photoresist is not paid so much attention nowadays.

18.4.4 Gases

The gas delivery system and its piping materials were originally transferred from the IC industry. However, when the processed glass area expanded by over 200 times (3,000 m^2/month in the Gen 1 and Gen 2 size plants and 600,000 m^2/month in the Gen 8 and Gen 10 size plants), the consumed gas volume was also extremely expanded, in which the conventional gas delivery system was no longer useful.

At first, on-site gas production and delivery were investigated, however it was solved with new transportation using high volume container. At the beginning, 0.3 L cylinders were used which over time increased to 1,000 L containers, as shown in Figure 18.19.

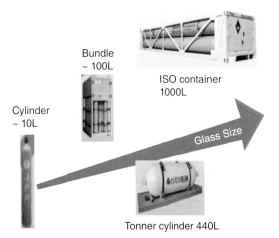

Figure 18.19 Evolution of specialty gas delivery system (*Source:* Provided by Taiyo Nippon Sanso Corporation).

Table 18.6 Evolution of electric power saving.

	G2	G3	G4	G5	G6	G8
Capacity	30k	30k	80k	72k	80k	90k
Annual Electric Power (MW)	180,000	224,000	336,000	254,000	350,000	513,000
Single Glass Area (m^2)	0.200	0.358	0.598	1.320	2.775	6.160
Total Glass Area (m^2/month)	6,000	10,725	47,872	95,040	222,000	554,400
Electricity (MW/month)	14,963	18,704	28,000	21,167	29,167	42,750
Electricity/Total Glass Area (MW/m^2)	2.494	1.744	0.585	0.223	0.131	0.077
Reduction Ratio	100.00%	69.93%	23.45%	8.93%	5.27%	3.09%

This increase was achieved by development of high volume containers and legal deregulation during transportation, and today gas production is centralized at the most suitable site and transported to the fab site. This has brought a big change to the industry in terms of fab location. LCD plant can be built up anywhere the transportation system is ready, and it concentrates the fab locations in four territories in the world: Japan, Korea, Taiwan and China. It also suggests that in the future new fabs may also relocate to other territories. Regarding bulk gas, on-site production is usual as in other industry.

The exhaust gas treatment system evolved from absorption removal to burning removal type. Since the burning type system tends to be compact enough, typically one-by-one installation directly accompanying the CVD equipment has been adopted as an effective countermeasure against gas leakage and residual accidents.

NF$_3$ gas is an inevitable material for cleaning the inside of vacuum chamber. It was introduced in the G3 period and led to drastic improvement of plasma damages on the TFT electrode, resulting in yield rate increased by a few %. Then all users in the world, including IC industry, have started to demand this gas and resulting in supply shortage and high prices. Presently, this issue was settled by price increases and enlargement of delivery capacity.

18.4.5 Electricity

The electricity delivery system for LCD plant is not so special, but just as referred in the previous "gas system" section, the energy consumption also drastically increased as the glass plate increased from Gen 1 to Gen 10. In case of the electricity supply system, such scaling-up was achieved by the regular approach without introduction of new system or technology. Installation of back-up power system for non-stop operation was also achieved by the conventional approach.

Table 18.6 shows the evolution of electricity savings from Gen 2 to Gen 8. The power consumption per unit glass area has been extremely and successfully reduced to 3.1% in Gen 8 relative to Gen 2, due to processing TACT reduction, power reduction in equipment and cleanroom operation, and cleanroom space area reduction.

18.5 PRODUCTION CONTROL SYSTEM

In a typical Gen 8 TFT-LCD plant with 90k sheets/month input, TFT processes have 38 steps and each step has 10 equipment units in average, resulting in 380 pieces of equipment under control during production. That typically means more than 1,000 processing units, including inspection tools for quality control and other auxiliary instruments, are under operation at a given time. Besides these, with additional control items including recipe and tool management (e.g., photomask, sputtering target, chemicals) the resulting number of controlling factors becomes very large. Furthermore, in some cases precise processing control is required. For instance, in exposure systems only a combination of specific machines is capable to achieve the critical performance, which is handled only by computer control being too complex for manual control. A CIM (computer-integrated manufacturing) system provides such complicated and highly sophisticated control capabilities, and it is served by IBM, HP, and other system vendors.

In the primitive Gen 1 or Gen 2 plant, computers were employed to control only the transportation with AGV and to convey the WIP (work-in-process) cassettes from stocker to loader/unloader at the front end of the process equipment, and then to the next station after processing is completed. The major purpose of computer system installation was essentially to manage transportation and keep a clean environment by excluding the chance of particle contamination introduced by human movement. It was a really tough work at that time. Thus, the starting point of CIM functionality was to provide transportation commands to cassettes in stand-by to be moved to the next process step. Later, in 1993, during the transferring phase from Gen 2 to Gen 2.5, the CIM system added new functionalities for "avoiding mistake" and "efficiently" in addition to transportation itself. The verification function between the process record of the specific cassette and applied equipment was further added and it contributed to avoiding process errors.

Then, efficiency improvement is done by collecting and monitoring the whole stock (inventory) status to minimize the number of WIPs and to maximize operation up-time. Further new functions were added to CIM for pursuing high operation efficiency, such as production scheduling to meet the customer requirement, high speed test lot flow to check the technical result in a short time, determination of PM (preventive maintenance) timing, negative impact minimization plan when equipment failure takes place, and so on.

Considering a typical Gen 8 fab, four cassette stands (ports) are included at the front end of each of the 380 process tools to reduce transportation loss, resulting in a total of 1,420 ports installed. The central control unit using CIM system is always monitoring the status of stocks in the standby cassette stocker, checking the waiting time of each cassette in the stocker and issuing the moving command at the best time. First-in/first-out is the base of such command.

Quality control and early failure discovery assistance functions have recently been added to the important role of CIM system. In TFT array production multiple process stations are used in parallel, as for photolithography and CVD, or multiple steps can be combined with inline connection as in photoresist stripping process using multiple in-line connected vessels, as shown in Figure 18.20. In any types of the process stations (chambers and vessels in the example in Figure 18.20), all the process information and history are stored in central CIM computer through the communication line. The process data for all the glass plates on

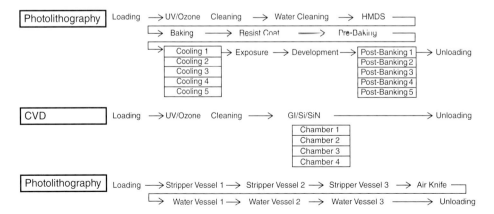

Figure 18.20 Process flows for photolithography, CVD, and stripper.

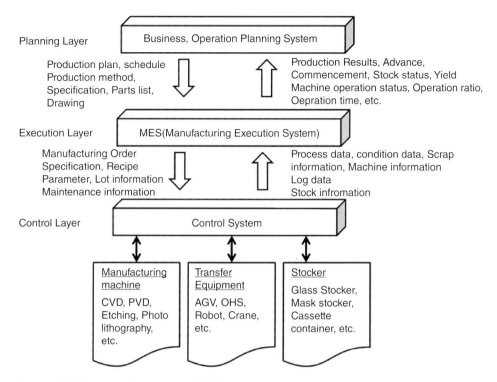

Figure 18.21 Conceptual structure of CIM [1].

process recipe, process station, processed time, and related equipment status information is accumulated in a database. Such data is the source of failure analysis when the reliability and quality problems happen, and is further helpful for yield improvement.

In a typical LCD plant, multiple types of LCDs with different panel sizes and different specifications are produced in parallel, using complex processing recipe involving multiple types of equipment shared by parallel flows. Only CIM can control such a complex production system. Figure 18.21 shows the conceptual structure of CIM system with three layers categorized into "1. Planning layer," "2. Execution layer," and "3. Control layer" where these three systems (layers) communicate with each other and control the production and the factory.

(1) **Planning layer**

Production schedule, production method, product specification, and so on, are planned in this layer and they are transferred to a production execution system. Then planning system receives the feedback data (production data, advance data, stock status, yield data) from the production execution system, and modifies the plan and transfer it to execution layer, if necessary.

(2) **Execution layer**

Based on the information from the planning system, the execution system determines the process and informs specifications, recipe, production sequence, and so on, to control system. Then the execution system receives process data, device data, log data, and so on, form control system.

(3) **Control layer**

The control system directly communicates with production machines, transportation equipment, stockers, and so on, controlling them with specifications, recipe, and production sequence received from the execution system. The control system collects production data, machine status data, log data, and inform these data to the execution system.

References

1 Controls and Definition & MES to Controls Data Flow Possibilities, MESA International White Paper No 3, (1995).

19

Green Manufacturing

YiLin Wei, Mona Yang, and Matt Chien

AU Optronics, Science Park, Hsinchu, Taiwan, ROC, 30078

19.1 INTRODUCTION

In the last 20 years, the TFT-LCD industry has seen major breakthroughs in technological development, product manufacturing, and market expansion. With increased product size and display resolution, the industry has also matured and expanded in fabrication plant (fab) size, total raw material consumption, and waste generation. The LCD manufacturing process involves many issues in different aspects such as fab construction and manufacturing machine selections, chemical management, water and electricity consumption, carbon emission, product design, end processing, raw material handling, and product delivery. Due to the complexity of supply chain and electronic product characteristics, enterprises endeavoring green manufacturing must not only consider the environmental impacts of the manufacture processes, but must also take into account the various cradle-to-gate and gate-to-grave impacts.

From a business perspective, any environmental issue should be carefully dealt with and reviewed by a cost assessment and subsequent benefits, such as tangible costs, intangible risks, and short- and long-term benefits. With increasing total resource volume to meet the demand of the LCD industries, life-cycle concepts of green manufacturing and environmental impact have become key issues in the last decade.

The relationship between the TFT-LCD display manufacturing and the environment comprises four dimensions:

- Fabrication plant (fab) design
- Product material uses
- Manufacturing features and green management
- Future challenges

These issues will be discussed in the following sections in more detail.

19.2 FABRICATION PLANT (FAB) DESIGN

19.2.1 Fab Features

To reduce the impact of particles on the yield rate, LCD panel must be manufactured in a clean room environment with positive pressure and controlled humidity and temperature. The clean room comprises three floors. The top floor is designed for the air returning system. The middle floor is for the production-line machines, called fab layer, and the ground floor is for the auxiliary machines, called sub-fab. A central utility building (CUB) located outside the clean room is designed for the support of utility for the manufacturing use in the fab. Given the specific demands of manufacturing, a building size of around 200 × 300 m is usually required for a production that meets economic scale. Therefore, a LCD fab consumes a considerably large area of land.

For those who have heard and wondered about the difference in fab generations, it's really quite simple. The major difference between the different generation fabs is the size of the mother glass, which can be cut into various sizes of panels. The next generation fabs have larger mother glass to either be cut into more pieces of panels to increase capacity and lower cost, or produce panels of larger size (e.g., LCD TV displays). From the birth of the industry in early 1990s, the early Japanese manufacturers began with generation one (Gen 1) production. The mother glass used in Gen 1 was approximately 30 × 40 cm, equivalent to the size of an open fashion magazine and could make one 15-inch panel. In 1996, the technology had already advanced to Gen 3.5 with the 60 × 72 cm mother glass size. Also, the gigantic mother glass of Gen 7.5 fab has dimensions of 195 × 225 cm to achieve the optimal cut for flat panels over 40 inches in size to produce large-sized LCD TV displays. The mother glass used in Gen 8.5 is 220 × 250 cm in size, equivalent to the size of a pool table. Also, the mother glass thickness has been decreasing over the years. Therefore, the new generation fabs require more advanced process technologies.

Compared to Gen 1 fab, which has a glass substrate measuring 30 × 40 cm, a Gen 8.5 fab, glass substrate is 2.2 × 2.5 m, which is a full 46-fold increase (see Figure 19.1). Substrate glass affects production machine size, which in turn affects the size and specifications of auxiliary machines, and subsequently increases the amount of raw materials and energy source needed and fab size. In addition, extensive human resource, material suppliers and complexity in manufacturing process are also increased along with the new generation fabs.

19.2.2 Green Building Design

To reduce negative environmental impact, factories are aiming for green building design and development:

- Greenbelt maintenance in site design: to protect existing and future development of open spaces and green space planning, as many native species are retained as possible without affecting fab configuration. Otherwise, attempt will be made to uproot and transplant the species to the green space on a new site.
- Reduce secondary damage in site development: achieve maximum earthwork balance when designing the new construction. Reduce transportation contamination from outward earth transportation or outside purchase. Adopt protective measures during construction to reduce environmental destruction from fugitive dust.
- Incorporate rainwater recycle design: to increase water resource usage, include a drainage system to retain rainwater for watering and relieve water runoff during heavy storms.

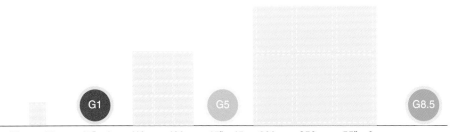

Figure 19.1 Comparison of glass substrate size and factory generation [1].

- Reduce heat island effect: in addition to adding rigidity or asphalt to roads that bear heavy weight, water permeable design is used for public roads such as pedestrian walkways and outdoor parking lots. In addition, more green space and plants are added to reduce heat island effect.
- Use recycled material for construction: given advances in material development, materials such as environmental protection coating, certified wood and water permeable pavers containing recycled glass are appropriate for the landscape design.

19.3 PRODUCT MATERIAL USES

19.3.1 Material Types and Uses

LCD materials can be divided into key categories such as metals, polymers, electronic components, glass and chemicals (see Figure 19.2). Metal is mainly used for the LCD backlight modules; metal frames (bazel) and

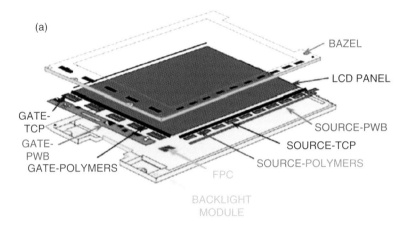

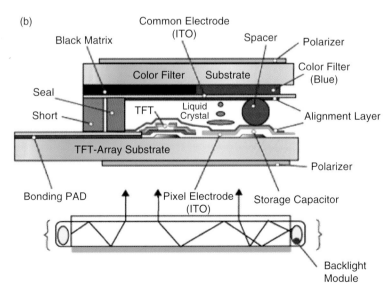

Figure 19.2 TFT-LCD panel assembly details (a) module assembly, and (b) LCD panel architecture overview [2].

screws are two major components for affixing the backlight modules to the front structure, and for protecting the LCD monitor from external impact; polymers are used in the polarizer, optical film of the backlight module, plastic frame, and cable jacket of the wirings; the lightness and thinness of the polymer allow the polarizer and optic film to be shaped into different forms of surface. Suitable surface patterns will help distribute and reflect light evenly, and reduce power consumption in LCD products.

To adjust polymer hardness to meet parts requirement, plasticizer is added to some of the polymers. In addition, the anisotropic conductive film (ACF) used for bonding the components are made with high-quality resin and conducting particles. Main electronic components include IC circuits, processors, resistors, capacitors, transistors, coupling devices, and diodes for connecting, driving, and adjusting the voltages. Glasses are mainly used to protect the RGB primary color pixels on the inside of the first layer of the LCD monitor without blocking light, and include plain glass, patterned glass, recycled glass, ground glass, and chemically tempered glass. Styrofoam (EPS) is commonly used as packaging material for LCD products.

19.3.2 Hazardous Substance Management

LCD technology development relies on material development and application. In particular, the huge increase in use of chemical elements in electronic products raises the concern of whether the materials are environmentally harmful. Since the 1990s, people have begun to realize that rapid industrial development is harming the very planet and natural environment that they depend on for survival. Some countries have realized that the rapid development of electronic products and their improper disposal may be environmentally hazardous, and have begun enacting legislative controls.

These regulations have a major impact on material selection for LCD products and LCD industry management policies. In 2001, when the SONY PS games violated the Dutch Cadmium Decree 1999, 2.5 million units of product were detained and the company was fined 17 million Euros, resulting in a huge financial and image loss for the company; thereafter, the electronic and electrical industry begun to understand the roles and social responsibility in environmental issues. Hence SONY began to actively construct its own hazardous substance management mechanism for their products, and the Restriction of Hazardous Substances Directive (RoHS) evolved from what was then considered a pioneer and model for hazardous substance management in the electronic and electrical industry. The RoHS restricts the use of six major chemical substances in imported or locally manufactured electronic and electrical products circulating in the European market, namely lead (Pb), mercury (Hg), cadmium (Cd), hexavalent chromium (Cr^{6+}), polybrominated biphenyls (PBB), and polybrominated diphenyl ethers (PBDE).

To ensure that products comply with RoHS standards, the International Organization for Standardization (ISO) 9001 quality management based International Electrotechnical Commission Quality (IECQ) QC080000 Hazardous Substance Process Management system is developed as a process management. Begun as a high-level management commitment, customer demand and international regulations, and transformed into product design development, component procurement, supply chain management, production process, testing and monitoring, and implementation of corresponding provisions for all resources used, this management system is now an important system for reinforcing non-hazardous products.

In addition, with the rapid development of chemical substances, international hazardous substance directives and standards continue to evolve and expand. To integrate and unify chemical controls for easy tracking and management, the European Commission established the Registration, Evaluation, and Authorization of Chemicals (REACH) system. Implemented in 2008, the system is a set of new chemical registration, evaluation and authorization mechanisms, and aims to integrate more than 40 existing chemical related directives and laws (see Figure 19.3). More than 30,000 chemical substances exceeding 1 ton in production will be controlled to increase chemical awareness, enhance safety, promote alternative test methods, encourage substituting highly hazardous substances with safer substances, and stimulate innovation (see Table 19.1).

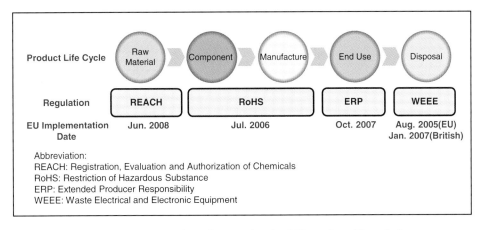

Figure 19.3 European environmental regulations related to LCD products life-cycle flow.

Table 19.1 Hazardous substances used in LCD production.

Part Group	Parts	Material Type	Possible Hazardous Substance	Function
TFT Panel	Glass	Glass	Antimony trioxide or Arsenic trioxide	Defoamer agent, fining agent
	LC	Chemical	Halogen compound	Polarity and Dipole Moments
	Color resist	Chemical	Halogen compound	Pigment for adjusting color
	Target (Al, Mo, Cu, ITO)	Metal	None	None
	Polarizer	Polymer	Dichloromethane	Solvent in process
			Boric acid	PVA stretching process
IC & ACF	IC	Electronic	Lead	Alloy, dielectronic ceramic or solder paste
	ACF	Polymer	Antimony trioxide	Hardener
PCBA & FPCA	PCB & FPC	Electronic	Antimony trioxide, halogen compound or phosphorus compound	Flame retardant
			Epoxy resin	Adhesive or molding compound
	IC	Electronic	Lead	Alloy, dielectronic ceramic or solder paste
			Epoxy resin	Adhesive or molding compound
	Capacitor	Electronic	Lead	Alloy, dielectronic ceramic or solder paste
	Resistor	Electronic	Lead	Alloy, dielectronic ceramic or solder paste
	Diode	Electronic	Lead	Alloy, dielectronic ceramic or solder paste
			Epoxy resin	Adhesive or molding compound
	Connector	Polymer	Antimony trioxide, halogen compound or phosphorus compound	Flame retardant

(Continued)

Table 19.1 (Continued)

Part Group	Parts	Material Type	Possible Hazardous Substance	Function
BLU	Optical films	Polymer	Antimony trioxide, halogen compound or phosphorus compound	Flame retardant
	Light guide	Polymer	None	None
	Light bar	Electronic	Lead	Solder paste
	LED	Electronic	Lead	Solder paste
	CCFL	Lamp	Mercury	Excitation source
	Bezel	Metal	Hexavalent chromium, Nickel	Anticorrosive coating
	Screw	Metal	Hexavalent chromium, Nickel	Anticorrosive coating
	Rubber	Polymer	PAHs	Impurity
	Ink	Chemical	PAHs	Impurity in solvent
	cable jacket	Polymer	PVC	Insulator
			Organic tin compound	Heat stabilizer
			Phthalates	Plasticizer
	Frame	Polymer	Antimony trioxide, halogen compound or phosphorus compound	Flame retardant
			Phthalates, phosphate ester	Plasticizer
			Organic tin compound	Catalyzer
			PC	Plastic
Packing	Packing material	Polymer	PET, EPP, EPO	Packing material

19.3.3 Material Hazard and Green Trend

In addition to price and performance, LCD material selection must take into account environmental and health risks. For example, to prevent rusting, nickel or chrome were used in the past to plate metals, and both are harmful to the human skin. Nowadays, zinc and trivalent chromium are the main substitutes for the harmful nickel and hexavalent chromium.

The main problem of existing polymer is the addition of flame retardant. To meet product safety requirements, polymer typically contains a flame retardant to achieve flame retardation. During waste disposal combustion, halogen flame retardants such as PBBs and PBDEs release dioxins, the toxin of the century, and phosphorus flame retardants such as TCEP, TXP, TDBPP, and TEPA are bio-accumulative and carcinogenic. These two categories of flame retardants are considered hazardous to humans and the environment, and have been replaced with chemical substances such as nitrogen, silicon, and metal hydroxides.

Electronic components containing lead are more environmentally polluting, and hence most electronic components today are unleaded. Other light sources commonly used a decade ago, such as cold cathode fluorescent lamp (CCFL) contain highly toxic mercury vapor, and have been replaced by light-emitting diode (LED). Moreover, in glass production, antimony trioxide or arsenic trioxide were often added for bubble removal or clarification, causing antimony or arsenic poisoning in those who come in frequent contact or inhalation. The process is now replaced with antimony and arsenic-free glass production.

19.3.4 Conflict Minerals Control

Conflict minerals refer to minerals such as gold (Au), tantalum (Ta), tin (Sn), and tungsten (W) that are mined under the conditions of armed conflict and human rights violations, especially in the Democratic Republic of the Congo and adjoining countries. These elements are widely used in information and communications technology products, and elements such as the Sn in indium tin oxide (ITO) are essential components in LCD.

In 2010, the United States passed the Dodd-Frank Wall Street Reform and Consumer Protection Act in Section 1502 of the financial reform bill. It requires the U.S. Securities and Exchange Commission (SEC) to formulate regulations on conflict minerals, and requires corporations to disclose whether conflict minerals are used in the manufacturing process. The proposal was announced in May 31, 2011, and all listed U.S. companies are required to comply with the standards by submitting a conflict mineral investigation report with their annual financial report to the SEC.

In addition to environmental issues, the latest material source control in supplier management is taking into account societal rights, giving rise to ore source traceability systems. Most current systems are based on the Organization for Economic Cooperation and Development (OECD) framework. To ensure non-conflict minerals in the production supply chain, the Global e-Sustainability Initiative (GeSi) of the Electronic Industry Citizenship Coalition (EICC) conflict mineral reporting guide is used for tracing whether smelting plants are certified by the Conflict-Free Smelter Program (CFSP) of the EICC.

19.4 MANUFACTURING FEATURES AND GREEN MANAGEMENT

19.4.1 The Manufacturing Processes

Using the TN-type LCD as example, the manufacture process includes stages such as array, color filter, cell, and module (see Figure 19.4). Different manufacture processes require different raw materials such as water, acid, dye, organic solvents, special chemicals, special gases, dry air, and inert gases. Some raw materials such as organic solvents are used directly in the manufacture phase. Some others are used indirectly to maintain manufacturing (e.g., it can be used to protect electrical plates from oxidation). Moreover, machine actuation or heating requires a lot of power, and to prevent power outages from interrupting production, plants are equipped with diesel and natural gas as emergency fuel for the generator. Hence the overall and environmental impact can be categorized into energy saving and carbon reduction in manufacturing, and pollution and emission reduction in end-process.

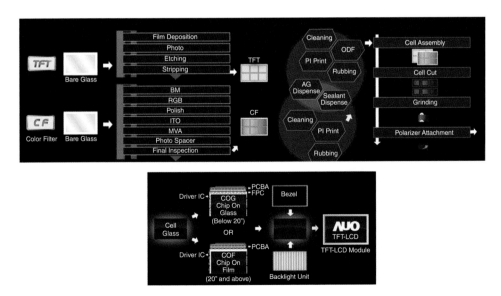

Figure 19.4 LCD module production process (array, call and module) [3].

19.4.2 Greenhouse Gas Inventory

Greenhouse gas has become a common international issue, and to understand the emission scenario within their organizations, the 2001 World Resources Institute (WRI) and the World Business Council for Sustainable Development (WBCSD) jointly published the first the Greenhouse Gas Protocol: A Corporate Accounting and Reporting Standard (Corporate Standard) inventory guide. The ISO also subsequently launched the ISO 14064 standard to provide industries with specific guidelines and instructions, and completed an inventory of its own volume of direct and indirect greenhouse gas emissions. Most LCD industries today adopt the GHG Protocol and the ISO 14064 for greenhouse gas inventory. Hence compared to the past, data for greenhouse gas emission volume is substantially more complete, which facilitates carbon reduction management.

The source of greenhouse gas in LCD industry can be classified into direct emissions from manufacturing, Scope 1; indirect emissions from energy, Scope 2; and other indirect emissions, Scope 3 (see Table 19.2). Main

Table 19.2 Greenhouse gas emissions from TFT-LCD production.

Category	Type	Response /Facility Type
Scope 1 Direct Greenhouse Emissions	Greenhouse gas emission from energy such as electricity, steam or other fossil fuels.	1. Diesel (e.g. generator, fire equipment, boiler and forklift) 2. Natural gas (e.g. zeolite turntable, boiler, kitchen, PFCs destruction facilities) 3. Liquefied petroleum gas (e.g. zeolite turntable, kitchen, PFCs destruction facilities)
	Greenhouse gas emissions from biological, physical or chemical processes	1. CF_4, SF_6, NF_3 (PFCs used in dry etching and CVD processes) 2. N_2O 3. CH_4 4. CO_2 (CF processes) 5. Organic waste gas emission
	Controlled raw materials, products, waste and transportation such as employee transportation	1. Gas (service vehicles) 2. Diesel (service vehicles)
	Fugitive greenhouse gas emissions	1. CO_2`HFC (Central fire facility) 2. CO_2 (Portable fire equipment) 3. SF_6 (high and medium voltage -GCB) 4. CH_4 (Anaerobic wastewater treatment) 5. Sewage (septic tank) 6. HFCs central air conditioning chiller, refrigerators, air conditioners, storage freezer, kitchen freezer, drinking fountains 7. HFCs machine cooling system
Scope 2 Indirect Emissions from Energy	Greenhouse gas emission from externally purchased electricity, heat, steam and other fossil fuels.	1. Electricity from externally purchased electricity
Scope 3 Other Indirect Emissions	Other indirect greenhouse gas emissions (e.g. employee business travel; third party products, raw materials or waste material shipping; sponsored activities, outsourced manufacturing and authorized distributors; facility waste material greenhouse gas emissions released outside the boundary of the emission source or facility; emissions from end-of-life stage of facility products and services; employee work commute; emissions from non-energy raw materials)	1. Fuel (employee business trip and travel) 2. Fuel (employee personal vehicle for commuting) 3. Wastes (waste incineration) 4. Fuel (Outsourced vehicle for commuting)

greenhouse gas emissions from the use of PFCs, N_2O, CO_2, and CH_4 during physical or chemical processes come from fossil fuels use in fixed facilities such as boilers, waste gas treatment equipment, and emergency generators. Moreover, fugitive greenhouse gas is also emitted by the combustion of raw materials, products, waste material, and employee transportation (such as service vehicles), portable extinguishers (CO_2), central air-conditioner chillers (HFCs), refrigerators (HFCs), high and medium voltage -GCB (SF_6), and septic tanks (CH_4). In addition, greenhouse gas is also emitted from externally purchased electricity.

19.4.3 Energy Saving in Manufacturing

About 70% to 90% of the greenhouse gases produced by LCD industries come from electricity use. Hence by adjusting the manufacture process and increasing efficiency, energy consumption can be reduced. Technological evolution have created feasible methods commonly used by the industry, such as simplifying manufacturing, reducing the power while machine in idle mode, installing frequency converter, using energy saving substitutes, recycling heat and steam, and replacing low-pressure faulty items. In addition, maintaining clean room temperature and humidity requires high-energy consumption air conditioning, and increasing chiller efficiency and adding refrigerant to enhance chiller conductivity are common measures for energy saving.

Moreover, the ISO 50001 Energy management has led to a growing trend of integrating energy data with smart metering, monitoring equipment, and big data management analysis to identify energy-saving potential opportunities and formulate action plans and objectives, and through the P-D-C-A cycle loops to continually improve the management of energy (see Figure 19.5).

19.4.4 Reduction of Greenhouse Gas from Manufacturing

The dry etch and chemical vapor deposition (CVD) equipment used in LCD manufacturing require the use of high global warming potential (GWP) fluorinated gases (PFCs) such as NF_3, SF_6 and CF_4.

To reduce the environmental impact of emissions, the earlier practices of directly releasing emissions into the environment has evolved into installing control facilities to manage emissions. Hence modern machines have incorporated such designs, and technology for control facilities has also evolved from earlier forms of

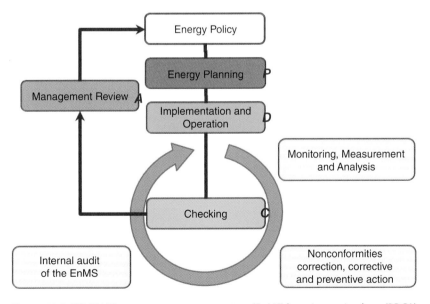

Figure 19.5 ISO 50001 energy management system (EnMS) keep improving loop (PDCA).

Figure 19.6 Emission reduction comparison.

electric heating to combustion. Correspondingly, the relative efficiency of manufacturing has increased to more than 90%. Other technologies such as enhanced recycling technology and material substitution are simultaneously under feasibility studies.

Other feasibility research on improving recycling technologies and alternative materials are also being conducted. PFC emission calculation is based on the World Display device Industry Cooperation Committee (WDICC) resolutions, and the Tier 2b calculation method, issued by the Intergovernmental Panel on Climate Change (IPCC) in 2006, has been in use since 2008. PFC has a high global warming potential (GWP), and the SF_6 used in dry etch facility and NF_3 used in chemical vapor deposition (CVD) facility during TFT-LCD manufacturing are PFCs. These gases are first disintegrated in the manufacturing facilities, and IPCC Tier 2b report indicates that the destruction rate can exceed 97% in chemical deposition facility and exceed 70% in dry etch facility. Since the above destruction occurs in the vacuum environment of the manufacturing facilities, the products do not carry any PFC, and the remaining gases from the manufacturing are treated in the local scrubber, which is capable of more than 90% removal rate. A schematic diagram of the PFC treatment is shown below (see Figure 19.6). The numbers in the diagram represent the PFC gas concentrations based on Tier 2b coefficients issued by the IPCC 2006.

To calculate the equivalent values for the various PFC greenhouse gas emissions, the destruction removal efficiency (DRE) in IPCC Tier 2b method for fueled combustion, plasma and high temperature catalytic end-gas treatment in local scrubbers are set at 90%, while other facilities such as water washing or electrical heating devices do not have DRE. The following chart is the IPCC list of DRE rate for various PFC treatment technologies (see Figure 19.7).

Tier 2a & 2b Default Efficiency Parameters for Electronics Industry Emission Reduction Technologies (a, b, e)

Emission Control Technology	CF_4	C_2F_6	CHF_3	C3F8	$c\text{-}C_4F_8$	NF_3[f]	SF_6
Destruction[c]	0.9	0.9	0.9	0.9	0.9	0.95	0.9
Capture/Recovery[d]	0.75	0.9	0.9	NT	NT	NT	0.9

a: Values are simple (unweighted) averages of destruction efficiencies for all abatement technologies. Emission factors do not apply to emission control technologies which cannot abate CF_4 at destruction or removal efficiency (DRE) ≥ 85 percent when CF4 is present as an input gas or by-product and all other FC gases at DRE ≥ 90 percent. If manufacturers use any other type of emission control technology, its destruction efficiency is 0 percent when using the Tier 2 methods.
b: Tier 2 emission control technology factors are applicable only to electrically heated, fuelled-combustion, plasma, and catalytic devices that
- are specifically designed to abate FCs,
- are used within the manufacturer's specified process window and in accordance with specified maintenance schedules and
- have been measured and has been confirmed under actual process conditions, using a technically sound protocol, which accounts for known measurement errors including, for example, CF_4 by-product formation during C_2F_6 as well as the effect of dilution, the use of oxygen or both in combustion abatement systems

c: Average values for fuelled combustion, plasma, and catalytic abatement technologies.
d: Average values for cryogenic and membrane capture and recovery technologies.
e: Vendor data verified by semiconductor manufacturers. Factors should only be used when an emission control technology is being utilised and maintained in accordance with abatement manufacturer specifications.
f: Use of NF_3 in the etch process is typically small compared to CVD. The aggregate emissions of NF_3 from etch and CVD under Tier 2b will usually not be greater than estimates made with Tier 2a or Tier 1 methods.
NT = not tested.

Figure 19.7 Default efficiency parameters for emission reduction [4].

19.4.5 Air Pollution and Control

In addition to emitting greenhouse gas, LCD manufacturing also produces acid gas, organic waste, dust, heat exhaust and other pollutants (see Table 19.3), but most organic waste generated by the machine stations comes from the use of solvents. The main control technology channels large-air-volume low-concentration exhaust into a higher specific adsorption surface area zeolite rotor (see Figure 19.8) to be concentrated, then desorbing exhaust and combust in an incinerator to reduce the discharge of air pollutants into the environment (see Figure 19.9). Compared to the past, current technology is more mature, and can maintain a stable 85% to 95% efficiency. In addition, recent introduction of monitoring has overcome the problem of complex components in LCD organic waste gas interfering with analyzers, and online continuous and real-time

Table 19.3 Environmental issues related to TFT LCD manufacturing process.

		Environmental Pollution Issues		
Materials	EQ	Air Pollution	Waste Water	Waste
Photo Resist/Thinner	Track	VOCs	COD	Waste Solvent/Sludge
Etching	Wet Etch	Acid Exhaust	Metal IN/Mo/Cu/Zn/Ni	Waste Cu Acid Waste Al Acid
$SiH_4/NF_3/Cl_2/SF_6$	CVD/Dry Etch	CVD Exhaust	F^-	Dust
Stripper	Stripper	VOCs	COD	Waste Stripper
Developer	Track/CVD/Wet	NH_3	NH_3-N	Sludge
Acid/Alkaline	Facility	Acid Exhaust	pH & Conductivity	Backwash
Heat Source	Equipment	General Exhaust	--	--

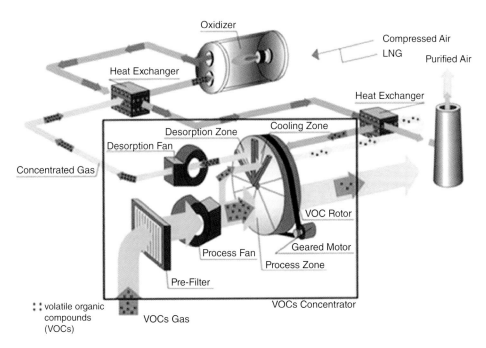

Figure 19.8 VOCs emission treatment [5].

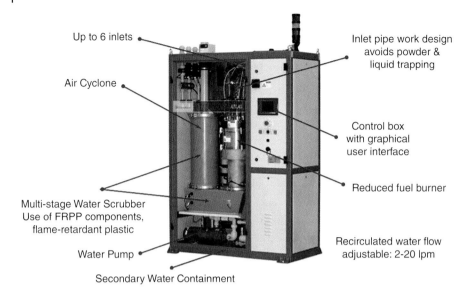

Figure 19.9 Details of combustion-type DRE facility [6].

monitoring are gradually being widely used. Exhaust gas with other properties are treated in scrubbing towers. With water as a medium, the gas forms a surface with high surface density raschig rings. Then through pH adjustment, acidic gases are absorbed by alkaline water, and alkaline gas by acidic water to achieve a certain pH balance. The resulting highly conductive water is then discharged into a waste water treatment system for further processing. In addition, dust exhaust contains fine suspension particles. To pre-remove the particulate contaminants, the exhaust gas is filtered through filter cloths within negative pressure environment in the bag house before being channeled into the wet scrubber.

19.4.6 Water Management and Emissions Control

LCD manufacturing requires multiple stages of pure water for cleaning the glass substrate, and reducing manufacturing water use and increasing water recovery are common methods actively adopted by industries to reduce industrial water demand. For example, developing more effective alternative pure water can improve the cleaning efficiency for glass substrates by about 30%, thereby reducing water use. To strengthen water shunting during manufacturing, water emitted from machines is divided into organic and inorganic waste, then further shunted according to concentration for effective water recycling. The shunting purpose is to increase recycling efficiency by treating similarly polluted waste water with applicable physical-chemical processes, thereby reducing the burden of end-process waste water treatment. In response to the impact of climate changes on water resource stability, water recycling technology in LCD manufacturing has improved significantly. In new factories, water shunting management for different departments are already in place during machine designing, and with facilities such as film processing system, about 80% to 90% of water can be recycled. Feasible technology is available to further increase waste water recycling to zero discharge; however, the processing cost is relatively high.

Waste water from LCD manufacturing can be divided into inorganic and organic waste (see Table 19.4). The inorganic wastes are treated with precipitation procedures such as coagulation-flocculation. Except for the more complex inorganic waste water containing fluorine ions, the environmental impact of inorganic waste water is limited. Organic waste water is more complex, especially its organic matter and ammonia components. Although modern technology can reduce the pollution through biological treatment, the culturing

Table 19.4 Wastewater categories and its optimization treatment.

Categories	Concentration Level	Pollutants	Treatment	Water Channel
Organic Wastewater	High	COD	Advanced Oxidation Technologies (AOP)	Effluent
	Medium to Low	NH_3-N	Biological Treatment & Membrane Bioreactor (MBR)	Effluent
Inorganic Waste water	Unrestricted	F^-/Cu In/Mo	Coagulation & Flocculation Electrochemistry Ion-Exchange Resin	Effluent
	Unrestricted	SO_4^{-2}/Cl^- PO_3^{-4}/Na^+	Neutralization Electro-dialysis	Effluent
Recycle Water	Very Low	SS/TOC	Filtration & Adsorption	Raw Water Tank

environment for the aerobic or anaerobic bacteria must be treated cautiously to prevent failure in the biological treatment system and significantly affecting the efficiency of the waste water treatment.

19.4.7 Waste Recycling and Reuse

Another LCD manufacturing burden on the environment is waste material, which primarily include waste liquid/solvent, waste material, waste packaging, and sludge from waste water treatment. Due to heightened environmental consciousness, waste recycling technology continues to evolve greatly and the waste-to-resource ratio has increased significantly to at least 50%. Most recycled resources come from waste solvents, which are purified through distillation for reuse, or downgraded from electronic use to industrial use, such as EBR. Another bulk of recycled resources come from sludge, which is heat treated into dried pellets for reuse as graded material, landfill, aggregates, or bricks. Next, wastes such as glass trimmings and plain glass can be used in concrete, cement bricks, or asphalt materials after debris removal and crushing. Some manufacturers use an additional continuous compression to produce compressed concrete, bricks, water permeable bricks, wall tiles, and retaining walls. Other wastes such as flue dust and hydrofluoric acid are chemically treated to produce industrial grade fluorine sodium silicate used in metallurgy (aluminum, beryllium), opalescent glass, vitrified ceramic frit, and wood preservatives. Treated aluminum etching solution can be distilled and used as industrial grade phosphoric acid, acetic acid, and nitric acid. Wastes from indium resin coating and indium dust are refined through electrolysis to obtain indium ingots for reuse. Utilizing the characteristic solubility of different pH, manufacturing waste fluid from the molybdenum resin desorption are dephosphorized and precipitated with calcium compounds. The usable residual solids from the centrifugal dehydration are calcium molybdate, which is calcinated into molybdenum oxide and used in steel making to increase ductility. Other resources such as packaging materials require more highly skilled recycling channels.

19.5 FUTURE CHALLENGES

Product end of life issues: future demands for LCD functions are evident, and user discard rate far exceeds the past. Hence future e-waste will be a major challenge and a bottle-neck is expected in the application of lifecycle concept to strengthen the environmental design of LCD. For example, to produce stable image quality in LCD, the various layers must be tightly bonded, which will lead to the obvious later difficulty of dismantling the electronic waste. In addition to RoHS and REACH, there are other worldwide directives for different requirements, such as the Electrical and Electronic Equipment Directive (WEEE, 2002/96/EC), Packaging and Packaging Waste Directive (PPW, 94/62/EC), and Energy-related Products Directive (ErP, 2009/125/EC).

However, the exercise of corporate product responsibility and execution of legislative power are issues that require continual attention.

Waste recycling bottleneck: most waste recycling technologies are already mature, but the market and relevant laws for recycled products are lagging, resulting in barriers to product flow. For example, sludge is typically treated and reused in the market as graded materials generally used in engineering or architecture. However de-sludged materials lack long-term stability and the government lacks clear requirements for end-product certification and control. Manufacturer's willingness to recycle other materials such as solvents, and the accepted quantity of recycling are affected by end customer and market demand for raw materials. Hence diversified recycling technologies must be developed for a single waste material to activate the market. In addition, application is required to develop new waste recycling methods, and manufacturers and organizations are deterred by the very time-consuming process. Organizations should also consider changing the nature of the waste to provide more avenues of waste-to-resource. Attention should be paid to reasonable feasible technologies for the small number of other valuable recyclables.

Tighter control of environmental regulations: although today's environmental engineering technology for reducing pollutant emission into air and water is more efficient and stable, traditional concepts of overall control and management general concept still prevail, such as emission standard requirements for total volatile organic compounds (VOCs) or total chemical oxygen demand (COD). The laws in most countries have yet to formulate their industrial control strategies for harmful substances and highly important environmental issues (e.g., PM2.5) in terms of environment and health impact dimensions. The community is expected to become increasingly concerned about environmental health risks. Micro-dose harmful substances will become the focus of attention, and hence their control mechanism and economically feasible processing technology should be developed in advance.

Supply chain management: the supply chain for LCD is varied and complex; and materials are subjected to supplier oligopoly, symbiotic clusters and standard differences. A comprehensive supply chain management that takes into account material transportation risks and the impact of production environment conditions and operational capacity is necessary to avoid supply interruption. Hence, more efforts should be invested into globalized and differentiated management, increased attention to environmental concern and the establishment of sustainable configurations. There is a need for green product international standards through green supply chain management by working together with suppliers. The focus is on the following directions:

- Implement material localization to reduce unnecessary cost of air or sea shipping, or encourage suppliers to produce locally.
- Encourage suppliers to import environmentally friendly renewable energy sources, and use environmentally friendly renewable wind power and combined heat and power system (CHPs).
- Increase reuse rate of raw materials and packaging, and recycle stripper and LC waste liquids for reuse.

In addition to focus on the value-added products, display industry should apply green manufacturing concept and its application to the process in order to keep this industry moving forward and provide customers with innovative, environmental friendly products. It is for sure that this industry will face the challenge from the coming technology revolution. However, it is important for display industry to think about green manufacturing process, green materials and green product, and this is also the way to be sustainable.

References

1 AU Optronics international documents, Size Matters (2016).
2 AU Optronics international documents, What is TFT-LCD (2016).
3 AU Optronics international documents, TFT-LCD Process Slide Show (2007).

4 H. S. Eggleston, L. Buendia, K. Miwa, T. Ngara, K. Tanabe, (eds), IPCC, 2006 IPCC Guidelines for National Greenhouse Gas Inventories, 3, Prepared by the National Greenhouse Gas Inventories Programme, IGES, Japan, (2006).
5 GCES, Rotor Concentrator Systems, http://www.gcesystems.com/rotor-concentrators.html (2016).
6 Edwards, Inward Fired Combustors, Edwards Ltd, Crawley (2016).

Index

a

Abnormal discharge   401, 405, 406
AC cathode   214, 217, 218
Acryl   282, 283
Active matrix organic light emitting diode (AMOLED)   13
  manufacturing   137, 138
  TV   115
Adobe RGB   43
Advanced excimer laser anneal (AELA)   117, 118
Air flow (clean rooms)   427, 428
Air-guide BLU   111
Air knife   301
Air pollution and control   451
AKT cluster PECVD   9
Al alloy   195
Alignment (Overlay)   293, 310
Alignment layer   61
AlNd alloy   201
Al_2O_3 (alumina)   201
Aluminium (Al)   209, 222, 232, 233
Amorphous carbon (a-C)   271
Amorphous silicon (a-Si)   13, 222, 253, 273, 276, 284
  defects   274, 276
  n^+ doped   257
  process   325
  TFT-LCD   13, 248
Anisotropic conductive film (ACF)   75
  attachment   76
Anode   206, 227, 228, 231
  PECVD   274, 275
Anodized Al   281, 282
AQUATRAN II   179
Aristo PVD tool   225, 226
Array process   394, 399, 402, 409
Array test   353, 399, 410
Automated ground vehicle (AGV)   422, 423, 424
Automatic Optical Inspection (AOI)   334, 379, 388

b

Back channel etch (BCE)   23
Backing plate   215, 216
Backlight unit (BLU)   87
Backplane   193
Backplane metallization   225, 232
Backside inspection   339
Band gap   207
Barix   170
Barrier film   173
Beveling   80
Binning   97
Black-matrix   44
Black matrix (BM) patterning   303, 305
BOE   7
Bright shot panel failure   406
BT.709   43
Buffer layer   93, 271
Buried particles (AOI)   332
By-product   401, 405

c

CAAC-IGZO   22
Calcium test   180
Capacitively coupled plasma (CCP)   273, 274, 275
Capital investment   394
Capture rate (AOI)   348
Carbon black   47
Carrier glass   174
Cascade rinse   316, 317
Cassette transportation system   420, 421, 423

Cathode, PECVD   274, 275
Ceiling height (clean rooms)   424, 425
Ceiling transportation system   422
Cell gap   401, 412
Cell process inspection   347
Charge generation layer (CGL)   152
Charge neutralizer   405
Cheese slab floor (clean rooms)   424
Chemical contamination   427, 431, 433
Chemical etching (dry)   323–325
Chemical life   317
Chemical vapor deposition (CVD)   395, 397, 401, 404
China Star Optoelectronics Technology (CSOT)   8
Chip on flex (COF)   73, 186, 198
   bonding   75, 77, 78
   replacement   389
Chip on glass (COG)   73
   bonding   81
Cholesteric liquid crystal film (CLCF)   108
CIE1931   41
CIE1976   41
Circular polarizer   186
Cleaning in-situ (PECVD)   274, 279, 280, 281
Cleanliness (clean rooms)   428, 433
Clean room   404, 405
Clean room design   423
Cluster   225
Cluster-type tool (equipment)   209, 212, 217, 219
CMOS process   27
Coating meniscus   291
Coefficient of thermal expansion (CTE)   175
   thermal expansion   202
Cold ablation repair   366, 376
Collision cascade   195
Color changing medium (CCM)   144
Color coordinate   41, 87
Color filter (CF)   39, 90, 153, 155
   patterning   287, 292, 303
Color filter on array (COA)   60
Color gamut   90
Color mixing   96
Color resist   289, 292, 305
Columnar   203
Column Spacer (CS)   46, 53
Computer-integrated manufacturing (CIM)   419, 423, 438
Computer simulation (clean rooms)   430, 431
Concentration management   317

Conflict minerals   446
Contact exposure   292, 297
Contact sensing   355
Contamination   289, 401, 409, 417
Continuous wave (CW) laser   117
Continuous wave laser crystallization (CLC)   117
Controlled super lateral growth (C-SLG)   120
Copper (Cu)   203
Corrosion   281
Cosmetic defect inspection   381
Cost competition   111
Cost of ownership (CoO)   211, 217
Coupling efficiency   88
Cover plastic window   188
Critical angle   98
Critical Dimension/Overlay (CD/O)   337, 349
Cross contamination   427
Cu alloys   209, 223, 224
Curing equipment   182
Cutting repair   367
Cylindrical rotating magnetron   229

d

Dam and fill   161, 166
Dangling bond   276
Data line   202
DC Cathode   210, 216, 217
Dead pixel (AOI)   333
Defect classification (AOI)   336
Defect detection sensitivity (AOI)   342, 344
Dehydration baking   289
Dehydrogenation   183, 260, 277
Deionized water (DIW)   289, 300
   consumption   315
Deposition rate   254
Device encapsulation   135
Diffuser   89, 106
Diffuser plate   110
Diffusion barrier layers   269
Diffusion path   169
Digital Cinema Initiative (DCI)   43
Digital mirror device (DMD) patterning   373
Dip etching   315, 316
Directional SLS (D-SLS)   120
Direct-lit BLU   87, 109
Dislocation density   92
Doctor blade roll   62
Double shielding method (double skinning)   433
Dow Corning OLED encapsulation   178

Downtime   395
Dry etching   319–325
DTI   2
Dual magnetron   236, 239
Dyad   176
Dye   49
Dynamic (in-line) magnetron sputtering layout   225
Dynamic Multi Cathode   216, 217

e

Edge-lit BLU   87
Edge Seal   161, 163
Electrical testing   352–353
Electricity delivery system (clean rooms)   437
Electroluminescent (EL)   90, 209
Electrostatic charging / discharging   405, 406
Electrostatic damage   404, 405
Emission control   452
Emission reduction   450
Encapsulation   135
End point detector (EPD)   323
End seal   67
Enhanced capacitive coupled plasma (ECCP) mode   322–326
Equipment cost   313
Etchability   237
Etch factor   313, 315
Etching   198
Etch stopper (ES)   24
European environmental regulations   445
Evaporation   196
Excimer laser   180
Excimer laser annealing (ELA)   17
Exposure image field   292
Exposure mask defects (AOI)   344
Exposure systems   287, 292
Extrusion   102

f

Fab layouts   5
Face seal   167
Fan filter unit (FFU)   420, 428, 431, 433
 energy saving   433
Fast steering mirror (FSM) patterning   373
Field-effect-transistor   2
Fine metal mask (FMM)   116, 129, 133, 144
Flat panel display (FPD)   209, 212, 213
Flexible display   173

Flexible OLED   136, 137, 173
Flexible OLED deposition process   183
Flexible printed circuit (FPC)   73
Flexible printed circuit board (FPCB)   97
Flexible substrate   173
Flexible touch solution   181
Flip chip   96
Floor structure (clean rooms)   424
FOG bonding   82
Foldable display   177
Foldable OLED   187
Foreign particles (AOI)   331
Four-mask count process   24
Free-form panel testing   361
Fringe-field switching (FFS)   59
Frit sealing   135, 162, 164
Fuchs-Sondheimer model   197, 198
Full contact probing   354
Functional defects inspection   379

g

GaN   92
Gas delivery system (clean rooms)   436
Gas flow diffuser   245
Gate/data (G/D) short   201
Gate on array (GOA) testing   359
GE OLED encapsulation   178
Glass breakage defects   344
Glass breaking clipping (chipping)   401
Glass edge seal   163
Glass size generations   396
Glass substrate sizes   5
Glow discharge   227
Grain boundary   196
Grating floor (clean rooms)   424
Gray level difference (AOI)   342
Green building design   442
Green gap   92
Green house gas   448
Ground Strap   246

h

Half-tone exposure   53, 394
Half-tone photoresist process   325
Hardness   188
Hazardous substance management   444
H content   254, 268
Heat dissipation   95
Hexamethyldisilazane (HMDS)   289

Hexamethyldisiloxane (HMDSO)   271
High density plasma   275
High-power impulse magnetron sputtering (HiPIMS)   225
High temperature PI film   175
Hillock   198
Historic review   2
Hollow cathode effect   243
Hollow cathode gradient   244
Homeotropic   61
Homogeneous   61
Hot pixel (AOI)   333
Hot spot   88
Hybrid encapsulation   176, 177
Hydrogen   273, 274, 276, 277, 283

i

IBM Japan   2
i-Components i-Barrier   178
Impressio (TM)   322
Imprinting   103
In-cell touch panel inspection   345
Indium-gallium-zinc-oxide (IGZO)   14, 204, 210, 218, 223, 232, 234, 263, 276, 283, 319, 324
Indium Tin Oxide (ITO)   53, 202, 236, 304, 307
  transparent films   238, 239
Indium Zinc Oxide (IZO)   53, 202, 237
Inductively coupled plasma (ICP)   275
  mode   320, 321, 323–328
InGaN   92
Injection molding   101
Inline AOI   350
Inline processing   287, 301, 308
In-plane switching (IPS)   4, 9, 54, 59, 304
In-process quality control (IPQC)   416, 417
Input quality control (IQC)   407, 416
In situ chamber cleaning   395, 405
Inter layer dielectric (ILD)   199
Ion assist etch   324
Ion bombardment   227
Ionizer   405
Ions (dry etching)   320, 321
ISO50001   449
Isotropic etching   311, 313

j

JI process   73, 74
Judgment (AOI)   337, 350
Just noticeable difference (JND)   385

l

Lamination of OLED protection film   185
Laser ablation repair   365
Laser anneal   276, 277
Laser chemical vapor deposition (LCVD) repair   369
Laser cutting   389, 406
Laser CVD   406
Laser engraving   103
Laser-induced forward transfer (LIFT) repair   371
Laser lift-off (LLO)   180, 185
Light-emitting diode (LED)   87
  package   95
Light extraction   94
Light extraction efficiency (LEE)   94
Light guide plate (LGP)   87, 98
Lightly doped drain (LDD)   28
Line defect locator (LDL)   375
Liquid crystal (LC)
  alignment layer   59
  cell fabrication   61, 399, 400, 402
  drop filling process   5, 9, 10, 402, 404
  filling process   65, 67, 68
Liquid crystal display (LCD)   13, 87, 209, 210, 218
  cleanroom fab   419
  desktop monitor   4
  module assembly   84
  TV   5
Liquid crystal module (LCM)   73, 88
  defect repair   388
  disassembly   391
  optical film rework   391
Load lock chamber   322
Location controlled crystallization (LCC)   117
Low density plasma   273
Low pressure chemical vapor deposition (LPCVD)   16
Low temperature poly silicon (LTPS)   13, 16, 115, 116, 209, 213, 217, 221, 258, 320
LTPS TFT dry etch process   326
Luminance   87
Luminance uniformity   89
Luminous efficacy   90
Luminous flux   87

m

MacAdam ellipses   42
Magnetron cathode   227
Magnetron sputtering   210, 211, 212, 225
Magnet-wobble   230

Maintenance cycle   313, 315
Major milestones   9
Manufacturing process   447
Material hazard   446
Material types and uses   443
Mayadas and Schatzkes (MS) model   197
Mean time between failures (MTBF)   395
Mean time to repair (MTTR)   395
Mega-trusses (clean rooms)   421
Metal can encapsulation   163
Metal frame   96
Metal-induced crystallization (MIC)   17
Metal ink deposition repair   370
Metal layer repair   364
Metal oxide   225, 234, 263
Metal shadow mask   133, 134
Metrology   337, 349
Micro-cavity effect (OLED)   129, 130
Microlens array (MLA)   104
Microstructure   196
Microwave-type reactor   274, 275, 282
Mill base (MBS)   49
Mirror projection scan system   296
Mitsubishi X-barrier   178
Mobility   253, 255
MOCON test   179
Module assembly (MA)   73, 399, 400
Moire formation   100
Mold frame   87
Molybdenum (Mo)   203, 233, 234
Momentum transfer   195
Motion picture quality   4
Moving cathode   219, 220, 221
Multi cathode   216, 217, 218
Multi-layer optical film (MOF)   108
Multi lens projection   296
Multi magnet cathode   216, 218
Multi-modality imaging (AOI)   341
Multi-panel assignment   394
Mura defects   346, 359, 401, 402

n

Narrow bezel   89, 177
National Television System Committee (NTSC)   43
NEC   3
Negative photoresist   289, 292, 306
Neutral zone shifting   186
Nitrogen trifluoride (NF_3)   247
Nodule   207

Non-reactive sputter processing   227
N/Si ratio   252
Numerical aperture   294, 296

o

O_2 ashing   302
Offline AOI   350
One drop filling (ODF)   67, 68, 394, 412
On-the-fly video grabbing (AOI)   340
Open cell   79
Open-circuit repair   369
Open/short defect   401, 403, 409
O_2 plasma   302
Optical band gap   253
Optical Density (OD)   45, 305, 309
Optical film   87
Optically Clear Adhesive (OCA)   186
Optical pattern   88
Optical pattern design (BLU)   99
Optical simulation   100
Organic light emitting diode (OLED)   205, 219, 223, 269
  deposition equipment   184
  encapsulation   161
  encapsulation:3M   178
  manufacturing   129
  yellow green stack   155
OS TFT   122
Outcoupling efficiency   94
Outgoing quality control (OQC)   416, 417
Overcoat (OC)   51, 149, 305
Overlay accuracy   292, 294
Oxide film   218, 223
Oxide semiconductor (OS)   14
Oxide TFT dry etching process   327
Oxygen vacancy   236, 267

p

Panel Defect   401
Panelization   394
Panel signal driving (testing)   353
Panel test   399, 400, 401, 402
PAN etch solution   198
Particle contamination   422
Particle defects (PECVD)   279, 280, 281, 282, 284
Pass-by   212, 214, 219
Pattern defects   329, 401
Patterned sapphire substrate   95
Patterning   288, 292, 303
PCBA repair   390

Peripheral inspection   346
PET film   104
Phase modulated excimer laser anneal (PM-ELA)   117
Phosphor   90
Photo-Acryl Compound (PAC)   56
Photo-alignment   64, 65
Photochemical ablation repair   366
Photochemistry   292
Photodecomposition   65, 66
Photodimerization   65, 66
Photoisomerization   65, 66
Photolithography   287
   resolution   292, 309
Photo mask   288, 292
Photoresist coating   288, 290, 303
Photoresist coating bead   290, 291
Photoresist development   300, 308
Photoresist layer repair   364, 372
Photoresist stripping   287, 302
Photoresist track system   287
Photo spacer (PS)   306, 404
Pigment   49
PiVot   225, 226, 230
Pixel defect   401, 406
Pixel patterning   133
Planar (magnetron sputter)   210, 211, 216, 218
Planar cathodes   227, 228
Planarization   51
Plasma   195
Plasma density   273, 274
Plasma enhanced chemical vapor deposition
      (PECVD)   15, 241, 273, 279
   precursors   273, 276, 279
Plasma etching (PE) mode   320, 321, 324–326
Plasma potential   274275
Plasma treatment   149
PMOS process   27, 28
Polarization recycling   108
Polarized UV light   65
Polarizer attachment   69, 81
Polarizer detach   390
Polycarbonate (PC)   100
Polyimide (PI)   61, 175
   coating and curing   181, 399, 402, 405
   rubbing   402
Poly(methyl methacrylate) (PMMA)   98
Poly-Si   16
Polyvinyl alcohol (PVA)   4, 9
Poor coating defects (AOI)   344

Positive photoresist   289, 292, 301
Precursor   260
Precursor dissociation   273, 275
Pressure sensitive adhesive (PSA)   178
Pre-tilt   61, 65
Printed circuit board assembly (PCBA)   388
Printed circuit board (PCB) bonding   79
Prism film   89, 107
Process capability index Cpk   409, 417
Process chamber   405
Process chamber (etching)   322
Process window   404
Process yield   399, 404, 406
Production control system   437
Productivity   292, 393, 397, 399
Productivity enhancement   5
Productivity-to-investment factor   397, 399
Projection exposure   292, 293
Projection scanning exposure system   294, 295
Protection film   89
Proximity exposure   287, 297

q

Quality control   393, 399, 406, 413, 416
Quantum dot (QD) film   108
Quantum efficiency   92
Quantum well   93
Quick rinse system   317, 318

r

Racetrack   227
Radiative recombination   93
Radical   321, 322
Railed ground vehicle (RGV)   422, 423
Raised floor (clean rooms)   424
Random failure   401
Rank mixing   97
Ray-tracing   100
RC delay   194
Reactive ion etching (RIE) mode   320, 321, 324–326
Reactive sputter   218, 223
Reactive sputter processing   227
Recrystallization   196
Re-deposition   228, 229
Reflection film   89
Reflective polarizer   108
Reflector   107
Remote plasma   279, 281
Remote plasma cleaning   405

Remote plasma source cleaning (RPSC)   247
Repairing   399, 404, 406, 409, 412, 415
Resin   96
Resistance (thin film properties)   209, 221, 222, 223
Resistivity   193, 258
Resonant cavity LED   95
Reticle   294
Return shaft (clean rooms)   427
Review image grabbing (AOI)   337, 340
Rework   406, 408
RF   210, 274, 275, 281, 284
RGBW   55
Roll printing   62
Roll stamping   103
Roll-to-roll processing   106
Roof-transportation (clean rooms)   420
Rotary Cathode (Magnetron Sputter)   211, 218, 225
Rotary cathode array   230
Rubbing alignment   61
Rubbing depth   63
Rubbing process   63
Running cost   311, 315, 317, 318

s

Sacrificial layer   180
Sapphire substrate   92
Saving of chemicals   436
Scattering   197
Screen printing   102
Sealant   59
Secondary electrons   227
Selective detection control (AOI)   344
Semi knock down (SKD)   79
Separated circulation (clean rooms)   427
Sequential lateral solidification (SLS)   117, 120
Sheet resistance   194
Short bar probing   354
Short-circuit repair   365
Sideways Static Deposition Method (PVD)   212
Signal driving probing   357
Si-H content   252
SiH, SiH_2, SiH_3, SiH_4 273, 274, 276, 278, 280
Silicone dispensing   78
Silicon nitride (SiNx) dielectric film   248, 269, 276
Silicon oxide (SiOx) dielectric film   259, 265, 276, 279, 282, 284
Silver nanowire   188
Single cathode   216
Single plate processing   394

Slit coating photolithography   25
Slit die coating   287, 290, 307
Small mask scanning (SMS)   145
Small panel JI process   80
Solid state laser   180
Source/drain layer testing   362
Spin coating   290, 291
Sputtering   195
Sputter yield   227
sRGB   43
Static (in-line or cluster) magnetron sputtering layout   225
Stepper   287, 293
Structure zone model   196
Stuck pixel (AOI)   333
Sub-fab (clean rooms)   420, 424
Substrate proximity handling   405
Super lateral growth (SLG)   20
Surface mounting technology (SMT)   97
Surface particles (AOI)   332
Susceptor   245
Symmorphix OLED encapsulation   178
Systematic failure   401
System on panel (SOP)   32

t

Tact time   314
Tape-automated bonding (TAB) connection   73, 402, 408, 413
Terra barrier   179
Tetraethylorthosilicate (TEOS)   276, 279
Tetramethyl ammonium hydroxide (TMAH)   289, 301
TFT-LCD plant   419, 423, 427, 437
Thermal ablation repair   366
Thermal evaporation   129
Thin film deposition   273, 282
Thin film encapsulation (TFE)   136, 168, 176, 185, 269
Thin-film transistor (TFT)   2, 39, 209, 215, 218, 221, 287, 292, 303
  array repair   363
  backplane   115, 173
  channel layer   273
  coplanar P-channel   117
  five-mask structure   26
  gate insulator   273, 279, 283, 284
  history   2
  insulated-gate   2
  inverted staggered   2
  passivation   234, 275, 276, 282

Threshold voltage (Vth)   283, 284, 285
Throughput   292, 296, 394, 397, 399
Ti/Al/Ti stack   206
Tilted transferring system   315, 316
Time constant   194
Ti-silicide   204
Titanium dioxide (TiO$_2$)   204
Titanium nitride (TiN)   204
Top emission OLED structure   130, 131, 132
Toshiba   2
Total circulation (clean rooms)   427
Total internal reflection (TIR)   98
Transfer chamber (etch)   322
Transparent conductive oxide (TCO)   206
Transparent oxide (TOS)   209, 210, 218, 219, 221
Turn-around time (TAT)   423
Twisted nematic (TN)   54, 59, 304
Two shot sequential lateral solidification (TS-SLS)   33, 120

u

Ultra-clean room   419
Ultra-high-resolution testing   357
Ultra low penetration air (ULPA)   420
Uniformity   394, 397, 401, 404, 412
University of Colorado OLED encapsulation   178
Uptime   395
Utilization   395
UV glue dispense   83
UV irradiation   287
UV LED   90

v

Vertical alignment (VA)   4, 9, 55, 304, 307
Vertical cluster or equipment   209, 213, 214, 215, 217, 219
Visual inspection   383, 384
Visual testing environment   385
Visual test patterns   387
Vitex   170
Vitex OLED encapsulation   176
Vitriflex OLED encapsulation   179
Volatile organic compounds (VOC)   451
Voltage image optical system (VIOS)   355

w

Waku (backlight light leakage)   386
Waste recycling and reuse   453
Water management   452
Water recycling   434
Water vapor transmission rate (WVTR)   135, 160, 176, 269, 282
Welding repair   368
Wet etch rate   253
White LED   90, 147
White OLED
  one-stacked   149, 150
  two-stacked   152, 153
  three-stacked   155
Wide viewing angle   4, 9
Wire grid polarizer (WGP)   108
Work function   206
Work in process (WIP)   395

y

Yield   393, 399, 403, 406, 413, 416
Yield optimization   395, 404
Young's modulus   200